Transfer Entropy

Transfer Entropy

Special Issue Editor

Deniz Gençağa

MDPI • Basel • Beijing • Wuhan • Barcelona • Belgrade

MDPI

Special Issue Editor
Deniz Gençağa
Antalya Bilim Universitesi
Turkey

Editorial Office
MDPI
St. Alban-Anlage 66
Basel, Switzerland

This is a reprint of articles from the Special Issue published online in the open access journal *Entropy* (ISSN 1099-4300) from 2013 to 2018 (available at: http://www.mdpi.com/journal/entropy/special_issues/transfer_entropy)

For citation purposes, cite each article independently as indicated on the article page online and as indicated below:

LastName, A.A.; LastName, B.B.; LastName, C.C. Article Title. *Journal Name* **Year**, *Article Number*, Page Range.

ISBN 978-3-03842-919-7 (Pbk)
ISBN 978-3-03842-920-3 (PDF)

Contents

About the Special Issue Editor

Deniz Gençağa received the Ph.D. degree in electrical and electronic engineering from Boğaziçi University in 2007. Same year, he joined SUNY at Albany, as a postdoctoral researcher. Between 2009 and 2011, he was a research associate at the National Oceanic and Atmospheric Administration Center of CUNY, USA. Until 2017, he took roles in interdisciplinary projects at universities, including the University of Texas and Carnegie Mellon University. Until 2016, he was the chair of IEEE Pittsburgh signal processing and control systems societies. Since 2017, he has been an Assistant Professor at the electrical and electronic engineering department of the Antalya Bilim University. He is a member of the editorial board of Entropy, inventor in two US patents, recipient of NATO research fellowship, general chair of the first ECEA and one of the organizers of MaxEnt 2007. His research interests include statistical signal processing, Bayesian inference, uncertainty modeling and causality.

Editorial

Transfer Entropy

Deniz Gençağa [ID]

Department of Electrical and Electronics Engineering, Antalya Bilim University, Antalya 07190, Turkey; deniz.gencaga@antalya.edu.tr

Received: 12 April 2018; Accepted: 13 April 2018; Published: 16 April 2018

Keywords: transfer entropy; causal relationships; entropy estimation; statistical dependency; nonlinear interactions; interacting subsystems; information-theory; Granger causality; mutual information; machine learning; data mining

Statistical relationships among the variables of a complex system reveal a lot about its physical behavior. Therefore, identification of the relevant variables and characterization of their interactions are crucial for a better understanding of a complex system. Correlation-based techniques have been widely utilized to elucidate the linear statistical dependencies in many science and engineering applications. However, for the analysis of nonlinear dependencies, information-theoretic quantities, such as Mutual Information (MI) and the Transfer Entropy (TE), have been proven to be superior. MI quantifies the amount of information obtained about one random variable, through the other random variable, and it is symmetric. As an asymmetrical measure, TE quantifies the amount of directed (time-asymmetric) transfer of information between random processes and therefore is related to the measures of causality.

In the literature, the Granger causality has been addressed in many fields, such as biomedicine, atmospheric sciences, fluid dynamics, finance, and neuroscience. Despite its success in the identification of couplings between the interacting variables, the use of structural models restricts its performance. Unlike Granger causality, TE is a quantity that is directly estimated from data and it does not suffer from such constraints. In the specific case of Gaussian distributed random variables, equivalence between TE and Granger causality has been proven.

The estimation of TE from data is a numerically challenging problem. Generally, this estimation depends on accurate representations of the probability distributions of the relevant variables. Histogram and kernel estimates are two common ways of estimating probability distributions from data. TE can be expressed in terms of other information-theoretic quantities, such as Shannon entropy and MI, which are functions of the probability distributions of the variables. Therefore, it is prone to errors due to the approximations of probability distributions. Moreover, many TE estimation techniques suffer from the bias effects arising from the algebraic sums of other information-theoretic quantities. Thus, bias correction has been an active research area for better estimation performance. Methods such as Symbolic TE and the Kraskov-Stögbauer-Grassberger (KSG) algorithm are among the other techniques used to estimate TE from data. The efficient estimation of TE is still an active research area.

Most of these techniques have been proposed to solve specific problems in diverse applications. Hence, a method proposed for the solution of one application might not be the best for another. This Special Issue has been organized to collect distinctive approaches in one publication, as a reference tool for the theory and applications of TE.

The contributions are categorized into two sections: the methods and the applications.

1. Methods and Theory

The first section begins with the presentation of a recipe to estimate the information flow in dynamical systems [1]. In their work, Gencaga et al. propose a Bayesian approach to estimate TE and apply a set of methods together as an accuracy cross-check to provide a reliable mathematical tool for any given dataset. The work of Zhu et al. [2] proposes a k-Nearest Neighbor approach to estimate TE and demonstrates its effectiveness as an extension of the KSG MI estimator.

The methodological section continues with the analytical derivations of the TE expressions for a class of non-Gaussian distributions. Here, Jafari-Mamaghani and Tyrcha [3] provide the expressions of TE in the cases of multivariate exponential, logistic, Pareto (Type I-IV), and Burr distributions. Next, Nichols et al. elaborate on the linearized TE for continuous and coupled second-order systems and they derive an analytical expression for time-delayed transfer entropy (TDTE) [4]. They conclude with an alternative interpretation of TE, which can be viewed as a measure of the ability of a given system component to predict the dynamics of another. Coupling between random processes is also explored by Hahs and Pethel [5], where the TE is computed over multiple time lags for multivariate Gaussian autoregressive processes. In two examples, they demonstrate the change in TE as a response of variations in the correlation and coupling coefficient parameters. The case of coupling dynamics with time-varying dependencies is investigated by Gómez-Herrero et al. [6] if access to an ensemble of independent repetitions of time series is available. They estimate combinations of entropies and detect time-varying information flow between dynamical systems using the ensemble members.

The relation between Granger causality and directed information theory is discussed next in the review paper of Amblard and Michel [7], in which they focus on conditional independence and causal influences between stochastic processes. In addition to the link between directed information and hypothesis testing, instantaneous dependencies are emphasized to be different than dependencies on past values.

The next two papers demonstrate two new interpretations of TE. First, motivated by the relativistic effects on the observation of information dynamics, Lizier and Mahoney bring a new explanation of a local framework for information dynamics [8]. Second, Prokopenko et al. present a thermodynamic interpretation of TE near equilibrium and emphasize the nuance between TE and causality [9]. The methodological section ends with the comparisons of Papana et al. where they study direct causality measures in multivariate time series by simulations. The authors compare measures such as the conditional Granger causality index, partial Granger causality index, partial directed coherence, partial TE, partial symbolic TE, and partial MI on mixed embedding. Simulations include stochastic and chaotic dynamical systems with different embedding dimensions and time series lengths [10].

2. Applications

In this section, we present six contributions on the applications of TE. In the first paper, Faes et al. introduce a tool for reliably estimating information transfer in physiological time series using compensated TE [11]. This tool provides a set of solutions to the problems arising from the high dimensionality and small sample size, which are frequently encountered in entropy estimations of cardiovascular and neurological time series. Next, Materassi et al. elucidate a different normalized TE and use it to detect the verse of energy flux transfer in a synthetic model of fluid turbulence, namely the Gledzer-Ohkitana-Yamada shell model [12]. They emphasize the superior performance compared to those of the traditional methods. Applications continue with a paper on network inference by Ai [13]. The author addresses a TE-based framework to quantify the relationships among topological measures and provides a general approach to infer a drive-response structure in a complex network. This work is followed by two financial applications. The first contribution is authored by Li et al., in which a TE-based method is developed to determine the interbank exposure matrix between banks and the stability of the Chinese banking system is evaluated by simulating the risk contagion process [14]. In the second application, Sandoval Jr. uses the stocks of the 197 largest companies in the world and explores their relationships using TE [15]. This Special Issue ends with the presentation of the theory and applications of the Liang-Kleeman information flow [16]. Here, Liang points out the importance of information flow as a potential measure of the cause and effect relation between dynamical events and presents applications on the Baker transformation, Henon map, truncated Burgers-Hopf system, and Langevin equation.

This Special Issue demonstrates the importance of information-theoretic quantities in the analysis of the statistical dependencies between the variables of a complex system. Unlike correlation and MI, TE is shown to be effective for the detection of directional interactions, which are closely related to cause and effect relationships. The examples demonstrate the difficulties in estimating information-theoretic quantities from data and present approaches to overcome these problems.

In this Special Issue, we have collected 16 outstanding papers by the experts in the field. I would like to express our special thanks to each researcher and anonymous referee for their invaluable contributions. I would also like to thank the Editor-in-Chief, Prof. Kevin H. Knuth, for his encouragement during the organization of this Special Issue. My grateful thanks are also extended to the members of the editorial board and the editorial assistants of the Entropy Journal for their support. Last, but not least, I would like to thank MDPI Books for giving me the opportunity to publish this Special Issue.

We are excited to present this Special Issue as a reference for the theory and applications of transfer entropy and we hope that this publication contributes to novelties in all disciplines of research and development.

Acknowledgments: We express our thanks to the authors of the above contributions, and to the journal *Entropy* and MDPI for their support during this work.

Conflicts of Interest: The author declares no conflict of interest.

References

1. Gencaga, D.; Knuth, K.; Rossow, W. A Recipe for the Estimation of Information Flow in a Dynamical System. *Entropy* **2015**, *17*, 438–470.
2. Zhu, J.; Bellanger, J.; Shu, H.; Le Bouquin Jeannès, R. Contribution to Transfer Entropy Estimation via the k-Nearest-Neighbors Approach. *Entropy* **2015**, *17*, 4173–4201.
3. Jafari-Mamaghani, M.; Tyrcha, J. Transfer Entropy Expressions for a Class of Non-Gaussian Distributions. *Entropy* **2014**, *16*, 1743–1755.
4. Nichols, J.; Bucholtz, F.; Michalowicz, J. Linearized Transfer Entropy for Continuous Second Order Systems. *Entropy* **2013**, *15*, 3186–3204.
5. Hahs, D.; Pethel, S. Transfer Entropy for Coupled Autoregressive Processes. *Entropy* **2013**, *15*, 767–788.
6. Gómez-Herrero, G.; Wu, W.; Rutanen, K.; Soriano, M.; Pipa, G.; Vicente, R. Assessing Coupling Dynamics from an Ensemble of Time Series. *Entropy* **2015**, *17*, 1958–1970.
7. Amblard, P.; Michel, O. The Relation between Granger Causality and Directed Information Theory: A Review. *Entropy* **2013**, *15*, 113–143.
8. Lizier, J.; Mahoney, J. Moving Frames of Reference, Relativity and Invariance in Transfer Entropy and Information Dynamics. *Entropy* **2013**, *15*, 177–197.
9. Prokopenko, M.; Lizier, J.; Price, D. On Thermodynamic Interpretation of Transfer Entropy. *Entropy* **2013**, *15*, 524–543.
10. Papana, A.; Kyrtsou, C.; Kugiumtzis, D.; Diks, C. Simulation Study of Direct Causality Measures in Multivariate Time Series. *Entropy* **2013**, *15*, 2635–2661.
11. Faes, L.; Nollo, G.; Porta, A. Compensated Transfer Entropy as a Tool for Reliably Estimating Information Transfer in Physiological Time Series. *Entropy* **2013**, *15*, 198–219.
12. Materassi, M.; Consolini, G.; Smith, N.; De Marco, R. Information Theory Analysis of Cascading Process in a Synthetic Model of Fluid Turbulence. *Entropy* **2014**, *16*, 1272–1286.
13. Ai, X. Inferring a Drive-Response Network from Time Series of Topological Measures in Complex Networks with Transfer Entropy. *Entropy* **2014**, *16*, 5753–5776.

14. Li, J.; Liang, C.; Zhu, X.; Sun, X.; Wu, D. Risk Contagion in Chinese Banking Industry: A Transfer Entropy-Based Analysis. *Entropy* **2013**, *15*, 5549–5564.
15. Sandoval, L. Structure of a Global Network of Financial Companies Based on Transfer Entropy. *Entropy* **2014**, *16*, 4443–4482.
16. Liang, X. The Liang-Kleeman Information Flow: Theory and Applications. *Entropy* **2013**, *15*, 327–360.

Article

A Recipe for the Estimation of Information Flow in a Dynamical System

Deniz Gencaga [1],*, Kevin H. Knuth [2] and William B. Rossow [1]

[1] NOAA-CREST, the City College of New York, New York, NY, 10031, USA; wbrossow@ccny.cuny.edu

[2] Depts. of Physics and Informatics, University at Albany (SUNY), Albany, NY 12222, USA; kknuth@albany.edu

* Author to whom correspondence should be addressed; d.gencaga@ieee.org; Tel.: +1-412-973-2241.

Academic Editor: J. Tenreiro Machado

Received: 7 February 2014; Accepted: 8 January 2015; Published: 19 January 2015

Abstract: Information-theoretic quantities, such as entropy and mutual information (MI), can be used to quantify the amount of information needed to describe a dataset or the information shared between two datasets. In the case of a dynamical system, the behavior of the relevant variables can be tightly coupled, such that information about one variable at a given instance in time may provide information about other variables at later instances in time. This is often viewed as a flow of information, and tracking such a flow can reveal relationships among the system variables. Since the MI is a symmetric quantity; an asymmetric quantity, called Transfer Entropy (TE), has been proposed to estimate the directionality of the coupling. However, accurate estimation of entropy-based measures is notoriously difficult. Every method has its own free tuning parameter(s) and there is no consensus on an optimal way of estimating the TE from a dataset. We propose a new methodology to estimate TE and apply a set of methods together as an accuracy cross-check to provide a reliable mathematical tool for any given data set. We demonstrate both the variability in TE estimation across techniques as well as the benefits of the proposed methodology to reliably estimate the directionality of coupling among variables.

Keywords: transfer entropy; information flow; statistical dependency; mutual information; Shannon entropy; information-theoretical quantities; Lorenz equations

PACS: 89.70.Cf; 02.50.-r; 89.70.-a; 05.10.-a; 02.50.Cw

1. Introduction

Complex dynamical systems consisting of nonlinearly coupled subsystems can be found in many application areas ranging from biomedicine [1] to engineering [2,3]. Teasing apart the subsystems and identifying and characterizing their interactions from observations of the system's behavior can be extremely difficult depending on the magnitude and nature of the coupling and the number of variables involved. In fact, the identification of a subsystem can be an ill-posed problem since the definition of strong or weak coupling is necessarily subjective.

The direction of the coupling between two variables is often thought of in terms of one variable driving another so that the values of one variable at a given time influence the future values of the other. This is a simplistic view based in part on our predilection for linear or "intuitively understandable" systems. In nonlinear systems, there may be mutual coupling across a range of temporal and spatial scales so that it is impossible to describe one variable as driving another without specifying the temporal and spatial scale to be considered.

Even in situations where one can unambiguously describe one variable as driving another, inferring the actual nature of the coupling between two variables from data can still be misleading since co-varying variables could reflect either a situation involving coupling where one variable drives another with a time delay or a situation where both variables are driven by an unknown third variable each with different time delays. While co-relation (we use the term co-relation to describe the situation where there is a relationship between the dynamics of the two variables; this is to be distinguished from correlation, which technically refers only to a second-order statistical relationship) cannot imply causality [4], one cannot have causality without co-relation. Thus co-relation can serve as a useful index for a potential causal interaction.

However, if past values of one variable enable one to predict future values of another variable, then this can be extremely useful despite the fact that the relationship may not be strictly causal. The majority of tests to identify and quantify co-relation depend on statistical tests that quantify the amount of information that one variable provides about another. The most common of these are based on linear techniques, which rely exclusively on second-order statistics, such as correlation analysis and Principal Component Analysis (PCA), which is called Empirical Orthogonal Functions (EOFs) in geophysical studies [5]. However, these techniques are insensitive to higher-order nonlinear interactions, which can dominate the behavior of a complex coupled dynamical system. In addition, such linear methods are generally applied by normalizing the data, which implies that they do not depend on scaling effects.

Information-theoretic techniques rely on directly estimating the amount of information contained in a dataset and, as such, rely not only on second-order statistics, but also on statistics of higher orders [6]. Perhaps most familiar is the Mutual Information (MI), which quantifies the amount of information that one variable provides about another variable. Thus MI can quantify the degree to which two variables co-relate. However, since it is a symmetric measure MI cannot distinguish potential directionality, or causality, of the coupling between variables [7].

The problem of finding a measure that is sensitive to the directionality of the flow of information has been widely explored. Granger Causality [8] was introduced to quantify directional coupling between variables. However, it is based on second-order statistics, and as such, it focuses on correlation, which constrains its relevance to linear systems. For this reason, generalizations to quantify nonlinear interactions between bi-variate time-series have been studied [9]. Schreiber proposed an information-theoretic measure called Transfer Entropy (TE) [7], which can be used to detect the directionality of the flow of information. Transfer Entropy, along with other information-based approaches, is included in the survey paper by Hlavackova-Schindler *et al.* [10] and differentiation between the information transfer and causal effects are discussed by Lizier and Propenko [11]. Kleeman presented both TE and time-lagged MI as applied to ensemble weather prediction [12]. In [13], Liang explored the information flow in dynamical systems that can be modeled by equations obtained by the underlying physical concepts. In such cases, the information flow has been analyzed by the evolution of the joint probability distributions using the Liouville equations and by the Fokker-Planck equations, in the cases of the deterministic and stochastic systems, respectively [13].

TE has been applied in many areas of science and engineering, such as neuroscience [1,14], structural engineering [2,3], complex dynamical systems [15,16] and environmental engineering [17,18]. In each of these cases, different approaches were used to estimate TE from the respective datasets. TE essentially quantifies the degree to which past information from one variable provides information about future values of the other variable based solely on the data without assuming any model regarding the dynamical relation of the variables or the subsystems. In this sense TE is a non-parametric method. The dependency of the current sample of a time series on its past values is formulated by k^{th} and l^{th} order Markov processes in Schreiber [7] to emphasize the fact that the current sample depends only on its k past values and the other process's past l values. There also exist parametric approaches where the spatio-temporal evolution of the dynamical system is explicitly modeled [15,16]. However, in many applications it is precisely this model that we would like to infer from the data. For this reason, we will focus on non-parametric methods.

Kaiser and Schreiber [19], Knuth *et al.* [20], and Ruddell and Kumar [17,18] have expressed the TE as a sum of Shannon entropies [21]. In [17,18], individual entropy terms were estimated from the data using histograms with bin numbers chosen using a graphical method. However, as we discuss in Appendix A1, TE estimates are sensitive to the number of bins used to form the histogram. Unfortunately, it is not clear how to optimally select the number of bins in order to optimize the TE estimate.

In the literature, various techniques have been proposed to efficiently estimate information-theoretic quantities, such as the entropy and MI. Knuth [22] proposed a Bayesian approach, implemented in Matlab and Python and known as the Knuth method, to estimate the probability distributions using a piecewise constant model incorporating the optimal bin-width estimated from data. Wolpert and Wolf [23] provided a successful Bayesian approach to estimate the mean and the variance of entropy from data. Nemenman *et al.* [24] utilized a mixture of Dirichlet distributions-based prior in their Bayesian Nemenman, Shafee, and Bialek (NSB) entropy estimator. In another study, Kaiser and Schreiber [19] give different expressions for TE as a summation and subtraction of various (conditional/marginal/joint) Shannon entropies and MI terms. However, it has been pointed out that summation and subtraction of information-theoretic quantities can result in large biases [25,26]. Prichard and Theiler [25] discuss the "bias correction" formula proposed by Grassberger [27] and conclude that it is better to estimate MI utilizing a "correlation integral" method by performing a kernel density estimation (KDE) of the underlying probability density functions (pdfs). KDE tends to produce a smoother pdf estimate from data points as compared to its histogram counterpart. In this method, a preselected distribution of values around each data point is averaged to obtain an overall, smoother pdf in the data range. This preselected distribution of values within a certain range, which is known as a "kernel", can be thought of as a window with a bandwidth [28]. Commonly-used examples of kernels include "Epanechnikov", "Rectangular", and "Gaussian" kernels. Prichard and Theiler showed that pdf models obtained by KDE can be utilized to estimate entropy [25] and other information theoretic quantities, such as the generalized entropy and the Time Lagged Mutual Information (TLMI), using the correlation integral and its approximation through the correlation sums [7]. In [25], Prichard and Theiler demonstrated that the utilization of correlation integrals corresponds to using a kernel that is far from optimal, also known as the "naïve estimator" described in [28]. It is also shown that the relationship between the correlation integral and information theoretic statistics allows defining "local" versions of many information theoretical quantities. Based on these concepts, Prichard and Theiler demonstrated the interactions among the components of a three-dimensional chaotic Lorenz model with a fractal nature [25]. The predictability of the dynamical systems, including the same Lorenz model have been explored by Kleeman in [29,30], where a practical approach for estimating entropy was developed for dynamical systems with non-integral information dimension.

In the estimation of information-theoretical quantities, the KDE approach requires estimation of an appropriate radius (aka bandwidth or rectangle kernel width) for the estimation of the correlation integral. In general cases, this can be accomplished by the Garassberger-Procaccia algorithm, as in [31,33]. In order to compute the TE from data using a KDE of the pdf, Sabesan and colleagues proposed a methodology to explore an appropriate region of radius values to be utilized in the estimation of the correlation sum [14].

The TE can be expressed as the difference between two relevant MI terms [19], which can be computed by several efficient MI estimation techniques using variable bin-width histograms. Fraser and Swinney [34] and Darbellay and Vajda [35] proposed adaptive partitioning of the observation space to estimate histograms with variable bin-widths thereby increasing the accuracy of MI estimation. However, problems can arise due to the subtraction of the two MI terms as described in [19] and explained in [25,26].

Another adaptive and more data efficient method was developed by Kraskov *et al.* [36] where MI estimations are based on *k*-nearest neighbor distances. This technique utilizes the estimation of smooth probability densities from the distances between each data sample point and its *k*-th nearest neighbor and as well as bias correction to estimate MI. It has been demonstrated [36] that no fine tuning of specific parameters is necessary unlike the case of the adaptive partitioning method of Darbellay and Vajda [35] and the efficiency of the method has been shown for Gaussian and three other non-Gaussian distributed data sets. Herrero *et al.* extended this technique to TE in [37] and this has been utilized in many applications where TE is estimated [38,40] due to its advantages.

We note that a majority of the proposed approaches to estimate TE rely on its specific parameter(s) that have to be selected prior to applying the procedure. However, there are no clear prescriptions available for picking these *ad hoc* parameter values, which may differ according to the specific application. Our main contribution is to synthesize three established techniques to be used together to perform TE estimation. With this composite approach, if one of the techniques does not agree with the others in terms of the direction of information flow between the variables, we can conclude that method-specific parameter values have been poorly chosen. Here, we propose using three methods to validate the conclusions drawn about the directions of the information flow between the variables, as we generally do not possess a priori facts about any physical phenomenon we explore.

In this paper, we propose an approach that employs efficient use of histogram based methods, adaptive partitioning technique of Darbellay and Vajda, and KDE based TE estimations, where fine tuning of parameters is required. We propose a Bayesian approach to estimate the width of the bins in a fixed bin-width histogram method to estimate the probability distributions from data.

In the rest of the paper, we focus on the demonstration of synthesizing three established techniques to be used together to perform TE estimation. As the TE estimation based on the *k*-th nearest neighbor approach of Kraskov *et al.* [36] is demonstrated to be robust to parameter settings, it does not require fine tunings to select parameter values. Thus it has been left for future exploration, as our main goal is to develop a strategy for the selection of parameters in the case of non-robust methods.

The paper is organized as follows. In Section 2, background material is presented on the three TE methods utilized. In Section 3, the performance of each method is demonstrated by applying it to both a linearly coupled autoregressive (AR) model and the Lorenz system equations [41] in both the chaotic and sub-chaotic regimes. The latter represents a simplified model of atmospheric circulation in a convection cell that exhibits attributes of non-linear coupling, including sensitive dependence on model parameter values that can lead to either periodic or chaotic variations. Finally conclusions are drawn in Section 4.

2. Estimation of Information-Theoretic Quantities from Data

The Shannon entropy:

$$H(X) = - \sum_{x \in X} p(x) \log p(x) \tag{1}$$

can be used to quantify the amount of information needed to describe a dataset [21]. It can be thought of as the average uncertainty for finding the system at a particular state "*x*" out of a possible set of states "*X*", where $p(x)$ denotes the probability of that state.

Another fundamental information-theoretic quantity is the mutual information (MI), which is used to quantify the information shared between two datasets. Given two datasets denoted by X and Y, the MI can be written as:

$$MI(X,Y) = \sum_{x \in X} \sum_{x \in Y} p(x,y) \log \frac{p(x,y)}{p(x)p(y)} \tag{2}$$

This is a special case of a measure called the Kullback-Leibler divergence, which in a more general form is given by:

$$D_{p||q} = \sum_{x \in X} p(x) \log \frac{p(x)}{q(x)} \tag{3}$$

which is a non-symmetric measure of the difference between two different probability distributions $p(x)$ and $q(x)$. We can see that in Equation (2), the MI represents the divergence between the joint distribution $p(x,y)$ of variables x and y and the product $p(x)p(y)$ of the two marginal distributions. The MI is a symmetric quantity and can be rewritten as a sum and difference of Shannon entropies by:

$$MI(X,Y) = H(X) + H(Y) - H(X,Y) \tag{4}$$

where $H(X,Y)$ is the joint Shannon entropy [21,42].

To define the transfer entropy (TE), we assume that there are two Markov processes such that the future value of each process either depends only on its past samples or on both its past samples and the past samples of the other process. Thus, the TE is defined as the ratio of the conditional distribution of one variable depending on the past samples of both processes *versus* the conditional distribution of that variable depending only on its own past values [7]. Thus the asymmetry of TE results in a differentiation of the two directions of information flow. This is demonstrated by the difference between Equation (5a), which defines the transfer entropy in the direction from X to Y and Equation (5b), which defines the transfer entropy in the direction from Y to X:

$$TE_{XY} = T\left(Y_{i+1} \middle| \mathbf{Y}_i^{(k)}, \mathbf{X}_i^{(l)}\right) = \sum_{y_{i+1}, \mathbf{y}_i^{(k)}, \mathbf{x}_i^{(l)}} p\left(y_{i+1}, \mathbf{y}_i^{(k)}, \mathbf{x}_i^{(l)}\right) \log_2 \frac{p\left(y_{i+1} \middle| \mathbf{y}_i^{(k)}, \mathbf{x}_i^{(l)}\right)}{p\left(y_{i+1} \middle| \mathbf{y}_i^{(k)}\right)} \tag{5a}$$

$$TE_{YX} = T\left(X_{i+1} \middle| \mathbf{X}_i^{(k)}, \mathbf{Y}_i^{(l)}\right) = \sum_{x_{i+1}, \mathbf{x}_i^{(k)}, \mathbf{y}_i^{(l)}} p\left(x_{i+1}, \mathbf{x}_i^{(k)}, \mathbf{y}_i^{(l)}\right) \log_2 \frac{p\left(x_{i+1} \middle| \mathbf{x}_i^{(k)}, \mathbf{y}_i^{(l)}\right)}{p\left(x_{i+1} \middle| \mathbf{x}_i^{(k)}\right)} \tag{5b}$$

where $\mathbf{x}_i^{(k)} = \{x_i, \ldots, x_{i-k+1}\}$ and $\mathbf{y}_i^{(l)} = \{y_i, \ldots, y_{i-l+1}\}$ are past states, and X and Y are k^{th} and l^{th} order Markov processes, respectively, such that X depends on the k previous values and Y depends on the l previous values. In the literature, k and l are also known as the embedding dimensions [33]. As an example, Equation (5b) describes the degree to which information about Y allows one to predict future values of X. Thus, the TE can be used as a measure to quantify the amount of information flow from the subsystem Y to the subsystem X. TE, as a conditional mutual information, can detect synergies between Y and $X^{(k)}$ in addition to removing redundancies [43,44]. In the following sections, we briefly introduce three methods used in the literature to estimate the quantities in Equation (5a) from data.

2.1. Fixed Bin-Width Histogram Approaches

To estimate the quantities in Equation (5a), conditional distributions are generally expressed in terms of their joint counterparts as in:

$$TE_{YX} = T\left(X_{i+1} \middle| \mathbf{X}_i^{(k)}, \mathbf{Y}_i^{(l)}\right) = \sum_{x_{i+1}, \mathbf{x}_i^{(k)}, \mathbf{y}_i^{(l)}} p\left(x_{i+1}, \mathbf{x}_i^{(k)}, \mathbf{y}_i^{(l)}\right) \log_2 \frac{p\left(x_{i+1}, \mathbf{x}_i^{(k)}, \mathbf{y}_i^{(l)}\right) p\left(\mathbf{x}_i^{(k)}\right)}{p\left(x_{i+1}, \mathbf{x}_i^{(k)}\right) p\left(\mathbf{x}_i^{(k)}, \mathbf{y}_i^{(l)}\right)} \tag{6}$$

In this sense, the TE estimation problem can be cast as a problem of density estimation from data. One of the most straightforward approaches to density estimation is based on histogram models [28,45]. However, histograms come with a free parameter—the number of bins. Unfortunately, the estimation of entropy-based quantities varies dramatically as the number of bins is varied. Numerous methods to identify the number of bins that optimally describes the density of a data set have been published [45,46]. However, most of these techniques assume that the underlying density is Gaussian. In this paper, we rely on a generalization of a method introduced by Knuth [20,22], which we refer to as the *Generalized Knuth method*. In this method, each of N observed data points is placed into one of M fixed-width bins, where the number of bins is selected utilizing a Bayesian paradigm. If the volume and the bin probabilities of each multivariate bin are denoted by V and π_i for the ith bin, respectively, then the likelihood of the data is given by the following multinomial distribution:

$$p(\mathbf{d}|M,\pi) = \left(\frac{M}{V}\right)^N \pi_1^{n_1} \pi_2^{n_2} \dots \pi_M^{n_M} \qquad (7)$$

where $\mathbf{d} = [d_1, d_2, \dots, d_N]$ denote the N observed data points, $n_1, n_2, \dots, n_N$ denote the number of data samples in each bin and $\pi = [\pi_1, \pi_2, \dots, \pi_M]$ denote the bin probabilities. Given M bins and the normalization condition that the integral of the probability density equals unity, we are left with M-1 bin probabilities, denoted by $\pi_1, \pi_2, \dots, \pi_{M-1}$. The normalization condition requires that $\pi_M = (1 - \sum_{i=1}^{M-1} \pi_i)$ [22]. The non-informative prior [20] is chosen to represent the bin probabilities:

$$p(\boldsymbol{\pi}|M) = \frac{\Gamma\left(\frac{M}{2}\right)}{\Gamma\left(\frac{1}{2}\right)^M} \left[\pi_1, \pi_2, \dots, \pi_{M-1}, \left(1 - \sum_{i=1}^{M-1} \pi_i\right)\right]^{\frac{-1}{2}} \qquad (8)$$

which is a Dirichlet prior conjugate to the multinomial likelihood function and Γ denotes the Gamma function [56]. The non-informative uniform prior models *a priori* belief regarding the number of bins where C denotes the maximum number of bins considered:

$$p(M) = \begin{cases} C^{-1}, & 1 \le M \le C \\ 0, & \text{otherwise} \end{cases} \qquad (9)$$

The posterior distribution of the bin probabilities and the bin numbers are given by Bayes theorem, which is written here as a proportionality where the Bayesian evidence, p(**d**) is the implicit proportionality constant:

$$p(\boldsymbol{\pi}, M|\mathbf{d}) \propto p(\boldsymbol{\pi}|M)p(M)p(\mathbf{d}|\boldsymbol{\pi}, M) \qquad (10)$$

Since the goal is to obtain the optimal number of constant-width bins one can marginalize over each of the bin probabilities resulting in the posterior of the bin number, which can be logarithmically written as follows [22]:

$$\log p(M|\mathbf{d}) = N \log M + \log \Gamma\left(\frac{M}{2}\right) - M \log \Gamma\left(\frac{1}{2}\right) - \log \Gamma\left(N + \frac{M}{2}\right) + \sum_{i=1}^{M} \log \Gamma\left(n_i + \frac{M}{2}\right) + K \qquad (11)$$

where K is a constant. To find the optimal number of bins, the mode of the posterior distribution in Equation (11) is estimated as follows:

$$\hat{M} = \max_{M} \{\log p(M|d)\} \qquad (12)$$

In Appendix II, we present the performance of entropy estimation based on the selection of the Dirichlet exponent, chosen as 0.5 in Equation (8). Below, we generalize this exponent of the Dirichlet prior to relax the constraint as follows:

$$p(\boldsymbol{\pi}|M) = \frac{\Gamma(\sum_{i=1}^{M} M\beta)}{\Gamma(\beta)^M} \left[\pi_1, \pi_2, \dots, \pi_{M-1}, \left(1 - \sum_{i=1}^{M-1} \pi_i\right)\right]^{\beta-1} \qquad (13)$$

In the literature, the prior in Equation (13) has been utilized to estimate the discrete entropy given by Equation (1), where *the number of bins are assumed to be known*, whereas here, we try to approximate a continuous pdf, thus the entropy, using a piecewise-constant model, where the number of bins is not known. In these publications, the main concern is to estimate Equation (1) as efficiently as possible for a small number of data samples. Different estimators have been named. For example, the assignment of β = 0.5 results in the Krichevsky-Trofimov estimator and the assignment of $\beta = \frac{1}{M}$ results in the Schurman-Grassberger estimator [23,24]. Here, we aim to approximate the continuous-valued differential entropy of a variable shown by using finite-precision data:

$$h(X) = -\int \hat{p}(x)\log\left[\frac{\hat{p}(x)}{m(x)}\right]dx \tag{14}$$

Using the same prior for the number of bins in Equation (9) and the procedures given by Equation (10) through Equation (12), the marginal posterior distribution of the bin numbers under the general Dirichlet prior Equation (13) is given by:

$$\log p(M|\mathbf{d}) = N \log M + \log \Gamma(M\beta) - M \log \Gamma(\beta) - \log \Gamma(N + M\beta) + \sum_{i=1}^{M} \log \Gamma(n_i + \beta) + K \tag{15}$$

Again, the point estimate for the optimal bin number can be found by identifying the mode of the above equation, that is, $\hat{M} = \max_{M}\{\log p(M|\mathbf{d})\}$ where $p(M \mid \mathbf{d})$ is obtained from Equation (15).

After the estimation of the optimal number of bins, the most important step is the accurate calculation of TE from the data. In [19], the TE is expressed as a summation of Shannon entropy terms:

$$TE_{YX} = T\left(X_{i+1}\middle|\mathbf{X}_i^{(k)},\mathbf{Y}_i^{(l)}\right) = H\left(\mathbf{X}_i^{(k)},\mathbf{Y}_i^{(l)}\right) - H\left(\mathbf{X}_i^{(k+1)},\mathbf{Y}_i^{(l)}\right) + H\left(\mathbf{X}_i^{(k+1)}\right) - H\left(\mathbf{X}_i^{(k)}\right) \tag{16}$$

where $\mathbf{X}_i^{(k)} = \{\mathbf{X}_i, \mathbf{X}_{i-1}, \ldots, \mathbf{X}_{i-k+1}\}$ denotes a matrix composed of k vectors [19] where $i = max(k,l)+1$. In other words, the latter representation can be interpreted as a concatenation of k column vectors in a *matrix*, where $\mathbf{X}_i = [x_i, x_{i-1}, \ldots, x_{i-S}]^T$, $\mathbf{X}_i = [x_{i-1}, x_{i-2}, \ldots, x_{i-S+1}]^T$ and S is the length of the column vector, defined as $S = max\ (length(X), length(Y))$. Here, $(\cdot)^T$ denotes transposition. Above, $H\left(\mathbf{X}_i^{(k)},\mathbf{Y}_i^{(l)}\right)$ is short for $H(\mathbf{X}_i, \mathbf{X}_{i-1}, \ldots, \mathbf{X}_{i-k+1}, \mathbf{Y}_i, \mathbf{Y}_{i-1}, \ldots, \mathbf{Y}_{i-l+1})$, where x denotes a particular value of the variable X and boldface is utilized to represent vectors. If $k = l = 1$ is selected, Equation (16) takes the following simplified form [20]:

$$TE_{YX} = T(X_{i+1}|X_i, Y_i) = H(X_i, Y_i) - H(X_{i+1}, X_i, Y_i) + H(X_{i+1}, X_i) - H(X_i) \tag{17}$$

In the best scenario, the above TE estimation requires the three-dimensional joint Shannon entropy, whereas its general expression in Equation (16) needs a $k+l+1$-dimensional entropy estimation.

According to our tests, if we use the prior in Equation (13), when $\beta = 10^{-10}$, posterior pdf estimates are biased significantly (see Appendix III), especially in high-dimensional problems. Thus, we aim to overcome this problem by using the generalized prior in Equation (13) for the Dirichlet prior with β values around 0.1. Using the generalized prior Equation (13), after selecting the number of bins by Equation (15), the mean value of the posterior bin height probability can be estimated by [22]:

$$\langle \pi_i \rangle = \frac{n_i + \beta}{N + M\beta}, k = 1, \ldots, M. \tag{18}$$

As the prior is Dirichlet and the likelihood function is multinomial-distributed, the posterior distribution of bin heights is Dirichlet-distributed with the mean given in Equation (18) above [22,47]. This allows us to sample from the Dirichlet posterior distribution of the bin heights to estimate the joint and marginal pdf's in the TE equations and then estimate their Shannon entropies and their uncertainties, too. The schematic in Figure 1 illustrates this procedure for estimating the entropies and their associated uncertainties.

11

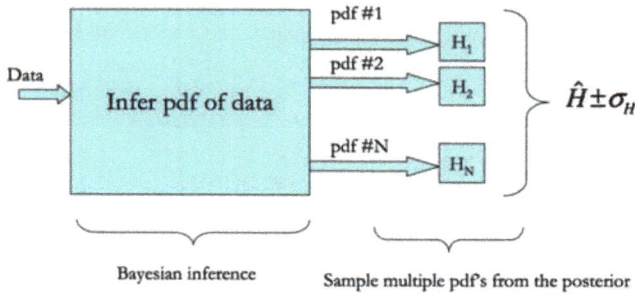

Figure 1. This schematic illustrates the procedure for estimating entropy as well as the uncertainty from data. First the number of bins is selected using the mode of the logarithm of the Dirichlet posterior in Equation (15). The Dirichlet posterior is then sampled resulting in multiple estimates of the pdf. The entropy of each pdf is estimated and the mean and standard deviation computed and reported.

As previously described, this method is known to produce biases, especially as higher dimensions are considered. There are a couple of reasons for this. As the pdf is modeled by a uniform distribution within a single bin, this corresponds to the maximum entropy for that bin. Additionally, Equation (18) tells us that, even if there is no data sample in a specific bin, an artificial amount β is added to the average bin probability. On the other hand, this addition mitigates the entropy underestimation encountered in the case of many empty bins, which is prevalent in higher dimensions. Moreover, the TE is estimated by the addition and subtraction of the marginal and joint Shannon entropies, as shown in Equation (16). Prichard and Theiler describe the artifacts originating from this summation procedure and advise using KDE methods instead [25]. However, before considering the KDE method, we discuss an alternative histogram method that has been proposed to overcome some of the drawbacks of the fixed-bin-width histogram approaches. In addition to the conjugate pairs of multinomial likelihood and Dirichlet prior model, the research topic of exploring other models has always been interesting. In addition to this conjugate pair, optimal binning in the case of other models, such as that of [24] including a mixture of Dirichlets provides a challenging research for optimal binning of data with the goal of efficient pdf estimation from the data.

2.2. Adaptive Bin-Width Histogram Approaches

The fixed bin-width histogram approaches are not very effective for estimating information-theoretic quantities from data due to the inaccurate filling of the bins with zero sampling frequency. Instead of generating a model based on bins with equal width, one can design a model consisting of bins with varying widths, determined according to a statistical criterion. Fraser and Swinney [34] and Darbellay and Vajda [35] proposed the adaptive partitioning of the observation space into cells using the latter approach and estimated the mutual information (MI) directly. Here, we will focus on the method proposed by Darbellay and Vajda. This approach relies on iteratively partitioning the cells on the observation space, based on a chi-square statistical test to ensure conditional independence of the proposed partitioned cells from the rest of the cells. We explain the details of this method schematically on Figure 2. Here, observation space of (X, Y) is shown by the largest rectangle. The partitioning of the observation space is done as follows:

1. Initially, we start with the largest rectangle containing all data samples.
2. Any cell containing less than two observations (data pairs) is not partitioned. The cell, which is partitioned into smaller blocks, is known as the parent cell; whereas each smaller block after partitioning is named as a child cell.

3. Every cell containing at least two observations is partitioned by dividing each one of its edges into two halves. It means four new cells are generated (according to the independence test, which will be described below).

4. In order to test whether we need to partition the upper cell (parent cell) into more cells (child cells), we rely on the Chi-Square test of independence, where the null hypothesis is phrased as follows:

H_0: Sample numbers N_1, N_2, N_3, N_4 in four child cells are similar (in other words, the sample distribution in the parent cell was uniform)

The Chi-Square (χ^2) test statistic for a 5% significance level with 3 degrees of freedom is given as follows:

$$T = \sum_{i=1}^{4} \left(\frac{\sum N_i}{4} - N_i \right)^2 \leq \chi^2_{95\%}(3) = 7.81 \tag{19}$$

If we happen to find that $T > 7.81$, we decide that the numbers of samples in each child cell are not similar and therefore we continue partitioning. Otherwise, we conclude that the numbers are similar and partitioning is stopped at this level. The data samples in this cell are used in the MI estimation.

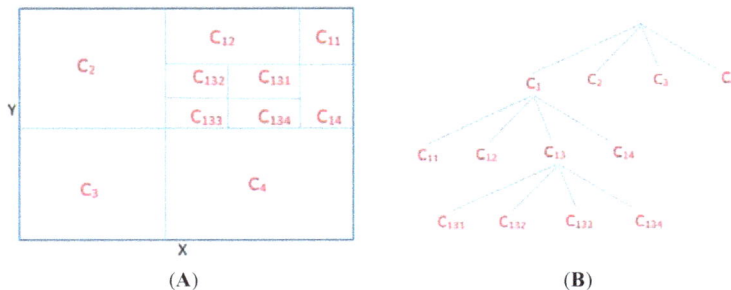

Figure 2. Illustration of the Adaptive Partitioning algorithm of Darbellay and Vajda (**A**) The observation space of two-dimensional data (X,Y) and its illustrative partitioning according to the independence test; (**B**) The corresponding tree showing the partitioning of each cell.

In this method, the level of statistical significance can be chosen according to the design, thus raising as *a parameter to be tuned* according to the application. After the partitioning is completed, the MI is estimated as shown below:

$$\hat{MI}_N(X,Y) = \sum_{i=1}^{m} \frac{N_i}{N} \log \frac{\frac{N_i}{N}}{\left(\frac{N_{x,i}}{N}\right)\left(\frac{N_{y,i}}{N}\right)} \tag{20}$$

where N denotes the total number of data samples with N_i showing the subset of these samples that fall into the ith cell, C_i, after the partitioning process is completed. Above, $N_{x,i}$ and $N_{y,i}$ represent the numbers of observations having the same x and y coordinates as observations in the cell C_i, respectively. The partitioning process is illustrated below using a similar discussion to that in [35]. The observation space is first divided into four child cells, namely C_1, C_2, C_3, C_4, to maintain equiprobable distributions in each cell. This forms the first set of branches shown in Figure 2. Then, according to the independence test, C_1 is divided into four cells whereas C_2, C_3, C_4 are retained to be included in the MI estimation and they are not divided into more child cells, forming the second layer of the partitioning tree shown in Figure 2. Finally, the third child cell of partition C_3 is divided into four child cells, namely C_{131}, C_{132}, C_{133} and C_{134}. In the last step, each cell is utilized in the MI estimation formula given by Equation (20). It should be noted that the partitioning is performed symbolically here for the sake of a better explanation without showing the actual data samples on the observation space, as done in [35].

As a result, finer resolution is used to describe larger MI regions and lower resolution is used for smaller MI regions [35]. Having estimated the MI from the data efficiently, the TE can be calculated using the expressions from [19] by:

$$TE_{YX} = T\left(X_{i+1}\middle|\mathbf{X}_i^{(k)}, \mathbf{Y}_i^{(l)}\right) = MI\left(X_{i+1}, \left[\mathbf{X}_i^{(k)}, \mathbf{Y}_i^{(l)}\right]\right) - MI\left(X_{i+1}, \mathbf{X}_i^{(k)}\right) \tag{21}$$

where $MI\left(X_{i+1}, \left[\mathbf{X}_i^{(k)}, \mathbf{Y}_i^{(l)}\right]\right)$ denotes the MI between X_{i+1} and the joint process denoted by $\left[\mathbf{X}_i^{(k)}, \mathbf{Y}_i^{(l)}\right]$ [19].

Because the MI is estimated more efficiently by this method, the overall TE estimation becomes less biased compared to the previous methods. However, the subtraction operation involved in Equation (21) can still produce a significant bias in the TE calculations. To overcome problems related to the addition and subtraction of information-theoretic quantities, KDE estimation methods have been utilized in the literature to estimate MI and redundancies [25], and TE in [49].

2.3. Kernel Density Estimation Methods

Kernel Density Estimation (KDE) is utilized to produce a smoothed pdf estimation using the data samples, which stands in contrast to the histogram model which has sharp edges resulting from a uniform distribution within each bin. In this method, a preselected distribution of values around each data sample is summed to obtain an overall, smoother pdf in the data range. This preselected distribution of values within a certain range is known as a "kernel". Some of the most commonly used kernels are "Epanechnikov", "Rectangular" and "Gaussian" [28]. Each kernel can be thought of as a window with a bandwidth or radius. Prichard and Theiler [25] showed that KDE can also be utilized to estimate entropy by the computation of the generalized correlation integral [7], which is approximated by the correlation sum. Even if a rectangular kernel is used, the resulting entropy estimation is more accurate compared to the histogram approach as discussed in [25]. In this method, entropies are estimated by first calculating the correlation sums through the Grassberger-Procaccia (GP) algorithm or some other effective procedure. Interested readers are referred to [31,33] for a detailed description of the algorithm. Here, the joint probabilities in the TE expression Equation (6) can be estimated from data by the following equation, which is known as the generalized correlation sum [7]:

$$p_\varepsilon\left(x_{i+1}, \mathbf{x}_i^{(k)}, \mathbf{y}_i^{(l)}\right) \cong \frac{1}{N} \sum_{\substack{m \\ i \neq m}}^{N} \Theta\left(\varepsilon - \left\| \begin{array}{c} x_{i+1} - x_{m+1} \\ \mathbf{x}_i^{(k)} - \mathbf{x}_m^{(k)} \\ \mathbf{y}_i^{(l)} - \mathbf{y}_m^{(l)} \end{array} \right\| \right) = C\left(x_{i+1}, \mathbf{x}_i^{(k)}, \mathbf{y}_i^{(l)}; \varepsilon\right) \tag{22}$$

where $\Theta(x > 0) = 1$; $\Theta(x < 0) = 0$ is the Heaviside function and ε is the radius around each data sample. In Equation (22), we count the number of neighboring data samples which are within ε distance. As a distance measure, the maximum norm, denoted $\|\cdot\|$, has been selected here, but the Euclidean norm could also be utilized. On the right-hand side of Equation (22), $C\left(x_{i+1}, \mathbf{x}_i^{(k)}, \mathbf{y}_i^{(l)}; \varepsilon\right)$ gives the mean probability that the states at two different indices (i and m) are within ε distance of each other. Using Equation (22), the TE can be expressed as [49,50]:

$$TE_{YX} = \left\langle \log_2 \frac{C\left(x_{i+1}, \mathbf{x}_i^{(k)}, \mathbf{y}_i^{(l)}; \varepsilon\right) C\left(\mathbf{x}_i^{(k)}; \varepsilon\right)}{C\left(x_{i+1}, \mathbf{x}_i^{(k)}; \varepsilon\right) C\left(\mathbf{x}_i^{(k)}, \mathbf{y}_i^{(l)}; \varepsilon\right)} \right\rangle \tag{23}$$

where $\langle . \rangle$ denotes the expectation [50].

The problem is that this method also has a free parameter, the radius value, ε, which must be selected to estimate the neighborhoods. Choosing this radius is similar to choosing the fixed bin width in a histogram. We utilize the Grassberger-Procaccia algorithm to plot log

ε *versus*$\log\left(C\left(x_{i+1}, \mathbf{x}_i^{(k)}, \mathbf{y}_i^{(l)}; \varepsilon\right)\right)$. The linear section along the resulting curve is used to select the radius ε. However, picking a radius value from any part of the linear section of the log ε *versus*$\log\left(C\left(x_{i+1}, \mathbf{x}_i^{(k)}, \mathbf{y}_i^{(l)}; \varepsilon\right)\right)$ curve appears to make the results sensitive over the broad range of possible values. This is explored in the following section and the benefit of exploring an appropriate radius range with the help of the embedding dimension selection is pointed out.

2.3.1. Selection of the Radius with the Embedding Dimensions in Kernel Density Estimation (KDE) Method

Here we explain a method to select the radius based on the above discussion in concert with the choice of the embedding dimensions k and l, based on the discussions in [14]. We demonstrate this procedure on a system consisting of a pair of linearly-coupled, autoregressive signals [51]:

$$
\begin{aligned}
y(i+1) &= 0.5y(i) + n_1(i) \\
x(i+1) &= 0.6x(i) + cy(i) + n_2(i)
\end{aligned}
\qquad
\begin{aligned}
n_1 &\sim \mathcal{N}(0,1) \\
n_2 &\sim \mathcal{N}(0,1) \\
c &\in [0.01, 1]
\end{aligned}
\tag{24}
$$

where $\mathcal{N}(\mu, \sigma)$ denotes the normal distribution with mean μ and standard deviation σ and the constant c denotes the coupling coefficient. First, we generate the log ε versus $\log\left(C\left(x_{i+1}, \mathbf{x}_i^{(k)}, \mathbf{y}_i^{(l)}; \varepsilon\right)\right)$ curve. The log ε versus $\log\left(C\left(x_{i+1}, \mathbf{x}_i^{(k)}, \mathbf{y}_i^{(l)}; \varepsilon\right)\right)$ curve is displayed in Figure 3 for different k values and $c = 1$.

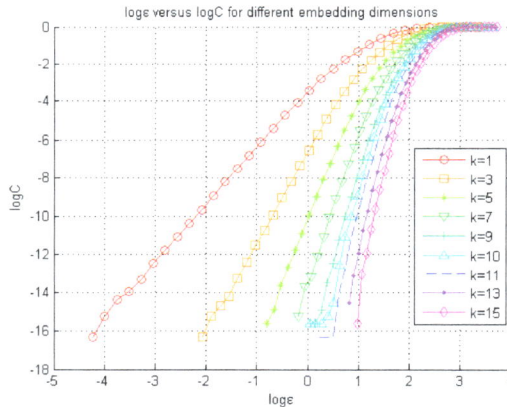

Figure 3. Exploration of the optimal radius for the KDE of a pdf using the Grassberger-Procaccia method. The figure illustrates the Correlation Sum, defined in Equation (22), estimated at different radius values represented by ε for the coupled AR model.

Here, $l = 1$ is selected [14]. It is known that the optimal radius lies in the linear region of these curves, where its logarithm is a point on the horizontal axis [14]. Above, we notice that the range of the radius values corresponding to the linear section of each curve varies significantly. As the k value increases, the linear region for each curve moves right, toward higher ε values [33]. With the increasing embedding dimensions, the embedding vectors include data, which are sampled with a lower frequency, *i.e.*, undersampling, leading to an increase in ε to achieve the same correlation sum obtained with a smaller radius. For example, a set of radius values within the range of $-3 \leq \log \varepsilon \leq 0$ provides the linear section of the log C curve for $k = 1$, whereas these values are not within the range of radius values used in forming the log C -log ε curve for an embedding dimension of $k = 10$. Thus, selection of an embedding dimension k first and then a radius value from the corresponding linear region on the curve can help us search for the radius in a more constrained and efficient way.

Entropy **2015**, *17*, 438–470

As seen in Figure 3, we end up with different radius ranges to select, based on the determination of the embedding dimension, k. Sabesan *et al.* [14] provide an approach to select the radius (ε) and k together.

According to [14,52], the embedding dimension, k, can be selected by considering the first local minimum of the Time-Lagged MI (TLMI) of the destination signal, followed by the determination of a radius value. The radius is selected such that it falls into the linear region of the curve for the corresponding k value, given in Figure 3. The k value, corresponding to the first local minima of MI(k), provides us with the time-lag k, where the statistical dependency between the current sample x_i and its k past value x_{i-k} is small. TLMI is defined by the following equation for the AR signal given in Equation (24):

$$MI(k) = \sum_x p(x_i, x_{i-k}) \log \frac{p(x_i, x_{i-k})}{p(x_i)p(x_{i-k})} \tag{25}$$

Below, we provide an estimate of the MI(k) of the AR signal, x_i, for different time lags $k \in [1, \ldots, 50]$. The adaptive partitioning algorithm of Darbellay and Vajda [35] was utilized to estimate the MI. As the MI is not bounded from above, we normalize its values between 0 and 1 as recommended in the literature, using the following formula [53]:

$$\lambda = \sqrt{1 - e^{-2MI}} \tag{26}$$

In Figure 4, we show the normalized MI for different lags after taking an ensemble of 10 members of the AR process x and utilizing an averaging to estimate MI.

Above, the first local minimum value of MI(k) is obtained at $k = 10$. Thus, we turn to Figure 3 to select a radius value on the linear region of the curve with the embedding dimension $k = 10$. This region can be described by the following values: $0.8 \leq \log \varepsilon \leq 1.4$. Thus, we can choose $k = 10$ and $\log \varepsilon = 0.85$ along with $l = 1$. If $k = l = 1$ is selected, the corresponding linear region on Figure 3 changes and a selection of $\log \varepsilon = -1$ can be chosen, instead. Once the radius and the embedding dimensions are determined, TE is estimated by Equation (23) using the correlation sums. These estimates are illustrated in Figure 5 for $k = l = 1$ and $k = 10, l = 1$; respectively.

In the next section, we will elaborate on the performance of the three methods used in TE estimation and emphasize the goal of our approach, which is to use all three methods together to fine tune their specific parameters.

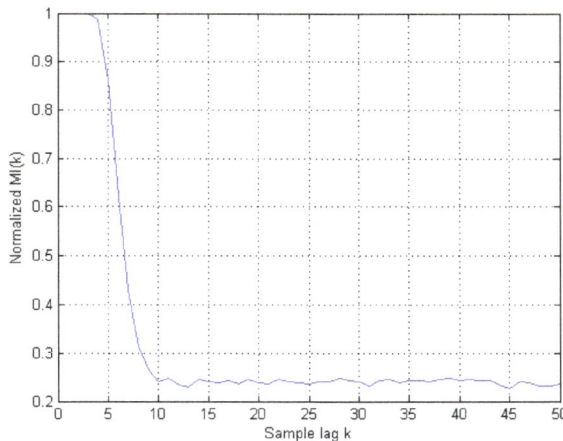

Figure 4. Ensemble averaged and normalized Time-lagged MI(k). As described in the text, the first local minima of the MI leads to the condition $k = 10$.

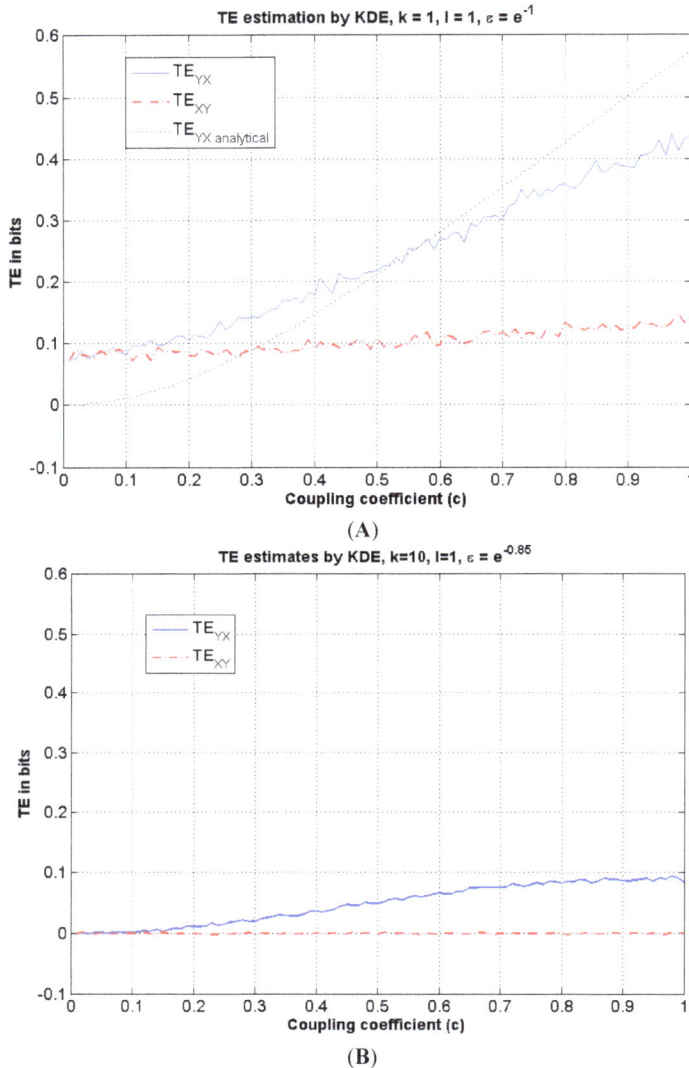

Figure 5. This figure illustrates TE estimation *versus* the coupling coefficient *c* in Equation (24) using the KDE method. (**A**) Both TE_{YX} (blue-solid) and TE_{XY} (red-dash dot) are estimated using the KDE method and illustrated along with the analytical solution (black-dotted) for $k = l = 1$. As there is no coupling from *X* to *Y*, analytically $TE_{XY} = 0$; (**B**) TE_{YX} (blue-solid) and TE_{XY} (red-dash dot) are estimated using the KDE method for $k = 10$, $l = 1$.

3. Experiments

In the preceding section, we described three different methods for estimating the TE from data, namely: the Generalized Knuth method, the adaptive bin-width histogram and the KDE method. We emphasized that we can compute different TE values by these three different methods, as the TE estimations depend on various factors, such as the value of the selected fixed bin-width, the bias resulting due to the subtraction and addition of various Shannon entropies, embedding dimensions and the value of the chosen KDE radius value. Due to this uncertainty in the TE estimations, we propose to use these three main techniques *together* to compute the TE values and to consistently identify the direction of relative information flows between two variables. With this approach, if one of the techniques does not agree with the others in terms of the direction of information flows between the variables, we determine that we need to fine tune the relevant parameters until all three methods agree with each other in the estimation of the NetTE direction between each pair of variables. The NetTE between two variables X and Y is defined to be the difference between TE_{XY} and TE_{YX}, which is defined as the difference of the TE magnitudes with opposite directions between X and Y:

$$NetTE_{XY} = \max(TE_{YX}, TE_{XY}) - \min(TE_{YX}, TE_{XY}) \qquad (27)$$

The NetTE allows us to compare the relative values of information flow in both directions and conclude which flow is larger than the other, giving a sense of main interaction direction between the two variables X and Y.

In order to use three methods together, we demonstrate our procedure on a synthetic dataset generated by a bivariate autoregressive model given by Equation (24). In Section 2.3.1, we have already described the KDE method using this autoregressive model example and we have explored different radius values in the KDE method by utilizing the Grassberger-Procaccia approach in conjunction with different selections of k values. In Section 3.1, we continue demonstrating the results using the same bivariate autoregressive model. We focus on the analysis of the adaptive partitioning and the Generalized Knuth methods. First, we analyze the performance of the adaptive partitioning method at a preferred statistical significance level. Then, we propose to investigate different β values to estimate the optimal fixed bin-width using Equation (15) in the Generalized Knuth method.

If an information flow direction consensus is not reached among the three methods, we try different values for the fine-tuning parameters until we get a consensus in the NetTE directions.

When each method has been fine-tuned to produce the same NetTE estimate, we conclude that the information flow direction has been correctly identified.

In Section 3.2, we apply our procedure to explore the information flow among the variables of the nonlinear dynamical system used by Lorenz to model an atmospheric convection cell.

3.1. Linearly-Coupled Bivariate Autoregressive Model

In this section, we apply the adaptive partitioning and the Generalized Knuth methods to estimate the TE among the processes defined by the same bivariate linearly-coupled autoregressive model (with variable coupling values) given by the equations in Equation (24). We demonstrate the performance of each TE estimation method using an ensemble of 10 members to average. The length of the synthetically generated processes is taken to be 1000 samples after eliminating the first 10,000 samples as the transient. For each method, TE estimations *versus* the value of coupling coefficients are shown in Figure 5—for both directions between processes X and Y. It should be noted that the process X is coupled to Y through coefficient c. Thus, there is no information flow from X to Y for this example, *i.e.*, $TE_{XY} = 0$ analytically. The analytical values of TE_{YX} have been obtained using the equations in [19] for $k = 1$ and $l = 1$. The performance of the three methods have been compared for the case of $k = 1$ and $l = 1$.

Below, TE is estimated for both directions using coupling values ranging from $c = 0.01$ to $c = 1$ in Equation (24). The information flows are consistently estimated to be in the same direction for all three methods, *i.e.*, $TE_{YX} \geq TE_{XY}$. If we compare the magnitudes of these TE estimates, we observe that the biases between the analytic solution and the TE_{YX} of the adaptive partitioning method, KDE and the Generalized Knuth method increase as the coefficient of the coupling in the autoregressive model increases to $c = 1$.

Above, we demonstrate the TE estimations using the KDE method with different embedding dimensions and different radius values. In Figure 5, we observe that the directions of each TE can be estimated correctly, *i.e.*, $TE_{YX} \geq TE_{XY}$. for the model given in Equation (24), demonstrating that we can obtain the same information flow directions, but with different bias values.

Below, results in Figure 5 are compared with the other two techniques for $k = l = 1$.

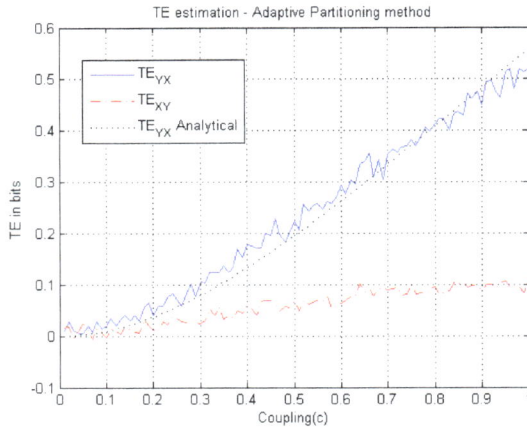

Figure 6. This figure illustrates TE estimation *versus* the coupling coefficient c in Equation (24) using the adaptive partitioning method. Both TE_{YX} (blue-solid) and TE_{XY} (red-dash dot) are estimated using the adaptive partitioning method and illustrated along with the analytical solution (black-dotted). As there is no coupling from X to Y, analytically $TE_{XY} = 0$. A statistical significance level of 5% has been utilized in the χ^2 test Equation (19) for a decision of partitioning with $k = l = 1$.

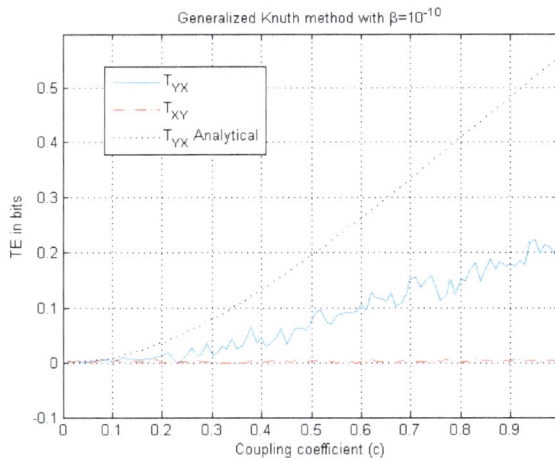

Figure 7. This figure illustrates TE estimation *versus* the coupling coefficient c in Equation (24) using the Generalized Knuth method. Both TE_{YX} (blue solid) and TE_{XY} (red-dash dot) are estimated for $\beta = 10^{-10}$ and illustrated along with the analytical solution (black dotted) where $k = l = 1$ is chosen.

When the magnitudes of the TE_{YX} estimates are compared in Figure 5 and , we observe bias both in TE_{YX} *and* TE_{XY}, whereas there is no bias in the TE_{XY} estimate in the Generalized Knuth method using $\beta = 10^{-10}$. On the other hand, the adaptive partitioning method provides the least bias for TE_{YX}

whereas KDE seems to produce larger bias for low coupling values and lower bias for high coupling values in Figure 5, compared to the Generalized Knuth method with $\beta = 10^{-10}$ in Figure 7.

For example, for $c = 1$, we note from the three graphs that the estimated transfer entropies are $TE_{YX} \cong 0.52$, $TE_{YX} \cong 0.43$, $TE_{YX} \cong 0.2$, for the adaptive partitioning, the KDE with $k = l = 1$ and the Generalized Knuth method with $\beta = 10^{-10}$, respectively. As the bias is the difference between the analytical value ($TE_{YX} = 0.55$ *for* $k = l = 1$) and the estimates, it obtains its largest value in the case of the Generalized Knuth method with $\beta = 10^{-10}$. On the other hand, we know that there is no information flow from the variable X to variable Y, *i.e.*, $TE_{XY} = 0$. This fact is reflected in Figure 7, but not in Figure 5 and where TE_{xy} is estimated to be non-zero, implying bias. As the same computation is also utilized to estimate TE_{YX} (in the other direction), we choose to analyze the NetTE, which equals the difference between TE_{YX} and TE_{XY}, which is defined in Equation (27). Before comparing the NetTE obtained by each method, we present the performance of the proposed Generalized Knuth method for different β values.

3.1.1. Fine-Tuning the Generalized Knuth Method

In this sub-section, we investigate the effect of β on the TE estimation bias in the case of the Generalized Knuth method. The piecewise-constant model of the Generalized Knuth method approaches a pure likelihood-dependent model, which has almost a constant value as β goes to zero in Equation (18). In this case, the mean posterior bin heights approach their frequencies in a bin, *i.e.*, $\langle \pi_i \rangle = \frac{n_i}{N}$. In this particular case, empty bins of the histogram cause large biases in entropy estimation, especially in higher dimensions as the data becomes sparser. This approach can only become unbiased asymptotically [54]. However, as shown in Equation (18), the Dirichlet prior with exponent β artificially fills each bin by an amount, β, reducing the bias problem. In Appendix III, Figure A.3 illustrates the effect of the free parameter β on the performance of the marginal and joint entropy estimates. We find that the entropy estimates fall within one to two standard deviations for $\beta \cong 0.1$. The performance degrades for much smaller and much larger β values. Figure 8 and illustrate less bias in TE_{YX} estimates for $\beta = 0.1$ and $\beta = 0.5$ unlike the case in shown in Figure 7 where we use $\beta = 10^{-10}$. However, the bias increases for low coupling values in these two cases. To illustrate the net effect of the bias, we explore NetTE estimates of Equation (27) for these cases in Section 3.1.2.

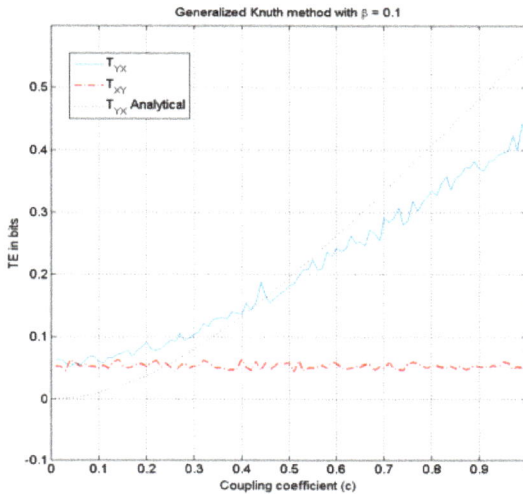

Figure 8. This figure illustrates TE estimation *versus* the coupling coefficient c in Equation (24) using the Generalized Knuth method method for $\beta = 0.1$, $k = l = 1$. These are illustrated along with the analytical solution.

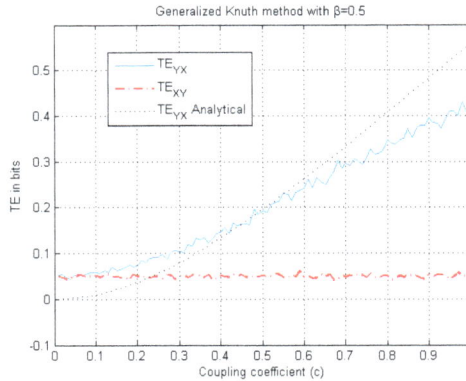

Figure 9. This figure illustrates TE estimation *versus* the coupling coefficient c in Equation (24) using the generalized piecewise-constant method (Knuth method) for $\beta = 0.5$, $k = l = 1$. These are illustrated along with the analytical solution.

3.1.2. Analysis of NetTE for the Bivariate AR Model

Since we are mainly interested in the direction of the information flow, we show that the estimation of the NetTE values exhibit more quantitative similarity among the methods for the case where $k = l = 1$ (Figure 10).

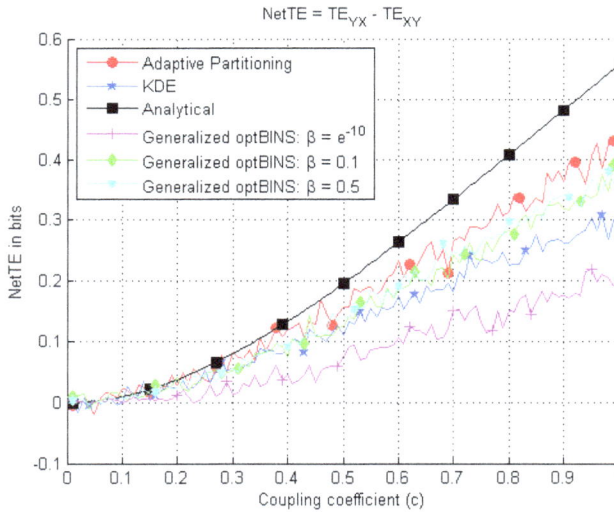

Figure 10. This figure illustrates the NetTE difference, given by Equation (27) between each pair of variables in Equation (24). Estimations are performed using all three methods and considering different β values in the case of the Generalized Knuth method.

In the KDE (Figure 5), Adaptive partitioning (Figure 6) and the Generalized Knuth method with $\beta = 0.1$ and $\beta = 0.5$, (Figures 8 and 9) a non-zero TE_{XY} is observed. The NetTE between the variables X and Y of the bivariate auroregressive model in Equation (24) still behaves similarly giving a net information flow in the direction of the coupling from Y to X as expected. Thus, in this case we find that the NetTE behaves in the same way, even though the individual TE estimates of each method have different biases. Above, we observe that the NetTE estimate of the adaptive partitioning outperforms the Generalized Knuth method with $\beta = 0.1$ and $\beta = 0.5$ and KDE. The largest bias in NetTE is achieved by the Generalized Knuth method with $\beta = 10^{-10}$. However, all methods agree that the information flow from Y to X is greater than that of X to Y, which is in agreement with the theoretical result obtained from Equation (24) using the equations in [19]. In the literature, the bias in the estimation has been obtained using surrogates of TE's estimated by shuffling the data samples [38]. These approaches will be explored in future work.

3.2. Lorenz System

In this section, the three methods of Section 2 are applied to a more challenging problem involving the detection of the direction of information flow among the three components of the Lorenz system, which is a simplified atmospheric circulation model that exhibits significant non-linear behavior. The Lorenz system is defined by a set of three coupled first-order differential equations [41]:

$$\begin{aligned} \frac{dX}{dt} &= \sigma(Y - X) \\ \frac{dY}{dt} &= -XZ + RX - Y \\ \frac{dZ}{dt} &= XY - bZ \end{aligned} \tag{28}$$

where $\sigma = 10, b = 8/3$, $R = 24$ (*sub − chaotic*) or $R = 28$ (*chaotic*). These equations derive from a simple model of an atmospheric convection cell, where the variables x, y, and z denote the convective velocity, vertical temperature difference and the mean convective heat flow, respectively. These equations are used to generate a synthetic time series, which is then used to test our TE estimation procedure. In the literature, the estimation of the TE of two Lorenz systems with nonlinear couplings have found applications in neuroscience [14,39,55]. Here, we explore the performance of our approach on a single Lorenz system which is not coupled to another one. Our goal is to estimate the interactions among the three variables of a single Lorenz system–not coupling from one system to another.

In our experiments, we tested the adaptive partitioning, KDE and Generalized Knuth methods in the case where the Rayleigh number, $R = 28$, which is well-known to result in chaotic dynamics and also for the sub-chaotic case where $R = 24$. For each variable, we generated 15,000 samples and used the last 5000 samples after the transient using a Runge-Kutta-based differential equation solver in MATLAB (ode45). Both in the chaotic and sub-chaotic cases, $\beta = 0.1$ was used at the Generalized Knuth method and a 5% significance level was selected in the adaptive partitioning method. Embedding dimensions of $k = l = 1$ have been selected in these two methods.

The embedding dimension values were implemented according to Section 2.3.2 at the KDE method: The log ε *versus* $\log\left(C\left(x_{i+1}, \mathbf{x}_i^{(k)}, \mathbf{y}_i^{(l)}; \varepsilon\right)\right)$ curves have been estimated for the chaotic and sub-chaotic cases.

In the chaotic case, the first minimum of TLMI was found to be at $k = 17$ and $\varepsilon = e^{-1}$ occured in the middle of the radii range of the linear part of the curve. The value of $l = 1$ was selected for both the chaotic and sub-chaotic cases. The curves for different k values have been illustrated in Figure 11 for the analysis of the interaction between X and Y. Similar curves have been observed for the analysis of the interactions between the other pairs in the model.

In the sub-chaotic case, values around $k = 15$ have been observed to provide the first local minimum of TLMI(k). However, the NetTE direction consistency cannot be obtained with the other two techniques, namely, the adaptive partitioning and the Generalized Knuth method. Therefore, as we propose in our method, k value has been fine-tuned along with the radius until we obtain consistency of NetTE directions among the three methods. Selection of $k = 3$, $l = 1$, $\varepsilon = e^{-2}$ has provided this consistency, where the NetTE directions are illustrated in Figure 15. Figure 12 illustrates log ε *versus* $\log\left(C\left(x_{i+1}, \mathbf{x}_i^{(k)}, \mathbf{y}_i^{(l)}; \varepsilon\right)\right)$ curves used in the selection of the appropriate region for ε, in the sub-chaotic case.

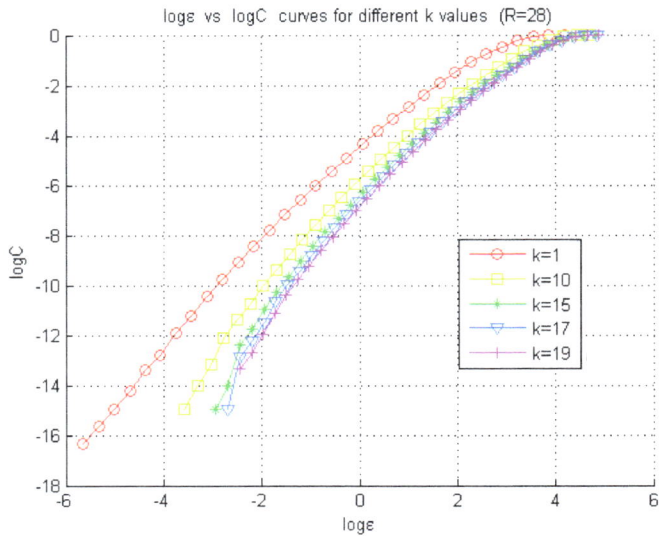

Figure 11. Exploration of the optimal radius for the KDE of a pdf using the Grassberger-Procaccia method. The figure illustrates the Correlation Sum Equation (22) estimated at different radius values represented by ε for the Lorenz model in the chaotic regime ($R = 28$).

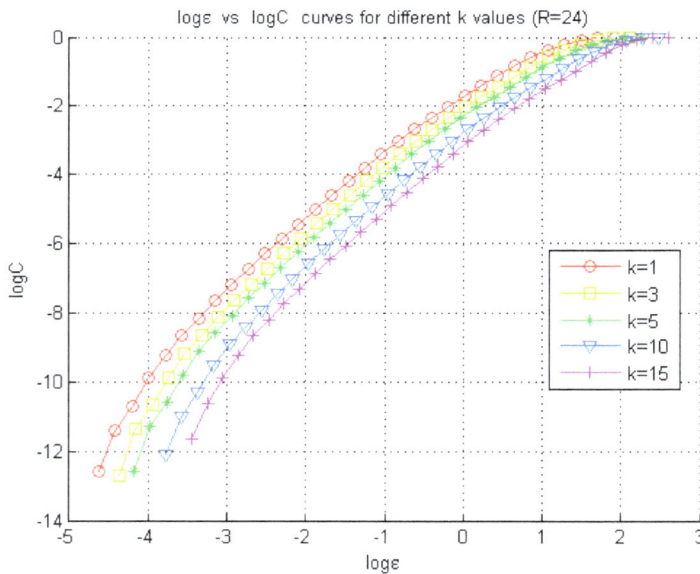

Figure 12. Exploration of the optimal radius for the KDE of a pdf using the Grassberger-Procaccia method. The figure illustrates the Correlation Sum Equation (22) estimated at different radius values represented by ε for the Lorenz model in the sub-chaotic regime ($R = 24$).

We estimated TE for both directions for each pair of variables (x,y), (x,z), and (y,z) using each of the three methods described in Section 2. Similar to the MI normalization of Equation (26) recommended in [53], we adapt the normalization for the NetTE as follows:

$$\delta_{XY} = \sqrt{1 - e^{-2(NetTE_{XY})}} \qquad (29)$$

where δ_{XY} denotes the normalized NetTE between variables X and Y, having values in the range of [0,1]. In Figures 13 and 14 , we illustrate the information flow between each pair of the Lorenz equation variables using both the un-normalized TE values obtained by the each of the three methods and the normalized NetTE estimates showing the net information flow between any pair of variables.

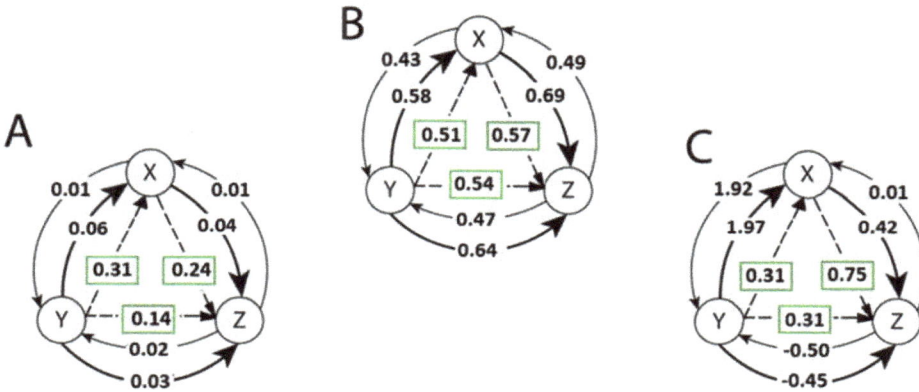

Figure 13. The *un-normalized* TE estimates between the variables of the Lorenz equations defined in Equation (28) for the chaotic case ($R = 28$) along with the normalized NetTE direction and magnitudes. Estimations were obtained using (**A**) Kernel Density Estimate method with $k = 17$, $l = 1$, $\varepsilon = e^{-1}$; (**B**) Generalized Knuth method method with $\beta = 0.1$, $k = l = 1$; and (**C**) Adaptive Partitioning method with 5% significance level and $k = l = 1$. Solid arrows denote the information flow (or TE) from X to Y or Y to X. Dashed lines show the direction of the *normalized* NetTE estimates.

Above, the *un-normalized* TE values are denoted by solid lines between each pair of variables. Also, the *normalized* NetTE estimates Equation (29) are illustrated with dashed lines. The direction of the NetTE has the same direction as the maximum of two un-normalized TE estimates between each pair, the magnitudes of which are shown in rectangles. For example, in the case of the adaptive partitioning method, the un-normalized TE values are estimated to be $TE_{YZ} = -0.45$ and $TE_{ZY} = -0.50$ between variables Y and Z, due to the biases originating from the subtraction used in Equation (21). However, the normalized NetTE is estimated to be $\delta_{YZ} = \sqrt{1 - e^{(-2(NetTE))}} = \sqrt{1 - e^{-2(-0.45 - (-0.5))}} = 0.31$ and shows a net information flow from variable Y to Z. Thus, we conclude that variable Y affects variable Z.

In Figure 14, we illustrate the estimates of TE's between each variable of the Lorenz Equation (28) in sub-chaotic regime with $R = 24$.

Above, we demonstrated the concept of our method: If the directions of information flows are not consistent with the three methods, then we can explore new parameter values to provide consistency in the directions. Above, for the selected parameters, the Generalized Knuth method and the adaptive partitioning provided consistent NetTE directions between the pairs of variables in the chaotic case. However, in the sub-chaotic case, we needed to explore a new parameter set for the KDE method as the NetTE directions were different than the other two consistent methods.

Based on the fact that the directions of the NetTE estimations obtained using each of the three methods agree, we conclude that information flow direction between the pairs of the Lorenz equation variables are as shown in Figure 15.

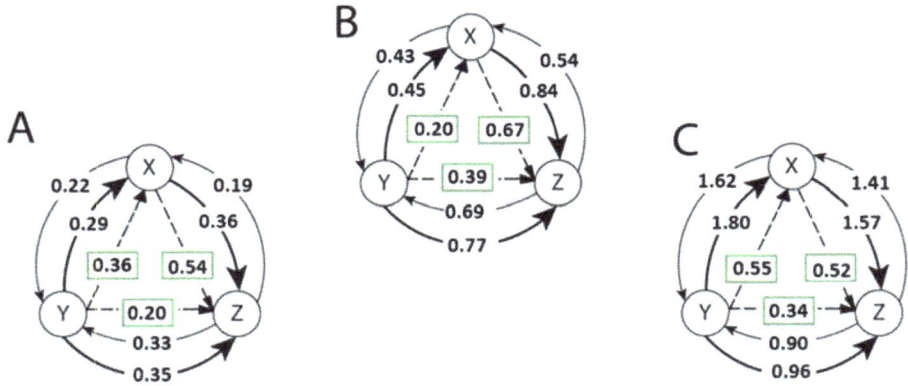

Figure 14. The *un-normalized* TE estimates between the variables of the Lorenz equations defined in Equation (28) for the sub-chaotic case ($R = 24$) along with the normalized NetTE direction and magnitudes. Estimations were obtained using: (**A**) Kernel Density Estimate method with $k = 3, l = 1$, $\varepsilon = e^{-2}$; (**B**) Generalized Knuth method where $\beta = 0.1, k = l = 1$; (**C**) Adaptive Partitioning method with 5% significance level and $k = l = 1$. Solid arrows denote the information flow (or TE) from X to Y or Y to X. Dashed lines illustrate the direction of the *normalized* NetTE estimates.

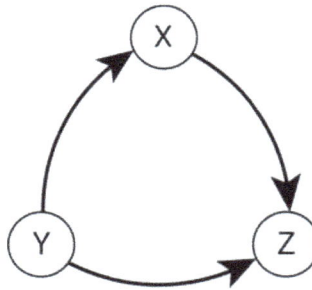

Figure 15. Information flow directions among the variables of the Lorenz equations, where X, Y, Z denote the velocity, temperature difference and the heat flow, respectively, in the case of the atmospheric convection roll model. These are also the NetTE directions, showing the larger influence among the bi-directional flows.

Note that these information flow directions are not only not obvious, but also not obviously obtainable, given the Lorenz system equations in Equation (28) despite the fact that these equations comprise a complete description of the system (sensitive dependence on initial conditions not withstanding). However, given the fact that this system of equations is derived from a well-understood physical system, one can evaluate these results based on the corresponding physics. In an atmospheric convection roll, it is known that both the velocity (X) and the heat flow (Z) are driven by the temperature difference (Y), and that it is the velocity (X) that mediates the heat flow (Z) in the system. This demonstrates that complex nonlinear relationships between different subsystems can be revealed by a TE analysis of the time series of the system variables. Furthermore, such an analysis reveals information about the system that is not readily accessible even with an analytic model, such as Equation (28), in hand.

4. Conclusions

Complex systems, such as the Earth's climate, the human brain, and a nation's economy, possess numerous subsystems, which not only interact in a highly nonlinear fashion, but also interact differently at different scales due to multiple feedback mechanisms. Analyzing these complex relationships in an attempt to better understand the physics underlying the observed behavior poses a serious challenge. Traditional methods, such as correlation analysis or PCA are inadequate due to the fact that they are designed for linear systems. TE has been demonstrated to be a potentially effective tool for complex systems consisting of nonlinearly-interacting subsystems due to its ability to estimate asymmetric information flow at different scales, which is indicative of cause and effect relationships. However, there are serious numerical challenges that need to be overcome before TE can be considered to be a dependable tool for identifying potential causal interactions. In response to this, we have developed a practical approach that involves utilizing three reasonably reliable estimation methods together. Instead of fine tuning the specific parameters of each method blindly, we find a working region where all three methods give the same direction of the information flow. In the case of collective agreement, we conclude that the individual tuning parameters for each method are near their optimal values. This was demonstrated on a bivariate linearly-coupled AR process as well as on the Lorenz system in both the chaotic and sub-chaotic regimes. Our success in deciphering the direction of information flow in the Lorenz system verified—not by the Lorenz system of differential equations—but rather by considering the known underlying physics suggests that this approach has significant promise in investigating and understanding the relationships among different variables in complex systems, such as the Earth's climate.

Appendix 1

In this Appendix we illustrate, via numerical simulation, the sensitivity of TE estimates on the number of bins used in a histogram model of a pdf. Consider the coupled autoregressive process:

$$
\begin{aligned}
y(i+1) &= 0.5y(i) + n_1(i) \\
x(i+1) &= 0.6x(i) + cy(i) + n_2(i)
\end{aligned}
\tag{A.1.1}
$$

where n_1 and n_2 are samples of zero mean and unit variance in Gaussian distributions, and c represents the coupling coefficient that couples the two time series equations for x and y. Here, $TE_{XY} = 0$, as the coupling direction is from Y to X (due to the coupling coefficient c). It was demonstrated by Kaiser and Schreiber ([19]) that the TE can be analytically solved for this system. By choosing the coupling coefficient to be $c = 0.5$, one finds $TE_{YX} = 0.2$. Numerical estimates of TE were performed by considering 11 datasets with the number of data points ranging from 10 to 1000. Eleven histograms were constructed for each dataset with the number of bins ranging from 2 to 100, and from these histograms the relevant Shannon entropies were computed. Figure A.1 illustrates the normalized TE_{YX} values, which are computed using the Shannon entropies in Equation (17), for each combination of N data points and M histogram bins considered. First, note that the estimated TE values range from below 0.1 to above 0.5 where the correct TE value is known to be 0.2 demonstrating that the TE estimates are highly dependent on both the number of data points and the number of bins. Second, note that there is no plateau where TE estimates remain approximately constant—not to mention correct—over a range of histogram bin numbers. For this reason, it is critical to select the correct number of bins in the histogram model of the pdf. However, this is made even more difficult since the entropy is a transform of the model of the pdf itself and therefore the number of bins required to produce the optimal model of the pdf will not be the same as the number of bins resulting in the optimal entropy estimate. This is explored in Appendix 2.

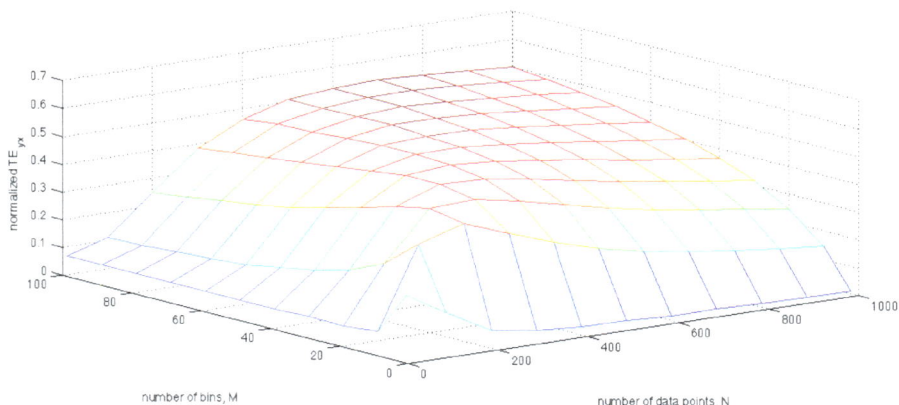

Figure A.1. This figure illustrates the numerically-estimated normalized transfer entropy, TE$_{YX}$, of the autoregressive system given in (A.1.1) as a function of varying numbers of data points and histogram bin numbers. To normalize TE, (29) was used as given in the text. Given that the correct value of the transfer entropy is TE$_{YX}$ = 0.2, this figure illustrates that the estimation of TE is extremely sensitive to the number of bins chosen for the histogram model of the pdf of the data. ($k = l = 1$).

Appendix 2

The differential entropy of a variable is estimated by Equation (14). However, due to the finite-precision of numerical calculations in a digital computer, the integral in Equation (14) is approximated by the following discrete summation:

$$h(X) \approx H(x) = \sum_{i=1}^{M} \hat{p}(x)[\log \hat{p}(x) - \log m(x)] \tag{A.2.1}$$

where M denotes the total number of bins used in the histogram and $\hat{p}(x)$ is the estimate of the continuous pdf of variable X. The Lebesgue measure, $m(x)$, used above is chosen to be the volume V of each bin. Equation (A.2.1) can easily be written for the joint entropies, where $\hat{p}(x)$ is replaced by its joint pdf counterpart and the volume is estimated for a multi-dimensional bin. In the one-dimensional case, the range of the variable x, is divided into $\hat{M}$ bins, which is selected to be optimal in Equation (12), and the volume is given by $m(x) = V = \left(\frac{\max(x) - \min(x)}{\hat{M}}\right)$. We show that the entropy calculation by (A.2.1) is biased.

The entropy of the standard Gaussian distribution $\mathcal{N}(0,1)$ with zero mean and unit variance can be analytically computed to be approximately 1.4189. We numerically generated 100 Gaussian-distributed datasets, each with 1000 data points, by sampling from $\mathcal{N}(0,1)$. Given these 100 datasets, we estimated the 100 corresponding entropy values using the Generalized Knuth method with $\beta = 0.5$. We found that 76% and 91% of the entropy estimates were within 1 or 2 standard deviations, respectively, of the true entropy value of 1.4189. However, this means that not every attempt at entropy estimation in this ensemble was successful.

We illustrate this with a specific data set that was found to lie outside of 76% percentile success rate. In this case, the optimal number of bins was estimated to be $M_{opt} = 11$ using Equation (12). In Figure A.2.1a,b, we illustrate the resulting histogram model of the pdf and the non-normalized log posterior probability of the number of bins in the model given the data.

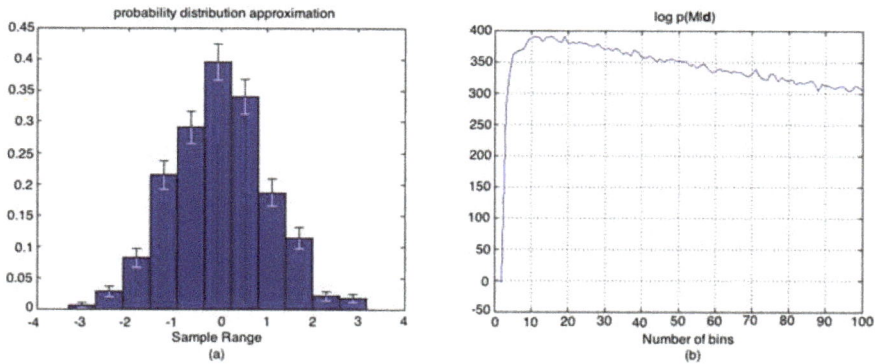

Figure A.2.1. (**a**) Histogram model of the pdf of the data set with error-bars on the bin heights; (**b**) The non-normalized log posterior probability of the number of bins in the model given the data.

In Figure A.2.2, we illustrate the entropy estimates for this data set as a function of the number of bins where the vertical bars denote one standard deviation from the mean.

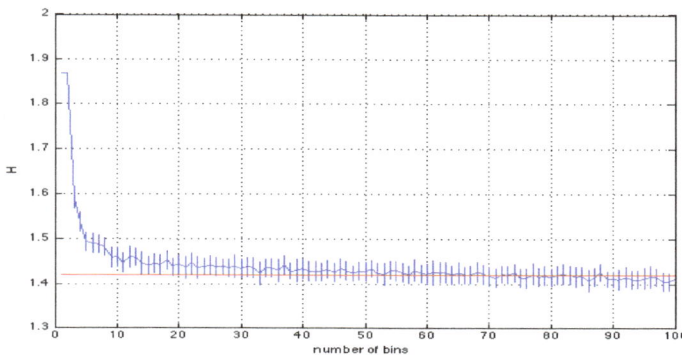

Figure A.2.2. Entropy estimate of a data-set outside of 76% percentile success rate (for one of the data-sets in the remaining 24% of 100 trials).

Figure A.2.2 shows that the true value of the entropy does not fall into the one standard deviation interval of the mean estimate using $M_{opt} = 11$, implying that the required number of bins is *different* for an optimal pdf model and an optimal entropy estimation. It is seen that $M = 19$ is the smallest number of bins where the entropy estimate falls within this interval and has a very close $\log p(M \,|\, \mathbf{d})$ value compared to that obtained for $M = 11$ in Figure A.2.1b.

Appendix 3

In Appendix 2, we estimated the entropy of a one-dimensional Gaussian variable using the Generalized Knuth method with the prior shown in Equation (8) and Equation (9). We notice that, even in the one-dimensional case, some of the entropy estimates lie outside the confidence intervals. If we estimate the joint entropy of two variables or more, the quality of the estimation decreases further due to empty bins. To overcome this problem, we proposed a different prior Equation (13) and computed the percentages of the relevant entropy estimates falling into 1 and 2 standard deviations (sigma's) within a total of 100 data-sets sampled from the same two-dimensional Gaussian distribution given by

$$N\left(\begin{bmatrix} 0 \\ 0 \end{bmatrix}, \begin{bmatrix} 1 & 0.5 \\ 0.5 & 1 \end{bmatrix}\right) versus\ \beta = [0.001, 0.01, 0.05, 0.1, 0.3, 0.5, 0.7, 1].$$ Figure A.3 illustrates these percentages as a function of different β values. Approximately 50% of the time, the marginal entropy estimate falls into the one-sigma interval for $\beta = 0.05$, and 80% of the time within the two-sigma interval (compare the first and second columns of Figure A.3). As a comparison, the corresponding statistics are approximately 10% for the marginal entropies falling into the one-sigma and 30% for marginal entropies falling into the two-sigma confidence intervals when we use the Krichevsky-Trofinov Dirichlet prior ($\beta = 0.5$), as in Equation (8) above. It is also observed that in both cases, the confidence interval statistics are lower for the joint entropies, due to the increase of the dimensionality of the space. As a result of this analysis, we observe the largest percentage of getting an entropy estimate within its one-sigma and two-sigma intervals from the true values take place for $\beta = 0.1$.

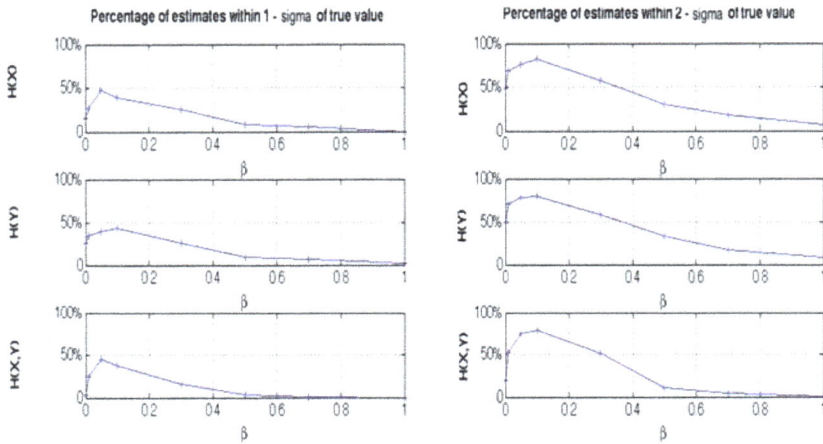

Figure A.3. Percentage performance (one- and two-standard deviation confidence intervals) of marginal and joint entropy estimates as a function of β.

Above, both joint and marginal Shannon entropies of 100 Gaussian-distributed data-sets are estimated using the Generalized Knuth method for the illustrated β values. Subfigures denote the percentage of estimates within one- and two- standard deviations from their analytical values.

Acknowledgments: The first author would like to thank Joseph Lizier for various discussions on TE during the beginning of this research. We also would like to thank three anonymous reviewers for their invaluable comments and suggestions. The first author would like to thank Petr Tichavsky for his code for the estimation of MI given at his web page. Also, we would like to thank NASA Cloud Modeling Analysis Initiative NASA GRANT NNX07AN04G for the support of this research.

Author Contributions: All authors conceived, designed and evaluated the experiments. Every author contributed to the preparation of the manuscript. All authors have read and approved the final manuscript.

Conflicts of Interest: The authors declare no conflict of interest.

References

1. Gourévitch, B.; Eggermont, J.J. Evaluating information transfer between auditory cortical neurons. *J. Neurophysiol.* **2007**, *97*, 2533–2543.
2. Overbey, L.A.; Todd, M.D. Dynamic system change detection using a modification of the transfer entropy. *J. Sound Vibr.* **2009**, *322*, 438–453.
3. Overbey, L.A.; Todd, M.D. Effects of noise on transfer entropy estimation for damage detection. *Mech. Syst. Signal Process.* **2009**, *23*, 2178–2191.

4. Pearl, J. *Causality: Models, Reasoning and Inference*; MIT Press: Cambridge, MA, USA, 2000; Volume 29.
5. Hannachi, A.; Jolliffe, I.T.; Stephenson, D.B. Empirical orthogonal functions and related techniques in atmospheric science: A review. *Int. J. Climatol.* **2007**, *27*, 1119–1152.
6. Principe, J.C.; Xu, D.; Fisher, J. Information theoretic learning. *Unsuperv. Adapt. Filter.* **2000**, *1*, 265–319.
7. Schreiber, T. Measuring information transfer. *Phys. Rev. Lett.* **2000**, *85*, 461–464.
8. Granger, C.W. Investigating causal relations by econometric models and cross-spectral methods. *Econometrica* **1969**, *37*, 424–438.
9. Ancona, N.; Marinazzo, D.; Stramaglia, S. Radial basis function approach to nonlinear Granger causality of time series. *Phys. Rev. E* **2004**, *70*, 056221.
10. Hlaváčková-Schindler, K.; Paluš, M.; Vejmelka, M.; Bhattacharya, J. Causality detection based on information-theoretic approaches in time series analysis. *Phys. Rep.* **2007**, *441*, 1–46.
11. Lizier, J.T.; Prokopenko, M. Differentiating information transfer and causal effect. *Eur. Phys. J. B* **2010**, *73*, 605–615.
12. Kleeman, R. Information flow in ensemble weather predictions. *J. Atmos. Sci.* **2007**, *64*, 1005–1016.
13. Liang, X.S.; The, Liang-Kleeman. Information Flow: Theory and Applications. *Entropy* **2013**, *15*, 327–360.
14. Sabesan, S.; Narayanan, K.; Prasad, A.; Iasemidis, L.D.; Spanias, A.; Tsakalis, K. Information flow in coupled nonlinear systems: Application to the epileptic human brain. *Data Mining Biomed.* **2007**, *7*, 483–503.
15. Majda, A.J.; Harlim, J. Information flow between subspaces of complex dynamical systems. *Proc. Natl. Acad. Sci. USA* **2007**, *104*, 9558–9563.
16. Liang, X.S.; Kleeman, R. Information transfer between dynamical system components. *Phys. Rev. Lett.* **2005**, *95*, 244101.
17. Ruddell, B.L.; Kumar, P. Ecohydrologic process networks: 1. Identification. *Water Resour. Res.* **2009**, *45*.
18. Ruddell, B.L.; Kumar, P. Ecohydrologic process networks: 2. Analysis and characterization. *Water Resour. Res.* **2009**, *45*, W03420.
19. Kaiser, A.; Schreiber, T. Information transfer in continuous processes. *Physica D* **2002**, *166*, 43–62.
20. Knuth, K.H.; Gotera, A.; Curry, C.T.; Huyser, K.A.; Wheeler, K.R.; Rossow, W.B. Revealing relationships among relevant climate variables with information theory, Proceedings of the Earth-Sun System Technology Conference (ESTC 2005), Adelphi, MD, USA, 27–30 June 2005.
21. Shannon, C.E.; Weaver, W. *The Mathematical Theory of Communication*; University of Illinois Press: Urbana-Champaign, IL, USA, 1949.
22. Knuth, K.H. Optimal data-based binning for histograms. *arXiv preprint physics/0605197.* This method has been implemented in Matlab and Python. 2006. Available online: http://knuthlab.rit.albany.edu/index.php/Products/Code & http://www.astroml.org/ accessed on 11 January 2015.
23. Wolpert, D.H.; Wolf, D.R. Estimating functions of probability distributions from a finite set of samples. *Phys. Rev. E* **1995**, *52*, 6841.
24. Nemenman, I.; Shafee, F. Bialek13, W. Entropy and Inference, Revisited. *Adv. Neur. Inf. Process. Syst.* **2002**, *1*, 471.
25. Prichard, D.; Theiler, J. Generalized redundancies for time series analysis. *Physica D* **1995**, *84*, 476–493.
26. Roulston, M.S. Estimating the errors on measured entropy and mutual information. *Physica D* **1999**, *125*, 285–294.
27. Grassberger, P. Finite sample corrections to entropy and dimension estimates. *Phys. Lett. A* **1988**, *128*, 369–373.
28. Silverman, B.W. *Density Estimation for Statistics and Data Analysis*; Chapman & Hall/CRC: London, UK, 1986.
29. Kleeman, R. Information theory and dynamical system predictability. *Entropy* **1986**, *13*, 612–649.
30. Kleeman, R. Measuring dynamical prediction utility using relative entropy. *J. Atmos. Sci.* **2002**, *59*, 2057–2072.
31. Grassberger, P.; Procaccia, I. Characterization of strange attractors. *Phys. Rev. Lett.* **1983**, *50*, 346–349.
32. Grassberger, P.; Procaccia, I. Measuring the strangeness of strange attractors. *Physica D* **1983**, *9*, 189–208.
33. Kantz, H.; Schreiber, T. *Nonlinear time Series Analysis*; Cambridge University Press: Cambridge, UK, 2003; Volume 7.
34. Fraser, A.M.; Swinney, H.L. Independent coordinates for strange attractors from mutual information. *Phys. Rev. A* **1986**, *33*, 1134.
35. Darbellay, G.A.; Vajda, I. Estimation of the information by an adaptive partitioning of the observation space. *IEEE Trans. Inf. Theory* **1999**, *45*, 1315–1321.

36. Kraskov, A.; Stögbauer, H.; Grassberger, P. Estimating mutual information. *Phys. Rev. E* **2004**, *69*, 066138.
37. Gomez-Herrero, G.; Wu, W.; Rutanen, K.; Soriano, M.C.; Pipa, G.; Vicente, R. *Assessing coupling dynamics from an ensemble of time series* **2010**, *arXiv:1008.0539*.
38. Vicente, R.; Wibral, M.; Lindner, M.; Pipa, G. Transfer entropy-a model-free measure of effective connectivity for the neurosciences. *J. Comput. Neurosci.* **2011**, *30*, 45–67.
39. Wibral, M.; Pampu, N.; Priesemann, V.; Siebenhühner, F.; Seiwert, H.; Lindner, M.; Lizier, J.T.; Vicente, R. Measuring Information-Transfer Delays. *PLoS One* **2013**, *8*.
40. Steeg, G.; Galstyan, V.A. Information-theoretic measures of influence based on content dynamics, Proceedings of the Sixth ACM International Conference on Web Search and Data Mining, Rome, Italy, 4–8 February 2013; pp. 3–12.
41. Lorenz, E.N. Deterministic nonperiodic flow. *J. Atmos. Sci.* **1963**, *20*, 130–141.
42. Cover, T.M.; Thomas, J.A. *Elements of Information Theory*; Wiley: Hoboken, NJ, USA, 2012.
43. Lizier, J.T.; Prokopenko, M.; Zomaya, A. Local information transfer as a spatiotemporal filter for complex systems. *Phys. Rev. E* **2008**, *77*, 026110.
44. Williams, P.L.; Beer, R.D. *Generalized Measures of Information Transfer*. 2011, arXiv:1102.1507. Available online: http://arxiv.org/abs/1102.1507 accessed on 9 January 2014.
45. Scott, D.W. On optimal and data-based histograms. *Biometrika* **1979**, *66*, 605–610.
46. Freedman, D.; Diaconis, P. On the histogram as a density estimator: L 2 theory. *Probab. Theory Relat. Fields* **1981**, *57*, 453–476.
47. Box, G.E.; Tiao, G.C. *Bayesian Inference in Statistical Analysis*; No. 622; Addison-Wesley: Boston, MA, USA, 1973.
48. Cellucci, C.J.; Albano, A.M.; Rapp, P.E. Statistical validation of mutual information calculations: Comparison of alternative numerical algorithms. *Phys. Rev. E* **2005**, *71*, 066208.
49. Kugiumtzis, D. Improvement of Symbolic Transfer Entropy, Proceedings of the 3rd International Conference on Complex Systems and Applications, Le Havre, France, 29 June–2 July 2009; Bertelle, C., Liu, X., Aziz-Alaoui, M.A., Eds.; pp. 338–342.
50. Prokopenko, M.; Lizier, J.T. Transfer Entropy and Transient Limits of Computation. *Sci. Rep.* **2014**, *4*, 5394.
51. Lungarella, M.; Ishiguro, K.; Kuniyoshi, Y.; Otsu, N. Methods for quantifying the causal structure of bivariate time series. *Int. J. Bifurc. Chaos* **2007**, *17*, 903–921.
52. Grassberger, P. Grassberger-Procaccia algorithm. *Scholarpedia* **2007**, *2*, 3043.
53. Dionisio, A.; Menezes, R.; Mendes, D.A. Mutual information: A measure of dependency for nonlinear time series. *Physica A* **2004**, *344*, 326–329.
54. Moddemeijer, R. On estimation of entropy and mutual information of continuous distributions. *Signal Process.* **1989**, *16*, 233–248.
55. Staniek, M.; Lehnertz, K. Symbolic Transfer Entropy. *Phys. Rev. Lett.* **2008**, *100*, 158101.
56. Abramowitz, M.; Stegun, I. *Handbook of Mathematical Functions*; Dover Publications: New York, NY, USA, 1972.

Article

Contribution to Transfer Entropy Estimation via the *k*-Nearest-Neighbors Approach

Jie Zhu [1,2,3], Jean-Jacques Bellanger [1,2], Huazhong Shu [3,4] and Régine Le Bouquin Jeannès [1,2,3,]*

[1] Institut National de la Santé Et de la Recherche Médicale (INSERM), U 1099, Rennes F-35000, France;
E-Mails: jie.zhu.1@etudiant.univ-rennes1.fr (J.Z.); jean-jacques.bellanger@univ-rennes1.fr (J.-J.B.)

[2] Université de Rennes 1, LTSI, Rennes F-35000, France

[3] Centre de Recherche en Information Biomédicale sino-français (CRIBs), Rennes F-35000, France

[4] Laboratory of Image Science and Technology (LIST), School of Computer Science and Engineering,
Southeast University, Nanjing 210018, China; E-Mail: shu.list@seu.edu.cn

* Author to whom correspondence should be addressed; E-Mail: regine.le-bouquin-jeannes@univ-rennes1.fr;
Tel.: +33-2-23236919; Fax: +33-2-23236917.

Academic Editor: Deniz Gencaga

Received: 31 December 2014 / Accepted: 10 June 2015 / Published: 16 June 2015

Abstract: This paper deals with the estimation of transfer entropy based on the *k*-nearest neighbors (*k*-NN) method. To this end, we first investigate the estimation of Shannon entropy involving a rectangular neighboring region, as suggested in already existing literature, and develop two kinds of entropy estimators. Then, applying the widely-used error cancellation approach to these entropy estimators, we propose two novel transfer entropy estimators, implying no extra computational cost compared to existing similar *k*-NN algorithms. Experimental simulations allow the comparison of the new estimators with the transfer entropy estimator available in free toolboxes, corresponding to two different extensions to the transfer entropy estimation of the Kraskov–Stögbauer–Grassberger (KSG) mutual information estimator and prove the effectiveness of these new estimators.

Keywords: entropy estimation; *k* nearest neighbors; transfer entropy; bias reduction

1. Introduction

Transfer entropy (TE) is an information-theoretic statistic measurement, which aims to measure an amount of time-directed information between two dynamical systems. Given the past time evolution of a dynamical system $\mathcal{A}$, TE from another dynamical system $\mathcal{B}$ to the first system $\mathcal{A}$ is the amount of Shannon uncertainty reduction in the future time evolution of $\mathcal{A}$ when including the knowledge of the past evolution of $\mathcal{B}$. After its introduction by Schreiber [1], TE obtained special attention in various fields, such as neuroscience [2–8], physiology [9–11], climatology [12] and others, such as physical systems [13–17].

More precisely, let us suppose that we observe the output $X_i \in \mathbb{R}$, $i \in \mathbb{Z}$, of some sensor connected to $\mathcal{A}$. If the sequence X is supposed to be an m-th order Markov process, *i.e.*, if considering subsequences $X_i^{(k)} = (X_{i-k+1}, X_{i-k+2}, \cdots, X_i)$, $k > 0$, the probability measure $\mathcal{P}_X$ (defined on measurable subsets of real sequences) attached to X fulfills the m-th order Markov hypothesis:

$$\forall i : \forall m' > m : \mathrm{d}\mathcal{P}_{X_{i+1}|X_i^{(m)}}\left(x_{i+1}|x_i^{(m)}\right) = \mathrm{d}\mathcal{P}_{X_{i+1}|X_i^{(m')}}\left(x_{i+1}|x_i^{(m')}\right), x_{i+1} \in \mathbb{R}, x_i^{(k)} \in \mathbb{R}^k, \quad (1)$$

then the past information $X_i^{(m)}$ (before time instant $i + 1$) is sufficient for a prediction of X_{i+k}, $k \geq 1$, and can be considered as an m-dimensional state vector at time i (note that, to know from X the hidden dynamical evolution of $\mathcal{A}$, we need a one-to-one relation between $X_i^{(m)}$ and the physical state of $\mathcal{A}$

at time i). For the sake of clarity, we introduce the following notation: $\left(X_i^p, X_i^-, Y_i^-\right), i = 1, 2, \ldots, N$, is an independent and identically distributed (IID) random sequence, each term following the same distribution as a random vector $(X^p, X^-, Y^-) \in \mathbb{R}^{1+m+n}$ whatever i (in X^p, X^-, Y^-, the upper indices "p" and "-" correspond to "predicted" and "past", respectively). This notation will substitute for the notation $\left(X_{i+1}, X_i^{(m)}, Y_i^{(n)}\right), i = 1, 2, \ldots, N$, and we will denote by $\mathcal{S}_{X^p, X^-, Y^-}$, $\mathcal{S}_{X^p, X^-}$, $\mathcal{S}_{X^-, Y^-}$ and $\mathcal{S}_{X^-}$ the spaces in which (X^p, X^-, Y^-), (X^p, X^-), (X^-, Y^-) and X^- are respectively observed.

Now, let us suppose that a causal influence exists from $\mathcal{B}$ on $\mathcal{A}$ and that an auxiliary random process $Y_i \in \mathbb{R}, i \in \mathbb{Z}$, recorded from a sensor connected to $\mathcal{B}$, is such that, at each time i and for some $n > 0$, $Y_i^- \triangleq Y_i^{(n)}$ is an image (not necessarily one-to-one) of the physical state of $\mathcal{B}$. The negation of this causal influence implies:

$$\forall (m > 0, n > 0) : \forall i : \mathrm{d}\mathcal{P}_{X_i^p|X_i^{(m)}}\left(x_i^p|x_i^{(m)}\right) = \mathrm{d}\mathcal{P}_{X_i^p|X_i^{(m)},Y_i^{(n)}}\left(x_i^p|x_i^{(m)},y_i^{(n)}\right). \tag{2}$$

If Equation (2) holds, it is said that there is an absence of information transfer from $\mathcal{B}$ to $\mathcal{A}$. Otherwise, the process X can be no longer considered strictly a Markov process. Let us suppose the joint process (X, Y) is Markovian, *i.e.*, there exist a given pair (m', n'), a transition function f and an independent random sequence $e_i, i \in \mathbb{Z}$, such that $[X_{i+1}, Y_{i+1}]^T = f\left(X_i^{(m')}, Y_i^{(n')}, e_{i+1}\right)$, where the random variable e_{i+1} is independent of the past random sequence $(X_j, Y_j, e_j), j \leq i$, whatever i. As $X_i = g\left(X_i^{(m)}, Y_i^{(n)}\right)$ where g is clearly a non-injective function, the pair $\left\{\left(X_i^{(m)}, Y_i^{(n)}\right), X_i\right\}, i \in \mathbb{Z}$, corresponds to a hidden Markov process, and it is well known that this observation process is not generally Markovian.

The deviation from this assumption can be quantified using the Kullback pseudo-metric, leading to the general definition of TE at time i:

$$\mathrm{TE}_{Y \to X, i} = \int_{\mathbb{R}^{m+n+1}} \log\left[\frac{\mathrm{d}\mathcal{P}_{X_i^p|X_i^-,Y_i^-}\left(x_i^p|x_i^-,y_i^-\right)}{\mathrm{d}\mathcal{P}_{X_i^p|X_i^-}\left(x_i^p|x_i^-\right)}\right] \mathrm{d}\mathcal{P}_{X_i^p,X_i^-,Y_i^-}\left(x_i^p,x_i^-,y_i^-\right), \tag{3}$$

where the ratio in Equation (3) corresponds to the Radon–Nikodym derivative [18,19] (*i.e.*, the density) of the conditional measure $\mathrm{d}\mathcal{P}_{X_i^p|X_i^-,Y_i^-}\left(\cdot|x_i^-,y_i^-\right)$ with respect to the conditional measure $\mathrm{d}\mathcal{P}_{X_i^p|X_i^-}\left(\cdot|x_i^-\right)$. Considering "log" as the natural logarithm, information is measured in natural units (nats). Now, given two observable scalar random time series X and Y with no *a priori* given model (as is generally the case), if we are interested in defining some causal influence from Y to X through TE analysis, we must specify the dimensions of the past information vectors X^- and Y^-, *i.e.*, m and n. Additionally, even if we impose them, it is not evident that all of the coordinates in $X_i^{(m)}$ and $Y_i^{(n)}$ will be useful. To deal with this issue, variable selection procedures have been proposed in the literature, such as uniform and non-uniform embedding algorithms [20,21].

If the joint probability measure $\mathcal{P}_{X_i^p,X_i^-,Y_i^-}\left(x_i^p,x_i^-,y_i^-\right)$ is derivable with respect to the Lebesgue measure μ^{n+m+1} in $\mathbb{R}^{1+n+m}$ (*i.e.*, if $\mathcal{P}_{X_i^p,X_i^-,Y_i^-}$ is absolutely continuous with respect to μ^{n+m+1}), then the pdf (joint probability density function) $p_{X_i^p,X_i^-,Y_i^-}\left(x_i^p,x_i^-,y_i^-\right)$ and also the pdf for each subset of $\left\{X_i^p, X_i^-, Y_i^-\right\}$ exist, and $\mathrm{TE}_{Y \to X, i}$ can then be written (see Appendix I):

$$\begin{aligned}\mathrm{TE}_{Y \to X, i} = &-E\left[\log\left(p_{X_i^-,Y_i^-}\left(X_i^-,Y_i^-\right)\right)\right] - E\left[\log\left(p_{X_i^p,X_i^-}\left(X_i^p,X_i^-\right)\right)\right] \\ &+ E\left[\log\left(p_{X_i^p,X_i^-,Y_i^-}\left(X_i^p,X_i^-,Y_i^-\right)\right)\right] + E\left[\log\left(p_{X_i^-}\left(X_i^-\right)\right)\right]\end{aligned} \tag{4}$$

or:

$$\mathrm{TE}_{Y \to X, i} = \mathcal{H}\left(X_i^-,Y_i^-\right) + \mathcal{H}\left(X_i^p,X_i^-\right) - \mathcal{H}\left(X_i^p,X_i^-,Y_i^-\right) - \mathcal{H}\left(X_i^-\right), \tag{5}$$

where $\mathcal{H}(U)$ denotes the Shannon differential entropy of a random vector U. Note that, if the processes Y and X are assumed to be jointly stationary, for any real function $g : \mathbb{R}^{m+n+1} \rightarrow \mathbb{R}$, the expectation $E\left[g\left(X_{i+1}, X_i^{(m)}, Y_i^{(n)}\right)\right]$ does not depend on i. Consequently, $\mathrm{TE}_{Y \rightarrow X, i}$ does not depend on i (and so can be simply denoted by $\mathrm{TE}_{Y \rightarrow X}$), nor all of the quantities defined in Equations (3) to (5). In theory, TE is never negative and is equal to zero if and only if Equation (2) holds.

According to Definition (3), TE is not symmetric, and it can be regarded as a conditional mutual information (CMI) [3,22] (sometimes also named partial mutual information (PMI) in the literature [23]). Recall that mutual information between two random vectors X and Y is defined by:

$$\mathcal{I}(X;Y) = \mathcal{H}(X) + \mathcal{H}(Y) - \mathcal{H}(X,Y), \tag{6}$$

and TE can be also written as:

$$\mathrm{TE}_{Y \rightarrow X} = \mathcal{I}\left(X^p, Y^- | X^-\right). \tag{7}$$

Considering the estimation $\widehat{\mathrm{TE}_{Y \rightarrow X}}$ of TE, $\mathrm{TE}_{Y \rightarrow X}$, as a function defined on the set of observable occurrences (x_i, y_i), $i = 1, \ldots, N$, of a stationary sequence (X_i, Y_i), $i = 1, \ldots, N$, and Equation (5), a standard structure for the estimator is given by (see Appendix B):

$$
\begin{aligned}
\widehat{\mathrm{TE}_{Y \rightarrow X}} &= \widehat{\mathcal{H}\left(X^-, Y^-\right)} + \widehat{\mathcal{H}\left(X^p, X^-\right)} - \widehat{\mathcal{H}\left(X^p, X^-, Y^-\right)} - \widehat{\mathcal{H}\left(X^-\right)} \\
&= -\frac{1}{N} \sum_{n=1}^{N} \widehat{\log\left(p_{U_1}\left(u_{1n}\right)\right)} - \frac{1}{N} \sum_{n=1}^{N} \widehat{\log\left(p_{U_2}\left(u_{2n}\right)\right)} + \frac{1}{N} \sum_{n=1}^{N} \widehat{\log\left(p_{U_3}\left(u_{3n}\right)\right)} \\
&\quad + \frac{1}{N} \sum_{n=1}^{N} \widehat{\log\left(p_{U_4}\left(u_{4n}\right)\right)},
\end{aligned} \tag{8}
$$

where U_1, U_2, U_3 and U_4 stand respectively for (X^-, Y^-), (X^p, X^-), (X^p, X^-, Y^-) and X^-. Here, for each n, $\widehat{\log\left(p_U\left(u_n\right)\right)}$ is an estimated value of $\log\left(p_U\left(u_n\right)\right)$ computed as a function $f_n\left(u_1, \ldots, u_N\right)$ of the observed sequence u_n, $n = 1, \ldots, N$. With the k-NN approach addressed in this study, $f_n\left(u_1, \ldots, u_N\right)$ depends explicitly only on u_n and on its k nearest neighbors. Therefore, the calculation of $\widehat{\mathcal{H}(U)}$ definitely depends on the chosen estimation functions f_n. Note that if, for N fixed, these functions correspond respectively to unbiased estimators of $\log\left(p\left(u_n\right)\right)$, then $\widehat{\mathrm{TE}_{Y \rightarrow X}}$ is also unbiased; otherwise, we can only expect that $\widehat{\mathrm{TE}_{Y \rightarrow X}}$ is asymptotically unbiased (for N large). This is so if the estimators of $\log\left(p_U\left(u_n\right)\right)$ are asymptotically unbiased.

Now, the theoretical derivation and analysis of the most currently used estimators $\widehat{\mathcal{H}(U)}(u_1, \ldots, u_N) = -\frac{1}{N} \sum_{n=1}^{N} \widehat{\log\left(p(u_n)\right)}$ for the estimation of $\mathcal{H}(U)$ generally suppose that $u_1, \ldots, u_N$ are N independent occurrences of the random vector U, i.e., $u_1, \ldots, u_N$ is an occurrence of an independent and identically distributed (IID) sequence $U_1, \ldots, U_N$ of random vectors $(\forall i = 1, \ldots, N : \mathcal{P}_{U_i} = \mathcal{P}_U)$. Although the IID hypothesis does not apply to our initial problem concerning the measure of TE on stationary random sequences (that are generally not IID), the new methods presented in this contribution are extended from existing ones assuming this hypothesis, without relaxing it. However, the experimental section will present results not only on IID observations, but also on non-IID stationary autoregressive (AR) processes, as our goal was to verify if some improvement can be nonetheless obtained for non-IID data, such as AR data.

If we come back to mutual information (MI) defined by Equation (6) and compare it with Equations (5), it is obvious that estimating MI and TE shares similarities. Hence, similarly to Equation (8) for TE, a basic estimation $\widehat{\mathcal{I}(X;Y)}$ of $\mathcal{I}(X;Y)$ from a sequence (x_i, y_i), $i = 1, \ldots, N$, of N independent trials is:

$$\widehat{\mathcal{I}(X;Y)} = -\frac{1}{N} \sum_{n=1}^{N} \widehat{\log\left(p_X\left(x_n\right)\right)} - \frac{1}{N} \sum_{n=1}^{N} \widehat{\log\left(p_Y\left(y_n\right)\right)} + \frac{1}{N} \sum_{n=1}^{N} \widehat{\log\left(p_{X,Y}\left(x_n, y_n\right)\right)}. \tag{9}$$

In what follows, when explaining the links among the existing methods and the proposed ones, we refer to Figure 1. In this diagram, a box identified by a number k in a circle is designed by box $\textcircled{k}$.

Improving performance (in terms of bias and variance) of TE and MI estimators (obtained by choosing specific estimation functions $\overline{\log{(p\,(\cdot))}}$ in Equations (8) and (9), respectively) remains an issue when applied on short-length IID (or non-IID) sequences [3]. In this work, we particularly focused on bias reduction. For MI, the most widely-used estimator is the Kraskov–Stögbauer–Grassberger (KSG) estimator [24,31], which was later extended to estimate transfer entropy, resulting in the k-NN TE estimator [25–27,32–35] (adopted in the widely-used TRENTOOL open source toolbox, Version 3.0). Our contribution originated in the Kozachenko–Leonenko entropy estimator summarized in [24] and proposed beforehand in the literature to get an estimation $\widehat{\mathcal{H}\,(X)}$ of the entropy $\mathcal{H}(X)$ of a continuously-distributed random vector X, from a finite sequence of independent outcomes x_i, $i = 1, \ldots, N$. This estimator, as well as another entropy estimator proposed by Singh *et al.* in [36] are briefly described in Section 2.1, before we introduce, in Section 4, our two new TE estimators based on both of them. In Section 2.2, Kraskov MI and standard TE estimators derived in literature from the Kozachenko–Leonenko entropy estimator are summarized, and the passage from a square to rectangular neighboring region to derive new entropy estimation is detailed in Section 3. Our methodology is depicted in Figure 1.

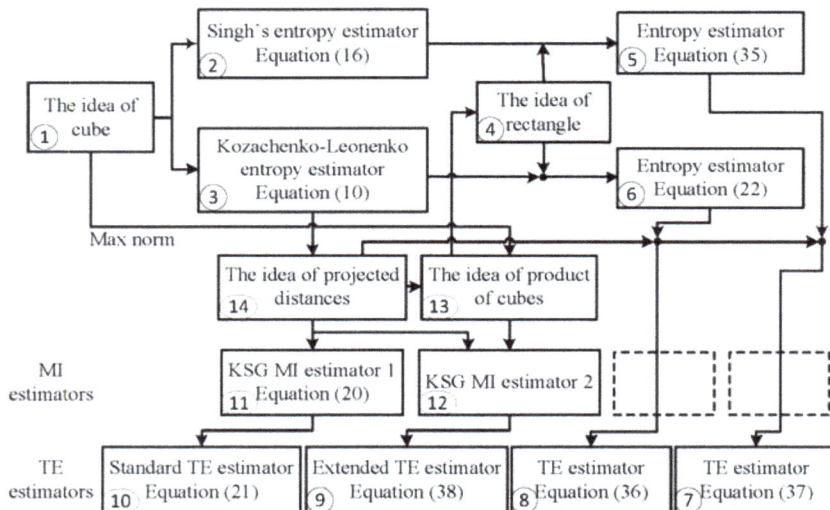

Figure 1. Concepts and methodology involved in k-nearest-neighbors transfer entropy (TE) estimation. Standard k-nearest-neighbors methods using maximum norm for probability density and entropy non-parametric estimation introduce, around each data point, a minimal (hyper-)cube (Box ①), which includes the first k-nearest neighbors, as is the case for two entropy estimators, namely the well-known Kozachenko–Leonenko estimator (Box ③) and the less commonly used Singh's estimator (Box ②). The former was used in [24] to measure mutual information (MI) between two signals X and Y by Kraskov *et al.*, who propose an MI estimator (Kraskov–Stögbauer–Grassberger (KSG) MI Estimator 1, Box ①) obtained by summing three entropy estimators (two estimators for the marginal entropies and one for the joint entropy). The strategy was to constrain the three corresponding (hyper-)cubes, including nearest neighbors, respectively in spaces $\mathcal{S}_X$, $\mathcal{S}_Y$ and $\mathcal{S}_{X,Y}$, to have an identical edge length (the idea of projected distances, Box ⑭) for a better cancellation of the three corresponding biases. The same approach was used to derive the standard TE estimator [25–29] (Box ⑩), which has been implemented in the TRENTOOL toolbox, Version 3.0. In [24], Kraskov *et al.* also suggested, for MI

estimation, to replace minimal (hyper-)cubes with smaller minimal (hyper-)rectangles equal to the product of two minimal (hyper-)cubes built separately in subspaces $\mathcal{S}_X$ and $\mathcal{S}_Y$ (KSG MI Estimator 2, Box ⑫) to exploit more efficiently the Kozachenko–Leonenko approach. An extended algorithm for TE estimation based on minimal (hyper-)rectangles equal to products of (hyper-)cubes was then proposed in [27] (extended TE estimator, Box ⑨) and implemented in the JIDT toolbox [30]. Boxes ⑩ and ⑨ are marked as "standard algorithm" and "extended algorithm". The new idea extends the idea of the product of cubes (Box ⑬). It consists of proposing a different construction of the neighborhoods, which are no longer minimal (hyper-)cubes, nor products of (hyper-)cubes, but minimal (hyper-)rectangles (Box ④), with possibly a different length for each dimension, to get two novel entropy estimators (Boxes ⑤ and ⑥), respectively derived from Singh's entropy estimator and the Kozachenko–Leonenko entropy estimator. These two new entropy estimators lead respectively to two new TE estimators (Box ⑦ and Box ⑧) to be compared with the standard and extended TE estimators.

2. Original k-Nearest-Neighbors Strategies

2.1. Kozachenko–Leonenko and Singh's Entropy Estimators for a Continuously-Distributed Random Vector

2.1.1. Notations

Let us consider a sequence x_i, $i = 1, \ldots, N$ in $\mathbb{R}^{d_X}$ (in our context, this sequence corresponds to an outcome of an IID sequence $X_1, \ldots, X_N$, such that the common probability distribution will be equal to that of a given random vector X). The set of the k nearest neighbors of x_i in this sequence (except for x_i) and the distance between x_i and its k-th nearest neighbor are respectively denoted by χ_i^k and $d_{x_i,k}$. We denote $\mathcal{D}_{x_i}\left(\chi_i^k\right) \subset \mathbb{R}^{d_X}$ a neighborhood of x_i in $\mathbb{R}^{d_X}$, which is the image of $\left(x_i, \chi_i^k\right)$ by a set valued map. For a given norm $\|\cdot\|$ on $\mathbb{R}^{d_X}$ (Euclidean norm, maximum norm, *etc.*), a standard construction $\left(x_i, \chi_i^k\right) \in \left(\mathbb{R}^{d_X}\right)^{k+1} \to \mathcal{D}_{x_i}\left(\chi_i^k\right) \subset \mathbb{R}^{d_X}$ is the (hyper-)ball of radius equal to $d_{x_i,k}$, *i.e.*, $\mathcal{D}_{x_i}\left(\chi_i^k\right) = \{x : \|x - x_i\| \leq d_{x_i,k}\}$. The (hyper-)volume (*i.e.*, the Lebesgue measure) of $\mathcal{D}_{x_i}\left(\chi_i^k\right)$ is then $v_i = \int_{\mathcal{D}_{x_i}(\chi_i^k)} \mathrm{d}x$ (where $\mathrm{d}x \triangleq \mathrm{d}\mu^{d_X}(x)$).

2.1.2. Kozachenko–Leonenko Entropy Estimator

The Kozachenko–Leonenko entropy estimator is given by (Box ③ in Figure 1):

$$\widehat{\mathcal{H}(X)}_{KL} = \psi(N) + \frac{1}{N}\sum_{i=1}^{N} \log(v_i) - \psi(k), \tag{10}$$

where v_i is the volume of $\mathcal{D}_{x_i}\left(\chi_i^k\right) = \{x : \|x - x_i\| \leq d_{x_i,k}\}$ computed with the maximum norm and $\psi(k) = \frac{\Gamma'(k)}{\Gamma(k)}$ denotes the digamma function. Note that using Equation (10), entropy is measured in natural units (nats).

To come up with a concise presentation of this estimator, we give hereafter a summary of the different steps to get it starting from [24]. First, let us consider the distance $d_{x_i,k}$ between x_i and its k-th nearest neighbor (introduced above) as a realization of the random variable $D_{x_i,k}$, and let us denote by $q_{x_i,k}(x)$, $x \in \mathbb{R}$, the corresponding probability density function (conditioned by $X_i = x_i$). Secondly, let us consider the quantity $h^{x_i}(\varepsilon) = \int_{\|u-x_i\| \leq \varepsilon/2} \mathrm{d}P_X(u)$. This is the probability mass of the (hyper-)ball with radius equal to $\varepsilon/2$ and centered on x_i. This probability mass is approximately equal to:

$$h^{x_i}(\varepsilon) \simeq p_X(x_i) \int_{\|\xi\| \leq \varepsilon/2} \mathrm{d}\mu^d(\xi) = p_X(x_i)\, c_d \varepsilon^d, \tag{11}$$

if the density function is approximately constant on the (hyper-)ball. The variable c_d is the volume of the unity radius d-dimensional (hyper-)ball in $\mathbb{R}^d$ ($c_d = 1$ with maximum norm). Furthermore, it can be established (see [24] for details) that the expectation $E\left[\log\left(h^{x_i}\left(D_{x_i,k}\right)\right)\right]$, where h^{x_i} is the random variable associated with h^{x_i}, $D_{X_i,k}$ (which must not be confused with the notation $\mathcal{D}_{x_i}\left(x_i^k\right)$ introduced previously) denotes the random distance between the k-th neighbor selected in the set of random vectors $\{X_k, 1 \leq k \leq N, k \neq i\}$, and the random point X_i is equal to $\psi(k) - \psi(N)$ and does not depend on $p_X\left(\cdot\right)$. Equating it with $E\left[\log\left(p_X\left(X_i\right)c_dD_{X_i,k}\right)\right]$ leads to:

$$\psi(k) - \psi(N) \simeq E\left[\log\left(p_X\left(X_i\right)\right)\right] + E\left[\log\left(c_dD_{X_i,k}^d\right)\right]$$
$$= -\mathcal{H}(X_i) + E\left[\log\left(V_i\right)\right]$$
(12)

and:

$$\mathcal{H}\left(X_i\right) \simeq \psi(N) - \psi(k) + E\left[\log\left(c_dD_{X_i,k}^d\right)\right].$$
(13)

Finally, by using the law of large numbers, when N is large, we get:

$$\mathcal{H}\left(X_i\right) \simeq \psi(N) - \psi(k) + \frac{1}{N}\sum_{i=1}^{N}\log\left(v_i\right)$$
$$= \widehat{\mathcal{H}\left(X\right)}_{KL},$$
(14)

where v_i is the realization of the random (hyper-)volume $V_i = c_dD_{x_i,k}^d$.

Moreover, as observed in [24], it is possible to make the number of neighbors k depend on i by substituting the mean $\frac{1}{N}\sum_{i=1}^{N}\psi(k_i)$ for the constant $\psi(k)$ in Equation (14), so that $\widehat{\mathcal{H}\left(X\right)}_{KL}$ becomes:

$$\widehat{\mathcal{H}\left(X\right)}_{KL} = \psi(N) + \frac{1}{N}\sum_{i=1}^{N}\left(\log\left(v_i\right) - \psi(k_i)\right).$$
(15)

2.1.3. Singh's Entropy Estimator

The question of k-NN entropy estimation is also discussed by Singh *et al.* in [36], where another estimator, denoted by $\widehat{\mathcal{H}(X)}_S$ hereafter, is proposed (Box ② in Figure 1):

$$\widehat{\mathcal{H}(X)}_S = \log(N) + \frac{1}{N}\sum_{i=1}^{N}\log\left(v_i\right) - \psi(k).$$
(16)

Using the approximation $\psi(N) \approx \log(N)$ for large values of N, the estimator given by Equation (16) is close to that defined by Equation (10). This estimator was derived by Singh *et al.* in [36] through the four following steps:

(1) Introduce the classical entropy estimator structure:

$$\widehat{\mathcal{H}(X)} \triangleq -\frac{1}{N}\sum_{i=1}^{N}\log\widehat{p_X\left(X_i\right)} = \frac{1}{N}\sum_{i=1}^{N}T_i,$$
(17)

where:

$$\widehat{p_X(x_i)} \triangleq \frac{k}{Nv_i}.$$
(18)

(2) Assuming that the random variables T_i, $i = 1,\ldots,N$ are identically distributed, so that $E\left[\widehat{\mathcal{H}(X)}\right] = E\left(T_1\right)$ (note that $E\left(T_1\right)$ depends on N, even if the notation does not make that explicit), compute the asymptotic value of $E\left(T_1\right)$ (when N is large) by firstly computing its asymptotic cumulative probability distribution function and the corresponding probability density p_{T_1}, and finally, compute the expectation $E\left(T_1\right) = \int_{\mathbb{R}} t p_{T_1}(t)\mathrm{d}t$.

(3) It appears that $E\left(T_1\right) = E\left[\widehat{\mathcal{H}(X)}\right] = \mathcal{H}(X) + B$ where B is a constant, which is identified with the bias.

(4)   Subtract this bias from $\widehat{\mathcal{H}(X)}$ to get $\widehat{\mathcal{H}(X)}_S = \widehat{\mathcal{H}(X)} - B$ and the formula given in Equation (16).

Note that the cancellation of the asymptotic bias does not imply that the bias obtained with a finite value of N is also exactly canceled. In Appendix C, we explain the origin of the bias for the entropy estimator given in Equation (17).

Observe also that, as for the Kozachenko–Leonenko estimator, it is possible to adapt Equation (16) if we want to consider a number of neighbors k_i depending on i. Equation (16) must then be replaced by:

$$\widehat{\mathcal{H}(X)}_S = \log(N) + \frac{1}{N} \sum_{i=1}^{N} \left(\log(v_i) - \psi(k_i) \right). \tag{19}$$

2.2. Standard Transfer Entropy Estimator

Estimating entropies separately in Equations (8) and (9) leads to individual bias values. Now, it is possible to cancel out (at least partially) the bias considering the algebraic sums (Equations (8) and (9)). To help in this cancellation, on the basis of Kozachenko–Leonenko entropy estimator, Kraskov *et al.* proposed to retain the same (hyper-)ball radius for each of the different spaces instead of using the same number k for both joint space $\mathcal{S}_{X,Y}$ and marginal spaces ($\mathcal{S}_X$ and $\mathcal{S}_Y$ spaces) [24,37], leading to the following MI estimator (Box ① in Figure 1):

$$\widehat{\mathcal{I}}_K = \psi(k) + \psi(N) - \frac{1}{N} \sum_{i=1}^{N} \left[\psi(n_{X,i} + 1) + \psi(n_{Y,i} + 1) \right], \tag{20}$$

where $n_{X,i}$ and $n_{Y,i}$ denote the number of points that strictly fall into the resulting distance in the lower-dimensional spaces $\mathcal{S}_X$ and $\mathcal{S}_Y$, respectively.

Applying the same strategy to estimate TE, the number of neighbors in the joint space $\mathcal{S}_{X^p,X^-,Y^-}$ is first fixed, then for each i, the resulting distance $\varepsilon_i \triangleq d_{(x_i^p,x_i^-,y_i^-),k}$ is projected into the other three lower dimensional spaces, leading to the standard TE estimator [25,27,28] (implementation available in the TRENTOOL toolbox, Version 3.0, Box ⑩ in Figure 1):

$$\widehat{TE}_{Y \to X SA} = \psi(k) + \frac{1}{N} \sum_{i=1}^{N} \left[\psi(n_{X^-,i} + 1) - \psi(n_{(X^-,Y^-),i} + 1) - \psi(n_{(X^p,X^-),i} + 1) \right], \tag{21}$$

where $n_{X^-,i}$, $n_{(X^-,Y^-),i}$ and $n_{(X^p,X^-),i}$ denote the number of points that fall into the distance ε_i from x_i^-, (x_i^-,y_i^-) and $\left(x_i^p, x_i^- \right)$ in the lower dimensional spaces $\mathcal{S}_{X^-}$, $\mathcal{S}_{X^-,Y^-}$ and $\mathcal{S}_{X^p,X^-}$, respectively. This estimator is marked as the "standard algorithm" in the experimental part.

Note that a generalization of Equation (21) was proposed in [28] to extend this formula to the estimation of entropy combinations other than MI and TE.

3. From a Square to a Rectangular Neighboring Region for Entropy Estimation

In [24], to estimate MI, as illustrated in Figure 2, Kraskov *et al.* discussed two different techniques to build the neighboring region to compute $\widehat{\mathcal{I}(X;Y)}$: in the standard technique (square $ABCD$ in Figure 2a,b), the region determined by the first k nearest neighbors is a (hyper-)cube and leads to Equation (20), and in the second technique (rectangle $A'B'C'D'$ in Figure 2a,b), the region determined by the first k nearest neighbors is a (hyper-)rectangle. Note that the TE estimator mentioned in the previous section (Equation (21)) is based on the first situation (square $ABCD$ in Figure 2a or 2b). The introduction of the second technique by Kraskov *et al.* was to circumvent the fact that Equation (15) was not applied rigorously to obtain the terms $\psi(n_{X,i} + 1)$ or $\psi(n_{Y,i} + 1)$ in Equation (20). As a matter of fact, for one of these terms, no point x_i (or y_i) falls exactly on the border of the (hyper-)cube $\mathcal{D}_{x_i}$ (or $\mathcal{D}_{y_i}$) obtained by the distance projection from the $\mathcal{S}_{X,Y}$ space. As clearly illustrated in Figure 2 (rectangle $A'B'C'D'$ in Figure 2a,b), the second strategy prevents that issue, since the border of the (hyper-)cube (in this case, an interval of $\mathbb{R}$) after projection from $\mathcal{S}_{X,Y}$ space to $\mathcal{S}_X$ space (cr $\mathcal{S}_Y$ space) contains one point. When the dimensions of $\mathcal{S}_X$ and $\mathcal{S}_Y$ are larger than one, this strategy leads to building an (hyper-)rectangle equal to the product of two (hyper-)cubes, one of them in $\mathcal{S}_X$ and the other one in $\mathcal{S}_Y$. If the maximum distance of the k-th NN in $\mathcal{S}_{X,Y}$ is obtained in one of the directions in $\mathcal{S}_X$, this maximum distance, after multiplying by two, fixes the size of the (hyper-)cube in $\mathcal{S}_X$. To obtain the size of the second (hyper-)cube (in $\mathcal{S}_Y$), the k neighbors in $\mathcal{S}_{X,Y}$ are first projected on $\mathcal{S}_Y$, and then, the largest of the distances calculated from these projections fixes the size of this second (hyper-)cube.

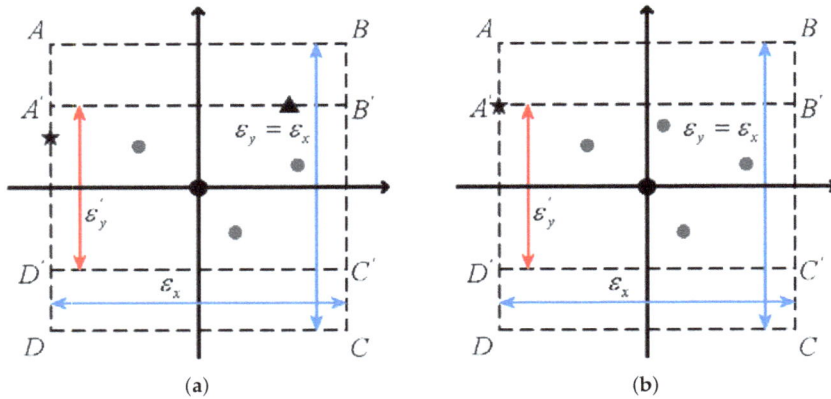

Figure 2. In this two-dimensional example, $k = 5$. The origin of the Cartesian axis corresponds to the current point x_i. Only the five nearest neighbors of this point, *i.e.*, the points in the set χ_i^k, are represented. The fifth nearest neighbor is symbolized by a star. The neighboring regions $ABCD$, obtained from the maximum norm around the center point, are squares, with equal edge lengths $\varepsilon_x = \varepsilon_y$. Reducing one of the edge lengths, ε_x or ε_y, until one point falls onto the border (in the present case, in the vertical direction), leads to the minimum size rectangle $A'B'C'D'$, where $\varepsilon_x \neq \varepsilon_y$. Two cases must be considered: (**a**) the fifth neighbor is not localized on a node, but between two nodes, contrary to (**b**). This leads to obtaining either two points (respectively the star and the triangle in (a)) or only one point (the star in(b)) on the border of $A'B'C'D'$. Clearly, it is theoretically possible to have more than two points on the border of $A'B'C'D'$, but the probability of such an occurrence is equal to zero when the probability distribution of the random points X_j is continuous.

In the remainder of this section, for an arbitrary dimension d, we propose to apply this strategy to estimate the entropy of a single multidimensional variable X observed in $\mathbb{R}^d$. This leads to introducing a d-dimensional (hyper-)rectangle centered on x_i having a minimal volume and including the set χ_i^k of neighbors. Hence, the rectangular neighboring is built by adjusting its size separately in each direction in the space $\mathcal{S}_X$. Using this strategy, we are sure that, in any of the d directions, there is at least one point on one of the two borders (and only one with probability one). Therefore, in this approach, the (hyper-)rectangle, denoted by $\mathcal{D}_{x_i}^{\varepsilon_1,\dots,\varepsilon_d}$, where the sizes $\varepsilon_1,\dots,\varepsilon_d$ in the respective d directions are completely specified from the neighbors set χ_i^k, is substituted for the basic (hyper-)square $\mathcal{D}_{x_i}\left(\chi_i^k\right) = \{x : \|x - x_i\| \leq d_{x_i,k}\}$. It should be mentioned that the central symmetry of the (hyper-)rectangle around the center point allows for reducing the bias in the density estimation [38] (*cf.* Equation (11) or (18)). Note that, when $k < d$, there must exist neighbors positioned on some vertex or edges of the (hyper-)rectangle. With $k < d$, it is impossible that, for any direction, one point falls exactly inside a face (*i.e.*, not on its border). For example, with $k = 1$ and $d > 1$, the first neighbor will be on a vertex, and the sizes of the edges of the reduced (hyper-)rectangle will be equal to twice the absolute value of its coordinates, whatever the direction.

Hereafter, we propose to extend the entropy estimators by Kozachenko–Leonenko and Singh using the above strategy before deriving the corresponding TE estimators and comparing their performance.

3.1. Extension of the Kozachenko–Leonenko Method

As indicated before, in [24], Kraskov *et al.* extended the Kozachenko–Leonenko estimator (Equations (10) and (15)) using the rectangular neighboring strategy to derive the MI estimator. Now, focusing on entropy estimation, after some mathematical developments (see Appendix D), we obtain another estimator of $\mathcal{H}(X)$, denoted by $\widehat{\mathcal{H}(X)}_K$ (Box ⑥ in Figure 1),

$$\widehat{\mathcal{H}(X)}_K = \psi(N) + \frac{1}{N}\sum_{i=1}^{N}\log{(v_i)} - \psi(k) + \frac{d-1}{k}. \tag{22}$$

Here, v_i is the volume of the minimum volume (hyper-)rectangle around the point x_i. Exploiting this entropy estimator, after substitution in Equation (8), we can derive a new estimation of TE.

3.2. Extension of Singh's Method

We propose in this section to extend Singh's entropy estimator by using a (hyper-)rectangular domain, as we did for the Kozachenko–Leonenko estimator extension introduced in the preceding section. Considering a d-dimensional random vector $X \in \mathbb{R}^d$ continuously distributed according to a probability density function p_X, we aim at estimating the entropy $\mathcal{H}(X)$ from the observation of a p_X distributed IID random sequence X_i, $i = 1, \ldots, N$. For any specific data point x_i and a fixed number k ($1 \leq k \leq N$), the minimum (hyper-)rectangle (rectangle $A'B'C'D'$ in Figure 2) is fixed, and we denote this region by $\mathcal{D}_{x_i}^{\varepsilon_1, \ldots, \varepsilon_d}$ and its volume by v_i. Let us denote ξ_i ($1 \leq \xi_i \leq \min(k,d)$) the number of points on the border of the (hyper-)rectangle that we consider as a realization of a random variable Ξ_i. In the situation described in Figure 2a,b, $\xi_i = 2$ and $\xi_i = 1$, respectively. According to [39] (Chapter 6, page 269), if $\mathcal{D}_{x_i}\left(\chi_i^k\right)$ corresponds to a ball (for a given norm) of volume v_i, an unbiased estimator of $p_X(x_i)$ is given by:

$$\widehat{p_X(x_i)} = \frac{k-1}{Nv_i}, i = 1, 2, \ldots, N. \tag{23}$$

This implies that the classical estimator $\widehat{p_X(x_i)} = \frac{k}{Nv_i}$ is biased and that presumably $\log\left(\frac{k}{Nv_i}\right)$ is also a biased estimation of $\log{(p_X(x_i))}$ for N large, as shown in [39].

Now, in the case $\mathcal{D}_{x_i}\left(\chi_i^k\right)$ is the minimal (*i.e.*, with minimal (hyper-)volume) (hyper-)rectangle $\mathcal{D}_{x_i}^{\varepsilon_1, \ldots, \varepsilon_d}$, including χ_i^k, more than one point can belong to the border, and a more general estimator $\widetilde{p_X(x_i)}$ of $p_X(x_i)$ can be *a priori* considered:

$$\widetilde{p_X(x_i)} = \frac{\tilde{k}_i}{Nv_i}, \tag{24}$$

where $\tilde{k}_i$ is some given function of k and ξ_i. The corresponding estimation of $\mathcal{H}(X)$ is then:

$$\widehat{\mathcal{H}(X)} = -\frac{1}{N}\sum_{i=1}^{N}\log{\left(\widetilde{p_X(x_i)}\right)} = \frac{1}{N}\sum_{i=1}^{N}t_i, \tag{25}$$

with:

$$t_i = \log{\left(\frac{Nv_i}{\tilde{k}_i}\right)}, i = 1, 2, \ldots, N, \tag{26}$$

t_i being realizations of random variables T_i and $\tilde{k}_i$ being realizations of random variables $\widetilde{K}_i$. We have:

$$\forall i = 1, \ldots, N : E\left[\widehat{\mathcal{H}(X)}\right] = E\left(T_i\right) = E\left(T_1\right). \tag{27}$$

Our goal is to derive $E\left[\widehat{\mathcal{H}(X)}\right] - \mathcal{H}(X) = E(T_1) - \mathcal{H}(X)$ for N large to correct the asymptotic bias of $\widehat{\mathcal{H}(X)}$, according to Steps (1) to (3), explained in Section 2.1.3. To this end, we must consider an asymptotic approximation of the conditional probability distribution $\mathcal{P}(T_1 \leq r|X_1 = x_1, \Xi_1 = \xi_1)$ before computing the asymptotic difference between the expectation $E[T_1] = E[E[T_1|X_1 = x_1, \Xi_1 = \xi_1]]$ and the true entropy $\mathcal{H}(X)$.

Let us consider the random Lebesgue measure V_1 of the random minimal (hyper-)rectangle $\mathcal{D}_{x_1}^{\epsilon_1,\ldots,\epsilon_d}$ (($\epsilon_1,\ldots,\epsilon_d$) denotes the random vector for which $(\epsilon_1,\ldots,\epsilon_d) \in \mathbb{R}^d$ is a realization) and the relation $T_1 = \log\left(\frac{NV_1}{\widetilde{K}_1}\right)$. For any $r > 0$, we have:

$$\mathcal{P}(T_1 > r|X_1 = x_1, \Xi_1 = \xi_1) = \mathcal{P}\left(\log\left(\frac{NV_1}{\widetilde{K}_1}\right) > r|X_1 = x_1, \Xi_1 = \xi_1\right)$$
$$= \mathcal{P}(V_1 > v_r|X_1 = x_1, \Xi_1 = \xi_1), \tag{28}$$

where $v_r = e^r\frac{\tilde{k}_1}{N}$, since, conditionally to $\Xi_1 = \xi_1$, we have $\widetilde{K}_1 = \tilde{k}_1$.

In Appendix E, we prove the following property.

Property 1. *For N large,*

$$\mathcal{P}(T_1 > r|X_1 = x_1, \Xi_1 = \xi_1) \simeq \sum_{i=0}^{k-\xi_1}\binom{N-\xi_1-1}{i}(p_X(x_1)v_r)^i(1 - p_X(x_1)v_r)^{N-\xi_1-1-i}. \tag{29}$$

The Poisson approximation (when $N \to \infty$ and $v_r \to 0$) of the binomial distribution summed in Equation (29) leads to a parameter $\lambda = (N - \xi_1 - 1)p_X(x_1)v_r$. As N is large compared to $\xi_1 + 1$, we obtain from Equation (26):

$$\lambda \simeq \tilde{k}_1 e^r p_X(x_1), \tag{30}$$

and we get the approximation:

$$\lim_{N\to\infty}\mathcal{P}(T_1 > r|X_1 = x_1, \Xi_1 = \xi_1) \simeq \sum_{i=0}^{k-\xi_1}\frac{[\tilde{k}_1 e^r p_X(x_1)]^i}{i!}e^{-\tilde{k}_1 e^r p_X(x_1)}. \tag{31}$$

Since $\mathcal{P}(T_1 \leq r|X_1 = x_1, \Xi_1 = \xi_1) = 1 - \mathcal{P}(T_1 > r|X_1 = x_1, \Xi_1 = \xi_1)$, we can get the density function of T_1, noted $g_{T_1}(r)$, by deriving $\mathcal{P}(T_1 \leq r|X_1 = x_1, \Xi_1 = \xi_1)$. After some mathematical developments (see Appendix F), we obtain:

$$g_{T_1}(r) = \mathcal{P}'(T_1 \leq r|X_1 = x_1, \Xi_1 = \xi_1)$$
$$= -\mathcal{P}'(T_1 > r|X_1 = x_1, \Xi_1 = \xi_1) \quad, r \in \mathbb{R}, \tag{32}$$
$$= \frac{[\tilde{k}_1 e^r p_X(x_1)]^{(k-\xi_1+1)}}{(k-\xi_1)!}e^{-\tilde{k}_1 e^r p_X(x_1)}$$

and consequently (see Appendix G for details),

$$\lim_{N\to\infty}E[T_1|X_1 = x_1, \Xi_1 = \xi_1] = \int_{-\infty}^{\infty} r\frac{[\tilde{k}_1 e^r p_X(x_1)]^{(k-\xi_1+1)}}{(k-\xi_1)!}e^{-\tilde{k}_1 e^r p_X(x_1)}dr$$
$$= \psi(k - \xi_1 + 1) - \log(\tilde{k}_1) - \log(p_X(x_1)). \tag{33}$$

Therefore, with the definition of differential entropy $\mathcal{H}(X_1) = E[-\log(p_X(X_1))]$, we have:

$$\lim_{N\to\infty}E[T_1] = \lim_{N\to\infty}E[E[T_1|X_1, \Xi_1]] = E\left[\psi(k - \Xi_1 + 1) - \log\left(\widetilde{K}_1\right)\right] + \mathcal{H}(X_1). \tag{34}$$

Thus, the estimator expressed by Equation (25) is asymptotically biased. Therefore, we consider a modified version, denoted by $\widehat{\mathcal{H}(X)}_{NS}$, obtained by subtracting an estimation of the bias $E\left[\psi(k - \Xi_1 + 1) - \log\left(\widetilde{K}_1\right)\right]$ given by the empirical mean $\frac{1}{N}\sum_{i=1}^{N}\psi(k - \xi_i + 1) + \frac{1}{N}\sum_{i=1}^{N}\log\left(\tilde{k}_i\right)$ (according to the large numbers law), and we obtain, finally (Box ⑤ in Figure 1):

$$\widehat{\mathcal{H}(X)}_{NS} = \frac{1}{N}\sum_{i=1}^{N}t_i - \frac{1}{N}\sum_{i=1}^{N}\psi(k - \xi_i + 1) + \frac{1}{N}\sum_{i=1}^{N}\log\left(\tilde{k}_i\right)$$

$$= \frac{1}{N}\sum_{i=1}^{N}\log\left(\frac{N v_i}{\tilde{k}_i}\right) - \frac{1}{N}\sum_{i=1}^{N}\psi(k - \xi_i + 1) + \frac{1}{N}\sum_{i=1}^{N}\log\left(\tilde{k}_i\right) \tag{35}$$

$$= \log(N) + \frac{1}{N}\sum_{i=1}^{N}\log\left(v_i\right) - \frac{1}{N}\sum_{i=1}^{N}\psi(k - \xi_i + 1).$$

In comparison with the development of Equation (22), we followed here the same methodology, except we take into account (through a conditioning technique) the influence of the number of points on the border.

We observe that, after cancellation of the asymptotic bias, the choice of the function of k and ξ_i to define $\tilde{k}_i$ in Equation (24) does not have any influence on the final result. In this way, we obtain an expression for $\widehat{\mathcal{H}(X)}_{NS}$, which simply takes into account the values ξ_i that could *a priori* influence the entropy estimation.

Note that, as for the original Kozachenko–Leonenko (Equation (10)) and Singh (Equation (16)) entropy estimators, both new estimation functions (Equations (22) and (35)) hold for any value of k, such that $k \ll N$, and we do not have to choose a fixed k while estimating entropy in lower dimensional spaces. Therefore, under the framework proposed in [24], we built two different TE estimators using Equations (22) and (35), respectively.

3.3. Computation of the Border Points Number and of the (Hyper-)Rectangle Sizes

We explain more precisely hereafter how to determine the numbers of points ξ_i on the border. Let us denote $x_i^j \in \mathbb{R}^d$, $j = 1, \ldots, k$, the k nearest neighbors of $x_i \in \mathbb{R}^d$, and let us consider the $d \times k$ array D_i, such that for any $(p, j) \in \{1, \ldots, d\} \times \{1, \ldots, k\}$, $D_i(p, j) = \left|x_i^j(p) - x_i(p)\right|$ is the distance (in $\mathbb{R}$) between the p-th component $x_i^j(p)$ of x_i^j and the p-th component $x_i(p)$ of x_i. For each p, let us introduce $J_i(p) \in \{1, \ldots, k\}$ defined by $D_i(p, J_i(p)) = \max\left(D_i(p, 1), \ldots, D_i(p, k)\right)$ and which is the value of the column index of D_i for which the distance $D_i(p, j)$ is maximum in the row number p. Now, if there exists more than one index $J_i(p)$ that fulfills this equality, we select arbitrarily the lowest one, hence avoiding the $\max(\cdot)$ function to be multi-valued. The MATLAB implementation of the max function selects such a unique index value. Then, let us introduce the $d \times k$ Boolean array B_i defined by $B_i(p, j) = 1$ if $j = J_i(p)$ and $B_i(p, j) = 0$, otherwise. Then:

(1) The d sizes ε_p, $p = 1, \ldots, d$ of the (hyper-)rectangle $\mathcal{D}_{x_i}^{\varepsilon_1, \ldots, \varepsilon_d}$ are equal respectively to $\varepsilon_p = 2D_i(p, J_i(p))$, $p = 1, \ldots, d$.

(2) We can define ξ_i as the number of non-null column vectors in B_i. For example, if the k-th nearest neighbor x_i^k is such that $\forall j \neq k, \forall p = 1, \ldots, d : \left|x_i^j(p) - x_i(p)\right| < \left|x_i^k(p) - x_i(p)\right|$, *i.e.*, when the k-th nearest neighbor is systematically the farthest from the central point x_i for each of the d directions, then all of the entries in the last column of B_i are equal to one, while all other entries are equal to zero: we have only one column including values different from zero and, so, only one point on the border ($\xi_i = 1$), which generalizes the case depicted in Figure 2b for $d = 2$.

N.B.: this determination of ξ_i may be incorrect when there exists a direction p, such that the number of indices j for which $D_i(p, j)$ reaches the maximal value is larger than one: the value of ξ_i obtained with our procedure can then be underestimated. However, we can argue that, theoretically, this case occurs with a probability equal to zero (because the observations are continuously distributed in the probability) and, so, it can be *a priori* discarded. Now, in practice, the measured quantification errors and the round off errors are unavoidable, and this probability will differ from zero (although remaining small when the aforesaid errors are small): theoretically distinct values $D_i(p,j)$ on the row p of D_i may be erroneously confounded after quantification and rounding. However, the $\max(\cdot)$ function then selects on row p only one value for $J_i(p)$ and, so, acts as an error correcting procedure. The fact that the maximum distance in the concerned p directions can then be allocated to the wrong neighbor index has no consequence for the correct determination of ξ_i.

4. New Estimators of Transfer Entropy

From an observed realization $\left(x_i^p, x_i^-, y_i^-\right) \in \mathcal{S}_{X^p, X^-, Y^-}$, $i = 1, 2, \ldots, N$ of the IID random sequence $\left(X_i^p, X_i^-, Y_i^-\right)$, $i = 1, 2, \ldots, N$ and a number k of neighbors, the procedure could be summarized as follows (distances are from the maximum norm):

(1) similarly to the MILCA [31] and TRENTOOL toolboxes [34], normalize, for each i, the vectors x_i^p, x_i^- and y_i^-;

(2) in joint space $\mathcal{S}_{X^p, X^-, Y^-}$, for each point $\left(x_i^p, x_i^-, y_i^-\right)$, calculate the distance $d_{\left(x_i^p, x_i^-, y_i^-\right), k}$ between $\left(x_i^p, x_i^-, y_i^-\right)$ and its k-th neighbor, then construct the (hyper-)rectangle with sizes $\varepsilon_1, \ldots, \varepsilon_d$ (d is the dimension of the vectors $\left(x_i^p, x_i^-, y_i^-\right)$), for which the (hyper-)volume is $v_{(X^p, X^-, Y^-),i} = \varepsilon_1 \times \ldots \times \varepsilon_d$ and the border contains $\xi_{(X^p, X^-, Y^-),i}$ points;

(3) for each point (x_i^p, x_i^-) in subspace $\mathcal{S}_{X^p, X^-}$, count the number $k_{(X^p, X^-),i}$ of points falling within the distance $d_{\left(x_i^p, x_i^-, y_i^-\right), k}$, then find the smallest (hyper-)rectangle that contains all of these points and for which $v_{(X^p, X^-),i}$ and $\xi_{(X^p, X^-),i}$ are respectively the volume and the number of points on the border; repeat the same procedure in subspaces $\mathcal{S}_{X^-, Y^-}$ and $\mathcal{S}_{X^-}$.

From Equation (22) (modified to k not constant for $\mathcal{S}_{X^-}$, $\mathcal{S}_{X^p, X^-}$ and $\mathcal{S}_{X^-, Y^-}$), the final TE estimator can be written as (Box ⑧ in Figure 1):

$$\widehat{\mathrm{TE}}_{Y \to X p1} = \frac{1}{N} \sum_{i=1}^{N} \log \frac{v_{(X^p, X^-),i} \cdot v_{(X^-, Y^-),i}}{v_{(X^p, X^-, Y^-),i} \cdot v_{X^-,i}}$$

$$+ \frac{1}{N} \sum_{i=1}^{N} \left(\psi(k) + \psi(k_{X^-,i}) - \psi(k_{(X^p, X^-),i}) - \psi(k_{(X^-, Y^-),i}) \right. \tag{36}$$

$$\left. + \frac{d_{X^p} + d_{X^-} - 1}{k_{(X^p, X^-),i}} + \frac{d_{X^-} + d_{Y^-} - 1}{k_{(X^-, Y^-),i}} - \frac{d_{X^p} + d_{X^-} + d_{Y^-} - 1}{k} - \frac{d_{X^-} - 1}{k_{X^-,i}} \right),$$

where $d_{X^p} = \dim\left(\mathcal{S}_{X^p}\right)$, $d_{X^-} = \dim\left(\mathcal{S}_{X^-}\right)$, $d_{Y^-} = \dim\left(\mathcal{S}_{Y^-}\right)$, and with Equation (35), it yields to (Box ⑦ in Figure 1):

$$\widehat{\mathrm{TE}}_{Y \to X p2} = \frac{1}{N} \sum_{i=1}^{N} \log \frac{v_{(X^p, X^-),i} \cdot v_{(X^-, Y^-),i}}{v_{(X^p, X^-, Y^-),i} \cdot v_{X^-,i}}$$

$$+ \frac{1}{N} \sum_{i=1}^{N} \left(\psi(k - \xi_{(X^p, X^-, Y^-),i} + 1) + \psi(k_{X^-,i} - \xi_{X^-,i} + 1) - \psi(k_{(X^p, X^-),i}) \right. \tag{37}$$

$$\left. - \xi_{(X^p, X^-),i} + 1) - \psi(k_{(X^-, Y^-),i} - \xi_{(X^-, Y^-),i} + 1) \right).$$

In Equations (36) and (37), the volumes $v_{(X^p, X^-),i}$, $v_{(X^-, Y^-),i}$, $v_{(X^p, X^-, Y^-),i}$, $v_{X^-,i}$ are obtained by computing, for each of them, the product of the edges lengths of the (hyper-)rectangle, *i.e.*, the product of d edges lengths, d being respectively equal to $d_{X^p} + d_{X^-}, d_{X^-} + d_{Y^-}, d_{X^p} + d_{X^-} + d_{Y^-}$ and d_{X^-}. In a given subspace and for a given direction, the edge length is equal to twice the largest distance between the corresponding coordinate of the reference point (at the center) and each of the corresponding coordinates of the k nearest neighbors. Hence a generic formula is $v_U = \prod_{j=1}^{\dim(\mathcal{U})} \varepsilon_{Uj}$, where U is one of the symbols $(X^p, X^-), (X^-, Y^-), (X^p, X^-, Y^-)$ and X^- and the ε_{Uj} are the edge lengths of the (hyper-)rectangle.

The new TE estimator $\widehat{\mathrm{TE}}_{Y \to Xp1}$ (Box ⑧ in Figure 1) can be compared with the extension of $\widehat{\mathrm{TE}}_{Y \to XSA}$, the TE estimator proposed in [27] (implemented in the JIDT toolbox [30]). This extension [27], included in Figure 1 (Box ⑨), is denoted here by $\widehat{\mathrm{TE}}_{Y \to XEA}$. The main difference with our $\widehat{\mathrm{TE}}_{Y \to Xp1}$ estimator is that our algorithm uses a different length for each sub-dimension within a variable, rather than one length for all sub-dimensions within the variable (which is the approach of the extended algorithm). We introduced this approach to make the tightest possible (hyper-)rectangle around the k nearest neighbors. $\widehat{\mathrm{TE}}_{Y \to XEA}$ is expressed as follows:

$$
\begin{aligned}
\widehat{\mathrm{TE}}_{Y \to XEA} = & \frac{1}{N} \sum_{i=1}^{N} (\psi(k) - \frac{2}{k} + \psi(l_{X^-,i}) - \psi(l_{(X^p,X^-),i}) \\
& + \frac{1}{l_{(X^p,X^-),i}} - \psi(l_{(X^-,Y^-),i}) + \frac{1}{l_{(X^-,Y^-),i}}).
\end{aligned}
\tag{38}
$$

In the experimental part, this estimator is marked as the "extended algorithm". It differs from Equation (36) in two ways. Firstly, the first summation on the right hand-side of Equation (36) does not exist. Secondly, compared with Equation (36), the numbers of neighbors $k_{X^-,i}$, $k_{(X^p,X^-),i}$ and $k_{(X^-,Y^-),i}$ included in the rectangular boxes, as explained in Section 3.1, are replaced respectively with $l_{X^-,i}$, $l_{(X^p,X^-),i}$ and $l_{(X^-,Y^-),i}$, which are obtained differently. More precisely, Step (2) in the above algorithm becomes:

(2') For each point (x_i^p, x_i^-) in subspace $\mathcal{S}_{X^p,X^-}$, $l_{(X^p,X^-),i}$ is the number of points falling within a (hyper-)rectangle equal to the Cartesian product of two (hyper-)cubes, the first one in $\mathcal{S}_{X^p}$ and the second one in $\mathcal{S}_{X^-}$, whose edge lengths are equal, respectively, to $d_{x_i^p}^{\max} = 2 \times \max \left\{ \left\| x_k^p - x_i^p \right\| : (x^p, x^-, y^-)_k \in \chi_{(x^p,x^-,y^-)_i}^k \right\}$ and $d_{x_i^-}^{\max} = 2 \times \max \left\{ \left\| x_k^- - x_i^- \right\| : (x^p, x^-, y^-)_k \in \chi_{(x^p,x^-,y^-)_i}^k \right\}$, *i.e.*, $l_{(X^p,X^-),i} = \mathrm{card} \left\{ \left(x_j^p, x_i^- \right) : j \in \{\{1,\ldots,N\} - \{i\}\} \ \& \ \left\| x_j^p - x_i^p \right\| \le d_{x_i^p}^{\max} \ \& \ \left\| x_j^- - x_i^- \right\| \le d_{x_i^-}^{\max} \right\}$. Denote by $v_{(X^p,X^-),i}$ the volume of this (hyper-)rectangle. Repeat the same procedure in subspaces $\mathcal{S}_{X^-,Y^-}$ and $\mathcal{S}_{X^-}$.

Note that the important difference between the construction of the neighborhoods used in $\widehat{\mathrm{TE}}_{Y \to XEA}$ and in $\widehat{\mathrm{TE}}_{Y \to Xp1}$ is that, for the first case, the minimum neighborhood, including the k neighbors, is constrained to be a Cartesian product of (hyper-)cubes and, in the second case, this neighborhood is a (hyper-)rectangle whose edge lengths can be completely different.

5. Experimental Results

In the experiments, we tested both Gaussian IID and Gaussian AR models to compare and validate the performance of the TE estimators proposed in the previous section. For a complete comparison, beyond the theoretical value of TE, we also computed the Granger causality index as a reference (as indicated previously, in the case of Gaussian signals TE and Granger causality index are equivalent up to a factor of two; see Appendix H). In each following figure, GCi/2 corresponds to the Granger causality index divided by two; TE estimated by the free TRENTOOL toolbox (corresponding to Equation (21)) is marked as the standard algorithm; that estimated by JIDT (corresponding to Equation (38)) is marked as the extended algorithm; TE_{p1} is the TE estimator given by Equation (36);

and TE$_{p2}$ is the TE estimator given by Equation (37). For all of the following results, the statistical means and the standard deviations of the different estimators have been estimated using an averaging on 200 trials.

5.1. Gaussian IID Random Processes

The first model we tested, named Model 1, is formulated as follows:

$$X_t = aY_t + bZ_t + W_t, \quad W_t \in \mathbb{R}, Y \in \mathbb{R}^{d_Y}, Z \in \mathbb{R}^{d_Z}, \tag{39}$$

where $Y_t \sim \mathcal{N}(0, C_Y)$, $Z_t \sim \mathcal{N}(0, C_Z)$, $W_t \sim \mathcal{N}(0, \sigma_W^2)$, the three processes Y, Z, and W being mutually independent. The triplet (X_t, Y_t, Z_t) corresponds to the triplet $\left(X_i^p, X_i^-, Y_i^-\right)$ introduced previously. C_U is a Toeplitz matrix with the first line equal to $[1, \alpha, \ldots, \alpha^{d_U-1}]$. For the matrix C_Y, we chose $\alpha = 0.5$, and for C_Z, $\alpha = 0.2$. The standard deviation σ_W was set to 0.5. The vectors a and b were such that $a = 0.1 * [1, 2, \ldots, d_Y]$ and $b = 0.1 * [d_Z, d_Z - 1, \ldots, 1]$. With this model, we aimed at estimating $\mathcal{H}(X|Y) - \mathcal{H}(X|Y, Z)$ to test if the knowledge of signals Y and Z could improve the prediction of X compared to only the knowledge of Y.

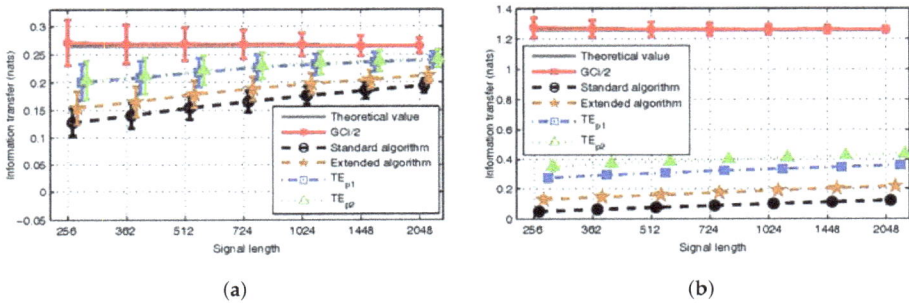

Figure 3. Information transfer from Z to X (Model 1) estimated for two different dimensions with $k = 8$. The figure displays the mean values and the standard deviations: (**a**) $d_Y = d_Z = 3$; (**b**) $d_Y = d_Z = 8$.

Results are reported in Figure 3 where the dimensions d_Y and d_Z are identical. We observe that, for a low dimension and a sufficient number of neighbors (Figure 3a), all TE estimators tend all the more to the theoretical value (around 0.26) that the length of the signals is large, the best estimation being obtained by the two new estimators. Compared to Granger causality, these estimators display a greater bias, but a lower variance. Due to the "curse of dimensionality", with an increasing dimension (see Figure 3b), it becomes much more difficult to obtain an accurate estimation of TE. For a high dimension, all estimators reveal a non-negligible bias, even if the two new estimators still behave better than the two reference ones (standard and extended algorithms).

5.2. Vectorial AR Models

In the second experiment, two AR models integrating either two or three signals have been tested. The first vectorial AR model (named Model 2) we tested was as follows:

$$\begin{cases} x_t = 0.45\sqrt{2}x_{t-1} - 0.9x_{t-2} - 0.6y_{t-2} + e_{x,t} \\ y_t = 0.6x_{t-2} - 0.175\sqrt{2}y_{t-1} + 0.55\sqrt{2}y_{t-2} + e_{y,t}. \end{cases} \tag{40}$$

The second vectorial AR model (named Model 3) was given by:

$$\begin{cases} x_t = -0.25x_{t-2} - 0.35y_{t-2} + 0.35z_{t-2} + e_{x,t} \\ y_t = -0.5x_{t-1} + 0.25y_{t-1} - 0.5z_{t-3} + e_{y,t} \\ z_t = -0.6x_{t-2} - 0.7y_{t-2} - 0.2z_{t-2} + e_{z,t}. \end{cases} \tag{41}$$

For both models, e_x, e_y and e_z denote realizations of independent white Gaussian noises with zero mean and a variance of 0.1. As previously, we display in the following figures not only the theoretical value of TE, but also the Granger causality index for comparison. In this experiment, the prediction orders m and n were equal to the corresponding regression orders of the AR models. For example, when estimating $TE_{Y \to X}$, we set $m = 2$, $n = 2$, and $\left(X_i^p, X_i^-, Y_i^- \right)$ corresponds to $\left(X_{i+1}, X_i^{(2)}, Y_i^{(2)} \right)$.

For Figures 4 and 5, the number k of neighbors was fixed to eight, whereas, in Figure 6, this number was set to four and three (respectively Figures 6a,b) to show the influence of this parameter. Figures 4 and 6 are related to Model 2, and Figure 5 is related to Model 3.

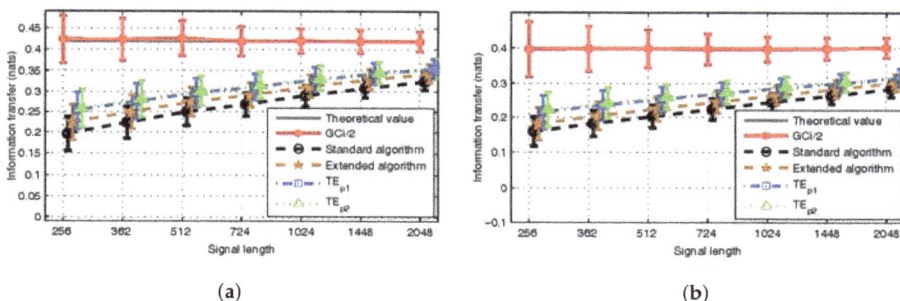

Figure 4. Information transfer (Model 2), mean values and standard deviations, $k = 8$. (**a**) From X to Y; (**b**) from Y to X.

As previously, for large values of k (*cf.* Figures 4 and 5), we observe that the four TE estimators converge towards the theoretical value. This result is all the more true when the signal length increases. As expected in such linear models, Granger causality outperforms the TE estimators at the expense of a slightly larger variance. Contrary to Granger causality, TE estimators are clearly more impacted by the signal length, even if their standard deviations remain lower. Here, again, when comparing the different TE estimators, it appears that the two new estimators achieve improved behavior compared to the standard and extended algorithms for large k.

In the scope of k-NN algorithms, the choice of k must be a tradeoff between the estimation of bias and variance. Globally, when the value of k decreases, the bias decreases for the standard and extended algorithms and for the new estimator TE_{p1}. Now, for the second proposed estimator TE_{p2}, it is much more sensitive to the number of neighbors (as can be seen when comparing Figures 4 and 6). As shown in Figures 3 to 5, the results obtained using TE_{p2} and TE_{p1} are quite comparable when the value of k is large ($k = 8$). Now, when the number of neighbors decreases, the second estimator we proposed, TE_{p2}, is much less reliable than all of the other ones (Figure 6). Concerning the variance, it remains relatively stable when the number of neighbors falls from eight to three, and in this case, the extended algorithm, which displays a slightly lower bias, may be preferred.

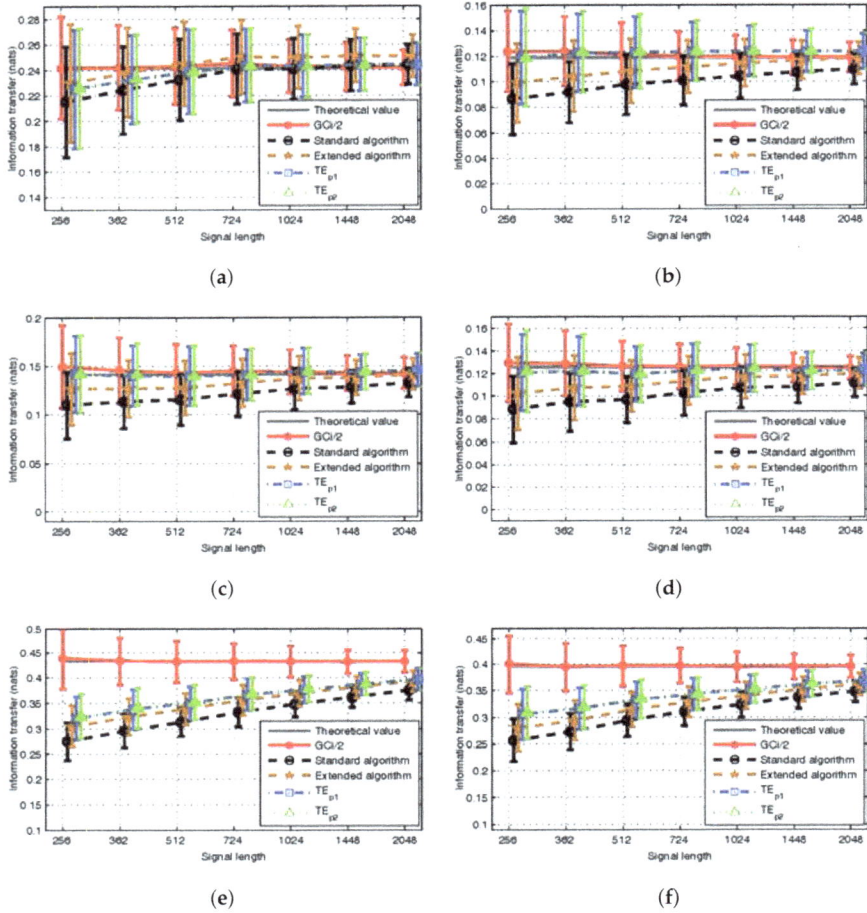

Figure 5. Information transfer (Model 3), mean values and standard deviations, $k = 8$. (**a**) From X to Y; (**b**) from Y to X; (**c**) from X to Z; (**d**) from Z to X; (**e**) from Y to Z; (**f**) from Z to Y.

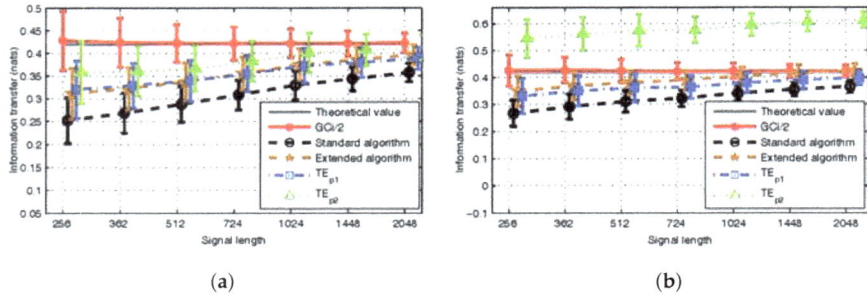

Figure 6. Information transfer from X to Y (Model 2), mean values and standard deviations: (**a**) $k = 4$; (**b**) $k = 3$.

When using $k = 8$, a possible interpretation of getting a lower bias with our algorithms could be that, once we are looking at a large enough number of k nearest neighbors, there is enough opportunity for the use of different lengths on the sub-dimensions of the (hyper-)rectangle to make a difference to the results, whereas with $k = 3$, there is less opportunity.

To investigate the impact on the dispersion (estimation error standard deviation) of (i) the estimation method and (ii) the number of neighbors, we display in Figures 7a,b the boxplots of the absolute values of the centered estimation errors (AVCE) corresponding to experiments reported in Figures 4a and 6b for a 1024-point signal length. These results show that neither the value of k, nor the tested TE estimator dramatically influence the dispersions. More precisely, we used a hypothesis testing procedure (two-sample Kolmogorov–Smirnov goodness-of-fit hypothesis, KSTEST2 in MATLAB) to test if two samples (each with 200 trials) of AVCE are drawn from the same underlying continuous population or not. The tested hypothesis corresponds to non-identical distributions and is denoted $H = 1$, and $H = 0$ corresponds to the rejection of this hypothesis. The confidence level was set to 0.05.

(1) Influence of the method:

    (a) Test between the standard algorithm and TE_{p1} in Figure 7a: $H = 0$, p-value = 0.69 → no influence

    (b) Test between the extended algorithm and TE_{p1} in Figure 7a: $H = 0$, p-value = 0.91 → no influence

    (c) Test between the standard algorithm and TE_{p1} in Figure 7b: $H = 0$, p-value = 0.081 → no influence

    (d) Test between the extended algorithm and TE_{p1} in Figure 7b: $H = 1$, p-value = 0.018 → influence exists.

(2) Influence of the neighbors' number k:

    (a) Test between $k = 8$ (Figure 7a) and $k = 3$ (Figure 7b) for the standard algorithm: $H = 0$, p-value = 0.97 → no influence

    (b) Test between $k = 8$ (Figure 7a) and $k = 3$ (Figure 7b) for TE_{p1}: $H = 0$, p-value = 0.97 → no influence.

For these six tested cases, the only case where a difference between distributions (and so, between the dispersions) corresponds to a different distribution is when comparing the extended algorithm and TE_{p1} in Figure 7b.

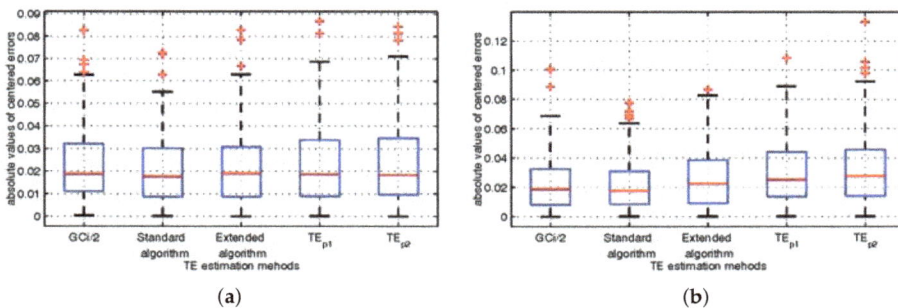

Figure 7. Box plots of the centered errors obtained with the five methods for Model 2, $X \rightarrow Y$: (**a**) $k = 8$ (corresponding to Figure 4a); (**b**) $k = 3$ (corresponding to Figure 6b).

6. Discussion and Summary

In the computation of k-NN based estimators, the most time-consuming part is the procedure of nearest neighbor searching. Compared to Equations (10) and (16), Equations (22) and (35) involve supplementary information, such as the maximum distance of the first k-th nearest neighbor in each dimension and the number of points on the border. However, most currently used neighbor searching algorithms, such as k-d tree (k-dimensional tree) and ATRIA (A TRiangle Inequality based Algorithm) [40], provide not only information on the k-th neighbor, but also on the first $(k-1)$ nearest neighbors. Therefore, in terms of computation cost, there is no significant difference among the three TE estimators (Boxes ⑦, ⑧, ⑨, ⑩ in Figure 1).

In this contribution, we discussed TE estimation based on k-NN techniques. The estimation of TE is always an important issue, especially in neuroscience, where getting large amounts of stationary data is problematic. The widely-used k-NN technique has been proven to be a good choice for the estimation of information theoretical measurement. In this work, we first investigated the estimation of Shannon entropy based on the k-NN technique involving a rectangular neighboring region and introduced two different k-NN entropy estimators. We derived mathematically these new entropy estimators by extending the results and methodology developed in [24] and [36]. Given the new entropy estimators, two novel TE estimators have been proposed, implying no extra computation cost compared to existing similar k-NN algorithm. To validate the performance of these estimators, we considered different simulated models and compared the new estimators with the two TE estimators available in the free TRENTOOL and JIDT toolboxes, respectively, and which are extensions of two Kraskov–Stögbauer–Grassberger (KSG) MI estimators, based respectively on (hyper-)cubic and (hyper-)rectangular neighborhoods.

Under the Gaussian assumption, experimental results showed the effectiveness of the new estimators under the IID assumption, as well as for time-correlated AR signals in comparison with the standard KSG algorithm estimator. This conclusion still holds when comparing the new algorithms with the extended KSG estimator. Globally, all TE estimators satisfactorily converge to the theoretical TE value, *i.e.*, to half the value of the Granger causality, while the newly proposed TE estimators showed lower bias for k sufficiently large (in comparison with the reference TE estimators) with comparable variances estimation errors.

As the variance remains relatively stable when the number of neighbors falls from eight to three, in this case, the extended algorithm, which displays a slightly lower bias, may be preferred.

Now, one of the new TE estimators suffered from noticeable error when the number of neighbors was small. Some experiments allowed us to verify that this issue already exists when estimating the entropy of a random vector: when the number of neighbors k falls below the dimension d, then the bias drastically increases. More details on this phenomenon are given in Appendix 6.

As expected, experiments with Model 1 showed that all three TE estimators under examination suffered from the "curse of dimensionality", which made it difficult to obtain accurate estimation of TE with high dimension data. In this contribution, we do not present the preliminary results that we obtained when simulating a nonlinear version of Model 1, for which the three variables X_t, Y_t and Z_t were scalar and their joint law was non-Gaussian, because a random nonlinear transformation was used to compute X_t from Y_t, Z_t. For this model, we computed the theoretical TE (numerically, with good precision) and tuned the parameters to obtain a strong coupling between X_t and Z_t. The theoretical Granger causality index was equal to zero. We observed the same issue as that pointed out in [41], *i.e.*, a very slow convergence of the estimator when the number of observations increases, and noticed that the four estimators $\widehat{TE}_{Y \to XSA}$, $\widehat{TE}_{Y \to XEA}$, $\widehat{TE}_{Y \to Xp1}$ and $\widehat{TE}_{Y \to Xp2}$, revealed very close performance. In this difficult case, our two methods do not outperform the existing ones. Probably, for this type of strong coupling, further improvement must be considered at the expense of an increasing computational complexity, as that proposed in [41].

This work is a first step in a more general context of connectivity investigation for neurophysiological activities obtained either from nonlinear physiological models or from clinical recordings. In this context, partial TE has also to be considered, and future work would address a comparison of the techniques presented in this contribution in terms of bias and variance. Moreover, considering the practical importance to know statistical distributions of the different TE estimators for independent channels, this point should be also addressed.

Author Contributions: Author Contributions

All authors have read and approved the final manuscript.

Appendix

A. Mathematical Expression of Transfer Entropy for Continuous Probability Distributions

Here, we consider that the joint probability measure $\mathcal{P}_{X_i^p, X_i^-, Y_i^-}$ is absolutely continuous (with respect to the Lebesgue measure in $\mathbb{R}^{m+n+1}$ denoted by μ^{m+n+1}) with the corresponding density:

$$p_{X_i^p, X_i^-, Y_i^-}\left(x_i^p, x_i^-, y_i^-\right) = \frac{d\mathcal{P}_{X_i^p, X_i^-, Y_i^-}\left(x_i^p, x_i^-, y_i^-\right)}{d\mu^{n+m+1}\left(x_i^p, x_i^-, y_i^-\right)}. \tag{42}$$

Then, we are sure that the two following conditional densities probability functions exist:

$$p_{X_i^p|X_i^-}\left(x_i^p|x_i^-\right) = \frac{d\mathcal{P}_{X_i^p|X_i^-}\left(x_i^p|x_i^-\right)}{d\mu^1\left(x_i^p\right)}$$

$$p_{X_i^p|X_i^-, Y_i^-}\left(x_i^p|x_i^-, y_i^-\right) = \frac{d\mathcal{P}_{X_i^p|X_i^-, Y_i^-}\left(x_i^p|x_i^-, y_i^-\right)}{d\mu^1\left(x_i^p\right)}. \tag{43}$$

and Equation (3) yields to:

$$\begin{aligned}
\mathrm{TE}_{Y \to X, i} &= \int_{\mathbb{R}^{m+n+1}} p_{X^p, X_i^-, Y_i^-}\left(x_i^p, x_i^-, y_i^-\right) \log \left[\frac{p_{X^p|X_i^-, Y_i^-}\left(x_i^p|x_i^-, y_i^-\right)}{p_{X^p|X_i^-}\left(x_i^p|x_i^-\right)}\right] dx_i^p dx_i^- y_i^- \\
&= \int_{\mathbb{R}^{m+n+1}} p_{X^p, X_i^-, Y_i^-}\left(x_i^p, x_i^-, y_i^-\right) \log \left[\frac{p_{X^p, X_i^-, Y_i^-}\left(x_i^p, x_i^-, y_i^-\right) p_{X_i^-}\left(x_i^-\right)}{p_{X_i^-, Y_i^-}\left(x_i^-, y_i^-\right) p_{X^p, X_i^-}\left(x_i^p, x_i^-\right)}\right] dx_i^p dx_i^- y_i^-.
\end{aligned} \tag{44}$$

Equation (44) can be rewritten:

$$\begin{aligned}
\mathrm{TE}_{Y \to X, i} &= -E\left[\log\left(p_{X_i^-, Y_i^-}\left(X_i^-, Y_i^-\right)\right)\right] - E\left[\log\left(p_{X_i^p, X_i^-}\left(X_i^p, X_i^-\right)\right)\right] \\
&\quad + E\left[\log\left(p_{X_i^p, X_i^-, Y_i^-}\left(X_i^p, X_i^-, Y_i^-\right)\right)\right] + E\left[\log\left(p_{X_i^-}\left(X_i^-\right)\right)\right].
\end{aligned} \tag{45}$$

B. Basic Structure of TE Estimators

From Equation (8), assuming that X and Y are jointly strongly ergodic leads to:

$$\begin{aligned}
\mathrm{TE}_{Y \to X} = \lim_{N \to \infty} \frac{1}{N} \sum_{i=1,\dots,N} &\left[-\log\left(p_{X_i^-, Y_i^-}\left(X_i^-, Y_i^-\right)\right) - \log\left(p_{X_i^p, X_i^-}\left(X_i^p, X_i^-\right)\right)\right. \\
&\left. + \log\left(p_{X_i^p, X_i^-, Y_i^-}\left(X_i^p, X_i^-, Y_i^-\right)\right) + \log\left(p_{X_i^-}\left(X_i^-\right)\right)\right],
\end{aligned} \tag{46}$$

where the convergence holds with probability one. Hence, as a function of an observed occurrence (x_i, y_i), $i = 1, \ldots, N$, of (X_i, Y_i), $i = 1, \ldots, N$, a standard estimation $\widehat{\text{TE}}_{Y \to X}$ of $\text{TE}_{Y \to X}$ is given by:

$$
\begin{aligned}
\widehat{\text{TE}}_{Y \to X} &= \widehat{\mathcal{H}\left(X^-, Y^-\right)} + \widehat{\mathcal{H}\left(X^p, X^-\right)} - \widehat{\mathcal{H}\left(X^p, X^-, Y^-\right)} - \widehat{\mathcal{H}\left(X^-\right)} \\
&= -\frac{1}{N} \sum_{n=1}^{N} \log\widehat{\left(p_{U_1}\left(u_{1n}\right)\right)} - \frac{1}{N} \sum_{n=1}^{N} \log\widehat{\left(p_{U_2}\left(u_{2n}\right)\right)} + \frac{1}{N} \sum_{n=1}^{N} \log\widehat{\left(p_{U_3}\left(u_{3n}\right)\right)} \\
&\quad + \frac{1}{N} \sum_{n=1}^{N} \log\widehat{\left(p_{U_4}\left(u_{4n}\right)\right)},
\end{aligned}
\tag{47}
$$

where U_1, U_2, U_3 and U_4 stand respectively for (X^-, Y^-), (X^p, X^-), (X^p, X^-, Y^-) and X^-.

C. The Bias of Singh's Estimator

Let us consider the equalities $E\left(T_1\right) = -E\left[\log\left(\widehat{p_X\left(X_1\right)}\right)\right] = -E\left[\log\left(\frac{k}{NV_1}\right)\right]$ where V_1 is the random volume for which v_1 is an outcome. Conditionally to $X_1 = x_1$, if we have $\frac{k}{NV_1} \xrightarrow[N \to \infty]{pr} p_X\left(x_1\right)$ (convergence in probability), then $E\left(T_1/X_1 = x_1\right) \xrightarrow[N \to \infty]{} -\log\left(p_X\left(x_1\right)\right)$, and by deconditioning, we obtain $E\left(T_1\right) \xrightarrow[N \to \infty]{} -E\left(\log\left(p_X\left(X_1\right)\right)\right) = \mathcal{H}(X)$. Therefore, if $\frac{k}{NV_1} \xrightarrow[N \to \infty]{pr} p_X\left(x_1\right)$, the estimation of $\mathcal{H}(X)$ is asymptotically unbiased. Here, this convergence in probability does not hold, even if we assume that $E\left(\frac{k}{NV_1}\right) \xrightarrow[N \to \infty]{} p_X\left(x_1\right)$ (one order mean convergence), because we do not have $\text{var}\left(\frac{k}{NV_1}\right) \xrightarrow[N \to \infty]{} 0$. The ratio $\frac{k}{NV_1}$ remains fluctuating when $N \to \infty$, because the ratio $\frac{\sqrt{\text{var}(V_1)}}{E(V_1)}$ does not tend to zero, even if V_1 tends to be smaller: when N increases, the neighborhoods become smaller and smaller, but continue to 'fluctuate'. This explains informally (see [37] for a more detailed analysis) why the naive estimator given by Equation (17) is not asymptotically unbiased. It is interesting to note that the Kozachenko–Leonenko entropy estimator avoids this problem, and so it does not need any bias subtraction.

D. Derivation of Equation (22)

As illustrated in Figure 2, for $d = 2$, there are two cases to be distinguished: (1) ε_x and ε_y are determined by the same point; (2) ε_x and ε_y are determined by distinct points.

Considering the probability density $q_{i,k}\left(\epsilon_x, \epsilon_y\right)$, $\left(\epsilon_x, \epsilon_y\right) \in \mathbb{R}^2$ of the pair of random sizes $\left(\epsilon_x, \epsilon_y\right)$ (along x and y, respectively), we can extend it to the case $d > 2$. Hence, let us denote by $q_{x_i, k}^d\left(\varepsilon_1, \ldots, \varepsilon_d\right)$, $\left(\varepsilon_1, \ldots, \varepsilon_d\right) \in \mathbb{R}^d$ the probability density (conditional to $X_i = x_i$) of the d-dimensional random vector whose d components are respectively the d random sizes of the (hyper-)rectangle built from the random k nearest neighbors, and denote by $h^{x_i}\left(\varepsilon_1, \ldots, \varepsilon_d\right) = \int_{u \in \mathcal{D}_{x_i}^{\varepsilon_1, \ldots, \varepsilon_d}} \mathrm{d}P_X\left(u\right)$ the probability mass (conditional to $X_i = x_i$) of the random (hyper-)rectangle $\mathcal{D}_{x_i}^{\varepsilon_1, \ldots, \varepsilon_d}$. In [24], the equality $E\left[\log\left(h^{x_i}\left(D_{x_i, k}\right)\right)\right] = \psi(k) - \psi(N)$ obtained for an (hyper-)cube is extended for the case $d > 2$ to:

$$
E\left[\log\left(h^{x_i}\left(\epsilon_1, \ldots, \epsilon_d\right)\right)\right] = \psi(k) - \frac{d-1}{k} - \psi(N).
\tag{48}
$$

Therefore, if p_X is approximately constant on $\mathcal{D}_{x_i}^{\varepsilon_1, \ldots, \varepsilon_d}$, we have:

$$
h^{x_i}\left(\varepsilon_1, \ldots, \varepsilon_d\right) \simeq v_i p_X\left(x_i\right),
\tag{49}
$$

where $v_i = \int_{\mathcal{D}_{x_i}^{\varepsilon_1, \ldots, \varepsilon_d}} \mathrm{d}\mu^d(\xi)$ is the volume of the (hyper-)rectangle, and we obtain:

$$
\log p_X(x_i) \approx \psi(k) - \psi(N) - \frac{d-1}{k} - \log\left(v_i\right).
\tag{50}
$$

Finally, by taking the experimental mean of the right term in Equation (50), we obtain an estimation of the expectation $E\left[\log p_X(X)\right]$, *i.e.,*:

$$\widehat{\mathcal{H}(X)} = -\psi(k) + \psi(N) + \frac{d-1}{k} + \frac{1}{N}\sum_{i=1}^{N}\log(v_i). \tag{51}$$

E. Proof of Property 1

Let us introduce the (hyper-)rectangle $\mathcal{D}_{x_1}^{\epsilon_1',\dots,\epsilon_d'}$ centered on x_1 for which the random sizes along the d directions are defined by $(\epsilon_1',\dots,\epsilon_d') = (\epsilon_1,\dots,\epsilon_d) \times \left(\frac{v_r}{\epsilon_1 \times \dots \times \epsilon_d}\right)^{1/d}$, so that $\mathcal{D}_{x_1}^{\epsilon_1',\dots,\epsilon_d'}$ and $\mathcal{D}_{x_1}^{\epsilon_1,\dots,\epsilon_d}$ are homothetic and $\mathcal{D}_{x_1}^{\epsilon_1',\dots,\epsilon_d'}$ has a (hyper-)volume constrained to the value v_r. We have:

$$\int_{x\in\mathcal{D}_{x_1}^{\epsilon_1,\dots,\epsilon_d}}d\mu^d(x) > v_r \Leftrightarrow \mathcal{D}_{x_1}^{\epsilon_1',\dots,\epsilon_d'} \subset \mathcal{D}_{x_1}^{\epsilon_1,\dots,\epsilon_d} \Leftrightarrow card\left\{x_j : x_j \in \mathcal{D}_{x_1}^{\epsilon_1',\dots,\epsilon_d'}\right\} \le k - \xi_1, \tag{52}$$

where the first equivalence (the inclusion is a strict inclusion) is clearly implied by the construction of $\mathcal{D}_{x_1}^{\epsilon_1',\dots,\epsilon_d'}$ and the second equivalence expresses the fact that the (hyper-)volume of $\mathcal{D}_{x_1}^{\epsilon_1,\dots,\epsilon_d}$ is larger than v_r if and only if the normalized domain $\mathcal{D}_{x_1}^{\epsilon_1',\dots,\epsilon_d'}$ does not contain more than $(k-\xi_1)$ points x_j (as ξ_1 of them are on the border of $\mathcal{D}_{x_1}^{\epsilon_1',\dots,\epsilon_d'}$, which is necessarily not included in $\mathcal{D}_{x_1}^{\epsilon_1',\dots,\epsilon_d'}$). These equivalences imply the equalities between conditional probability values:

$$\mathcal{P}\left(T_1 > r|X_1 = x_1, \Xi_1 = \xi_1\right) = \mathcal{P}\left(\log\left(\frac{NV_1}{\widetilde{K}_1}\right) > r|X_1 = x_1, \Xi_1 = \xi_1\right)$$
$$= \mathcal{P}\left(V_1 > v_r|X_1 = x_1, \Xi_1 = \xi_1\right) \tag{53}$$
$$= \mathcal{P}\left(card\left\{X_j : X_j \in \mathcal{D}_{x_1}^{\epsilon_1',\dots,\epsilon_d'}\right\} \le k - \xi_1\right).$$

Only $(N-1-\xi_1)$ events $\left\{X_j : X_j \in \mathcal{D}_{x_1}^{\epsilon_1',\dots,\epsilon_d'}\right\}$ are to be considered, because the variable X_1 and the ξ_1 variable(s) on the border of $\mathcal{D}_{x_1}^{\epsilon_1,\dots,\epsilon_d}$ must be discarded. Moreover, these events are independent. Hence, the probability value in (53) can be developed as follows:

$$\mathcal{P}\left(T_1 > r|X_1 = x_1, \Xi_1 = \xi_1\right) \simeq \sum_{i=0}^{k-\xi_1}\binom{N-\xi_1-1}{i}\left(\mathcal{P}\left(X \in \mathcal{D}_{x_1}^{\epsilon_1',\dots,\epsilon_d'}\right)\right)^i$$
$$\left(1 - \mathcal{P}\left(X \in \mathcal{D}_{x_1}^{\epsilon_1',\dots,\epsilon_d'}\right)\right)^{N-\xi_1-1-i}. \tag{54}$$

If $p_X(x_1)$ is approximately constant on $\mathcal{D}_{x_1}^{\epsilon_1',\dots,\epsilon_d'}$, we have $\mathcal{P}\left(X \in \mathcal{D}_{x_1}^{\epsilon_1',\dots,\epsilon_d'}\right) \simeq p_X(x_1)v_r$ (note that the randomness of $(\epsilon_1',\dots,\epsilon_d')$ does not influence this approximation as the (hyper-)volume of $\mathcal{D}_{x_1}^{\epsilon_1',\dots,\epsilon_d'}$ is imposed to be equal to v_r). Finally, we can write:

$$\mathcal{P}\left(T_1 > r|X_1 = x_1, \Xi_1 = \xi_1\right) \simeq \sum_{i=0}^{k-\xi_1}\binom{N-\xi_1-1}{i}(p_X(x_1)v_r)^i(1 - p_X(x_1)v_r)^{N-\xi_1-1-i}. \tag{55}$$

F. Derivation of Equation (32)

With $\mathcal{P}(T_1 \le r | X_1 = x_1, \Xi_1 = \xi_1) = 1 - \mathcal{P}(T_1 > r | X_1 = x_1, \Xi_1 = \xi_1)$, we take the derivative of $\mathcal{P}(T_1 \le r | X_1 = x_1, \Xi_1 = \xi_1)$ to get the conditional density function of T_1:

$$
\begin{aligned}
&\mathcal{P}'(T_1 \le r | X_1 = x_1, \Xi_1 = \xi_1) \\
&= -\mathcal{P}'(T_1 > r | X_1 = x_1, \Xi_1 = \xi_1) \\
&= -\left[\sum_{i=0}^{k-\xi_1} \frac{[\tilde{k}_1 p_X(x_1) e^r]^i}{i!} e^{-\tilde{k}_1 p_X(x_1) e^r} \right]' \\
&= -\sum_{i=0}^{k-\xi_1} \left(\left[\frac{[\tilde{k}_1 p_X(x_1) e^r]^i}{i!} \right]' e^{-\tilde{k}_1 p_X(x_1) e^r} + \frac{[\tilde{k}_1 p_X(x_1) e^r]^i}{i!} \left[e^{-\tilde{k}_1 p_X(x_1) e^r} \right]' \right) \\
&= -\sum_{i=0}^{k-\xi_1} \left(\frac{i[\tilde{k}_1 p_X(x_1) e^r]^{i-1}(\tilde{k}_1 p_X(x_1) e^r)}{i!} e^{-\tilde{k}_1 p_X(x_1) e^r} + \frac{[\tilde{k}_1 p_X(x_1) e^r]^i}{i!} e^{-\tilde{k}_1 p_X(x_1) e^r}(-\tilde{k}_1 p_X(x_1) e^r) \right) \\
&= -\sum_{i=0}^{k-\xi_1} e^{-\tilde{k}_1 p_X(x_1) e^r} \left(\frac{[\tilde{k}_1 p_X(x_1) e^r]^i}{(i-1)!} - \frac{[\tilde{k}_1 p_X(x_1) e^r]^{i+1}}{i!} \right).
\end{aligned}
\tag{56}
$$

Defining:

$$
a(i) = \frac{[\tilde{k}_1 p_X(x_1) e^r]^i}{(i-1)!} \quad \text{and} \quad a(0) = 0,
\tag{57}
$$

we have:

$$
\begin{aligned}
\mathcal{P}'(T_1 \le r) &= -\sum_{i=0}^{k-\xi_1} e^{-\tilde{k}_1 p_X(x_1) e^r} (a(i) - a(i+1)) \\
&= -e^{-\tilde{k}_1 p_X(x_1) e^r} (a(0) - a(k - \xi_1 + 1)) \\
&= e^{-\tilde{k}_1 p_X(x_1) e^r} a(k - \xi_1 + 1) \\
&= \frac{[\tilde{k}_1 p_X(x_1) e^r]^{(k-\xi_1+1)}}{(k-\xi_1)!} e^{-\tilde{k}_1 p_X(x_1) e^r}.
\end{aligned}
\tag{58}
$$

G. Derivation of Equation (33)

$$
\begin{aligned}
\lim_{n \to \infty} E(T_1 | X_1 = x_1) &= \int_{-\infty}^{\infty} r \frac{[\tilde{k}_1 p_X(x_1) e^r]^{(k-\xi_1+1)}}{(k-\xi_1)!} e^{-\tilde{k}_1 p_X(x_1) e^r} \, dr \\
&= \int_0^{\infty} [\log(z) - \log(\tilde{k}_1) - \log p_X(x_1)] \frac{z^{k-\xi_1}}{(k-\xi_1)!} e^{-z} \, dz \\
&= \frac{1}{\Gamma(k - \xi_1 + 1)} \int_0^{\infty} \left[\log(z) z^{k-\xi_1} e^{-z} \right] dz - \log(\tilde{k}_1) - \log p_X(x_1) \\
&= \frac{1}{\Gamma(k - \xi_1 + 1)} \int_0^{\infty} \left[\log(z) z^{(k-\xi_1+1)-1} e^{-z} \right] dz - \log(\tilde{k}_1) - \log p_X(x_1) \\
&= \frac{\Gamma'(k - \xi_1 + 1)}{\Gamma(k - \xi_1 + 1)} - \log(\tilde{k}_1) - \log p_X(x_1) \\
&= \psi(k - \xi_1 + 1) - \log(\tilde{k}_1) - \log p_X(x_1).
\end{aligned}
\tag{59}
$$

H. Transfer Entropy and Granger Causality

TE can be considered as a measurement of the degree to which the history Y^- of the process Y disambiguates the future X^p of X beyond the degree to how its history X^- disambiguates this future [22]. It is an information theoretic implementation of Wiener's principle of observational causality. Hence, TE reveals a natural relation to Granger causality. As is well known, Granger causality emphasizes the concept of reduction of the mean square error of the linear prediction of X_i^p when adding Y_i^- to X_i^- by introducing the Granger causality index:

$$\mathrm{GC}_{Y \to X} = \log \left[\frac{\mathrm{var}\left(lpe_{X_i^p | X_i^-}\right)}{\mathrm{var}\left(lpe_{X_i^p | X_i^-, Y_i^-}\right)} \right], \tag{60}$$

where $lpe_{X_i^p | U}$ is the error when predicting linearly X_i^p from U. TE is framed in terms of the reduction of the Shannon uncertainty (entropy) of the predictive probability distribution. When the probability distribution of $\left(X_i^p, X_i^-, Y_i^-\right)$ is assumed to be Gaussian, TE and Granger causality are entirely equivalent, up to a factor of two [42]:

$$\mathrm{TE}_{Y \to X} = \frac{1}{2} \mathrm{GC}_{Y \to X}. \tag{61}$$

Consequently, in the Gaussian case, TE can be easily computed from a statistical second order characterization of $\left(X_i^p, X_i^-, Y_i^-\right)$. This Gaussian assumption obviously holds when the processes Y and X are jointly normally distributed and, more particularly, when they correspond to a Gaussian autoregressive (AR) bivariate process. In [42], Barnett *et al.* discussed the relation between these two causality measures, and this work bridged information-theoretic methods and autoregressive ones.

I. Comparison between Entropy Estimators

Figure 8 displays the values of entropy for a Gaussian d-dimensional vector as a function of the number of neighbors k, for $d = 3$ in Figure 8a and $d = 8$ in Figure 8b, obtained with different estimators. The theoretical entropy value is compared with its estimation from the Kozachenko–Leonenko reference estimator (Equation (10), red circles), its extension (Equation (22), black stars) and the extension of Singh's estimator (Equation (35), blue squares). It appears clearly that, for the extended Singh's estimator, the bias (true value minus estimated value) increases drastically when the number of neighbors decreases under a threshold slightly lower than the dimension d of the vector. This allows us to interpret some apparently surprising results obtained with this estimator in the estimation of TE, as reported in Figure 6b. TE estimation is a sum of four separate vector entropy estimations, $\widehat{\mathrm{TE}_{Y \to X}} = \mathcal{H}\left(X^-, Y^-\right) + \mathcal{H}\left(X^p, X^-\right) - \mathcal{H}\left(X^p, X^-, Y^-\right) - \mathcal{H}\left(X^-\right)$. Here, the dimensions of the four vectors are $d\left(X^-, Y^-\right) = m + n = 4$, $d\left(X^p, X^-\right) = 1 + m = 3$, $d\left(X^p, X^-, Y^-\right) = 1 + m + n = 5$, $d\left(X^-\right) = m = 2$, respectively. Note that, if we denote by X_{M2} and Y_{M2} the two components in Model 2, the general notation (X^p, X^-, Y^-) corresponds to $\left(Y_{M2}^p, Y_{M2}^-, X_{M2}^-\right)$, because in Figure 6b, the analyzed direction is $X \to Y$ and not the reverse. We see that, when considering the estimation of $\mathcal{H}\left(X^p, X^-, Y^-\right)$, we have $d = 5$ and $k = 3$, which is the imposed neighbors number in the global space. Consequently, from the results shown in Figure 8, we can expect that in Model 2, the quantity $\mathcal{H}\left(X^p, X^-, Y^-\right)$ will be drastically underestimated. For the other components $\widehat{\mathcal{H}\left(X^-, Y^-\right)}$, $\widehat{\mathcal{H}\left(X^p, X^-\right)}$, $\widehat{\mathcal{H}\left(X^-\right)}$, the numbers of neighbors to consider are generally larger than three (as a consequence of Kraskov's technique, which introduces projected distances) and $d \leq 5$, so that we do not expect any underestimation of these terms. Therefore, globally, when summing the four entropy estimations, the resulting positive bias observed in Figure 6b is understandable.

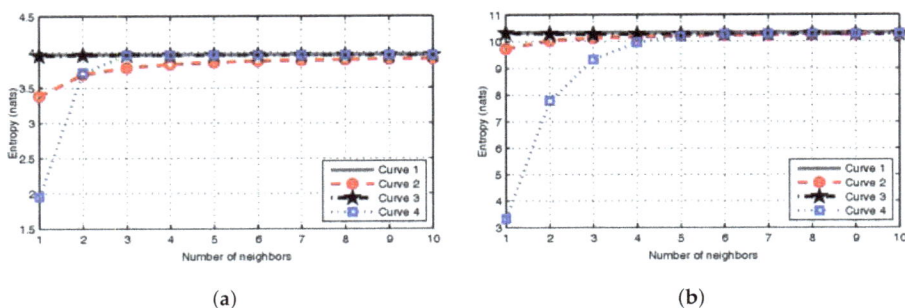

Figure 8. Comparison between four entropy estimators: (**a**) $d = 3$; (**b**) $d = 8$. The covariance matrix of the signals is a Toeplitz matrix with first line $\beta^{[0:d-1]}$, where $\beta = 0.5$. "Curve 1" stands for the true value; "Curve 2", "Curve 3" and "Curve 4" correspond to the values of entropy obtained using respectively Equations (10), (22) and (35).

Conflicts of Interest: The authors declare no conflict of interest.

References

1. Schreiber, T. Measuring information transfer. *Phys. Rev. Lett.* **2000**, *85*, doi:10.1103/PhysRevLett.85.461.
2. Gourévitch, B.; Eggermont, J.J. Evaluating information transfer between auditory cortical neurons. *J. Neurophysiol.* **2007**, *97*, 2533–2543.
3. Hlaváčková-Schindler, K.; Paluš, M.; Vejmelka, M.; Bhattacharya, J. Causality detection based on information-theoretic approaches in time series analysis. *Phys. Rep.* **2007**, *441*, 1–46.
4. Sabesan, S.; Narayanan, K.; Prasad, A.; Iasemidis, L.; Spanias, A.; Tsakalis, K. Information flow in coupled nonlinear systems: Application to the epileptic human brain. In *Data Mining in Biomedicine*; Springer: Berlin/Heidelberg, Germany, 2007; pp. 483–503.
5. Ma, C.; Pan, X.; Wang, R.; Sakagami, M. Estimating causal interaction between prefrontal cortex and striatum by transfer entropy. *Cogn. Neurodyn.* **2013**, *7*, 253–261.
6. Vakorin, V.A.; Krakovska, O.A.; McIntosh, A.R. Confounding effects of indirect connections on causality estimation. *J. Neurosci. Methods* **2009**, *184*, 152–160.
7. Yang, C.; Le Bouquin Jeannes, R.; Bellanger, J.J.; Shu, H. A new strategy for model order identification and its application to transfer entropy for EEG signals analysis. *IEEE Trans. Biomed. Eng.* **2013**, *60*, 1318–1327.
8. Zuo, K.; Zhu, J.; Bellanger, J.J.; Jeannès, R.L.B. Adaptive kernels and transfer entropy for neural connectivity analysis in EEG signals. *IRBM* **2013**, *34*, 330–336.
9. Faes, L.; Nollo, G. Bivariate nonlinear prediction to quantify the strength of complex dynamical interactions in short-term cardiovascular variability. *Med. Biol. Eng. Comput.* **2006**, *44*, 383–392.
10. Faes, L.; Nollo, G.; Porta, A. Non-uniform multivariate embedding to assess the information transfer in cardiovascular and cardiorespiratory variability series. *Comput. Biol. Med.* **2012**, *42*, 290–297.
11. Faes, L.; Nollo, G.; Porta, A. Information-based detection of nonlinear Granger causality in multivariate processes via a nonuniform embedding technique. *Phys. Rev. E* **2011**, *83*, 051112.
12. Runge, J.; Heitzig, J.; Petoukhov, V.; Kurths, J. Escaping the curse of dimensionality in estimating multivariate transfer entropy. *Phys. Rev. Lett.* **2012**, *108*, 258701.
13. Duan, P.; Yang, F.; Chen, T.; Shah, S.L. Direct causality detection via the transfer entropy approach. *IEEE Trans. Control Syst. Technol.* **2013**, *21*, 2052–2066.
14. Bauer, M.; Thornhill, N.F.; Meaburn, A. Specifying the directionality of fault propagation paths using transfer entropy. In Proceedings of the 7th International Symposium on Dynamics and Control of Process Systems (DYCOPS 7), Cambridge, MA, USA, 7–9 July 2004; pp. 203–208.

15. Bauer, M.; Cox, J.W.; Caveness, M.H.; Downs, J.J.; Thornhill, N.F. Finding the direction of disturbance propagation in a chemical process using transfer entropy. *IEEE Trans. Control Syst. Technol.* **2007**, *15*, 12–21.
16. Kulp, C.; Tracy, E. The application of the transfer entropy to gappy time series. *Phys. Lett. A* **2009**, *373*, 1261–1267.
17. Overbey, L.; Todd, M. Dynamic system change detection using a modification of the transfer entropy. *J. Sound Vib.* **2009**, *322*, 438–453.
18. Gray, R.M. *Entropy and Information Theory*; Springer: Berlin/Heidelberg, Germany, 2011.
19. Roman, P. *Some Modern Mathematics for Physicists and Other Outsiders: An Introduction to Algebra, Topology, and Functional Analysis*; Elsevier: Amsterdam, The Netherlands, 2014.
20. Kugiumtzis, D. Direct-coupling information measure from nonuniform embedding. *Phys. Rev. E* **2013**, *87*, 062918.
21. Montalto, A.; Faes, L.; Marinazzo, D. MuTE: A MATLAB toolbox to compare established and novel estimators of the multivariate transfer entropy. *PLoS One* **2014**, *9*, e109462.
22. Paluš, M.; Komárek, V.; Hrnčíř, Z.; Štěrbová, K. Synchronization as adjustment of information rates: Detection from bivariate time series. *Phys. Rev. E* **2001**, *63*, 046211.
23. Frenzel, S.; Pompe, B. Partial mutual information for coupling analysis of multivariate time series. *Phys. Rev. Lett.* **2007**, *99*, 204101.
24. Kraskov, A.; Stögbauer, H.; Grassberger, P. Estimating mutual information. *Phys. Rev. E* **2004**, *69*, 066138.
25. Vicente, R.; Wibral, M.; Lindner, M.; Pipa, G. Transfer entropy—A model-free measure of effective connectivity for the neurosciences. *J. Comput. Neurosci.* **2011**, *30*, 45–67.
26. Lindner, M.; Vicente, R.; Priesemann, V.; Wibral, M. TRENTOOL: A Matlab open source toolbox to analyse information flow in time series data with transfer entropy. *BMC Neurosci.* **2011**, *12*, 119.
27. Wibral, M.; Vicente, R.; Lindner, M. Transfer Entropy in Neuroscience. In *Directed Information Measures in Neuroscience*; Springer: Berlin/Heidelberg, Germany, 2014; pp. 3–36.
28. Gómez-Herrero, G.; Wu, W.; Rutanen, K.; Soriano, M.C.; Pipa, G.; Vicente, R. Assessing coupling dynamics from an ensemble of time series. **2010**, arXiv:1008.0539.
29. Vlachos, I.; Kugiumtzis, D. Nonuniform state-space reconstruction and coupling detection. *Phys. Rev. E* **2010**, *82*, 016207.
30. Lizier, J.T. JIDT: An information-theoretic toolkit for studying the dynamics of complex systems. **2014**, arXiv:1408.3270.
31. MILCA Toolbox. Available online: http://www.ucl.ac.uk/ion/departments/sobell/Research/RLemon/MILCA/MILCA (accessed on 11 June 2015).
32. Wibral, M.; Rahm, B.; Rieder, M.; Lindner, M.; Vicente, R.; Kaiser, J. Transfer entropy in magnetoencephalographic data: Quantifying information flow in cortical and cerebellar networks. *Prog. Biophys. Mol. Biol.* **2011**, *105*, 80–97.
33. Wollstadt, P.; Martínez-Zarzuela, M.; Vicente, R.; Díaz-Pernas, F.J.; Wibral, M. Efficient transfer entropy analysis of non-stationary neural time series. *PLoS One* **2014**, *9*, e102833.
34. Wollstadt, P.; Lindner, M.; Vicente, R.; Wibral, M.; Pampu, N.; Martinez-Zarzuela, M. Trentool Toolbox. Available online: www.trentool.de (accessed on 11 June 2015).
35. Wibral, M.; Pampu, N.; Priesemann, V.; Siebenhühner, F.; Seiwert, H.; Lindner, M.; Lizier, J.T.; Vicente, R. Measuring information-transfer delays. *PLoS One* **2013**, *8*, e55809.
36. Singh, H.; Misra, N.; Hnizdo, V.; Fedorowicz, A.; Demchuk, E. Nearest neighbor estimates of entropy. *Am. J. Math. Manag. Sci.* **2003**, *23*, 301–321.
37. Zhu, J.; Bellanger, J.J.; Shu, H.; Yang, C.; Jeannès, R.L.B. Bias reduction in the estimation of mutual information. *Phys. Rev. E* **2014**, *90*, 052714.
38. Fukunaga, K.; Hostetler, L. Optimization of k nearest neighbor density estimates. *IEEE Trans. Inf. Theory* **1973**, *19*, 320–326.
39. Fukunaga, K. *Introduction to Statistical Pattern Recognition*; Academic Press: Waltham, MA, USA, 1990.
40. Merkwirth, C.; Parlitz, U.; Lauterborn, W. Fast nearest-neighbor searching for nonlinear signal processing. *Phys. Rev. E* **2000**, *62*, 2089–2097.

Entropy **2015**, *17*, 4173–4201

41. Gao, S.; Steeg, G.V.; Galstyan, A. Efficient Estimation of Mutual Information for Strongly Dependent Variables. In Proceedings of the 18th International Conference on Artificial Intelligence and Statistics (AISTATS), San Diego, CA, USA, 9–12 May 2015; pp. 277–286.
42. Barnett, L.; Barrett, A.B.; Seth, A.K. Granger causality and transfer entropy are equivalent for Gaussian variables. *Phys. Rev. Lett.* **2009**, *103*, 238701.

MDPI

Article

Transfer Entropy Expressions for a Class of Non-Gaussian Distributions

Mehrdad Jafari-Mamaghani [1,2,*] and Joanna Tyrcha [1]

[1] Department of Mathematics, Stockholm University, SE-106 91 Stockholm, Sweden;
    E-Mail: joanna@math.su.se
[2] Center for Biosciences, Department of Biosciences and Nutrition, Karolinska Institutet,
    SE-141 83 Huddinge, Sweden

* Author to whom correspondence should be addressed; E-Mail: mjm@math.su.se; Tel.: +46-8-164507.

Received: 17 January 2014; in revised form: 10 March 2014 / Accepted: 18 March 2014 / Published: 24 March 2014

Abstract: Transfer entropy is a frequently employed measure of conditional co-dependence in non-parametric analysis of Granger causality. In this paper, we derive analytical expressions for transfer entropy for the multivariate exponential, logistic, Pareto (type $\mathcal{I} - \mathcal{IV}$) and Burr distributions. The latter two fall into the class of fat-tailed distributions with power law properties, used frequently in biological, physical and actuarial sciences. We discover that the transfer entropy expressions for all four distributions are identical and depend merely on the multivariate distribution parameter and the number of distribution dimensions. Moreover, we find that in all four cases the transfer entropies are given by the same decreasing function of distribution dimensionality.

Keywords: Granger causality; information theory; transfer entropy; multivariate distributions; power-law distributions

---

1. Introduction

Granger causality is a well-known concept based on dynamic co-dependence [1]. In the framework of Granger causality, the cause precedes and contains unique information about the effect. The concept of Granger causality has been applied in a wide array of scientific disciplines from econometrics to neurophysiology, from sociology to climate research (see [2,3] and references therein), and most recently in cell biology [4].

Information theory has increasingly become a useful complement to the existing repertoire of methodologies in mathematical statistics [5,6]. Particularly, in the area of Granger causality, transfer entropy [7], an information theoretical measure of co-dependence based on Shannon entropy, has been applied extensively in non-parametric analysis of time-resolved causal relationships. It has been shown that (conditional) mutual information measured in *nats* and transfer entropy coincide in definition [8–10]. Moreover, for Gaussian-distributed variables, there is a tractable equivalence by a factor of two between transfer entropy and a linear test statistic for Granger causality [11]. Although similar equivalences for non-Gaussian variables have been given in [8], it should be remarked that such equivalences cannot be generalized to non-Gaussian distributions as the linear models underlying the construction of linear test statistics for Granger causality are rendered invalid under assumptions of non-Gaussianity.

The aim of this paper is to present closed-form expressions for transfer entropy for a number of non-Gaussian, unimodal, skewed distributions used in the modeling of occurrence rates, rare events and 'fat-tailed' phenomena in biological, physical and actuarial sciences [12]. More specifically, we will derive expressions for transfer entropy for the multivariate exponential, logistic, Pareto (type $\mathcal{I} - \mathcal{IV}$) and Burr distributions. As for real-world applications, the exponential distribution is the naturally occurring distribution for describing inter-arrival times in a homogeneous Poisson process. In a similar manner, the exponential distribution can be used to model many other change of state scenarios in continuous settings, e.g., time until the occurrence of an accident given certain specifications. The logistic distribution is of great utility given its morphological similarity to the Gaussian distribution and is frequently used to model Gaussian-like phenomena in the presence of thicker distribution tails. The Pareto distribution (in either of its forms) is used in modeling of size related phenomena such as size of incurred casualties in non-life insurance, size of meteorites, and size of trafficked files over the Internet. The Burr distribution is another distribution used in non-life insurance to model incurred casualties, as well as in econometrics where it is used to model income distribution.

The specific choice of these distributions is contingent upon the existence of unique expressions for the corresponding probability density functions and Shannon entropy expressions. A counter-example is given by the multivariate gamma distribution, which although derived in a number of tractable formats under certain preconditions [12,13], lacks a unique and unequivocal multivariate density function and hence a unique Shannon entropy expression.

Another remark shall be dedicated to stable distributions. Such distributions are limits of appropriately scaled sums of independent and identically distributed variables. The general tractability of distributions with this property lies in their "attractor" behavior and their ability to accommodate skewness and heavy tails. Other than the Gaussian distribution (stable by the Central Limit Theorem), the Cauchy-Lorentz distribution and the Lévy distribution are considered to be the only stable distributions that can be expressed analytically. However, the latter lacks analytical expressions for Shannon entropy in the multivariate case. Expressions for Shannon entropy and transfer entropy for the multivariate Gaussian distribution have been derived in [14] and [11], respectively. Expressions for Shannon entropy and transfer entropy for the multivariate Cauchy-Lorentz distribution can be found in the Appendix.

As a brief methodological introduction, we will go through a conceptual sketch of Granger causality, the formulation of the linear models underlying the above-mentioned test statistic, and the definition of transfer entropy before deriving the expressions for our target distributions.

2. Methods

Employment of Granger causality is common practice within cause-effect analysis of dynamic phenomena where the cause temporally precedes the effect and where the information embedded in the cause about the effect is unique. Formulated using probability theory, under H_0, given k lags and the random variables A and B and the set of all other random variables C in any arbitrary system, B is said to not Granger-cause A at observation index t, if

$$H_0 : A_t \perp \{B_{t-1}, \ldots, B_{t-k}\} | \{A_{t-1}, \ldots, A_{t-k}, C_{t-1}, \ldots, C_{t-k}\} \tag{1}$$

where $\perp$ denotes probabilistic independence. Henceforth, for the sake of convenience, we implement the following substitutions: $X = A_t, Y = \{B\}_{t-1}^{t-k}$ and $Z = \{A, C\}_{t-1}^{t-k}$. It is understood that all formulations in what follows are compatible with any multivariate setting. Thus, one can parsimoniously reformulate the hypothesis in Equation (1) as:

$$H_0 : X \perp Y | Z \tag{2}$$

The statement above can be tested by comparing the two conditional probability densities: $f_{X|Z}$ and $f_{X|YZ}$ [15].

Entropy **2014**, *16*, 1743–1755

2.1. Linear Test Statistics

In parametric analysis of Granger causality, techniques of linear regression have been the dominant choice. Under fulfilled assumptions of ordinary least squares' regression and stationarity, the hypothesis in Equation (2), can be tested using the following models:

$$H_0 : X = \beta_1 + Z\beta_2 + \epsilon \tag{3}$$

$$H_1 : X = \gamma_1 + Z\gamma_2 + Y\gamma_3 + \eta \tag{4}$$

where the β and γ terms are the regression coefficients, and the residuals ϵ and η are independent and identically distributed following a centered Gaussian $N(0, \sigma^2)$. Traditionally, the F-distributed Granger-Sargent test [1], equivalent to the structural Chow test [16], has been used to examine the statistical significance of the reduction in residual sum of squares in the latter model compared to the former. In this study however, we will focus on the statistic $G(X, Y|Z) = \ln \left(\text{Var}_\epsilon / \text{Var}_\eta \right)$ [11,17]. This statistic is χ^2-distributed under the null hypothesis, and non-central χ^2-distributed under the alternate hypothesis. There are two types of multivariate generalizations of $G(X, Y|Z)$; one by means of *total variance*, using the trace of covariance matrices [18], and one by *generalized variance*, using the determinant of covariance matrices [11,17]. For a thorough discussion on the advantages of either measure we refer the reader to [18,19]. Choosing the latter extension, the test statistic in $G(X, Y|Z)$ can be reformulated as:

$$
\begin{aligned}
G(X, Y|Z) &= \ln \left(\frac{|\Sigma_\epsilon|}{|\Sigma_\eta|} \right) \\
&= \ln \left(\frac{|\Sigma_{XZ}| \cdot |\Sigma_{YZ}|}{|\Sigma_Z| \cdot |\Sigma_{XYZ}|} \right)
\end{aligned}
\tag{5}
$$

where the last equality follows the scheme presented in [11].

2.2. Transfer Entropy

Transfer entropy, a non-parametric measure of co-dependence is identical to (conditional) mutual information measured in *nats* (using the natural logarithm). Mutual information is a basic concept, based on the most fundamental measure in information theory, the Shannon entropy, or, more specifically, the differential Shannon entropy in the case of continuous distributions. The differential Shannon entropy of a random variable S with a continuous probability density f_S with support on $\mathcal{S}$ is

$$H(S) \equiv -\mathrm{E}[\log_b f_S] = -\int_{\mathcal{S}} f_S \log_b f_S ds \tag{6}$$

where b is the base of the logarithm determining the terms in which the entropy is measured; $b = 2$ for *bits* and $b = e$ for *nats* [14,20]. The transfer entropy for the hypothesis in Equation (2) is defined as [7]:

$$
\begin{aligned}
T(Y \to X|Z) &= H(X|Z) - H(X|Y, Z) \\
&= H(X, Z) - H(Z) + H(Y, Z) - H(X, Y, Z)
\end{aligned}
\tag{7}
$$

Interestingly, for Gaussian variables one can show that $G(X, Y|Z) = 2 \cdot T(Y \to X|Z)$ [11]. Naturally, such equivalences fail when using other types of distributions that do not meet the requirements of linear models used to construct $G(X, Y|Z)$.

In the following, we shall look at closed-form expressions for transfer entropy for the multivariate exponential, logistic, Pareto (type $\mathcal{I} - \mathcal{IV}$) and Burr distributions. Before deriving the results, it should be noted that all marginal densities of the multivariate density functions in this study are distributed according to the same distribution; *i.e.*, the marginal densities of a multivariate exponential density are themselves exponential densities.

3. Results

In this section we will derive the expression for transfer entropy for the multivariate exponential distribution. The remaining derivations follow an identical scheme and are presented in the Appendix. The differential Shannon entropy expressions employed in this study can be found in [21].

The multivariate exponential density function for a d-dimensional random vector $\mathbf{S}$ is:

$$f_{\mathbf{S}} = \prod_{i=1}^{d} \frac{\alpha + i - 1}{\theta_i} \exp\left(\frac{s_i - \lambda_i}{\theta_i}\right) \left[\sum_{i=1}^{d} \exp\left(\frac{s_i - \lambda_i}{\theta_i}\right) - d + 1\right]^{-(\alpha+d)} \tag{8}$$

where $\mathbf{S} \in \mathbb{R}^d$, $s_i > \lambda_i$, $\theta_i > 0$ for $i = 1, ..., d$ and $\alpha > 0$. For the multivariate exponential distribution the differential Shannon entropy of $\mathbf{S}$ is:

$$H(\mathbf{S}) = -\sum_{i=1}^{d} \log\left(\frac{\alpha + i - 1}{\theta_i}\right) + (\alpha + d)\sum_{i=1}^{d} \frac{1}{\alpha + i - 1} - \frac{d}{\alpha} \tag{9}$$

Thus, transfer entropy for a set of multivariate exponential variables can be formulated as:

$$
\begin{aligned}
T(Y \to X|Z) =\, & H(X,Z) - H(Z) + H(Y,Z) - H(X,Y,Z) \\
=\, & -\sum_{i=1}^{d_X} \log\left(\frac{\alpha+i-1}{\theta_i^{(X)}}\right) - \sum_{i=1}^{d_Z} \log\left(\frac{\alpha+d_X+i-1}{\theta_i^{(Z)}}\right) \\
& + (\alpha+d_X+d_Z)\left(\sum_{i=1}^{d_X}\frac{1}{\alpha+i-1} + \sum_{i=1}^{d_Z}\frac{1}{\alpha+d_X+i-1}\right) \\
& - \frac{d_X+d_Z}{\alpha} + \sum_{i=1}^{d_Z}\log\left(\frac{\alpha+i-1}{\theta_i^{(Z)}}\right) - (\alpha+d_Z)\left(\sum_{i=1}^{d_Z}\frac{1}{\alpha+i-1}\right) \\
& + \frac{d_Z}{\alpha} - \sum_{i=1}^{d_Z}\log\left(\frac{\alpha+i-1}{\theta_i^{(Z)}}\right) - \sum_{i=1}^{d_Y}\log\left(\frac{\alpha+d_Z+i-1}{\theta_i^{(Y)}}\right) \\
& + (\alpha+d_Z+d_Y)\left(\sum_{i=1}^{d_Z}\frac{1}{\alpha+i-1} + \sum_{i=1}^{d_Y}\frac{1}{\alpha+d_Z+i-1}\right) \\
& - \frac{d_Z+d_Y}{\alpha} + \sum_{i=1}^{d_X}\log\left(\frac{\alpha+i-1}{\theta_i^{(X)}}\right) + \sum_{i=1}^{d_Z}\log\left(\frac{\alpha+d_X+i-1}{\theta_i^{(Z)}}\right) \\
& + \sum_{i=1}^{d_Y}\log\left(\frac{\alpha+d_X+d_Z+i-1}{\theta_i^{(Y)}}\right) - (\alpha+d_X+d_Z+d_Y) \\
& \left(\sum_{i=1}^{d_X}\frac{1}{\alpha+i-1} + \sum_{i=1}^{d_Z}\frac{1}{\alpha+d_X+i-1} + \sum_{i=1}^{d_Y}\frac{1}{\alpha+d_X+d_Z+i-1}\right) \\
& + \frac{d_X+d_Z+d_Y}{\alpha}
\end{aligned}
\tag{10}
$$

which, after simplifications, reduces to

$$
\begin{aligned}
T(Y \to X|Z) = & \sum_{i=1}^{d_Y}\log\left(1 + \frac{d_X}{\alpha+d_Z+i-1}\right) \\
& - d_Y\left[\sum_{i=1}^{d_X}\frac{1}{\alpha+i-1} + \sum_{i=1}^{d_Z}\left(\frac{1}{\alpha+d_X+i-1} - \frac{1}{\alpha+i-1}\right)\right]
\end{aligned}
$$

$$+ (\alpha + d_Z + d_Y) \sum_{i=1}^{d_Y} \frac{1}{\alpha + d_Z + i - 1}$$

$$- (\alpha + d_X + d_Z + d_Y) \sum_{i=1}^{d_Y} \frac{1}{\alpha + d_X + d_Z + i - 1} \tag{11}$$

where d_X represents the number of dimensions in X, and where α is the multivariate distribution parameter. As stated previously, the expression in Equation (11) holds for the multivariate logistic, Pareto (type $\mathcal{I} - \mathcal{IV}$) and Burr distributions as proven in the Appendix. For the specific case of $d_X = d_Y = d_Z = 1$, the transfer entropy expression reduces to:

$$T(Y \to X|Z) = \log \left(\frac{\alpha + 2}{\alpha + 1} \right) - \frac{1}{\alpha + 2} \tag{12}$$

In any regard, $T(Y \to X|Z)$ depends only on the number of involved dimensions and the parameter α. The latter parameter, α, operates as a multivariate distribution feature and does not have a univariate counterpart. This result indicates that the value assigned to the conditional transfer of information from the cause to the effect decreases with increasing values of α. However, the impact of the multivariate distribution parameter α in this decrease, shrinks rather rapidly as the numbers of dimensions increase.

4. Conclusions

The distributions discussed in this paper are frequently subject to the modeling of natural phenomena, and utilized frequently within biological, physical and actuarial engineering. Events distributed according to any of the discussed distributions are not suitable for analysis using linear models and require non-parametric models of analysis or transformations where feasible.

The focus of this paper has been on non-parametric modeling of Granger causality using transfer entropy. Our results show that the expressions for transfer entropy for the multivariate exponential, logistic, Pareto (type $\mathcal{I} - \mathcal{IV}$) and Burr distributions coincide in definition and are dependent on the multivariate distribution parameter α, and the number of dimensions. In other words, the transfer entropy expressions are independent of other parameters of the multivariate distributions.

As underlined by our result, the value of transfer entropy depends in a declining manner on the multivariate distribution parameter α as the number of dimensions increase.

Acknowledgments: The authors wish to thank John Hertz for insightful discussions and feedback. MJM has been supported by the Magnusson's Fund at the Royal Swedish Academy of Sciences and the European Union's Seventh Framework Programme (FP7/2007-2013) under grant agreement #258068, EU-FP7-Systems Microscopy NoE. MJM and JT have been supported by the Swedish Research Council grant #340-2012-6011.

Author Contributions: Mehrdad Jafari-Mamaghani and Joanna Tyrcha designed, performed research and analyzed the data; Mehrdad Jafari-Mamaghani wrote the paper. All authors read and approved the final manuscript.

Conflicts of Interest: The authors declare no conflicts of interest.

A. Appendix

A.1. Multivariate Logistic Distribution

The multivariate logistic density function for a d-dimensional random vector $\mathbf{S}$ is:

$$f_{\mathbf{S}} = \prod_{i=1}^{d} \frac{\alpha + i - 1}{\theta_i} \exp \left(-\frac{s_i - \lambda_i}{\theta_i} \right) \left[\sum_{i=1}^{d} \exp \left(-\frac{s_i - \lambda_i}{\theta_i} \right) + 1 \right]^{-(\alpha + d)} \tag{13}$$

Entropy **2014**, *16*, 1743–1755

with $\mathbf{S} \in \mathbb{R}^d$, $\theta_i > 0$ for $i = 1, ..., d$ and $\alpha > 0$. For the multivariate logistic distribution the differential Shannon entropy of $\mathbf{S}$ is:

$$H(\mathbf{S}) = -\sum_{i=1}^{d} \log\left(\frac{\alpha + i - 1}{\theta_i}\right) + (\alpha + d)\Psi(\alpha + d) - \alpha\Psi(\alpha) - d\Psi(1) \tag{14}$$

where $\Psi(s) = \frac{d}{ds}\ln\Gamma(s)$ is the digamma function. Thus, the transfer entropy for the multivariate logistic distribution can be formulated as:

$$
\begin{aligned}
T(Y \to X|Z) =& H(XZ) - H(Z) + H(YZ) - H(XYZ) \\
=& -\sum_{i=1}^{d_X} \log\left(\frac{\alpha + i - 1}{\theta_i^{(X)}}\right) - \sum_{i=1}^{d_Z} \log\left(\frac{\alpha + d_X + i - 1}{\theta_i^{(Z)}}\right) \\
& + (\alpha + d_X + d_Z)\Psi(\alpha + d_X + d_Z) - \alpha\Psi(\alpha) - (d_X + d_Z)\Psi(1) \\
& + \sum_{i=1}^{d_Z} \log\left(\frac{\alpha + i - 1}{\theta_i^{(Z)}}\right) \\
& - (\alpha + d_Z)\Psi(\alpha + d_Z) + \alpha\Psi(\alpha) + d_Z\Psi(1) \\
& - \sum_{i=1}^{d_Z} \log\left(\frac{\alpha + i - 1}{\theta_i^{(Z)}}\right) - \sum_{i=1}^{d_Y} \log\left(\frac{\alpha + d_Z + i - 1}{\theta_i^{(Y)}}\right) \\
& + (\alpha + d_Z + d_Y)\Psi(\alpha + d_Z + d_Y) - \alpha\Psi(\alpha) - (d_Z + d_Y)\Psi(1) \\
& + \sum_{i=1}^{d_X} \log\left(\frac{\alpha + i - 1}{\theta_i^{(X)}}\right) + \sum_{i=1}^{d_Z} \log\left(\frac{\alpha + d_X + i - 1}{\theta_i^{(Z)}}\right) \\
& - \sum_{i=1}^{d_Y} \log\left(\frac{\alpha + d_X + d_Z + i - 1}{\theta_i^{(Y)}}\right) \\
& - (\alpha + d_X + d_Z + d_Y)\Psi(\alpha + d_X + d_Z + d_Y) \\
& + \alpha\Psi(\alpha) + (d_X + d_Z + d_Y)\Psi(1)
\end{aligned}
\tag{15}
$$

which, after simplifications, using the identity

$$\Psi(\alpha + d) = \Psi(\alpha) + \sum_{i=1}^{d} \frac{1}{\alpha + i - 1} \tag{16}$$

reduces to

$$
\begin{aligned}
T(Y \to X|Z) =& \sum_{i=1}^{d_Y} \log\left(1 + \frac{d_X}{\alpha + d_Z + i - 1}\right) \\
& - d_Y \left[\sum_{i=1}^{d_X} \frac{1}{\alpha + i - 1} + \sum_{i=1}^{d_Z}\left(\frac{1}{\alpha + d_X + i - 1} - \frac{1}{\alpha + i - 1}\right)\right] \\
& + (\alpha + d_Z + d_Y)\sum_{i=1}^{d_Y} \frac{1}{\alpha + d_Z + i - 1} \\
& - (\alpha + d_X + d_Z + d_Y)\sum_{i=1}^{d_Y} \frac{1}{\alpha + d_X + d_Z + i - 1}
\end{aligned}
\tag{17}
$$

A.2. Multivariate Pareto Distribution

The multivariate Pareto density function of type $\mathcal{I} - \mathcal{IV}$ for a d-dimensional random vector $\mathbf{S}$ is:

$$f_{\mathbf{S}} = \prod_{i=1}^{d} \frac{\alpha + i - 1}{\gamma_i \theta_i} \left(\frac{s_i - \mu_i}{\theta_i} \right)^{(1/\gamma_i)-1} \left(1 + \sum_{i=1}^{d} \left(\frac{s_i - \mu_i}{\theta_i} \right)^{1/\gamma_i} \right)^{-(\alpha+d)} \tag{18}$$

with $\mathbf{S} \in \mathbb{R}^d$, $s_i > \mu_i$, $\gamma_i > 0$ and $\theta_i > 0$ for $i = 1, ..., d$ and $\alpha > 0$. Other types of the multivariate Pareto density function are obtained as follows:

- Pareto $\mathcal{III}$ by setting $\alpha = 1$ in Equation (18).
- Pareto $\mathcal{II}$ by setting $\gamma_i = 1$ in Equation (18).
- Pareto $\mathcal{I}$ by setting $\gamma_i = 1$ and $\mu_i = \theta_i$ in Equation (18).

For the multivariate Pareto distribution in Equation (18) the differential entropy of $\mathbf{S}$ is:

$$H(\mathbf{S}) = -\sum_{i=1}^{d} \log\left(\frac{\alpha + i - 1}{\gamma_i \theta_i} \right) + (\alpha+d)\left[\Psi(\alpha+d) - \Psi(\alpha) \right] - \left[\Psi(1) - \Psi(\alpha) \right]\left(d - \sum_{i=1}^{d} \gamma_i \right) \tag{19}$$

Thus, the transfer entropy for the multivariate Pareto density function of type $\mathcal{I} - \mathcal{IV}$ can be formulated as:

$$
\begin{aligned}
T(Y \to X|Z) =& H(X,Z) - H(Z) + H(Y,Z) - H(X,Y,Z) \\
=& -\sum_{i=1}^{d_X} \log\left(\frac{\alpha + i - 1}{\gamma_i^{(X)}\theta_i^{(X)}} \right) - \sum_{i=1}^{d_Z} \log\left(\frac{\alpha + d_X + i - 1}{\gamma_i^{(Z)}\theta_i^{(Z)}} \right) \\
& + (\alpha + d_X + d_Z)\left[\Psi(\alpha + d_X + d_Z) - \Psi(\alpha) \right] \\
& - \left[\Psi(1) - \Psi(\alpha) \right]\left(d_X + d_Z - \sum_{i=1}^{d_X} \gamma_i^{(X)} - \sum_{i=1}^{d_Z} \gamma_i^{(Z)} \right) \\
& + \sum_{i=1}^{d_Z} \log\left(\frac{\alpha + i - 1}{\gamma_i^{(Z)}\theta_i^{(Z)}} \right) - (\alpha + d_Z)\left[\Psi(\alpha + d_Z) - \Psi(\alpha) \right] \\
& + \left[\Psi(1) - \Psi(\alpha) \right]\left(d_Z - \sum_{i=1}^{d_Z} \gamma_i^{(Z)} \right) \\
& - \sum_{i=1}^{d_Z} \log\left(\frac{\alpha + i - 1}{\gamma_i^{(Z)}\theta_i^{(Z)}} \right) - \sum_{i=1}^{d_Y} \log\left(\frac{\alpha + d_Y + i - 1}{\gamma_i^{(Y)}\theta_i^{(Y)}} \right) \\
& + (\alpha + d_Z + d_Y)\left[\Psi(\alpha + d_Z + d_Y) - \Psi(\alpha) \right] \\
& - \left[\Psi(1) - \Psi(\alpha) \right]\left(d_Z + d_Y - \sum_{i=1}^{d_Z} \gamma_i^{(Z)} - \sum_{i=1}^{d_Y} \gamma_i^{(Y)} \right) \\
& + \sum_{i=1}^{d_X} \log\left(\frac{\alpha + i - 1}{\gamma_i^{(X)}\theta_i^{(X)}} \right) + \sum_{i=1}^{d_Z} \log\left(\frac{\alpha + d_X + i - 1}{\gamma_i^{(Z)}\theta_i^{(Z)}} \right) \\
& + \sum_{i=1}^{d_Y} \log\left(\frac{\alpha + d_X + d_Z + i - 1}{\gamma_i^{(Y)}\theta_i^{(Y)}} \right) \\
& - (\alpha + d_X + d_Z + d_Y)\left[\Psi(\alpha + d_X + d_Z + d_Y) - \Psi(\alpha) \right] \\
& + \left[\Psi(1) - \Psi(\alpha) \right]\left(d_X + d_Z + d_Y - \sum_{i=1}^{d_X} \gamma_i^{(X)} - \sum_{i=1}^{d_Z} \gamma_i^{(Z)} - \sum_{i=1}^{d_Y} \gamma_i^{(Y)} \right)
\end{aligned}
\tag{20}
$$

which, after simplifications, reduces to

$$
\begin{aligned}
T(Y \to X|Z) ={}& \sum_{i=1}^{d_Y} \log\left(1 + \frac{d_X}{\alpha + d_Z + i - 1}\right) \\
& - d_Y \left[\sum_{i=1}^{d_X} \frac{1}{\alpha + i - 1} + \sum_{i=1}^{d_Z}\left(\frac{1}{\alpha + d_X + i - 1} - \frac{1}{\alpha + i - 1}\right)\right] \\
& + (\alpha + d_Z + d_Y) \sum_{i=1}^{d_Y} \frac{1}{\alpha + d_Z + i - 1} \\
& - (\alpha + d_X + d_Z + d_Y) \sum_{i=1}^{d_Y} \frac{1}{\alpha + d_X + d_Z + i - 1}
\end{aligned}
\tag{21}
$$

A.3. Multivariate Burr Distribution

The multivariate Burr density function for a d-dimensional random vector $\mathbf{S}$ is:

$$
f_{\mathbf{S}} = \prod_{i=1}^{d} (\alpha + i - 1) p_i c_i s_i^{c_i - 1} \left(1 + \sum_{j=1}^{d} p_j s_j^{c_j - 1}\right)^{-(\alpha + d)}
\tag{22}
$$

with $\mathbf{S} \in \mathbb{R}^d$, $s_i > 0$, $c_i > 0$, $d_i > 0$ for $i = 1, ..., n$ and $\alpha > 0$. For the multivariate Burr distribution the differential entropy of $\mathbf{S}$ is:

$$
\begin{aligned}
H(\mathbf{S}) ={}& - \sum_{i=1}^{d} \log(\alpha + i - 1) + (\alpha + d)\left[\Psi(\alpha + d) - \Psi(\alpha)\right] - \sum_{i=1}^{d} \log\left(c_i \sqrt[c_i]{p_i}\right) \\
& + \left[\Psi(\alpha) - \Psi(1)\right]\left(\sum_{i=1}^{d} \frac{c_i - 1}{c_i}\right)
\end{aligned}
\tag{23}
$$

Thus, the transfer entropy for the multivariate Burr distribution can be formulated as:

$$
\begin{aligned}
T(Y \to X|Z) ={}& H(XZ) - H(Z) + H(YZ) - H(XYZ) \\
={}& - \sum_{i=1}^{d_X} \log(\alpha + i - 1) - \sum_{i=1}^{d_Z} \log(\alpha + d_X + i - 1) \\
& + (\alpha + d_X + d_Z)\left[\Psi(\alpha + d_X + d_Z) - \Psi(\alpha)\right] \\
& - \sum_{i=1}^{d_X} \log\left(c_i^{(X)} \sqrt[c_i^{(X)}]{p_i^{(X)}}\right) - \sum_{i=1}^{d_Z} \log\left(c_i^{(Z)} \sqrt[c_i^{(Z)}]{p_i^{(Z)}}\right) \\
& + \left[\Psi(\alpha) - \Psi(1)\right]\left(\sum_{i=1}^{d_X} \frac{c_i^{(X)} - 1}{c_i^{(X)}} + \sum_{i=1}^{d_Z} \frac{c_i^{(Z)} - 1}{c_i^{(Z)}}\right) \\
& + \sum_{i=1}^{d_Z} \log(\alpha + i - 1) - (\alpha + d_Z)\left[\Psi(\alpha + d_Z) - \Psi(\alpha)\right] \\
& + \sum_{i=1}^{d_Z} \log\left(c_i^{(Z)} \sqrt[c_i^{(Z)}]{p_i^{(Z)}}\right) - \left[\Psi(\alpha) - \Psi(1)\right]\left(\sum_{i=1}^{d_Z} \frac{c_i^{(Z)} - 1}{c_i^{(Z)}}\right) \\
& - \sum_{i=1}^{d_Z} \log(\alpha + i - 1) - \sum_{i=1}^{d_Y} \log(\alpha + d_Z + i - 1) \\
& + (\alpha + d_Z + d_Y)\left[\Psi(\alpha + d_Z + d_Y) - \Psi(\alpha)\right] \\
& - \sum_{i=1}^{d_Z} \log\left(c_i^{(Z)} \sqrt[c_i^{(Z)}]{p_i^{(Z)}}\right) - \sum_{i=1}^{d_Y} \log\left(c_i^{(Y)} \sqrt[c_i^{(Y)}]{p_i^{(Y)}}\right)
\end{aligned}
$$

$$+ \left[\Psi(\alpha) - \Psi(1)\right] \left(\sum_{i=1}^{d_Z} \frac{c_i^{(Z)} - 1}{c_i^{(Z)}} + \sum_{i=1}^{d_Y} \frac{c_i^{(Y)} - 1}{c_i^{(Y)}}\right)$$

$$+ \sum_{i=1}^{d_X} \log(\alpha + i - 1) + \sum_{i=1}^{d_Z} \log(\alpha + d_X + i - 1)$$

$$+ \sum_{i=1}^{d_Y} \log(\alpha + d_X + d_Y + i - 1)$$

$$- (\alpha + d_X + d_Z + d_Y) \left[\Psi(\alpha + d_X + d_Z + d_Y) - \Psi(\alpha)\right]$$

$$+ \sum_{i=1}^{d_X} \log\left(c_i^{(X)} \sqrt[c_i^{(X)}]{p_i^{(X)}}\right) + \sum_{i=1}^{d_Z} \log\left(c_i^{(Z)} \sqrt[c_i^{(Z)}]{p_i^{(Z)}}\right)$$

$$+ \sum_{i=1}^{d_Y} \log\left(c_i^{(Y)} \sqrt[c_i^{(Y)}]{p_i^{(Y)}}\right)$$

$$- \left[\Psi(\alpha) - \Psi(1)\right] \left(\sum_{i=1}^{d_X} \frac{c_i^{(X)} - 1}{c_i^{(X)}} + \sum_{i=1}^{d_Z} \frac{c_i^{(Z)} - 1}{c_i^{(Z)}} + \sum_{i=1}^{d_Y} \frac{c_i^{(Y)} - 1}{c_i^{(Y)}}\right) \qquad (24)$$

which, after simplifications, reduces to

$$T(Y \to X|Z) = \sum_{i=1}^{d_Y} \log\left(1 + \frac{d_X}{\alpha + d_Z + i - 1}\right)$$

$$- d_Y \left[\sum_{i=1}^{d_X} \frac{1}{\alpha + i - 1} + \sum_{i=1}^{d_Z} \left(\frac{1}{\alpha + d_X + i - 1} - \frac{1}{\alpha + i - 1}\right)\right]$$

$$+ (\alpha + d_Z + d_Y) \sum_{i=1}^{d_Y} \frac{1}{\alpha + d_Z + i - 1}$$

$$- (\alpha + d_X + d_Z + d_Y) \sum_{i=1}^{d_Y} \frac{1}{\alpha + d_X + d_Z + i - 1} \qquad (25)$$

B. Appendix

B.1. Multivariate Cauchy-Lorentz Distribution

The multivariate Cauchy-Lorentz density function for a d-dimensional random vector $\mathbf{S}$ is:

$$f_{\mathbf{S}} = \frac{\Gamma(\frac{1+d}{2})}{\sqrt{\pi^{1+d}}} \left(1 + s_1^2 + s_2^2 + \ldots + s_d^2\right)^{-\frac{1+d}{2}} \qquad (26)$$

for $\mathbf{S} \in \mathbb{R}^d$. Interestingly, Equation (26) is equivalent to the multivariate t-distribution with one degree of freedom, zero expectation, and an identity covariance matrix [21]. For the case of $d = 1$, Equation (26) reduces to the univariate Cauchy-Lorentz density function [22]. The differential entropy of $\mathbf{S}$ is:

$$H(\mathbf{S}) = -\log\left(\frac{\Gamma\left(\frac{1+d}{2}\right)}{\sqrt{\pi^{1+d}}}\right) + \frac{1+d}{2}\left[\Psi\left(\frac{1+d}{2}\right) - \Psi\left(\frac{1}{2}\right)\right] \qquad (27)$$

Thus, the transfer entropy $T(Y \to X|Z)$ for the multivariate Cauchy-Lorentz distribution can be formulated as:

$$T(Y \to X|Z) = H(XZ) - H(Z) + H(YZ) - H(XYZ)$$

$$- \log \left(\frac{\Gamma \left(\frac{1+d_X+d_Z}{2} \right)}{\sqrt{\pi^{1+d_X+d_Z}}} \right)$$

$$+ \frac{1+d_X+d_Z}{2} \left[\Psi \left(\frac{1+d_X+d_Z}{2} \right) - \Psi \left(\frac{1}{2} \right) \right]$$

$$+ \log \left(\frac{\Gamma \left(\frac{1+d_Z}{2} \right)}{\sqrt{\pi^{1+d_Z}}} \right) - \frac{1+d_Z}{2} \left[\Psi \left(\frac{1+d_Z}{2} \right) - \Psi \left(\frac{1}{2} \right) \right]$$

$$- \log \left(\frac{\Gamma \left(\frac{1+d_Y+d_Z}{2} \right)}{\sqrt{\pi^{1+d_Y+d_Z}}} \right)$$

$$+ \frac{1+d_Y+d_Z}{2} \left[\Psi \left(\frac{1+d_Y+d_Z}{2} \right) - \Psi \left(\frac{1}{2} \right) \right]$$

$$+ \log \left(\frac{\Gamma \left(\frac{1+d_X+d_Y+d_Z}{2} \right)}{\sqrt{\pi^{1+d_X+d_Y+d_Z}}} \right)$$

$$- \frac{1+d_X+d_Y+d_Z}{2} \left[\Psi \left(\frac{1+d_X+d_Y+d_Z}{2} \right) - \Psi \left(\frac{1}{2} \right) \right] \tag{28}$$

which, after simplifications, using the identity in Equation (16), reduces to

$$T(Y \to X|Z) = \log \left(\frac{\Gamma \left(\frac{1+d_Z}{2} \right) \Gamma \left(\frac{1+d_X+d_Y+d_Z}{2} \right)}{\Gamma \left(\frac{1+d_X+d_Z}{2} \right) \Gamma \left(\frac{1+d_Y+d_Z}{2} \right)} \right) + \frac{1+d_X+d_Z}{2} \zeta \left(\frac{d_X+d_Z}{2} \right)$$

$$- \frac{1+d_Z}{2} \zeta \left(\frac{d_Z}{2} \right) + \frac{1+d_Y+d_Z}{2} \zeta \left(\frac{d_Y+d_Z}{2} \right)$$

$$- \frac{1+d_X+d_Y+d_Z}{2} \zeta \left(\frac{d_X+d_Y+d_Z}{2} \right) \tag{29}$$

where

$$\zeta(a) = \sum_{i=1}^{a} \frac{1}{i - 0.5} \tag{30}$$

is obtained after a simplification of the digamma function.

References

1. Granger, C.W.J. Investigating causal relations by econometric models and cross-spectral methods. *Econometrica* **1969**, *37*, 424–438.
2. Hlaváčková-Schindler, K.; Paluš, M.; Vejmelka, M.; Bhattacharya, J. Causality detection based on information-theoretic approaches in time series analysis. *Phys. Rep.* **2007**, *441*, 1–46.
3. Guo, S.; Ladroue, C.; Feng, J. Granger Causality: Theory and Applications. In *Frontiers in Computational and Systems Biology*; Springer: Berlin/Heidelberg, Germany, 2010; pp. 83–111.
4. Lock, J.G.; Jafari-Mamaghani, M.; Shafqat-Abbasi, H.; Gong, X.; Tyrcha, J.; Strömblad, S. Plasticity in the macromolecular-scale causal networks of cell migration. *PLoS One* **2014**, *9*, e90593.
5. Soofi, E.S. Principal information theoretic approaches. *J. Am. Stat. Assoc.* **2000**, *95*, 1349–1353.
6. Soofi, E.S.; Zhao, H.; Nazareth, D.L. Information measures. *Wiley Interdiscip. Rev. Comput. Stat.* **2010**, *2*, 75–86.

Entropy **2014**, *16*, 1743–1755

7.   Schreiber, T.   Measuring information transfer.   *Phys. Rev. Lett.* **2000**, *85*, doi:http://dx.doi.org/10.1103/PhysRevLett.85.461 .

8.   Hlaváčková-Schindler, K. Equivalence of Granger causality and transfer entropy: A generalization. *Appl. Math. Sci.* **2011**, *5*, 3637–3648.

9.   Seghouane, A.-K.; Amari, S. Identification of directed influence: Granger causality, Kullback-Leibler divergence, and complexity. *Neural Comput.* **2012**, *24*, 1722–1739.

10. Jafari-Mamaghani, M. Non-parametric analysis of Granger causality using local measures of divergence. *Appl. Math. Sci.* **2013**, *7*, 4107–4136.

11. Barnett, L.; Barrett, A.B.; Seth, A.K. Granger causality and transfer entropy are equivalent for Gaussian variables. *Phys. Rev. Lett.* **2009**, *103*, 238701.

12. Johnson, N.L.; Kotz, S.; Balakrishnan, N. *Continuous Multivariate Distributions, Models and Applications*; Volume 1; Wiley: New York, NY, USA, 2002.

13. Furman, E. On a multivariate gamma distribution. *Stat. Probab. Lett.* **2008**, *78*, 2353–2360.

14. Cover, T.M.; Thomas, J.A. *Elements of information theory*; Wiley: New York, NY, USA, 1991.

15. Florens, J.P.; Mouchart, M. A note on noncausality. *Econometrica* **1982**, *50*, 583–591.

16. Chow, G.C. Tests of equality between sets of coefficients in two linear regressions. *Econometrica* **1960**, *28*, 591–605.

17. Geweke, J. Measurement of linear dependence and feedback between multiple time series. *J. Am. Stat. Assoc.* **1982**, *77*, 304–313.

18. Ladroue, C.; Guo, S.; Kendrick, K.; Feng, J. Beyond element-wise interactions: Identifying complex interactions in biological processes. *PLoS One* **2009**, *4*, e6899.

19. Barrett, A.B.; Barnett, L.; Seth, A.K. Multivariate Granger causality and generalized variance. *Phys. Rev. E* **2010**, *81*, 041907.

20. Shannon, C.E. A mathematical theory of communication. *Bell Syst. Tech. J.* **2001**, *1*, 3–55.

21. Zografos, K.; Nadarajah, S. Expressions for Rényi and Shannon entropies for multivariate distributions. *Stat. Prob. Lett.* **2005**, *71*, 71–84.

22. Abe, S.; Rajagopal, A.K. Information theoretic approach to statistical properties of multivariate Cauchy-Lorentz distributions. *J. Phys. A* **2001**, *34*, doi:10.1088/0305-4470/34/42/301.

entropy

MDPI

Article

Linearized Transfer Entropy for Continuous Second Order Systems

Jonathan M. Nichols [1,*], Frank Bucholtz [1] and Joe V. Michalowicz [2,†]

[1] U.S. Naval Research Laboratory, Optical Sciences Division, Washington DC 20375, USA;
E-Mail: frank.bucholtz@nrl.navy.mil

[2] U.S. Naval Research Laboratory, Optical Science Division, Washington DC 20375, USA

[†] Permanent Address: Global Strategies Group, Crofton, MD 21114, USA;
E-Mail: georgiamsa@yahoo.com

[*] Author to whom correspondence should be addressed; E-Mail: jonathan.nichols@nrl.navy.mil.

Received: 22 May 2013; in revised form: 5 July 2013/Accepted: 18 July 2013/Published: 7 August 2013

Abstract: The transfer entropy has proven a useful measure of coupling among components of a dynamical system. This measure effectively captures the influence of one system component on the transition probabilities (dynamics) of another. The original motivation for the measure was to quantify such relationships among signals collected from a nonlinear system. However, we have found the transfer entropy to also be a useful concept in describing linear coupling among system components. In this work we derive the analytical transfer entropy for the response of coupled, second order linear systems driven with a Gaussian random process. The resulting expression is a function of the auto- and cross-correlation functions associated with the system response for different degrees-of-freedom. We show clearly that the interpretation of the transfer entropy as a measure of "information flow" is not always valid. In fact, in certain instances the "flow" can appear to switch directions simply by altering the degree of linear coupling. A safer way to view the transfer entropy is as a measure of the ability of a given system component to predict the dynamics of another.

Keywords: transfer entropy; joint entropy; coupling

1. Introduction

One of the biggest challenges in the modeling and analysis of dynamical systems is understanding coupling mechanisms among different system components. Whether one is studying coupling on a small scale (e.g., neurons in a biological system) or large scale (e.g. coupling among widely separated geographical locations due to climate), understanding the functional form, strength, and/or direction of the coupling between two or more system components is a non-trivial task. However, this understanding is necessary if we are to build accurate models of the coupled system and make predictions (our ultimate goal). Accurately assessing the functional form of the coupling is beyond the scope of this work. To do so would require positing various models for a particular coupled system and then testing the predictive power of those models against observed data. Rather, the focus here is on understanding the strength and direction of the coupling among two system components. This task can be accomplished by forming a general hypothesis about what it means for two system components to be coupled, and then testing that hypothesis against observation. It is in this framework that the transfer entropy is operates.

The transfer entropy (TE) is a scalar measure designed to capture both the magnitude and direction of coupling among two components of a dynamical system. This measure was posed initially for data described by discrete probability distributions [1] and was later extended to continuous random variables [2]. By construction, this measure quantifies a general definition of coupling that is appropriate for both linear and nonlinear systems. Moreover, TE is defined in such a way as to provide insight into the direction of the coupling (is component *A* driving component *B* or *vice-versa*?). Since its introduction, the TE has been applied to a diverse set of systems, including biological [1,3], chemical [4], economic [5], structural [6,7], and climate [8]. A number of papers in the Neurosciences also have focused on the TE as a useful way to draw inference about coupling [9–11]. In each case the TE provided information about the system that traditional linear measures of coupling (e.g., cross-correlation) could not.

The TE has also been linked to other concepts of coupling such as "Wiener-Granger Causality". in fact, for the class of systems studied in this work the TE can be shown to be entirely equivalent to measures of Granger causality [12]. Linkages to other models and concepts of dynamical coupling such as conditional mutual information [13] and Dynamic Causal Modeling (DCM) [14], are also possible for certain special cases. The connectivity model assumed by DCM is fundamentally nonlinear (specifically bilinear), however as the degree of nonlinearity decreases the form of the DCM model approaches that of the model studied here.

Although the TE was designed as a way to gain insight into nonlinear system coupling, we have found the TE to be quite useful in the study of linear systems as well. In this special case, analytical expressions for the TE are possible and can be used to provide useful insight into the behavior of the TE. Furthermore, unlike in the general case, the linearized TE can be easily estimated from observed data. This work is therefore devoted to the understanding of TE as applied to coupled, driven linear systems. Specifically, we consider coupling among components of a general, second order linear structural system driven by a Gaussian random process. The particular model studied is used to describe numerous phenomena, including structural dynamics, electrical circuits, heat transfer, *etc.* [15]. As such, it presents an opportunity to better understand the properties of the TE for a broad class of dynamical systems. Section 1 develops the general analytical expression for the TE in terms of the covariance matrices associated with different combinations of system response data. Section 2 specifies the general model under study and derives the TE for the model response data. Sections 3 and 4 present results and concluding remarks.

2. Mathematical Development

In what follows we assume that we have observed the signals $x_i(t_n)$, $i = 1 \cdots M$ as the output of a dynamical system and that we have sampled these signals at times t_n, $n = 1 \cdots N$. The system is assumed to be appropriately modeled as a mixture of deterministic and stochastic components, hence we choose to model each sampled value $x_i(t_n)$ as a random variable X_{in}. That is to say, for any particular observation time t_n we can define a function $P_{X_{in}}(x_i(t_n))$ that assigns a probability to the event that $X_{in} < x_i(t_n)$. We further assume that these are continuous random variables and that we may also define the probability density function (PDF) $p_{X_{in}}(x(t_n)) = dP_{X_{in}}/dx_n$.

The vector of random variables $\mathbf{X}_i \equiv (X_{i1}, X_{i2}, \cdots, X_{iN})$ defines a random process and will be used to model the ith signal $\mathbf{x}_i \equiv x_i(t_n)$, $n = 1 \cdots N$. Using this notation,we can also define the joint PDF $p_{\mathbf{X}_i}(\mathbf{x}_i)$ which specifies the probability of observing such a sequence. In this work we further assume that the random processes are strictly stationary, that is to say the joint PDF obeys $p_{\mathbf{X}_i}(x_i(t_1), x_i(t_2), \cdots, x_i(t_N)) = p_{\mathbf{X}_i}(x_i(t_1 + \tau), x_i(t_2 + \tau), \cdots, x_i(t_N + \tau))$ *i.e.* the joint PDF is invariant to a fixed temporal shift τ.

The joint probability density functions are models that predict the likelihood of observing a particular sequence of values. These same models can be extended to include dynamical effects by including conditional probability, $p_{X_{in}}(x_i(t_n)|x_i(t_{n-1}))$, which can be used to specify the probability of observing the value $x_i(t_n)$ given that we have already observed $x_i(t_{n-1})$. The idea that knowledge of past observations changes the likelihood of future events is certainly common in dynamical systems. A dynamical system whose output is a repeating sequence of $010101 \cdots$ is equally likely to be in state 0 or state 1 (probability 0.5) if the system is observed at a randomly chosen time. However, if we know the value at $t_1 = 0$ the value $t_2 = 1$ is known with probability 1. This concept lies at the heart of the $\mathcal{P}^{th}$ order Markov model, which by definition obeys

$$
\begin{aligned}
p_{X_i}(x_i(t_{n+1})|x_i(t_n),& x_i(t_{n-1}), x_i(t_{n-2}), \cdots, x_i(t_{n-\mathcal{P}})) = \\
& p_{X_i}(x_i(t_{n+1})|x_i(t_n), x_i(t_{n-1}), x_i(t_{n-2}), \cdots, x_i(t_{n-\mathcal{P}}), x_i(t_{n-\mathcal{P}-1}), \cdots) \\
& \equiv p_{X_i}(x_i(t_n)^{(1)}|x_i(t_n)^{(\mathcal{P})}).
\end{aligned}
\tag{1}
$$

That is to say, the probability of the random variable attaining the value $x_i(t_{n+1})$ is conditional on the previous $\mathcal{P}$ values only. The shorthand notation used here specifies relative lags/advances as a superscript.

Armed with this notation we consider the work of Kaiser and Schreiber [2] and define the continuous transfer entropy between processes $\mathbf{X}_i$ and $\mathbf{X}_j$ as

$$
\begin{aligned}
TE_{j\to i}(t_n) = \int_{\mathbb{R}^{\mathcal{P}+\mathcal{Q}+1}} & p_{X_i}\left(x_i(t_n)^{(1)}|\mathbf{x}_i^{(\mathcal{P})}(t_n), \mathbf{x}_j^{(\mathcal{Q})}(t_n)\right) \\
& \times \log_2\left(\frac{p_{X_i}(x_i(t_n)^{(1)}|\mathbf{x}_i^{(\mathcal{P})}(t_n), \mathbf{x}_j^{(\mathcal{Q})}(t_n))}{p_{\mathbf{x}_i}(x_i(t_n)^{(1)}|\mathbf{x}_i^{(\mathcal{P})})}\right) dx_i(t_n^{(1)})d\mathbf{x}_i(t_n)^{(\mathcal{P})}d\mathbf{x}_j(t_n)^{(\mathcal{Q})}
\end{aligned}
\tag{2}
$$

where $\int_{\mathbb{R}^N}$ is used to denote the N-dimensional integral over the support of the random variables. By definition, this measure quantifies the ability of the random process $\mathbf{X}_j$ to predict the dynamics of the random process $\mathbf{X}_i$. To see why, we can examine the argument of the logarithm. In the event that the two random processes are *not* coupled, the dynamics will obey the Markov model in the denominator of Equation (2). However, should $\mathbf{X}_j$ carry added information about the transition probabilities of $\mathbf{X}_i$, the numerator is a better model. The transfer entropy is effectively mapping the difference between these hypotheses to the scalar $TE_{j\to i}(t_n)$. In short, the transfer entropy measures deviations from the hypothesis that the dynamics of $\mathbf{X}_i$ can be described entirely by its own past history and that no new information is gained by considering the dynamics of system $\mathbf{X}_j$.

Two simplifications are possible which will aid in the evaluation of Equation (2). First, recall that we assumed the processes were stationary such that the joint probability distributions are invariant to the particular temporal location t_n at which they are evaluated (only relative lags between observations matter). Hence, in what follows we may drop this index from the notation, *i.e.*, $TE_{j\to i}(t_n) \to TE_{j\to i}$. Secondly, we may use the law of conditional probability and expand Equation (2) as

$$
\begin{aligned}
TE_{j\to i} = & \int_{\mathbb{R}^{\mathcal{P}+\mathcal{Q}+1}} p_{X_i^{(1)}\mathbf{X}_i\mathbf{X}_j}\left(x_i^{(1)}, \mathbf{x}_i^{(\mathcal{P})}, \mathbf{x}_j^{(\mathcal{Q})}\right) \log_2\left(p_{X_i^{(1)}\mathbf{X}_i\mathbf{X}_j}(x_i^{(1)}, \mathbf{x}_i^{(\mathcal{P})}, \mathbf{x}_j^{(\mathcal{Q})})\right) \\
& \times dx_i^{(1)}d\mathbf{x}_i^{(\mathcal{P})}d\mathbf{x}_j^{(\mathcal{Q})} \\
& - \int_{\mathbb{R}^{\mathcal{P}+\mathcal{Q}}} p_{\mathbf{X}_i\mathbf{X}_j}\left(\mathbf{x}_i^{(\mathcal{P})}, \mathbf{x}_j^{(\mathcal{Q})}\right) \log_2\left(p_{\mathbf{X}_i\mathbf{X}_j}(\mathbf{x}_i^{(\mathcal{P})}, \mathbf{x}_j^{(\mathcal{Q})})\right) d\mathbf{x}_i^{(\mathcal{P})}d\mathbf{x}_j^{(\mathcal{Q})} \\
& - \int_{\mathbb{R}^{\mathcal{P}+1}} p_{X_i^{(1)}\mathbf{X}_i}\left(x_i^{(1)}, \mathbf{x}_i^{(\mathcal{P})}\right) \log_2\left(p_{X_i^{(1)}\mathbf{X}_i}(x_i^{(1)}, \mathbf{x}_i^{(\mathcal{P})})\right) dx_i^{(1)}d\mathbf{x}_i^{(\mathcal{P})} \\
& + \int_{\mathbb{R}^{\mathcal{P}}} p_{\mathbf{X}_i}\left(\mathbf{x}_i^{(\mathcal{P})}\right) \log_2\left(p_{\mathbf{x}_i}(\mathbf{x}_i^{(\mathcal{P})})\right) d\mathbf{x}_i^{(\mathcal{P})} \\
= & -h_{X_i^{(1)}\mathbf{X}_i^{(\mathcal{P})}\mathbf{X}_j^{(\mathcal{Q})}} + h_{\mathbf{X}_i^{(\mathcal{P})}\mathbf{X}_j^{(\mathcal{Q})}} + h_{X_i^{(1)}\mathbf{X}_i^{(\mathcal{P})}} - h_{\mathbf{X}_i^{(\mathcal{P})}}
\end{aligned}
\tag{3}
$$

where the terms $h_{\mathbf{X}} = -\int_{\mathbb{R}^M} p_{\mathbf{X}}(\mathbf{x})\log_2(p(\mathbf{x}))\,d\mathbf{x}$ are the joint differential entropies associated with the $M-$dimensional random variable $\mathbf{X}$. In the next section we evaluate Equation (3) among the outputs of a second-order linear system driven with a jointly Gaussian random process.

Entropy **2013**, *15*, 3186–3204

3. Transfer Entropy (TE) for Second Order Linear Systems

3.1. Time-Delayed TE

The only multivariate probability distribution that readily admits an analytical solution for the differential entropies is the jointly Gaussian distribution. Consider the general case of the two data vectors $\mathbf{x} \in \mathbb{R}^N$ and the $\mathbf{y} \in \mathbb{R}^M$. The jointly Gaussian model for these data vectors is

$$p_{\mathbf{XY}}(\mathbf{x}, \mathbf{y}) = \frac{1}{(2\pi)^{(N+M)/2}|C_{\mathbf{XY}}|^{1/2}} e^{-\frac{1}{2}\mathbf{x}^T C_{\mathbf{XY}}^{-1} \mathbf{y}} \tag{4}$$

where $C_{\mathbf{XY}}$ is the $N \times M$ covariance matrix and $|\cdot|$ takes the determinant. Substituting Equation (4) into the expression for the corresponding differential entropy yields

$$h_{\mathbf{XY}} = -\int_{\mathbb{R}^{M \times N}} p_{\mathbf{XY}}(\mathbf{x}, \mathbf{y}) \log_2\left(p(\mathbf{x}, \mathbf{y})\right) dxdy$$
$$= \frac{1}{2}\log_2\left(|C_{\mathbf{XY}}|\right). \tag{5}$$

Therefore, assuming that both random processes $\mathbf{X}_i$ and $\mathbf{X}_j$ are jointly Gaussian distributed, we may substitute Equation (4) into Equation (3) for each of the differential entropies yielding

$$TE_{j \to i} = \frac{1}{2}\log_2\left(\frac{|C_{X_i^{(P)} X_j^{(Q)}}||C_{X_i^{(1)} X_i^{(P)}}|}{|C_{X_i^{(1)} X_i^{(P)} X_j^{(Q)}}||C_{X_i}|}\right). \tag{6}$$

For $\mathcal{P}, Q$ large the needed determinants become difficult to compute. We therefore employ a simplification to the model that retains the spirit of the transfer entropy, but that makes an analytical solution more tractable. In our approach, we set $\mathcal{P} = Q = 1$ *i.e.*, both random processes are assumed to follow a first order Markov model. However, we allow the time interval between the random processes to vary, just as is typically done for the mutual information and/or linear cross-correlation functions [6]. Specifically, we model $X_i(t)$ as the first order Markov model $p_{X_i}(x_i(t_n + \Delta_t)|x_i(t_n))$ and use the TE to consider the alternative $p_{X_i}(x_i(t_n + \Delta_t)|x_i(t_n), x_j(t_n + \tau))$. Note that in anticipation of dealing with measured data, sampled at constant time interval Δ_t, we have made the replacement $t_{n+1} = t_n + \Delta_t$. Although we are only using first order Markov models, by varying the time delay τ we can explore whether or not the random variable $X_j(t_n + \tau)$ carries information about the transition probability $p_{X_i}(x_i(t_n + \Delta_t)|x_i(t_n))$. Should consideration of $x_j(t_n + \tau)$ provide no additional knowledge about the dynamics of $x_i(t_n)$ the transfer entropy will be zero, rising to some positive value should $x_j(t_n + \tau)$ carry information not possessed in $x_j(t_n)$.

In what follows we refer to this particular form of the TE as the time-delayed transfer entropy, or, TDTE. In this simplified situation the needed covariance matrices are

$$C_{X_i X_j}(\tau) = \begin{bmatrix} E[(x_i(t_n) - \bar{x}_i)^2] & E[(x_i(t_n) - \bar{x}_i)(x_j(t_n + \tau) - \bar{x}_j)] \\ E[(x_j(t_n + \tau) - \bar{x}_j)(x_i(t_n) - \bar{x}_i)] & E[(x_j(t_n + \tau) - \bar{x}_j)^2] \end{bmatrix}$$

$$C_{X_i^{(1)} X_i X_j}(\tau) = \begin{bmatrix} E[(x_i(t_n + \Delta_t) - \bar{x}_i)^2] & E[(x_i(t_n + \Delta_t) - \bar{x}_i)(x_i(t_n) - \bar{x}_i)] \\ E[(x_i(t_n) - \bar{x}_i)(x_i(t_n + \Delta_t) - \bar{x}_i)] & E[(x_i(t_n) - \bar{x}_i)^2] \\ E[(x_j(t_n + \tau) - \bar{x}_j)(x_i(t_n + \Delta_t) - \bar{x}_i)] & E[(x_j(t_n + \tau) - \bar{x}_j)(x_i(t_n) - \bar{x}_i)] \end{bmatrix}$$

$$\begin{aligned} & E[(x_i(t_n + \Delta_t) - \bar{x}_i)(x_j(t_n + \tau) - \bar{x}_j)] \\ & E[(x_i(t_n) - \bar{x}_i)(x_j(t_n + \tau) - \bar{x}_j)] \\ & E[(x_j(t_n + \tau) - \bar{x}_j)^2] \end{aligned}$$

$$C_{X_i^{(1)} X_i} = \begin{bmatrix} E[(x_i(t_n + \Delta_t) - \bar{x}_i)^2] & E[(x_i(t_n + \Delta_t) - \bar{x}_i)(x_i(t_n) - \bar{x}_i)] \\ E[(x_i(t_n) - \bar{x}_i)(x_i(t_n + \Delta_t) - \bar{x}_i)] & E[(x_i(t_n) - \bar{x}_i)^2] \end{bmatrix} \tag{7}$$

and $C_{X_i X_i} = E[(x_i(t_n) - \bar{x}_i)^2] \equiv \sigma_i^2$ is simply the variance of the random process X_i and $\bar{x}_i$ its mean. The assumption of stationarity also allows to write $E[(x_i(t_n + \Delta_t) - \bar{x}_i)^2] = \sigma_i^2$ and $E[(x_j(t_n + \tau) - \bar{x}_j)^2] = \sigma_j^2$. Making these substitutions into Equation (6) yields the expression

$$TE_{j \to i}(\tau) = \frac{1}{2} \log_2 \left[\frac{\left(1 - \rho_{ii}^2(\Delta_t)\right)\left(1 - \rho_{ij}^2(\tau)\right)}{1 - \rho_{ij}^2(\tau) - \rho_{ij}^2(\tau - \Delta_t) - \rho_{ii}^2(\Delta_t) + 2\rho_{ii}(\Delta_t)\rho_{ij}(\tau)\rho_{ij}(\tau - \Delta_t)} \right] \tag{8}$$

where we have defined particular expectations in the covariance matrices using the shorthand $\rho_{ij}(\tau) \equiv E[(x_i(t_n) - \bar{x}_i)(x_j(t_n + \tau) - \bar{x}_j)]/\sigma_i\sigma_j$. This particular quantity is referred to in the literature as the cross-correlation function [16]. Note that the covariance matrices are positive-definite matrices and that the determinant of a positive definite matrix is positive [17]. Thus the quantity inside the logarithm will always be positive and the logarithm will exist.

Now, the hypothesis that the TE was designed to test is whether or not past values of the process X_j carry information about the transition probabilities of the second process X_i. Thus, if we are to keep with the original intent of the measure we would only consider $\tau < 0$. However, this restriction is only necessary if one implicitly assumes a non-zero TE means X_j is *influencing* the transition $p_{X_i}(x_i(t_n + \Delta_t)|x_i(t_n))$ as opposed to simply carrying additional information *about* the transition. Again, this latter statement is a more accurate depiction of what the TE is really quantifying and we have found it useful to consider both negative and positive delays τ in trying to understand coupling among system components.

It is also interesting to note the bounds of this function. Certainly for constant signals (*i.e.* $x_i(t_n)$, $x_j(t_n)$ are single-valued for all time) we have $\rho_{X_i X_i}(\Delta_t) = \rho_{X_i X_j}(\tau) = 0 \; \forall \; \tau$ and the transfer entropy is zero for any choice of time-scales τ defining the Markov processes. Knowledge of X_j does not aid in forecasting X_i simply because the transition probability in going from $x_i(t_n)$ to $x_i(t_n + \Delta_t)$ is always unity. Likewise, if there is no coupling between system components we have $\rho_{X_i X_j}(\tau) = 0$ and the TDTE becomes $TE_{j \to i}(\tau) = \frac{1}{2} \log_2 \left[\frac{1 - \rho_{X_i X_i}^2(\Delta_t)}{1 - \rho_{X_i X_i}^2(\Delta_t)} \right] = 0$. At the other extreme, for *perfectly* coupled systems *i.e.* $X_i = X_j$, consider $\tau \to 0$. In this case, we have $\rho_{X_i X_j}^2(\tau) \to 1$, and $\rho_{X_i X_j}(\tau - \Delta_t) \to \rho_{X_i X_i}(-\Delta_t) = \rho_{X_i X_i}(\Delta_t)$ (in this last expression we have noted the symmetry of the function $\rho_{X_i X_i}(\tau)$ with respect to the time-delay). The transfer entropy then becomes

$$TE_{j \to i}(0) = \frac{1}{2} \log_2 \left[\frac{0}{0} \right] \to 0 \tag{9}$$

and the random process X_j at $\tau = 0$ is seen to carry no *additional* information about the dynamics of X_i simply due to the fact that in this special case we have $p_{X_i}(x_i(t_n + \Delta_t)|x_i(t_n)) = p_{X_i}(x_i(t_n + \Delta_t)|x_i(t_n), x_i(t_n))$. These extremes highlight the care that must be taken in interpreting the transfer entropy. Because the TDTE is zero for both the perfectly coupled and uncoupled case we must not interpret the measure to quantify the coupling strength between two random processes. Rather, the TDTE measures the additional information provided by one random process about the dynamics of another.

We should point out that the average mutual information function can resolve the ambiguity in the TDTE as a measure of coupling strength. For two Gaussian random processes the time-delayed mutual information is known to be $I_{X_i X_j}(\tau) = -\frac{1}{2} \log_2 \left[1 - \rho_{ij}^2(\tau)\right]$. Hence, for perfect coupling $I_{X_i X_j}(0) \to \infty$ whereas for uncoupled systems $I_{X_i X_j}(0) \to 0$. Estimating both time-delayed mutual information and transfer entropies can therefore permit stronger inference about dynamical coupling.

3.2. Analytical Cross-Correlation Function

To fully define the TDTE, the auto- and cross-correlation functions $\rho_{ii}(T)$, $\rho_{ij}(T)$ are required. They are derived here for a general class of linear system found frequently in the modeling and analysis of physical processes. Consider the system

$$\mathbf{M}\ddot{\mathbf{x}}(t) + \mathbf{C}\dot{\mathbf{x}}(t) + \mathbf{K}\mathbf{x}(t) = \mathbf{f}(t) \tag{10}$$

where $\mathbf{x}(t) \equiv (x_1(t), x_2(t), \cdots, x_M(t))^T$ is the system's response to the forcing function(s) $\mathbf{f}(t) \equiv (f_1(t), f_2(t), \cdots, f_M(t))^T$ and $\mathbf{M}$, $\mathbf{C}$, $\mathbf{K}$ are $M \times M$ constant coefficient matrices that capture the system's physical properties. Thus, we are considering a second-order, constant coefficient, $M-$degree-of-freedom (DOF) linear system. It is assumed that we may measure the response of this system at any of the DOFs and/or the forcing functions.

One physical embodiment of this system is shown schematically in Figure 1. Five masses are coupled together via restoring elements k_i (springs) and dissipative elements, c_i (dash-pots). The first mass is fixed to a boundary while the driving force is applied at the end mass. If the response data $\mathbf{x}(t)$ are each modeled as a stationary random process we may use the analytical TDTE to answer questions about shared information between any two masses. We can explore this relationship as a function of coupling strength and also which particular mass response data we choose to analyze.

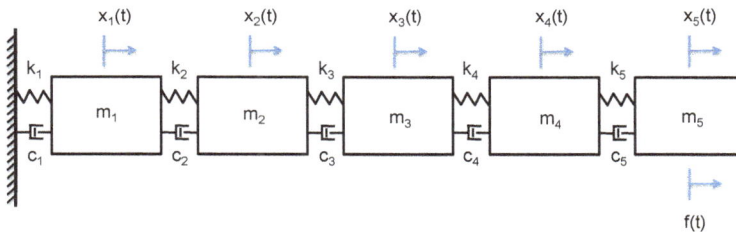

Figure 1. Physical system modeled by Equation (10). Here, an $M = 5$ DOF structure is represented by masses coupled together via both restoring and dissipative elements. Forcing is applied at the end mass.

However, before proceeding we require a general expression for the cross-correlation between any two DOFs, i, $j \in [1, M]$. In other words, we require the expectation $E[x_i(n)x_j(n + T)]$ for any combination of i, j. Such an expression can be obtained by first transforming coordinates. Let $x(t) = \mathbf{u}\eta(t)$ where the matrix $\mathbf{u}$ contain the non-trivial solutions to the eigen-value problem $|\mathbf{M}^{-1}\mathbf{K} - \omega_i^2\mathbf{I}|\mathbf{u}_i = 0$ as its columns [18]. Here the eigen-values are the natural frequencies of the system, denoted ω_i, $i = 1 \cdots M$. Making the above coordinate transformation, substituting into Equation (10) and then pre-multiplying both sides by $\mathbf{u}^T$ allows the equations of motion to be uncoupled and written separately as

$$\ddot{\eta}_i(t) + 2\zeta_i\omega_i\dot{\eta}_i(t) + \omega_i^2\eta_i(t) = \mathbf{u}_i^T\mathbf{f}(t) \equiv q_i(t). \tag{11}$$

where the eigenvectors have been normalized such that $\mathbf{u}^T\mathbf{M}\mathbf{u} = \mathbf{I}$ (the identity matrix). In the above formulation we have also made the assumption that $\mathbf{C} = \alpha\mathbf{K}$ *i.e.*, the dissipative coupling $\mathbf{C}\dot{\mathbf{x}}(t)$ is of the same form as the restoring term, albeit scaled by the constant $\alpha << 1$ (*i.e.*, a lightly damped system). To obtain the form shown in Equation (11) we introduce the dimensionless damping coefficient $\zeta_i = \frac{\alpha}{2}\omega_i$.

The general solution to these un-coupled, linear equations is well-known [18] and can be written as the convolution

$$\eta_i(t) = \int_0^\infty h_i(\theta) q_i(t - \theta) d\theta \tag{12}$$

where $h(\theta)$ is the impulse response function

$$h_i(\theta) = \frac{1}{\omega_{di}} e^{-\zeta_i \omega_i \theta} \sin(\omega_{di}\theta) \tag{13}$$

and $\omega_{di} \equiv \omega_i \sqrt{1 - \zeta_i^2}$. In general terms, we therefore have

$$x_i(t) = \sum_{l=1}^M u_{il}\eta_l(t)$$

$$= \int_0^\infty \sum_{l=1}^M u_{il} h_l(\theta) q_l(t - \theta) d\theta \tag{14}$$

If we further consider the excitation $\mathbf{f}(t)$ to be a zero-mean random process, so too will be $q_l(t)$. Using this model, we may construct the covariance

$$E[x_i(t)x_j(t + \tau)] =$$

$$E\left[\int_0^\infty \int_0^\infty \sum_{l=1}^M \sum_{m=1}^M u_{il}u_{jm} h_l(\theta_1) h_m(\theta_2) q_l(t - \theta_1) q_m(t + \tau - \theta_2) d\theta_1 d\theta_2 \right]$$

$$= \int_0^\infty \int_0^\infty \sum_{l=1}^M \sum_{m=1}^M u_{il}u_{jm} h_l(\theta_1) h_m(\theta_2) E[q_l(t - \theta_1) q_m(t + \tau - \theta_2)] d\theta_1 d\theta_2 \tag{15}$$

which is a function of the eigen-vectors $\mathbf{u}_i$, the impulse response function $h(\cdot)$ and the covariance of the modal forcing matrix. Knowledge of this covariance matrix can be obtained from knowledge of the forcing covariance matrix $R_{F_l F_m}(\tau) \equiv E[f_l(t)f_m(t + \tau)]$. Recalling that

$$q_l(t) = \sum_{p=1}^M u_{lp} f_p(t) \tag{16}$$

we write

$$E[q_l(t - \theta_1) q_m(t + \tau - \theta_2)] = \sum_{p=1}^M \sum_{q=1}^M u_{lq}u_{mp} E[f_q(t - \theta_1) f_p(t + \tau - \theta_2)] \tag{17}$$

It is assumed that the random vibration inputs are uncorrelated, *i.e.* $E[f_q(t)f_p(t)] = 0 \; \forall \; q \neq p$, with variance $\sigma_{F_p}^2 = E[f_p(t)f_p(t)]$. Thus, the above can therefore be simplified as

$$E[q_l(t - \theta_1) q_m(t + \tau - \theta_2)] = \sum_{p=1}^M u_{lp}u_{mp} E[f_p(t - \theta_1) f_p(t + \tau - \theta_2)] \tag{18}$$

The most common linear models assume the input is applied at a single DOF, i.e. $f_p(t)$ is non-zero only for $p = P$. For a load applied at DOF P, the auto-covariance becomes

$$E[x_i(t)x_j(t+\tau)] = \int_0^\infty \int_0^\infty \sum_{l=1}^M \sum_{m=1}^M u_{il}u_{jm}u_{lP}u_{mP}h_l(\theta_1)h_m(\theta_2)E[f_P(t-\theta_1)f_P(t+\tau-\theta_2)]d\theta_1 d\theta_2$$

$$= \sum_{l=1}^M \sum_{m=1}^M u_{il}u_{jm}u_{lP}u_{mP} \int_0^\infty h_l(\theta_1) \int_0^\infty h_m(\theta_2)E[f_P(t-\theta_1)f_P(t+\tau-\theta_2)]d\theta_2 d\theta_1. \tag{19}$$

The inner integral can be further evaluated as

$$\int_0^\infty h_m(\theta_2)E[f_P(t-\theta_1)f_P(t+\tau-\theta_2)]d\theta_2 = \int_0^\infty h_m(\theta_2) \int_{-\infty}^\infty S_{FF}(\omega)e^{i\omega(\tau-\theta_2+\theta_1)}d\omega d\theta_2. \tag{20}$$

Note that we have re-written the forcing auto-covariance as the inverse Fourier transform of the associated power spectral density function, denoted $S_{FF}(\omega)$, via the well-known Wiener-Khinchine relation [16]. We have already assumed the forcing is comprised of independent, identically distributed values, in which case the forcing power spectral density $S_{FF}(\omega) = const \ \forall \omega$. Denoting this constant $S_{FF}(0)$, we note that the Fourier Transform of a constant is simply $\int_{-\infty}^\infty S_{FF}(0) \times e^{i\omega t}dt = S_{FF}(0) \times \delta(t)$, hence our integral becomes

$$\int_0^\infty h_m(\theta_2)E[f_P(t-\theta_1)f_P(t+\tau-\theta_2)]d\theta_2$$

$$= \int_0^\infty h_m(\theta_2)S_{FF}(0)\delta(\tau-\theta_2+\theta_1)d\theta_2 = h(\tau+\theta_1)S_{FF}(0). \tag{21}$$

Returning to Equation (19) we have

$$E[x_i(t)x_j(t+\tau)] = \int_0^t \sum_{l=1}^M \sum_{m=1}^M u_{lP}u_{mP}u_{il}u_{jm}h_l(\theta_1)h_m(\theta_1+\tau)S_{FF}(0)d\theta_1. \tag{22}$$

At this point we can simplify the expression by carrying out the integral.

Substituting the expression for the impulse response in Equation (13), the needed expectation in Equation (22) becomes [19,20]

$$R_{X_iX_j}(\tau) = \frac{S_{FF}(0)}{4} \sum_{l=1}^M \sum_{m=1}^M u_{lP}u_{mP}u_{il}u_{jm} \left[A_{lm}e^{-\zeta_m\omega_m\tau}\cos(\omega_{dm}\tau) + B_{lm}e^{-\zeta_m\omega_m\tau}\sin(\omega_{dm}\tau) \right] \tag{23}$$

where

$$A_{lm} = \frac{8\left(\omega_l\zeta_l + \omega_m\zeta_m\right)}{\omega_l^4 + \omega_m^4 + 4\omega_l^3\omega_m\zeta_l\zeta_m + 4\omega_m^3\omega_l\zeta_l\zeta_m + 2\omega_m^2\omega_l^2\left(-1 + 2\zeta_l^2 + 2\zeta_m^2\right)}$$

$$B_{lm} = \frac{4\left(\omega_l^2 + 2\omega_l\omega_m\zeta_l\zeta_m + \omega_m^2\left(-1 + 2\zeta_m^2\right)\right)}{\omega_{dm}\left(\omega_l^4 + \omega_m^4 + 4\omega_l^3\omega_m\zeta_l\zeta_m + 4\omega_m^3\omega_l\zeta_l\zeta_m + 2\omega_m^2\omega_l^2\left(-1 + 2\zeta_l^2 + 2\zeta_m^2\right)\right)} \tag{24}$$

We can further normalize this function to give

$$\rho_{ij}(\tau) = R_{ij}(\tau)/\sqrt{R_{ii}(0)R_{jj}(0)} \tag{25}$$

for the normalized auto- and cross-correlation functions.

It will also prove instructive to study the TDTE between the drive and response. This requires $R_{X_i F_P}(\tau) \equiv E[x_i(t)f_P(t+\tau)]$. Following the same procedure as above results in the expression

$$R_{X_i F_P}(\tau) = \begin{cases} \mathbf{S}_{FF}(0)\sum_{m=1}^{M} u_{im}u_{mP}h_m(-\tau) & : \quad \tau \leq 0 \\ 0 & : \quad \tau > 0 \end{cases} \tag{26}$$

Normalizing by the variance of the random process X_i and assuming $\sigma_{F_P}^2 = 1$ yields the needed correlation function $\rho_{if}(\tau)$. This expression may be substituted into the expression for the transfer entropy to yield the TDTE between drive and response. At this point we have completely defined the analytical TDTE for a broad class of second order linear systems. The behavior of this function is described next. Before concluding this section we note that it also may be possible to derive expressions for the TDTE for different types of forcing functions. Impulse excitation and also non-Gaussian inputs where the marginal PDF can be described as a polynomial transformation of a Gaussian random variable (see e.g., [21]) are two such possibilities.

4. Behavior of the TDTE

Before proceeding with an example, we first require a means of estimating $TE_{j \to i}(\tau)$ from observed data. Assume we have recorded the signals $x_i(n\Delta_t)$, $x_j(n\Delta_t)$, $n = 1 \cdots N$ with a fixed sampling interval Δ_t. In order to estimate the TDTE we require a means of estimating the normalized correlation functions $\rho_{ij}(\tau)$ which can be substituted into Equation (8). While different estimators of correlation functions exist (see e.g., [16]), we use a frequency domain estimator. This estimator relies on the assumption that the observed data are the output of an ergodic (therefore stationary) random process. If we further assume that the correlation functions are absolute integrable, e.g., $\int |R_{ij}(\tau)d\tau| < \infty$, the Wiener-Khinchin Theorem tells us that the cross-spectral density and cross-covariance functions are related via Fourier transform as [16].

$$\int_{-\infty}^{\infty} E[x_i(t)x_j(t+\tau)]e^{-i2\pi f\tau}d\tau = S_{X_j X_i}(f) \equiv \lim_{T \to \infty} E\left[\frac{X_i^*(f)X_j(f)}{2T}\right]. \tag{27}$$

where $X_i(f)$ denotes the Fourier transform of the signal $x_i(t)$. One approach is to therefore estimate the spectral density $\hat{S}_{X_j X_i}(f)$ and then inverse Fourier transform to give $\hat{R}_{X_j X_i}(\tau)$. We further rely on the ergodic theorem of Birkhoff ([22]) which (when applied to probability) allows one to write expectations defined over multiple realizations to be well-approximated temporally averaging over a finite number of samples. More specifically, we divide the temporal sequences $x_i(n)$, $x_j(n)$, $n = 1 \cdots N$ into S segments of length N_s (possibly) overlapping by L points. Taking the discrete Fourier transform of each segment, e.g., $X_{is}(k) = \sum_{n=0}^{N_s-1} x_i(n+sN_s-L)e^{-i2\pi kn/N_s}$, $s = 0 \cdots S-1$ and averaging gives the estimator

$$\hat{S}_{X_j X_i}(k) = \frac{\Delta_t}{N_s S}\sum_{s=0}^{S-1}\hat{X}_{is}^*(k)\hat{X}_{js}(k) \tag{28}$$

at discrete frequency k. This quantity is then inverse discrete Fourier transformed to give

$$\hat{R}_{X_i X_j}(n) = \sum_{k=0}^{N_s-1}\hat{S}_{X_j X_i}(k)e^{i2\pi kn/S}. \tag{29}$$

Finally, we may normalize the estimate to give the cross-correlation coefficient

$$\hat{\rho}_{X_i X_j}(n) = \hat{R}_{X_i X_j}(n)/\sqrt{\hat{R}_{X_i X_i}(0)\hat{R}_{X_j X_j}(0)}. \tag{30}$$

This estimator is asymptotically consistent and unbiased and can therefore be substituted into Equation (8) to produce very accurate estimates of the TE (see examples to follow). In the general (nonlinear) case, kernel density estimators are typically used but are known to be poor in many cases, particularly when data are scarce (see e.g., [6,23]). We also point out that for this study stationarity (and ergodicity) only up to second order (covariance) is required. In general the TDTE is a function of all joint moments hence higher-order ergodicity must be assumed.

As an example, consider a five-DOF system governed by Equation (10), where:

$$
\mathbf{M} = \begin{bmatrix} m_1 & 0 & 0 & 0 & 0 \\ 0 & m_2 & 0 & 0 & 0 \\ 0 & 0 & m_3 & 0 & 0 \\ 0 & 0 & 0 & m_4 & 0 \\ 0 & 0 & 0 & 0 & m_5 \end{bmatrix}
$$

$$
\mathbf{C} = \begin{bmatrix} c_1 + c_2 & -c_2 & 0 & 0 & 0 \\ -c_2 & c_2 + c_3 & -c_3 & 0 & 0 \\ 0 & -c_3 & c_3 + c_4 & -c_4 & 0 \\ 0 & 0 & -c_4 & c_4 + c_5 & -c_5 \\ 0 & 0 & 0 & -c_5 & c_5 \end{bmatrix}
$$

$$
\mathbf{K} = \begin{bmatrix} k_1 + k_2 & -k_2 & 0 & 0 & 0 \\ -k_2 & k_2 + k_3 & -k_3 & 0 & 0 \\ 0 & -k_3 & k_3 + k_4 & -k_4 & 0 \\ 0 & 0 & -k_4 & k_4 + k_5 & -k_5 \\ 0 & 0 & 0 & -k_5 & k_5 \end{bmatrix}
$$

(31)

are constant coefficient matrices commonly used to describe structural systems. In this case, these particular matrices describe the motion of a cantilevered structure where we assume a joint normally distributed random process applied at the end mass, i.e. $\mathbf{f}(t) = (0,0,0,0,\mathcal{N}(0,1))$. In this first example we examine the TDTE between response data collected from two different points on the structure. We fix $m_i = 0.01\ kg$, $c_i = 0.1\ N \cdot s/m$, and $k_i = 10\ N/m$ for each of the $i = 1 \cdots 5$ degrees of freedom (thus we are using $\alpha = 0.01$ in the modal damping model $\mathbf{C} = \alpha \mathbf{K}$). The system response data $x_i(n\Delta_t)$, $n = 1 \cdots 2^{15}$ to the stochastic forcing is then generated via numerical integration. For simulation purposes we used a time-step of $\Delta_t = 0.01\ s$ which is sufficient to capture all five of the system natural frequencies (the lowest of which is $\omega_1 = 9.00$ rad/s). Based on these parameters, we generated the analytical expressions $TE_{3\to2}(\tau)$ and $TE_{2\to3}(\tau)$ and also $TE_{5\to1}(\tau)$ and $TE_{1\to5}(\tau)$ for illustrative purposes. These are shown in Figure 2 along with the estimates formed using the Fourier transform-based procedure. In forming the estimates we used $L = 0$, $S = 2^3$, $N_s = 2^{12}$, resulting in low bias and variance, and providing curves that are in very close agreement with theory.

With Figure 2 in mind, first consider negative delays only where $\tau < 0$. Clearly, the further the random variable $X_j(t_n + \tau)$ is from $X_i(t_n)$, the less information it carries about the probability of X_i transitioning to a new state Δ_t seconds into the future. This is to be expected from a stochastically driven system and accounts for the decay of the transfer entropy to zero for large $|\tau|$. However, we also see periodic returns to the point $TE_{j\to i}(\tau) = 0$ for even small temporal separation. Clearly this is a reflection of the periodicity observed in second order linear systems. In fact, for this system the dominant period of oscillation is $2\pi/\omega_1 = 0.698$ seconds. It can be seen that the argument of the logarithm in Equation (8) periodically reaches a minimum value of unity at precisely half this period, thus we observe zeros of the TDTE at times $(i-1) \times \pi/\omega_1$, $i = 1 \cdots$. In this case the TDTE is going to zero *not* because the random variables $X_j(t_n + \tau)$, $X_i(t_n)$ are unrelated, but because knowledge of one allows us to exactly predict the position of the other (no additional information is present). We believe this is likely to be a feature of most systems possessing an underlying periodicity and is one reason why using the TE as a measure of coupling must be done with care.

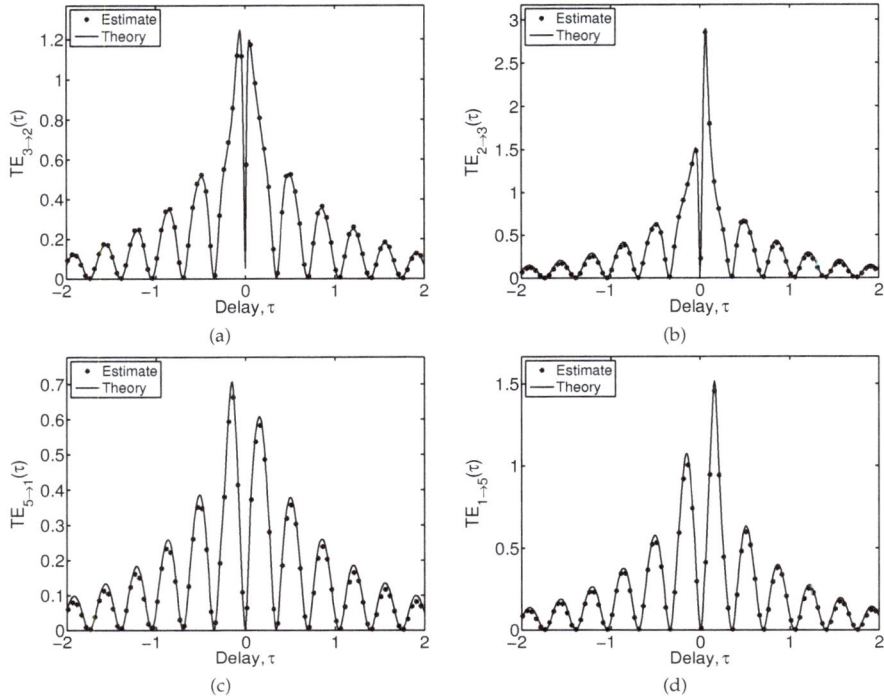

Figure 2. Time delay transfer entropy between masses two and three (top row) and one and five (bottom row) of a 5 DOF system driven at mass, $P = 5$.

One possible way to eliminate this feature is to condition the measure on more of the signal's past history. In fact, several papers (see e.g., [9,13]) mention the importance of conditioning on the full state vector $X_j(t_n - \tau_1), X_j(t_n - \tau_2), \cdots, X_j(t_n - \tau_d)$ where d is the dimensionality (in a loose sense, the number of dynamical degrees of freedom) of the random process X_j. Building in more past history would almost certainly remove the oscillations as some of the past observations would always be providing additional predictive power. However, building in more history significantly complicates the ability to derive closed-form expressions. Moreover, for this simple linear system the basic envelope of the TDTE curves would not likely be effected by altering the model in this way.

We also point out that values of the TDTE are non-zero for positive delays as well. Again, so long as we interpret the TE as a measure of predictive power this makes sense. That is to say, future values X_j can aid in predicting the current dynamics of X_i. Interestingly, the asymmetry in the TE peaks near $\tau = 0$ may provide the largest clue as to the location of the forcing signal. Consistently we have found that the TE is larger for negative delays when mass closest the driven end plays the role of X_j; conversely it is larger for positive delays when the mass furthest from the driven end plays this role. So long as the coupling is bi-directional, results such as those shown in Figure 2 can be expected in general.

However, the situation is quite different if we consider the case of uni-directional coupling. For example, we may consider $TE_{f \rightarrow i}(\tau)$, *i.e.* the TDTE between the forcing signal and response variable i. This is a particularly interesting case as, unlike in previous examples, there is no feedback from DOF i to the driving signal. Figure 3 shows the TDTE between drive and response and clearly highlights the directional nature of the coupling. Past values of the forcing function clearly help in predicting the dynamics of the response. Conversely, future values of the forcing say nothing about transition probabilities for the mass response simply because the mass has not "seen" that information yet. Thus, for uni-directional coupling, the TDTE can easily diagnose whether X_j is driving X_i or vice-versa. It can also be noticed from these plots that the drive signal is not that much help in predicting the response as the TDTE is much smaller in magnitude that when computed between masses. We interpret this to mean that the response data are dominated by the physics of the structure (e.g., the structural modes), which is information not carried in the drive signal. Hence, the drive signal offers little in the way of additional predictive power. While the drive signal puts energy into the system, it is not very good at predicting the response. It should also be pointed out that the kernel density estimation techniques are not able to capture these small values of the TDTE. The error in such estimates is larger than these subtle fluctuations. Only the "linearized" estimator is able to capture the fluctuations in the TDTE for small ($O(10^{-2})$) values.

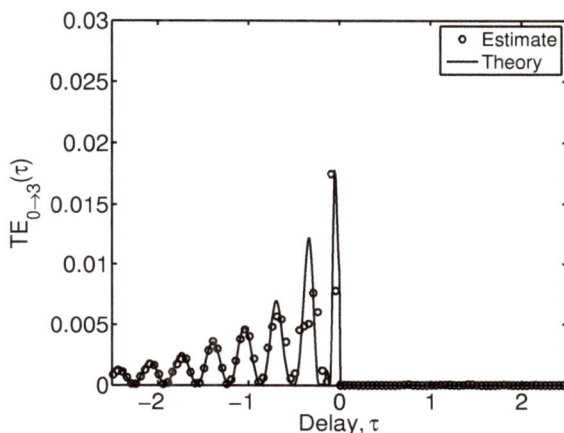

Figure 3. Time delay transfer entropy between the forcing (denoted as DOF "0") and mass three for the 5 DOF system driven at mass, $P = 5$. The plot is consistent with the interpretation of information moving from the forcing to mass three.

It has been suggested that the main utility of the TE is to, given a sequence of observations, assess the direction of information flow in a coupled system. More specifically, one computes the difference $TE_{i \to j} - TE_{j \to i}$ with a positive difference suggesting information flow from i to j (negative differences indicating the opposite) [2,4]. In the system modeled by Equation (10) one would heuristically understand the information as flowing from the drive signal to the response. This is certainly reinforced by Figure 3. However, by extension it might seem probable that information would similarly flow from the mass closest the drive signal to the mass closest the boundary (e.g., DOF 5 to DOF 1).

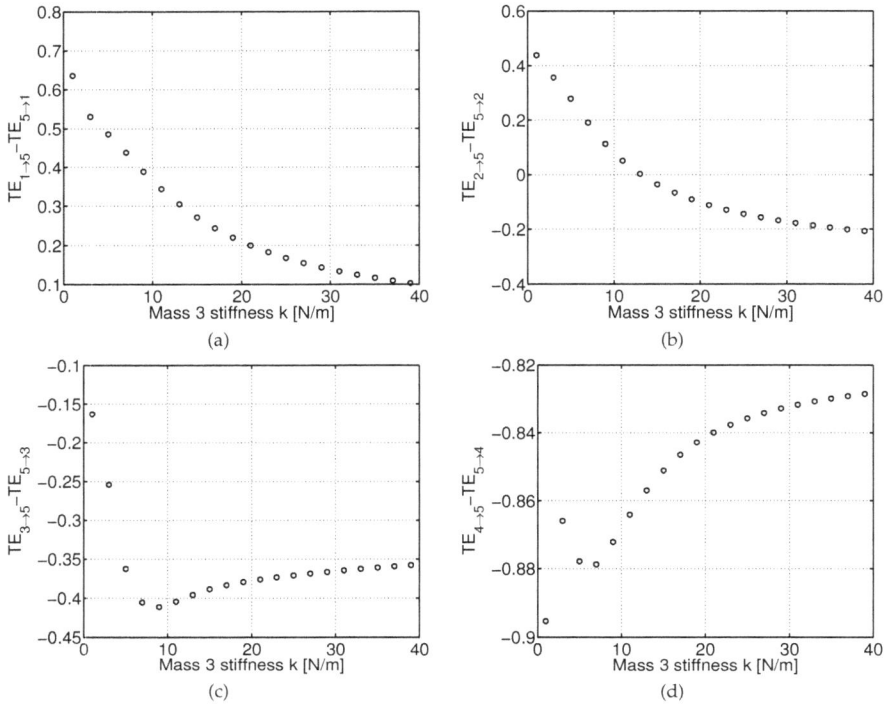

Figure 4. Difference in time delay transfer entropy between the driven mass five and each other DOF as a function of k_3. A positive difference indicates $TE_{i \rightarrow j} > TE_{j \rightarrow i}$ and is commonly used to indicate that information is moving from mass i to mass j. Based on this interpretation, negative values indicate information moving from the driven end to the base; positive values indicate the opposite. Even for this linear system, choosing different masses in the analysis can produce very different results. In fact, $TE_{2 \rightarrow 5} - TE_{5 \rightarrow 2}$ implies a different direction of information transfer, depending on the strength of the coupling, k_3

We test this hypothesis as a function of the coupling strength between masses. Fixing each stiffness and damping coefficient to the previously used values, we vary k_3 from 1 N/m to 40 N/m and examine the quantity $TE_{i \rightarrow j} - TE_{j \rightarrow i}$ evaluated at τ^*, taken as the delay at which the TDTE reaches its maximum. Varying k_3 slightly alters the dominant period of the response. By accounting for this shift we eliminate the possibility of capturing the TE at one of its nulls (see Figure 2). For example, in Figure 2 we see that $\tau^* = -0.15$ in the plot of $TE_{3 \rightarrow 2}(\tau)$. Figure 4 shows the difference in TDTE as a function of the coupling strength. The result is non-intuitive if one assumes information would move from driven end toward the non-driven end of the system. For certain DOFs this interpretation holds, for others, it does not. Herein lies the difficulty in interpreting the TE when bi-directional coupling exists. This was also pointed out by Schreiber [1] who noted "Reducing the analysis to the identification of a "drive" and a "response" may not be useful and could even be misleading". The above results certainly reinforce this statement.

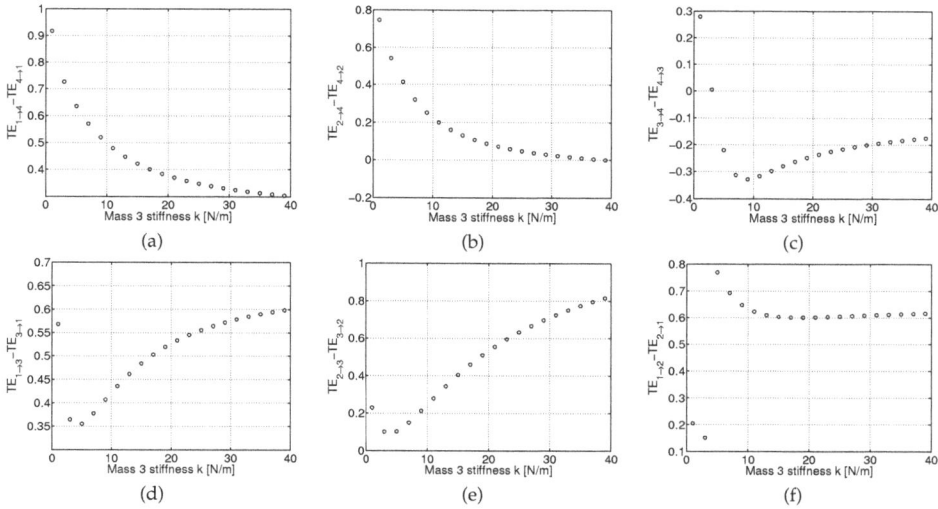

Figure 5. Difference in time-delayed transfer entropy (TDTE) among different combinations of masses. By the traditional interpretation of TE, negative values indicate information moving from the driven end to the base; positive values indicate the opposite.

Rather than being viewed as a measure of information flow, we find it more useful to interpret the difference measure as simply one of predictive power. That is to say, does knowledge of system j help predict system i more so than i helps predict j. This is a slightly different question. Our analysis suggests that if X_i and X_j are both near the driven end but with DOF i the closer of the two , then knowledge of $\mathbf{X}_j$ is of more use in predicting $\mathbf{X}_i$ than vice-versa. This interpretation also happens to be consistent with the notion of information moving from the driven end toward the base. However as i and j become de-coupled (physically separated) it appears the reverse is true. The random process $\mathbf{X}_i$ is better at predicting $\mathbf{X}_j$ than $\mathbf{X}_j$ is in predicting $\mathbf{X}_i$. Thus, for certain pairs of masses information seems to be traveling from the base toward the drive. One possible explanation is that because the mass $\mathbf{X}_i$ is further removed from the drive signal it is strongly influenced by the vibration of each of the other masses. By contrast, a mass near the driven end is strongly influenced only by the drive signal. Because the dynamics $\mathbf{X}_i$ are influenced heavily by the structure (as opposed to the drive), $\mathbf{X}_i$ does a good job in helping to predict the dynamics everywhere. The main point of this analysis is that the difference in TE is not at all an unambiguous measure of the direction of information flow.

To further explore this question, we have repeated this numerical experiment for all possible combinations of masses. These results are displayed in Figure 5 where the same basic phenomenology is observed. If both masses being analyzed are near the driven end, the mass closest the drive is a better predictor of the one that is further away. However again, as i and j become decoupled the reverse is true. Our interpretation is that the further the process is removed from the drive signal, the more it is dominated by the other mass dynamics and the boundary conditions. Because such a process is strongly influenced by the other DOFs, it can successfully predict the motion for these other DOFs.

It is also interesting to note how the strength, and even directionality (sign) of the difference in TDTE changes with variations in a single stiffness element. Depending on the value of k_3 we see changes in which of the two masses is a better predictor. In some cases we even see zero TDTE difference, implying that the dynamics of the constituent signals are equally useful in predicting one another. Again, this does not support our intuitive notion of what it means for information to travel through a structural system. Only in the case of uni-directional coupling can we unambiguously use the TE to indicate directionality of information transport.

Entropy **2013**, *15*, 3186–3204

One of the strengths of our analysis is that these conclusions are not influenced by estimation error. In studying heart and breath rate interactions, for example, the ambiguity in information flow was assigned to difficulties in the estimation process [2]. We have shown here that even when estimation error is not a factor the ambiguity remains. We would imagine a similar result would hold for more complex systems, however such systems are beyond our ability to develop analytical expressions. The difference in TDTE is, however, a useful indicator of which system component carries the most predictive power about the rest of the system dynamics.

In short, the TDTE can be a very useful descriptor of system dynamics and coupling among system components. However any real understanding is only likely to be obtained in the context of a particular system model, or class of models (e.g., linear). Absent physical insight into the process that generates the observations, understanding results of a TDTE analysis can be challenging at best.

However, it is perhaps worth mentioning that the expressions derived here might permit inference about the general form of the underlying "linearized" system model. Different linear system models yield different expressions for $\rho_{ij}(\tau)$, hence different expressions for the TDTE. One could then conceivably use estimates of the TDTE as a means to select among this class of models given observed data. Whether or not the TDTE is of use in the context of model selection remains to be seen.

5. Conclusions

In this work we have derived an analytical expression for the time-delayed transfer entropy (TDTE) among components of a broad class of second order linear systems driven by a jointly Gaussian input. This solution has proven particularly useful in understanding the behavior of the TDTE as a measure of dynamical coupling. In particular, when the coupling is uni-directional, we have found the TDTE to be an unambiguous indicator of the direction of information flow in a system. However, for bi-directional coupling the situation is significantly more complicated, even for linear systems. We have found that a heuristic understanding of information flow is not always accurate. For example, one might expect information to travel from the driven end of a system toward the non-driven end. In fact, we have shown precisely the opposite to be true. Simply varying a linear stiffness element can cause the apparent direction of flow to change. It would seem a safer interpretation is that a positive difference in the transfer entropy between two system components tells the practitioner which component has the greater predictive power.

Acknowledgments: The authors would like to thank the Naval Research Laboratory for providing funding for this work.

Conflicts of Interest: The authors declare no conflict of interest.

References

1. Schreiber, T. Measuring information transfer. *Phys. Rev. Lett.* **2000**, *85*, 461–464.
2. Kaiser, A.; Schreiber, T. Information transfer in continuous processes. *Phys. D: Nonlinear Phenom.* **2002**, *166*, 43–62.
3. Moniz, L.J.; Cooch, E.G.; Ellner, S.P.; Nichols, J.D.; Nichols, J.M. Application of information theory methods to food web reconstruction. *Ecol. Model.* **2007**, *208*, 145–158.
4. Bauer, M.; Cox, J.W.; Caveness, M.H.; Downs, J.J.; Thornhill, N.F. Finding the direction of disturbance propagation in a chemical process using transfer entropy. *IEEE Trans. Control Syst. Technol.* **2007**, *15*, 12–21.
5. Marschinski, R.; Kantz, H. Analysing the information flow between financial time series: An improved estimator for transfer entropy. *Eur. Phys. J. B* **2002**, *30*, 275–281.
6. Nichols, J.M. Examining structural dynamics using information flow. *Probab. Eng. Mech.* **2006**, *21*, 420–433.
7. Nichols, J.M.; Seaver, M.; Trickey, S.T.; Salvino, L.W.; Pecora, D.L. Detecting impact damage in experimental composite structures: An information-theoretic approach. *Smart Mater. Struct.* **2006**, *15*, 424–434.
8. Moniz, L.J.; Nichols, J.D.; Nichols, J.M. Mapping the information landscape: Discerning peaks and valleys for ecological monitoring. *J. Biol. Phys.* **2007**, *33*, 171–181.

9. Vicente, R.; Wibral, M.; Lindner, M.; Pipa, G. Transfer entropy-a model-free measure of effective connectivity for the neurosciences. *J. Comput. Neurosci.* **2001**, *30*, 45–67.

10. Wibral, M.; Rahm, B.; Rieder, M.; Lindner, M.; Vicente, R.; Kaiser, J. Transfer entropy in magnetoencephalographic data: Quantifying information flow in cortical and cerebellar networks. *Progr. Biophys. Mol. Biol.* **2011**, *105*, 80–97.

11. Vakorin, V.A.; Krakovska, O.A.; McIntosh, A.R. Confounding effects of indirect connections on causality estimation. *J. Neurosci. Method.* **2009**, *184*, 152–160.

12. Barnett, L.; Barrett, A.B.; Seth, A.K. Granger causality and transfer entropy are equivalent for gaussian variables. *Phys. Rev. Lett.* **2009**, *103*, 238701.

13. Palus, M.; Vejmelka, M. Directionality of coupling from bivariate time series: How to avoid false causalities and missed connections. *Phys. Rev. E* **2007**, *75*, 056211.

14. Friston, K.J.; Harrison, L.; Penny, W. Dynamic causal modelling. *NeuroImage* **2003**, *19*, 1273–1302.

15. Greenberg, M.D. *Advanced Engineering Mathematics*; Prentice-Hall, Inc.: Englewood Cliffs, NJ, USA, 1988.

16. Bendat, J.S.; Piersol, A.G. *Random Data Analysis and Measurement Procedures*, 3rd ed.; Wiley & Sons: New York, NY, USA, 2000.

17. Kay, S.M. *Fundamentals of Statistical Signal Processing: Volume I, Estimation Theory*; Prentice Hall: New Jersey, NJ, USA, 1993.

18. Meirovitch, L. *Introduction to Dynamics and Control*, 1st ed.; Wiley & Sons.: New York, NY, USA, 1985.

19. Crandall, S.H.; Mark, W.D. *Random Vibration in Mechanical Systems*; Academic Press: New York, NY, USA, 1963.

20. Benaroya, H. *Mechanical Vibration: Analysis, Uncertainties, and Control*; Prentice Hall: New Jersey, NJ, USA, 1998.

21. Nichols, J.M.; Olson, C.C.; Michalowicz, J.V.; Bucholtz, F. The bispectrum and bicoherence for quadratically nonlinear systems subject to non-gaussian inputs. *IEEE Trans. Signal Process.* **2009**, *57*, 3879–3890.

22. Birkhoff, G.D. Proof of the ergodic theorem. *Proc. Natl. Acad. Sci. USA* **1931**, *17*, 656–660.

23. Hahs, D.W.; Pethel, S.D. Distinguishing Anticipation from Causality: Anticipatory Bias in the Estimation of Information Flow. *Phys. Rev. Lett.* **2011**, *107*, 128701.

Article

Transfer Entropy for Coupled Autoregressive Processes

Daniel W. Hahs [1] and Shawn D. Pethel [2,*]

[1] Torch Technologies, Inc. Huntsville, AL 35802, USA; dan.hahs@torchtechnologies.com
[2] U.S. Army, Redstone Arsenal, Huntsville, AL 35898, USA

* Author to whom correspondence should be addressed; shawn.d.pethel.civ@mail.mil; Tel.: 256-842-9734; Fax: 256-842-2507.

Received: 26 January 2013; in revised form: 13 February 2013; Accepted: 19 February 2013; Published: 25 February 2013

Abstract: A method is shown for computing transfer entropy over multiple time lags for coupled autoregressive processes using formulas for the differential entropy of multivariate Gaussian processes. Two examples are provided: (1) a first-order filtered noise process whose state is measured with additive noise, and (2) two first-order coupled processes each of which is driven by white process noise. We found that, for the first example, increasing the first-order AR coefficient while keeping the correlation coefficient between filtered and measured process fixed, transfer entropy increased since the entropy of the measured process was itself increased. For the second example, the minimum correlation coefficient occurs when the process noise variances match. It was seen that matching of these variances results in minimum information flow, expressed as the sum of transfer entropies in both directions. Without a match, the transfer entropy is larger in the direction away from the process having the larger process noise. Fixing the process noise variances, transfer entropies in both directions increase with the coupling strength. Finally, we note that the method can be generally employed to compute other information theoretic quantities as well.

Keywords: transfer entropy; autoregressive process; Gaussian process; information transfer

---

1. Introduction

Transfer entropy [1] quantifies the information flow between two processes. Information is defined to be flowing from system X to system Y whenever knowing the past states of X reduces the uncertainty of one or more of the current states of Y above and beyond what uncertainty reduction is achieved by only knowing the past Y states. Transfer entropy is the mutual information between the current state of system Y and one or more past states of system X, conditioned on one or more past states of system Y. We will employ the following notation. Assume that data from two systems X and Y are simultaneously available at k timestamps: $t_{n-k+2\,:\,n+1} \equiv \{t_{n-k+2}, t_{n-k+2}, ..., t_n, t_{n+1}\}$. Then we express transfer entropies as:

$$TE_{x \to y}^{(k)} = I(y_{n+1}; x_{n-k+2\,:\,n}|y_{n-k+2\,:\,n}) = H(y_{n+1}|y_{n-k+2\,:\,n}) - H(y_{n+1}|y_{n-k+2\,:\,n}, x_{n-k+2\,:\,n}) \quad (1)$$

$$TE_{y \to x}^{(k)} = I(x_{n+1}; y_{n-k+2\,:\,n}|x_{n-k+2\,:\,n}) = H(x_{n+1}|x_{n-k+2\,:\,n}) - H(x_{n+1}|x_{n-k+2\,:\,n}, y_{n-k+2\,:\,n}). \quad (2)$$

Each of the two transfer entropy values $TE_{x \to y}$ and $TE_{y \to x}$ is nonnegative and both will be positive (and not necessarily equal) when information flow is bi-directional. Because of these properties, transfer entropy is useful for detecting causal relationships between systems generating measurement time series. Indeed, transfer entropy has been shown to be equivalent, for Gaussian variables, to Granger causality [2]. Reasons for caution about making causal inferences in some situations using transfer entropy, however, are discussed in [3–6]. A formula for normalized transfer entropy is provided in [7].

The contribution of this paper is to explicitly show how to compute transfer entropy over a variable number of time lags for autoregressive (AR) processes driven by Gaussian noise and to gain insight into the meaning of transfer entropy in such processes by way of two example systems: (1) a first-order AR process X = {x_n} with its noisy measurement process Y = {y_n}, and (2) a set of two mutually-coupled AR processes. Computation of transfer entropies for these systems is a worthwhile demonstration since they are simple models that admit intuitive understanding. In what follows we first show how to compute the covariance matrix for successive iterates of the example AR processes and then use these matrices to compute transfer entropy quantities based on the differential entropy expression for multivariate Gaussian random variables. Plots of transfer entropies *versus* various system parameters are provided to illustrate various relationships of interest.

Note that Kaiser and Schreiber [8] have previously shown how to compute information transfer metrics for continuous-time processes. In their paper they provide an explicit example, computing transfer entropy for two linear stochastic processes where one of the processes is autonomous and the other is coupled to it. To perform the calculation for the Gaussian processes the authors utilize expressions for the differential entropy of multivariate Gaussian noise. In our work, we add to this understanding by showing how to compute these quantities analytically for higher time lags. We now provide a discussion of differential entropy, the formulation of entropy appropriate to continuous-valued processes as we are considering.

2. Differential Entropy

The entropy of a continuous-valued process is given by its differential entropy. Recall that the entropy of a discrete-valued random variable is given by the *Shannon entropy* $H = -\sum_i p_i \log p_i$ (we shall always choose log base 2 so that entropy will be expressed in units of bits) where p_i is the probability of the ith outcome and the sum is over all possible outcomes.

Following [9] we derive the appropriate expression for differential entropies for conditioned and unconditioned continuous-valued random variables. When a process X is continuous-valued we may approximate it as a discrete-value process by identifying $p_i = f_i \Delta x$ where f_i is the value of the pdf at the ith partition point and Δx is the refinement of the partition. We then obtain:

$$
\begin{aligned}
H(X) &= -\sum_i p_i \log p_i \\
&= -\sum_i f_i \Delta x \log f_i \Delta x \\
&= -\sum_i f_i \Delta x (\log f_i + \log \Delta x) \\
&= -\sum_i f_i \log f_i \Delta x - \sum_i \log \Delta x f_i \Delta x \\
&= -\int f \log f \, dx - \log \Delta x \int f \, dx \\
&= h(X) - \log \Delta x
\end{aligned}
\tag{3}
$$

Note that since the X process is continuous-valued, then, as $\Delta x \to 0$, we have H(X) $\to$ + infinity. Thus, for continuous-valued processes, the quantity h(X), when itself defined and finite, is used to represent the entropy of the process. This quantity is known as the *differential entropy* of random process X.

Closed-form expressions for the differential entropy of many distributions are known. For our purposes, the key expression is the one for the (unconditional) multivariate normal distribution [10].

Let the probability density function of the n-dimensional random vector x be denoted f(x), then the relevant expressions are:

$$f(\overline{x}) = \frac{\exp\left[-\frac{1}{2}(\overline{x}-\overline{\mu})^T C^{-1}(\overline{x}-\overline{\mu})\right]}{(2\pi)^{\frac{n}{2}}[\det C]^{\frac{1}{2}}}$$
$$h(\overline{x}) = -\int f(\overline{x})\log[f(\overline{x})]dx \qquad (4)$$
$$= \frac{1}{2}\log\left[(2\pi e)^n \det C\right]$$

where detC is the determinant of matrix C, the covariance of x. In what follows, this expression will be used to compute differential entropy of unconditional and conditional normal probability density functions. The case for conditional density functions warrants a little more discussion.

Recall that the relationships between the joint and conditional covariance matrices, C_{XY} and $C_{Y|X}$, respectively, of two random variables X and Y (having dimensions n_x and n_y, respectively) are given by:

$$C_{XY} = \text{cov}\left(\begin{bmatrix} X \\ Y \end{bmatrix}\right) = \begin{bmatrix} \Sigma_{11} & \Sigma_{12} \\ \Sigma_{21} & \Sigma_{22} \end{bmatrix} \qquad (5)$$

$$\text{cov}[Y|X=x] = C_{Y|X} = \Sigma_{22} - \Sigma_{21}\Sigma_{11}^{-1}\Sigma_{12}.$$

Here blocks Σ_{11} and Σ_{22} have dimensions n_x by n_x and n_y by n_y, respectively. Now, using Leibniz's formula, we have that:

$$\det C_{XY} = \det\begin{bmatrix} \Sigma_{11} & \Sigma_{12} \\ \Sigma_{21} & \Sigma_{22} \end{bmatrix} = \det\Sigma_{11}\det\left(\Sigma_{22} - \Sigma_{21}\Sigma_{11}^{-1}\Sigma_{12}\right) = \det C_X \det C_{Y|X}. \qquad (6)$$

Hence the conditional differential entropy of Y, given X, may be conveniently computed using:

$$\begin{aligned} h(Y|X) &= \frac{1}{2}\log\left[(2\pi e)^{n_y}\det C_{Y|X}\right] \\ &= \frac{1}{2}\log\left[(2\pi e)^{n_y}\frac{\det C_{XY}}{\det C_X}\right] \\ &= \frac{1}{2}\log\left[(2\pi e)^{n_x+n_y}\det C_{XY}\right] - \frac{1}{2}\log\left[(2\pi e)^{n_x}\det C_X\right] \\ &= h(X,Y) - h(X). \end{aligned} \qquad (7)$$

This formulation is very handy as it allows us to compute many information-theoretic quantities with ease. The strategy is as follows. We define $C^{(k)}$ to be the covariance of two random processes sampled at k consecutive timestamps $\{t_{n-k+2}, t_{n-k+1}, \ldots, t_n, t_{n+1}\}$. We then compute transfer entropies for values of k up to k sufficiently large to ensure that their valuations do not change significantly if k is further increased. For our examples, we have found k = 10 to be more than sufficient. A discussion of the importance of considering this sufficiency is provided in [11].

3. Transfer Entropy Computation Using Variable Number of Timestamps

We wish to consider two example processes each of which conforms to one of the two model systems having the general expressions:

$$(1)\quad \begin{cases} x_{n+1} = a_0 x_n + a_1 x_{n-1} + \cdots + a_m x_{n-m} + w_n \\ y_{n+1} = c_{-1} x_{n+1} + v_n \\ v_n \sim N(0,R), w_n \sim N(0,Q) \end{cases} \qquad (8)$$

and:

$$(2)\quad \begin{cases} x_{n+1} = a_0 x_n + a_1 x_{n-1} + \cdots + a_m x_{n-m} + b_0 y_n + b_1 y_{n-1} + \cdots + b_j y_{n-j} + w_n \\ y_{n+1} = c_0 x_n + c_1 x_{n-1} + \cdots + c_m x_{n-m} + d_0 y_n + d_1 y_{n-1} + \cdots + d_j y_{n-j} + v_n \\ v_n \sim N(0,R), w_n \sim N(0,Q). \end{cases} \qquad (9)$$

Here, v_n and w_n are zero mean uncorrelated Gaussian noise processes having variances R and Q, respectively. For system stability, we require the model poles to lie within the unit circle. The first model is of a filtered process noise X one-way coupled to an instantaneous, but noisy measurement process Y. The second model is a two-way coupled pair of processes, X and Y.

Transfer entropy (as defined by Schreiber [1]) considers the flow of information from past states (*i.e.*, state values having, timetags $t_{n-k+2:n} \equiv \{t_{n-k+2}, t_{n-k+2}, ..., t_n\}$) of one process to the present (t_{n+1}) state of another process. However, note that in the first general model (measurement process) there is an explicit flow of information from the present state of the X process; x_{n+1} determines the present state of the Y process y_{n+1} (assuming c_{-1} is not zero). To fully capture the information transfer from the X process to the current state of the Y process we must identify the correct causal states [4]. For the measurement system, the causal states include the current (present) state. This state is not included in the definition of transfer entropy, being a mutual information quantity conditioned on only past states. Hence, for the purpose of this paper, we will temporarily define a quantity, "information transfer," similar to transfer entropy, except that the present of the driving process, x_{n+1}, will be lumped in with the past values of the X process: $x_{n-k+2}:x_n$. For the first general model there is no information transferred from the Y to the X process. We define the (non-zero) information transfer from the X to the Y process (based on data from k timetags) as:

$$IT^{(k)}_{x \to y} = I(y_{n+1}; x_{n-k+2:n+1}|y_{n-k+2:n}) = H(y_{n+1}|y_{n-k+2:n}) - H(y_{n+1}|y_{n-k+2:n}, x_{n-k+2:n+1}). \quad (10)$$

The major contribution of this paper is to show how to analytically compute transfer entropy for AR Gaussian processes using an iterative method for computing the required covariance matrices. Computation of information transfer is additionally presented to elucidate the power of the method when similar information quantities are of interest and to make the measurement example more interesting. We now present a general method for computing the covariance matrices required to compute information-theoretic quantities for the AR models above. Two numerical examples follow.

To compute transfer entropy over a variable number of multiple time lags for AR processes of the general types shown above, we compute its block entropy components over multiple time lags. By virtue of the fact that the processes are Gaussian we can avail ourselves of analytical entropy expressions that depend only on the covariance of the processes. In this section we show how to analytically obtain the required covariance expressions starting with the covariance for a single time instance. Taking expectations, using the AR equations, we obtain the necessary statistics to characterize the process. Representing these expectation results in general, the process covariance matrix $C^{(1)}(t_n)$ corresponding to a single timestamp, t_n, is:

$$C^{(1)}(t_n) \equiv \mathrm{cov}\left(\begin{bmatrix} x_n \\ y_n \end{bmatrix}\right) = \begin{bmatrix} E\left[x_n^2\right] & E[x_n y_n] \\ E[y_n x_n] & E\left[y_n^2\right] \end{bmatrix}. \quad (11)$$

To obtain an expanded covariance matrix, accounting for two time instances (t_n and t_{n+1}), we compute the additional expectations required to fill in the matrix $C^{(2)}(t_n)$:

$$C^{(2)}(t_n) \equiv \mathrm{cov}\left(\begin{bmatrix} x_n \\ y_n \\ x_{n+1} \\ y_{n+1} \end{bmatrix}\right) = \begin{bmatrix} E\left[x_n^2\right] & E[x_n y_n] & E[x_n x_{n+1}] & E[x_n y_{n+1}] \\ E[x_n y_n] & E\left[y_n^2\right] & E[x_{n+1} y_n] & E[y_n y_{n+1}] \\ E[x_n x_{n+1}] & E[x_{n+1} y_n] & E\left[x_{n+1}^2\right] & E[x_{n+1} y_{n+1}] \\ E[x_n y_{n+1}] & E[y_n y_{n+1}] & E[x_{n+1} y_{n+1}] & E\left[y_{n+1}^2\right] \end{bmatrix}. \quad (12)$$

Because the process is stationary, we may write:

$$C^{(2)}(t_n) = C^{(2)} = \begin{bmatrix} \Sigma_{11} & \Sigma_{12} \\ \Sigma_{21} & \Sigma_{22} \end{bmatrix} \quad (13)$$

where:

$$
\begin{aligned}
\Sigma_{11} &\equiv \begin{bmatrix} E\left[x_n^2\right] & E\left[x_n y_n\right] \\ E\left[x_n y_n\right] & E\left[y_n^2\right] \end{bmatrix} \\
\Sigma_{12} &\equiv \begin{bmatrix} E\left[x_n x_{n+1}\right] & E\left[x_n y_{n+1}\right] \\ E\left[x_{n+1} y_n\right] & E\left[y_n y_{n+1}\right] \end{bmatrix} \\
\Sigma_{21} &= \Sigma_{12}^T \\
\Sigma_{22} &= \Sigma_{11}.
\end{aligned}
\tag{14}
$$

Thus we have found the covariance matrix $C^{(2)}$ required to compute block entropies based on two timetags or, equivalently, one time lag. Using this matrix the single-lag transfer entropies may be computed.

We now show how to compute the covariance matrices corresponding to any finite number of time stamps. Define vector $\bar{z}_n = \begin{bmatrix} x_n \\ y_n \end{bmatrix}$. Using the definitions above, write the matrix $C^{(2)}$ as a block matrix and, using standard formulas, compute the conditional mean and covariance C_c of $\bar{z}_{n+1}$ given $\bar{z}_n$:

$$
\begin{aligned}
C^{(2)} = \mathrm{cov}\left(\begin{bmatrix} \bar{z}_n \\ \bar{z}_{n+1} \end{bmatrix}\right) &= E\left(\begin{bmatrix} \bar{z}_n \\ \bar{z}_{n+1} \end{bmatrix}\begin{bmatrix} \bar{z}_n & \bar{z}_{n+1} \end{bmatrix}\right) = \begin{bmatrix} \Sigma_{11} & \Sigma_{12} \\ \Sigma_{21} & \Sigma_{22} \end{bmatrix} \\
E[\bar{z}_{n+1}|\bar{z}_n = \bar{z}] &= E[\bar{z}_n] + \Sigma_{21}\Sigma_{11}^{-1}[\bar{z} - E[\bar{z}_n]] \\
&= \bar{\mu}_{\bar{z}} + \Sigma_{21}\Sigma_{11}^{-1}[\bar{z} - \bar{\mu}_{\bar{z}}] \\
C_c \equiv \mathrm{cov}[\bar{z}_{n+1}|\bar{z}_n = \bar{z}] &= \Sigma_{22} - \Sigma_{21}\Sigma_{11}^{-1}\Sigma_{12}.
\end{aligned}
\tag{15}
$$

Note that the expected value of the conditional mean is zero since the mean of the $\bar{z}_n$ process, $\bar{\mu}_{\bar{z}}$, is itself zero.

With these expressions in hand, we note that we may view propagation of the state $\bar{z}_n$ to its value $\bar{z}_{n+1}$ at the next timestamp as accomplished by the recursion:

$$
\begin{aligned}
\bar{z}_{n+1} &= \bar{\mu}_z + D(\bar{z}_n - \bar{\mu}_z) + S\bar{u}_n : \bar{u}_n \sim N(0_2, I_2) \\
D &\equiv \Sigma_{21}\Sigma_{11}^{-1} \\
C_c &\equiv SS^T \equiv \Sigma_{22} - \Sigma_{21}\Sigma_{11}^{-1}\Sigma_{12}.
\end{aligned}
\tag{16}
$$

Here S is the principal square root of the matrix C_c. It is conveniently computed using the inbuilt Matlab function *sqrtm*. To see analytically that the recursion works, note that using it we recover at each timestamp a process having the correct mean and covariance:

$$
E\{\bar{z}_{n+1}|\bar{z}_n = \bar{z}\} = E\{\bar{\mu}_z + D(\bar{z}_n - \bar{\mu}_z) + S\bar{u}_n|\bar{z}_n = \bar{z}\} = \bar{\mu}_z + D(\bar{z} - \bar{\mu}_z)
\tag{17}
$$

and:

$$
\bar{z}_{n+1} - E\{\bar{z}_{n+1}|\bar{z}_n = \bar{z}\} = \bar{\mu}_z + D(\bar{z}_n - \bar{\mu}_z) + S\bar{u}_n - (\bar{\mu}_z + D(\bar{z} - \bar{\mu}_z)) = S\bar{u}_n + D(\bar{z}_n - \bar{z})
$$

$$
\begin{aligned}
\mathrm{cov}(\bar{z}_{n+1}|\bar{z}_n = \bar{z}) &= E\left\{[\bar{z}_{n+1} - E\{\bar{z}_{n+1}|\bar{z}_n = \bar{z}\}][\bar{z}_{n+1} - E\{\bar{z}_{n+1}|\bar{z}_n = \bar{z}\}]^T\Big|\bar{z}_n = \bar{z}\right\} \\
&= E\left\{[S\bar{u}_n + D(\bar{z}_n - \bar{z})][S\bar{u}_n + D(\bar{z}_n - \bar{z})]^T\Big|\bar{z}_n = \bar{z}\right\} \\
&= E\left\{[S\bar{u}_n][S\bar{u}_n]^T\right\} = SE\{\bar{u}_n\bar{u}_n^T\}S^T = SS^T.
\end{aligned}
\tag{18}
$$

Thus, because the process is Gaussian and fully specified by its mean and covariance, we have verified that the recursive representation yields consistent statistics for the stationary AR system. Using the above insights, we may now recursively compute the covariance matrix $C^{(k)}$ for a variable number (k) of timestamps. Note that $C^{(k)}$ has dimensions of $2k \times 2k$. We denote 2×2 blocks of $C^{(k)}$ as $C^{(k)}_{ij}$

for i, j = 1,2, ..., k , where $C^{(k)}_{ij}$ is the 2-by-2 block of $C^{(k)}$ consisting of the four elements of $C^{(k)}$ that are individually located in row $2i - 1$ or $2i$ and column $2j - 1$ or $2j$.

The above recursion is now used to compute the block elements of $C^{(3)}$. Then each of these block elements is, in turn, expressed in terms of block elements of $C^{(2)}$. These calculations are shown in detail below where we have also used the fact that the mean of the z_n vector is zero:

$$C^{(3)}_{ij} = C^{(2)}_{ij} : i = 1,2; \; j = 1,2$$
$$\bar{z}_{n+2} = D\bar{z}_{n+1} + S\bar{u}_{n+1}$$
$$= D[D\bar{z}_n + S\bar{u}_n] + S\bar{u}_{n+1} = D^2\bar{z}_n + DS\bar{u}_n + S\bar{u}_{n+1} \tag{19}$$

$$C^{(3)}_{13} = E[\bar{z}_n\bar{z}^T_{n+2}] = E\left[\bar{z}_n\left(D^2\bar{z}_n + DS\bar{u}_n + S\bar{u}_{n+1}\right)^T\right] = \Sigma_{11}\left[D^2\right]^T$$
$$C^{(3)}_{31} = \left[C^{(3)}_{13}\right]^T \tag{20}$$

$$C^{(3)}_{23} = E[\bar{z}_{n+1}\bar{z}^T_{n+2}] = E\left[(D\bar{z}_n + S\bar{u}_n)\left(D^2\bar{z}_n + DS\bar{u}_n + S\bar{u}_{n+1}\right)^T\right]$$
$$= D\Sigma_{11}\left[D^2\right]^T + C_c D^T = DC^{(3)}_{13} + C_c D^T \tag{21}$$
$$C^{(3)}_{32} = \left[C^{(3)}_{23}\right]^T$$

$$C^{(3)}_{33} = E[\bar{z}_{n+2}\bar{z}^T_{n+2}] = E\left[\left(D^2\bar{z}_n + DS\bar{u}_n + S\bar{u}_{n+1}\right)\left(D^2\bar{z}_n + DS\bar{u}_n + S\bar{u}_{n+1}\right)^T\right]$$
$$= D^2\Sigma_{11}\left[D^2\right]^T + DC_c D^T + C_c = DC^{(3)}_{23} + C_c. \tag{22}$$

By continuation of this calculation to larger timestamp blocks (k > 3), we find the following pattern that can be used to extend (augment) $C^{(k-1)}$ to yield $C^{(k)}$. The pattern consists of setting most of the augmented matrix equal to that of the previous one, and then computing two additional rows and columns for $C^{(k)}$, k > 2, to fill out the remaining elements. The general expressions are:

$$C^{(k)}_{m,n} = C^{(k-1)}_{m,n} : m, n = 1, 2, ..., k-1$$
$$C^{(k)}_{1k} = \Sigma_{11}\left[D^{k-1}\right]^T$$
$$C^{(k)}_{ik} = DC^{(k)}_{i-1,k} + C_c\left[D^{k-i}\right]^T : i = 2, 3, ... k \tag{23}$$
$$C^{(k)}_{ki} = \left[C^{(k)}_{ik}\right]^T : i = 1, 2, ..., k.$$

At this point in the development we have shown how to compute the covariance matrix:

$$C^{(k)} = \text{cov}\left(\bar{z}^{(k)}\right) = \text{cov}\left(\left[\begin{array}{cccccc} x_n & y_n & x_{n+1} & y_{n+1} & \cdots & x_{n+k-1} \quad y_{n+k-1} \end{array}\right]^T\right) \tag{24}$$

Since the system is linear and the process noise w_n and measurement noise v_n are white zero-mean Gaussian noise processes, we may express the joint probability density function for the 2k variates as:

$$f\left(\bar{z}^{(k)}\right) = pdf\left(\bar{z}^{(k)}\right) = pdf\left(\left[\begin{array}{cccccc} x_n & y_n & x_{n+1} & y_{n+1} & \cdots & x_{n+k-1} \quad y_{n+k-1} \end{array}\right]\right) = \frac{\exp\left\{-\frac{1}{2}\left[\bar{z}^{(k)}\right]^T\left[C^{(k)}\right]^{-1}\left[\bar{z}^{(k)}\right]\right\}}{(2\pi)^{\frac{n}{2}}\left(\det[C^{(k)}]\right)^{\frac{1}{2}}} \tag{25}$$

Note that the mean of all 2k variates is zero.

Finally, to obtain empirical confirmation of the equivalence of the covariance terms obtained using the original AR system and its recursive representation, numerical simulations were conducted. Using the example 1 system (below) 500 sequences were generated each of length one million. For each sequence the $C^{(3)}$ covariance was computed. The error for all $C^{(3)}$ matrices was then averaged, assuming that the $C^{(3)}$ matrix calculated using the method based on the recursive representation was the true value. The result was that for each of the matrix elements, the error was less than 0.0071% of its true value. We are now in position to compute transfer entropies for a couple of illustrative examples.

4. Example 1: A One-Way Coupled System

For this example we consider the following system:

$$
\begin{aligned}
x_{n+1} &= a x_n + w_n : w_n \sim N(0, Q) \\
y_{n+1} &= h_c x_{n+1} + v_n : v_n \sim N(0, R)
\end{aligned}
\tag{26}
$$

Parameter hc specifies the coupling strength of the Y process to the first-order AR process X, and R and Q are their respective (wn and vn) zero-mean Gaussian process noise variances. For stability, we require $|a| < 1$. Comparing to the first general representation given above, we have $m = 0$, $a_0 = a$, and $c_{-1} = h_c a$. The system models filtered noise x_n and a noisy measurement, y_n, of x_n. Thus the x_n sequence represents a hidden process (or model) which is observable by way of another sequence, y_n. We wish to examine the behavior of transfer entropy as a function of the correlation ρ between x_n and y_n. One might expect that the correlation ρ between x_n and y_n to be proportional of the degree of information flow; however, we will see that the relationship between transfer entropy and correlation is not quite that simple.

Both the X and Y processes have zero mean. Computing the joint covariance matrix $C^{(1)}$ for x_n and y_n and their correlation we obtain:

$$
\begin{aligned}
Var(x_n) &= \frac{Q}{1 - a^2} \\
Var(y_n) &= h_c^2 Var(x_n) + R \\
E(x_n y_n) &= h_c Var(x_n) \\
\rho &\equiv \frac{E(x_n y_n)}{\sqrt{Var(x_n) Var(y_n)}}
\end{aligned}
\tag{27}
$$

Hence the process covariance matrix $C^{(1)}$ corresponding to a single timestamp, t_n is:

$$
C^{(1)} \equiv \mathrm{cov}\left(\begin{bmatrix} x_n \\ y_n \end{bmatrix}\right) = \begin{bmatrix} Var(x_n) & h Var(x_n) \\ h_c Var(x_n) & h_c^2 Var(y_n) + R \end{bmatrix}.
\tag{28}
$$

In order to obtain an expanded covariance matrix, accounting for two time instances (t_n and t_{n+1}) we compute the additional expectations required to fill in the matrix $C^{(2)}$:

$$
C^{(2)} \equiv \mathrm{cov}\left(\begin{bmatrix} x_n \\ y_n \\ x_{n+1} \\ y_{n+1} \end{bmatrix}\right) = \begin{bmatrix}
Var(x_n) & h_c Var(x_n) & a Var(x_n) & h_c a Var(x_n) \\
h_c Var(x_n) & h_c^2 Var(x_n) + R & h_c a Var(x_n) & h_c^2 a Var(x_n) \\
a Var(x_n) & h_c a Var(x_n) & Var(x_n) & h_c Var(x_n) \\
h_c a Var(x_n) & h_c^2 a Var(x_n) & h_c Var(x_n) & h_c^2 Var(x_n) + R
\end{bmatrix}.
\tag{29}
$$

Thus we have found the covariance matrix $C^{(2)}$ required to compute block entropies based on a single time lag. Using this matrix the single-lag transfer entropies may be computed. Using the recursive process described in the previous section we can compute $C^{(1\circ)}$. We have found that using higher lags does not change the entropy values significantly.

To aid the reader in understanding the calculations required to compute transfer entropies using higher time lags, it is worthwhile to compute transfer entropy for a single lag. We first define transfer entropy using general notation indicating the partitioning of the X and Y sequences in to past and future $\left(\overleftarrow{x}, \overrightarrow{x}\right)$ and $\left(\overleftarrow{y}, \overrightarrow{y}\right)$, respectively. We then compute transfer entropy as a sum of block entropies:

$$
\begin{aligned}
TE_{x \to y} &= I\left(\overleftarrow{x}; \overrightarrow{y} \middle| \overleftarrow{y}\right) = h\left(\overleftarrow{x} \middle| \overleftarrow{y}\right) + h\left(\overrightarrow{y} \middle| \overleftarrow{y}\right) - h\left(\overleftarrow{x}; \overrightarrow{y} \middle| \overleftarrow{y}\right) \\
&= \left[h\left(\overleftarrow{x}, \overleftarrow{y}\right) - h\left(\overleftarrow{y}\right)\right] + \left[h\left(\overleftarrow{y}, \overrightarrow{y}\right) - h\left(\overleftarrow{y}\right)\right] - \left[h\left(\overleftarrow{x}, \overleftarrow{y}, \overrightarrow{y}\right) - h\left(\overleftarrow{y}\right)\right] \\
&= h\left(\overleftarrow{x}, \overleftarrow{y}\right) + h\left(\overleftarrow{y}, \overrightarrow{y}\right) - h\left(\overleftarrow{y}\right) - h\left(\overleftarrow{x}, \overleftarrow{y}, \overrightarrow{y}\right).
\end{aligned}
\tag{30}
$$

Similarly:

$$TE_{y->x} = I\left(\overleftarrow{y}; \overrightarrow{x} \middle| \overleftarrow{x}\right) = h\left(\overleftarrow{x}, \overleftarrow{y}\right) + h\left(\overleftarrow{x}, \overrightarrow{x}\right) - h\left(\overleftarrow{x}\right) - h\left(\overleftarrow{x}, \overleftarrow{y}, \overrightarrow{x}\right) \tag{31}$$

The Y states have no influence on the X sequence in this example. Hence $TE_{y\to x} = 0$. Since we are here computing transfer entropy for a single lag (*i.e.*, two time tags t_n and t_{n+1}) we have:

$$TE^{(2)}_{x->y} = I(x_n; y_{n+1}|y_n) = h(x_n, y_n) + h(y_n, y_{n+1}) - h(y_n) - h(x_n, y_n, y_{n+1}) \tag{32}$$

By substitution of the expression for the differential entropy of each block we obtain:

$$\begin{aligned}
TE^{(2)}_{x->y} &= \tfrac{1}{2}\log\left[(2\pi e)^2 \det C^{(2)}_{[1,2],[1,2]}\right] + \tfrac{1}{2}\log\left[(2\pi e)^2 \det C^{(2)}_{[2,4],[2,4]}\right] - \\
&\quad \tfrac{1}{2}\log\left[(2\pi e)^1 \det C^{(2)}_{[2],[2]}\right] - \tfrac{1}{2}\log\left[(2\pi e)^3 \det C^{(2)}_{[1,2,4],[1:,2,4]}\right] \\
&= \tfrac{1}{2}\log\left[\frac{\det C^{(2)}_{[1,2],[1,2]}\det C^{(2)}_{[2,4],[2,4]}}{\det C^{(2)}_{[2],[2]}\det C^{(2)}_{[1,2,4],[1,2,4]}}\right].
\end{aligned} \tag{33}$$

For this example, note from the equation for y_{n+1} that state x_{n+1} is a causal state of X influencing the value of y_{n+1}. In fact, it is the most important such state. To capture the full information that is transferred from the X process to the Y process over the course of two time tags we need to include state x_{n+1}. Hence we compute the information transfer from $x \to y$ as:

$$IT^{(2)}_{x->y} = I(x_n, x_{n+1}; y_{n+1}|y_n) = h(x_n, x_{n+1}, y_n) + h(y_n, y_{n+1}) - h(y_n) - h(x_n, x_{n+1}, y_n, y_{n+1}) \tag{34}$$

$$\begin{aligned}
IT^{(2)}_{x->y} &= \tfrac{1}{2}\log\left[(2\pi e)^3 \det C^{(2)}_{[1,2,3],[1,2,3]}\right] + \tfrac{1}{2}\log\left[(2\pi e)^2 \det C^{(2)}_{[2,4],[2,4]}\right] - \\
&\quad \tfrac{1}{2}\log\left[(2\pi e)^1 \det C^{(2)}_{[2],[2]}\right] - \tfrac{1}{2}\log\left[(2\pi e)^4 \det C^{(2)}_{[1:4],[1:4]}\right] \\
&= \tfrac{1}{2}\log\left[\frac{\det C^{(2)}_{[1:,2,3],[1,2,3]}\det C^{(2)}_{[2,4],[2,4]}}{\det C^{(2)}_{[2],[2]}\det C^{(2)}_{[1:4],[1:4]}}\right].
\end{aligned} \tag{35}$$

Here the notation $\det C^{(2)}_{[i],[i]}$ indicates the determinant of the matrix composed of the rows and columns of $C^{(2)}$ indicated by the list of indices i shown in the subscripted brackets. For example, $\det C^{(2)}_{[1:4],[1:4]}$ is the determinant of the matrix formed by extracting columns {1, 2, 3, 4} and rows {1, 2, 3, 4} from matrix $C^{(2)}$. In later calculations we will use slightly more complicated-looking notation. For example, $\det C^{(10)}_{[2:2:20],[2:2:20]}$ is the determinant of the matrix formed by extracting columns {2, 4 , . . . , 18, 20} and the same-numbered rows from matrix $C^{(1\circ)}$. (Note $C^{(k)}_{[i],[i]}$ is not the same as $C^{(k)}_{ii}$ as used in Section 3).

It is interesting to note that a simplification in the expression for information transfer can be obtained by writing the expression for it in terms of conditional entropies:

$$IT^{(2)}_{x->y} = I(x_n, x_{n+1}; y_{n+1}|y_n) = h(y_{n+1}|y_n) - h(y_{n+1}|x_n, y_n, x_{n+1}) \tag{36}$$

From the fact that $y_{n+1} = x_{n+1} + v_{n+1}$ we see immediately that:

$$h(y_{n+1}|x_n, y_n, x_{n+1}) = h(v_{n+1}) = \frac{1}{2}\log(2\pi e R). \tag{37}$$

Hence we may write:

$$
\begin{aligned}
IT^{(2)}_{x->y} &= h(y_{n+1}|y_n) - h(y_{n+1}|x_n, y_n, x_{n+1}) \\
&= \tfrac{1}{2}\log\left[\frac{2\pi e \det C^{(2)}_{[2,4],[2,4]}}{\det C^{(2)}_{[2],[2]}}\right] - \tfrac{1}{2}\log[2\pi e R] \\
&= \tfrac{1}{2}\log\left[\frac{\det C^{(2)}_{[2,4],[2,4]}}{R\det C^{(2)}_{[2],[2]}}\right].
\end{aligned}
\tag{38}
$$

To compute transfer entropy using nine lags (ten timestamps) assume that we have already computed $C^{(10)}$ as defined above. We partition the sequence $\{\bar{z}^T_{n+i}\}^9_{i=0} = \{x_n, y_n, x_{n+1}, y_{n+1}, x_{n+2}, y_{n+2}, x_{n+3}, y_{n+3}, x_{n+4}, y_{n+4}, x_{n+5}, y_{n+5}, x_{n+6}, y_{n+6}, x_{n+7}, y_{n+7}, x_{n+8}, y_{n+8}, x_{n-9}, y_{n+9}\}$ into three subsets:

$$
\begin{aligned}
\overleftarrow{x} &\equiv \{x_n, x_{n+1}, \dots, x_{n+8}\} \\
\overleftarrow{y} &\equiv \{y_n, y_{n+1}, \dots, y_{n+8}\} \\
\overrightarrow{y} &\equiv \{y_{n+9}\}.
\end{aligned}
\tag{39}
$$

Now, using these definitions, and substituting in expressions for differential block entropies we obtain:

$$
\begin{aligned}
TE^{(10)}_{x->y} &= I\left(\overleftarrow{x};\overrightarrow{y}\,\middle|\,\overleftarrow{y}\right) = h\left(\overleftarrow{x},\overleftarrow{y}\right) + h\left(\overleftarrow{y},\overrightarrow{y}\right) - h\left(\overleftarrow{y}\right) - h\left(\overleftarrow{x},\overleftarrow{y},\overrightarrow{y}\right) \\
&= \tfrac{1}{2}\log\left[(2\pi e)^{18}\det C^{(10)}_{[1:18],[1:18]}\right] + \tfrac{1}{2}\log\left[(2\pi e)^{10}\det C^{(10)}_{[2:2:20],[2:2:20]}\right] - \\
&\quad \tfrac{1}{2}\log\left[(2\pi e)^9\det C^{(10)}_{[2:2:18],[2:2:18]}\right] - \tfrac{1}{2}\log\left[(2\pi e)^{19}\det C^{(10)}_{[1:18,20],[1:18,20]}\right] \\
&= \tfrac{1}{2}\log\left[\frac{\det C^{(10)}_{[1:18],[1:18]}\det C^{(10)}_{[2:2:20],[2:2:20]}}{\det C^{(10)}_{[2:2:18],[2:2:18]}\det C^{(10)}_{[1:18,20],[1:18,20]}}\right].
\end{aligned}
\tag{40}
$$

Similarly:

$$
IT^{(10)}_{x->y} = h\left(\overrightarrow{y}\,\middle|\,\overleftarrow{y}\right) - h\left(\overrightarrow{y}\,\middle|\,\overleftarrow{y},\overleftarrow{x},x_{n+1}\right) = \frac{1}{2}\log\left[\frac{\det C^{(10)}_{[2:2:20],[2:2:20]}}{R\det C^{(10)}_{[2:2:18],[2:2:18]}}\right].
\tag{41}
$$

As a numerical example we set $h_c = 1$, $Q = 1$, and for three different values of a (0.5, 0.7 and 0.9) we vary R so as to scan the correlation ρ between the x and y processes between the values of 0 and 1.

In Figure 1 it is seen that for each value of parameter a there is a peak in the transfer entropy $TE^{(k)}_{x\to y}$. As the correlation ρ between x_n and y_n increases from a low value the transfer entropy increases since the amount of information shared between y_{n+1} and x_n is increasing. At a critical value of ρ transfer entropy peaks and then starts to decrease. This decrease is due to the fact that at high values of ρ the measurement noise variance R is small. Hence y_n becomes very close to equaling x_n so that the amount of information gained (about y_{n+1}) by learning x_n, given y_n, becomes small. Hence $h(y_{n+1} \mid y_n) - h(y_{n+1} \mid y_n, x_n)$ is small. This difference is $TE^{(2)}_{x\to y}$.

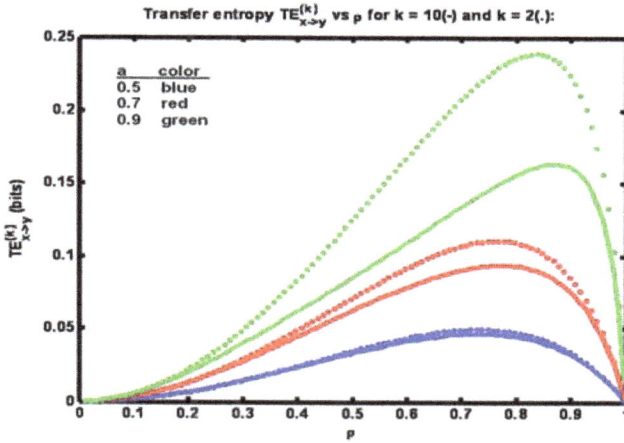

Figure 1. Example 1: Transfer entropy $TE^{(k)}_{x \to y}$ *versus* correlation coefficient ρ for three values of parameter a (see legend). Solid trace: $k = 10$, dotted trace: $k = 2$.

The relationship between ρ and R is shown in Figure 2. Note that when parameter a is increased, a larger value of R is required to maintain ρ at a fixed value. Also, in Figure 1 we see the effect of including more timetags in the analysis. When k is increased from 2 to 10 transfer entropy values fall, particularly for the largest value of parameter a. It is known that entropies decline when conditioned on additional variables. Here, transfer entropy is acting similarly. In general, however, transfer entropy, being a mutual information quantity, has the property that conditioning could make it increase as well [12].

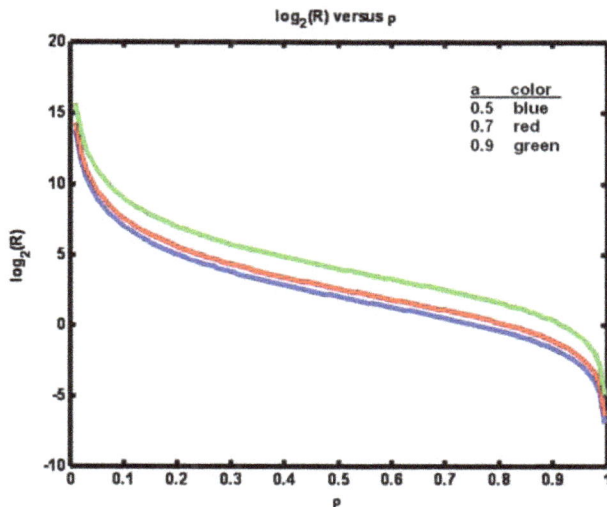

Figure 2. Example 1: Logarithm of R *versus* ρ for three values of parameter a (see legend).

The observation that the transfer entropy decrease is greatest for the largest value of parameter a is perhaps due to the fact that the entropy of the X process is itself greatest for the largest a value and therefore has more sensitivity to an increase in X data availability (Figure 3).

Entropy **2013**, *15*, 767–786

From Figure 1 it is seen that as the value of parameter *a* is increased, transfer entropy is increased for a fixed value of ρ. The reason for this increase may be gleaned from Figure 3 where it is clear that the amount of information contained in the x process, H_X, is greater for larger values of *a*. Hence more information is available to be transferred at the fixed value of ρ when *a* is larger. In the lower half of Figure 3 we see that as ρ increases the entropy of the Y process, H_Y, approaches the value of H_X. This result is due to the fact that the mechanism being used to increase ρ is to decrease R. Hence as R drops close to zero y_n looks increasingly identical to x_n (since $h_c = 1$).

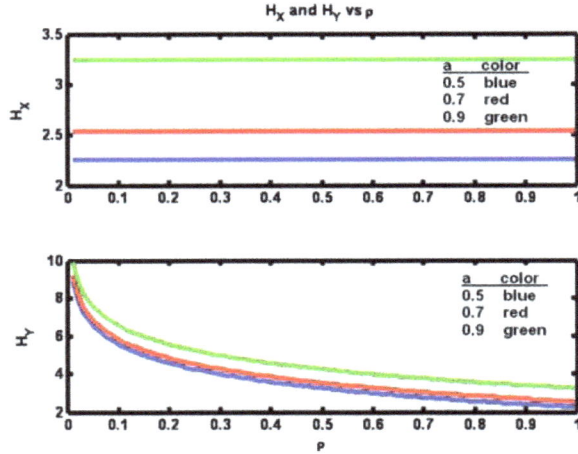

Figure 3. Example 1: Process entropies H_X and H_Y *versus* correlation coefficient ρ for three values of parameter *a* (see legend).

Figure 4 shows information transfer $IT^{(k)}{}_{x\to y}$ plotted *versus* correlation coefficient ρ. Now note that the trend is for information transfer to increase as ρ is increased over its full range of values. °

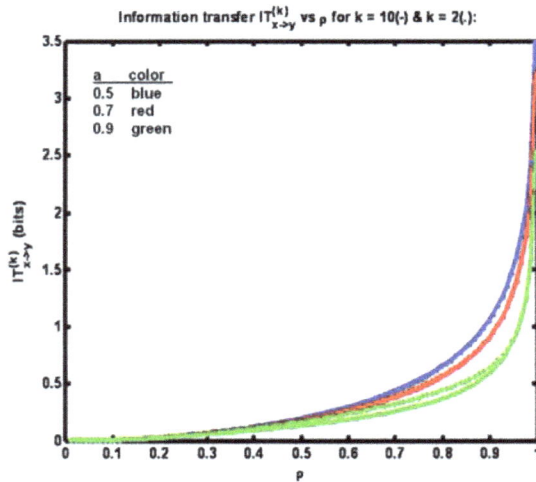

Figure 4. Example 1: Information transfer $IT^{(k)}{}_{x\to y}$ *versus* correlation coefficient ρ for three different values of parameter *a* (see legend) for k = 10 (solid trace) and k = 2 (dotted trace).

This result is obtained since as ρ is increased y_{n+1} becomes increasingly correlated with x_{n+1}. Also, for a fixed ρ, the lowest information transfer occurs for the largest value of parameter a. We obtain this result since at the higher a values x_n and x_{n+1} are more correlated. Thus the benefit of learning the value of y_{n+1} through knowledge of x_{n+1} is relatively reduced, given that y_n (itself correlated with x_n) is presumed known. Finally, we have $IT^{(10)}_{x \to y} < IT^{(2)}_{x \to y}$ since conditioning the entropy quantities comprising the expression for information transfer with more state data acts to reduce their difference. Also, by comparison of Figures 2 and 4, it is seen that information transfer is much greater than transfer entropy. This relationship is expected since information transfer as defined herein (for $k = 2$) is the amount of information that is gained about y_{n+1} from learning x_{n+1} and x_n, given that y_n is already known. Whereas transfer entropy (for $k = 2$) is the information gained about y_{n+1} from learning only x_n, given that y_n is known. Since the state y_{n+1} in fact equals x_{n+1}, plus noise, learning x_{n+1} is highly informative, especially when the noise variance is small (corresponding to high values of ρ). The difference between transfer entropy and information transfer therefore quantifies the benefit of learning x_{n+1}, given that x_n and y_n are known (when the goal is to determine y_{n+1}).

Figure 5 shows how information transfer varies with measurement noise variance R. As R increases the information transfer decreases since measurement noise makes determination of the value of y_{n+1} from knowledge of x_n and x_{n+1} less accurate. Now, for a fixed R, the greatest value for information transfer occurs for the greatest value of parameter a. This is the opposite of what we obtained for a fixed value of ρ as shown in Figure 4. The way to see the rationale for this is to note that, for a fixed value of information transfer, R is highest for the largest value of parameter a. This result is obtained since larger values of a yield the most correlation between states x_n and x_{n+1}. Hence, even though the measurement y_{n+1} of x_{n+1} is more corrupted by noise (due to higher R), the same information transfer is achieved nevertheless, because x_n provides a good estimate of x_{n+1} and, thus, of y_{n+1}.

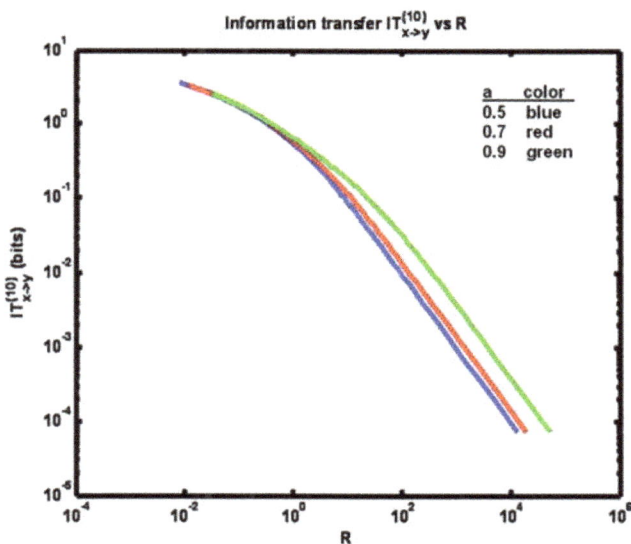

Figure 5. Example 1: Information transfer $IT^{(10)}_{x \to y}$ *versus* measurement error variance R for three different values of parameter a (see legend).

5. Example 2: Information-theoretic Analysis of Two Coupled AR Processes.

In example 1 the information flow was unidirectional. We now consider a bidirectional example achieved by coupling two AR processes. One question we may ask in such a system is how transfer entropies change with variations in correlation and coupling coefficient parameters. It might be anticipated that increasing either of these quantities will have the effect of increasing information flow and thus transfer entropies will increase.

The system is defined by the equations:

$$
\begin{aligned}
x_{n+1} &= ax_n + by_n + w_n : w_n \sim N(0, Q) \\
y_{n+1} &= cx_n + dy_n + v_n : v_n \sim N(0, R).
\end{aligned}
\tag{42}
$$

For stability, we require that the eigenvalues of the constant matrix $\begin{bmatrix} a & b \\ c & d \end{bmatrix}$ lie in the unit circle. The means of processes X and Y are zero. The terms w_n and v_n are the X and Y processes noise terms respectively. Using the following definitions:

$$
\begin{aligned}
\lambda_0 &\equiv 1 + ad - bc \\
\lambda_1 &\equiv 1 - ad - bc \\
\psi_a &\equiv (1 - ad)(1 - a^2) - bc(1 + a^2) \\
\psi_d &\equiv (1 - ad)(1 - d^2) - bc(1 + d^2) \\
\tau &\equiv \psi_a \psi_d - b^2 c^2 \lambda_0^2 \\
\eta_{x1} &\equiv \lambda_1 \psi_d / \tau \\
\eta_{x2} &\equiv b^2 \lambda_0 \lambda_1 / \tau \\
\eta_{y1} &\equiv c^2 \lambda_0 \lambda_1 / \tau \\
\eta_{y2} &\equiv \lambda_1 \psi_a / \tau
\end{aligned}
\tag{43}
$$

we may solve for the correlation coefficient ρ between x_n and y_n to obtain:

$$
\begin{bmatrix} Var(x_n) \\ Var(y_n) \end{bmatrix} = \begin{bmatrix} \eta_{x1} & \eta_{x2} \\ \eta_{y1} & \eta_{y2} \end{bmatrix} \begin{bmatrix} Q \\ R \end{bmatrix}.
\tag{44}
$$

$$
\begin{aligned}
C_{[xy]} &\equiv cov\left(\begin{bmatrix} x_n \\ y_n \end{bmatrix} \right) = \begin{bmatrix} Var(x_n) & \xi \\ \xi & Var(y_n) \end{bmatrix} \\
\xi &\equiv E[x_n y_n] = \frac{b(d\psi_a + abc\lambda_0)R + c(a\psi_d + bcd\lambda_0)Q}{\psi_a \psi_d - b^2 c^2 \lambda_0^2} \\
\rho &= \frac{\xi}{\sqrt{Var(x_n)Var(y_n)}}.
\end{aligned}
\tag{45}
$$

Now, as we did previously in example 1 above, compute the covariance $C^{(2)}$ of the variates obtained at two consecutive timestamps to yield:

$$
C^{(2)} \equiv cov\left(\begin{bmatrix} x_n \\ y_n \\ x_{n+1} \\ y_{n+1} \end{bmatrix} \right) = \begin{bmatrix} Var(x_n) & \xi & aVar(x_n) + b\xi & cVar(x_n) + d\xi \\ \xi & Var(y_n) & bVar(y_n) + a\xi & dVar(y_n) + c\xi \\ aVar(x_n) + b\xi & bVar(y_n) + a\xi & Var(x_n) & \xi \\ cVar(x_n) + d\xi & dVar(y_n) + c\xi & \xi & Var(y_n) \end{bmatrix}.
\tag{46}
$$

At this point the difficult part is done and the same calculations can be made as in example 1 to obtain $C^{(k)}$; $k = 3,4, \ldots, 10$ and transfer entropies. For illustration purposes, we define the parameters of the system as shown below, yielding a symmetrically coupled pair of processes. To generate a family of curves for each transfer entropy we choose a fixed coupling term ε from a set of four values. We set Q = 1000 and vary R so that ρ varies from about 0 to 1. For each ρ value we compute the transfer entropies. The relevant system equations and parameters are:

$$\begin{aligned}
x_{n+1} &= \left(\tfrac{1}{2} - \varepsilon\right)x_n + \varepsilon y_n + w_n : w_n \sim N(0, Q) \\
y_{n+1} &= \varepsilon x_n + \left(\tfrac{1}{2} - \varepsilon\right)y_n + v_n : w_n \sim N(0, R) \\
\varepsilon &\in \{0.1, 0.2, 0.3, 0.4\} \\
Q &= 1000.
\end{aligned} \tag{47}$$

Hence, we make the following substitutions to compute $C^{(2)}$:

$$\begin{aligned}
a &= \left(\tfrac{1}{2} - \varepsilon\right) \\
b &= \varepsilon \\
c &= \varepsilon \\
d &= \left(\tfrac{1}{2} - \varepsilon\right).
\end{aligned} \tag{48}$$

For each parameter set $\{\varepsilon, Q, R\}$ there is a maximum possible ρ, ρ_∞ obtained by taking the limit as R→ ∞ of the expression for ρ given above. Doing so, we obtain:

$$\rho_\infty = \frac{\phi_1 \phi_2 + \phi_3}{\sqrt{\phi_1(\phi_1 \mu_1 + 1)}} \tag{49}$$

where:

$$\begin{aligned}
\phi_1 &\equiv \frac{2ab^2 d + b^2 \lambda_1}{(1 - a^2 - b^2 \mu_1)\lambda_1 - 2ab(ac + bd\mu_1)} \\
\phi_2 &\equiv \frac{ac + bd\mu_1}{\lambda_1} \\
\phi_2 &\equiv \frac{bd}{\lambda_1}
\end{aligned} \tag{50}$$

$$\begin{aligned}
\lambda_1 &\equiv 1 - ad - bc \\
\mu_1 &\equiv \frac{c^2 \lambda_1 + 2ac^2 d}{(1 - d^2)\lambda_1 - 2bcd^2}.
\end{aligned} \tag{51}$$

There is a minimum value of ρ also. The corresponding value for R, R_{min}, was found by means of the inbuilt Matlab program *fminbnd*. This program is designed to find the minimum of a function in this case $\rho(a, b, c, d, R, Q))$ with respect to one parameter (in this case R) starting from an initial guess (here, R = 500). The program returns the minimum functional value (ρ_{min}) and the value of the parameter at which the minimum is achieved (R_{min}). After identifying R_{min} a set of R values were computed so that the corresponding set of ρ values spanned from ρ_{min} to the maximum ρ_∞ in fixed increments of Δρ (here equal to 0.002). This set of R values was generated using the iteration:

$$R_{new} = R_{old} + \Delta R = R_{old} + \left.\left(\frac{\partial \rho}{\partial R}\right)^{-1}\right|_{R=R_{old}} \Delta \rho \tag{52}$$

For the four selections of parameter ε we obtain the functional relationships shown in Figure 6.

From Figure 6 we see that for a fixed ε, increasing R increases (or decreases) ρ depending on whether R is less than (or greater than) Q (Q = 1000). Note that large increases in R > Q are required to marginally increase ρ when ρ nears its maximum value. The reason that the minimum ρ value occurs when Q equals R is because whenever they are unequal one of the processes dominates the other, leading to increased correlation. Also, note that if R << Q, then increasing ε will cause ρ to decrease since increasing the coupling will cause the variance of the y process $Var(y_n)$, a term appearing in the denominator of the expression for ρ, to increase. If Q << R, a similar result is obtained when ε is increased.

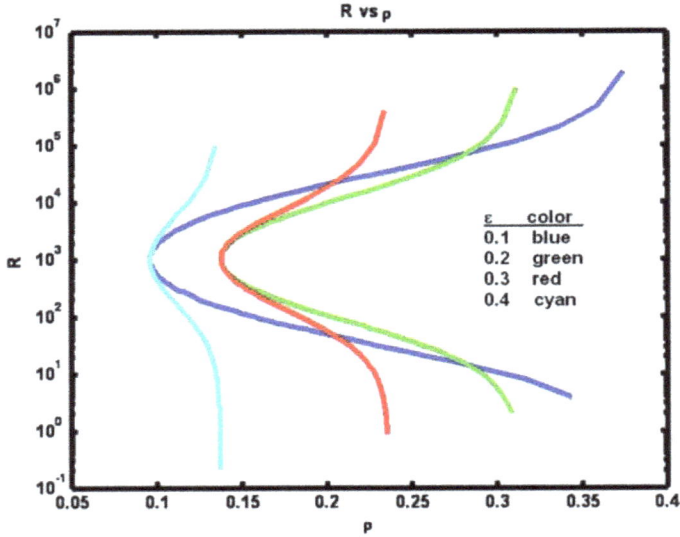

Figure 6. Example 2: Process noise variance R *versus* correlation coefficient ρ for a set of ε parameter values (see figure legend).

Transfer entropies in both directions are shown in Figure 7. Fixing ε, we note that as R is increased from a low value both ρ and $TE_{x->y}$ initially decrease while $TE_{y->x}$ increases. Ther. for further increases of R, ρ reaches a minimum value then begins to increase, while $TE_{x \to y}$ continues to decrease and $TE_{y \to x}$ continues to increase.

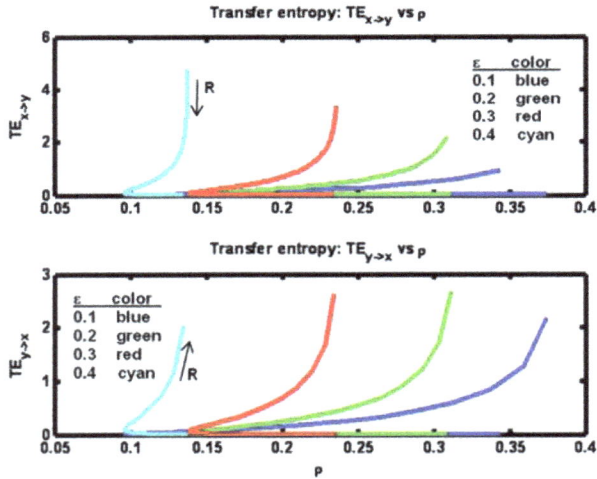

Figure 7. Example 2: Transfer entropy values *versus* correlation ρ for a set of ε parameter values (see figure legend). Arrows indicate direction of increasing R values.

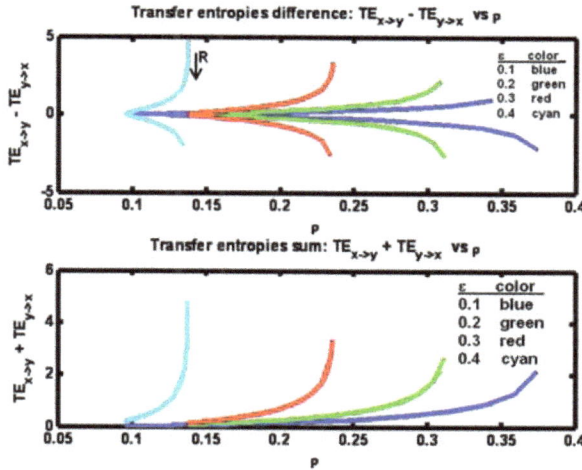

Figure 8. Example 2: Transfer entropies difference ($TE_{x->y} - TE_{y->x}$) and sum ($TE_{x->y} + TE_{y->x}$) *versus* correlation ρ for a set of ε parameter values (see figure legend). Arrow indicates direction of increasing R values.

By plotting the difference $TE_{x\to y} - TE_{y\to x}$ in Figure 8 we see the symmetry that arises as R increases from a low value to a high value. What is happening is that when R is low, the X process dominates the Y process so that $TE_{x\to y} > TE_{y\to x}$. As R increases, the two entropies equilibrate. Then, as R rises above Q, the Y process dominates giving $TE_{x\to y} < TE_{y\to x}$. The sum of the transfer entropies shown in Figure 8 reveal that the total information transfer is minimal at the minimum value of ρ and increases monotonically with ρ. The minimum value for ρ in this example occurs when the process noise variances Q and R are equal (matched). Figure 9 shows the changes in the transfer entropy values explicitly as a function of R. Clearly, when R is small (as compared to Q = 1000), $TE_{x\to y} > TE_{y\to x}$. Also it is clear that at every fixed value of R, both transfer entropies are higher at the larger values for the coupling term ε.

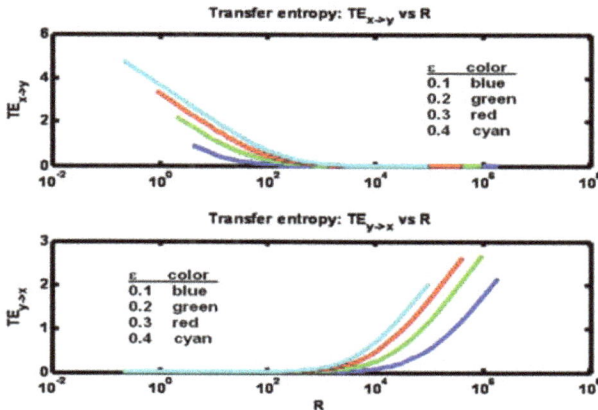

Figure 9. *Example 2:* Transfer entropies $TE_{x\to y}$ and $TE_{y\to x}$ *versus* process noise variance R for a set of ε parameter values (see figure legend).

Another informative view is obtained by plotting one transfer entropy value *versus* the other as shown in Figure 10.

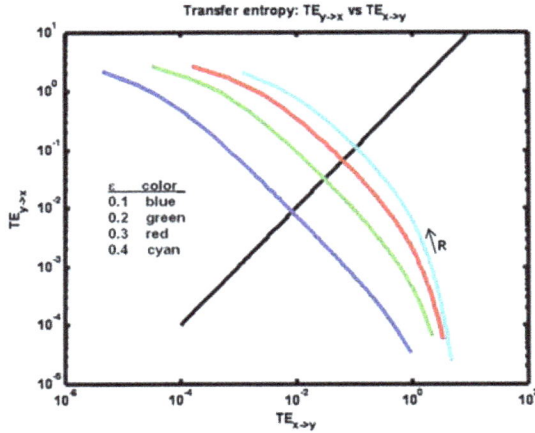

Figure 10. Example 2: Transfer entropy $TE_{x->y}$ plotted *versus* $TE_{y->x}$ for a set of ε parameter values (see figure legend). The black diagonal line indicates locations where equality obtains. Arrow indicates direction of increasing R values.

Here it is evident how $TE_{y \to x}$ increases from a value less than $TE_{x \to y}$ to a value greater than $TE_{x \to y}$ as R increases. Note that for higher coupling values ε this relative increase is more abrupt.

Finally, we consider the sensitivity of the transfer entropies to the coupling term ε. We reprise example system 2 where now ε is varied in the interval $(0, \frac{1}{2})$ and three values of R (somewhat arbitrarily selected to provide visually appealing figures to follow) are considered:

$$
\begin{aligned}
x_{n+1} &= \left(\tfrac{1}{2} - \varepsilon_x\right)x_n + \varepsilon_x y_n + w_n : w_n \sim N(0, Q) \\
y_{n+1} &= \varepsilon_y x_n + \left(\tfrac{1}{2} - \varepsilon_y\right)y_n + v_n : w_n \sim N(0, R) \\
R &\in \left\{10^0, 10^3, 10^4\right\} \\
Q &= 10^3.
\end{aligned}
\tag{53}
$$

Figure 11 shows the relationship between ρ and ε, where $\varepsilon_x = \varepsilon_y = \varepsilon$ for the three R values. Note that for the case R = Q the relationship is symmetric around $\varepsilon = \frac{1}{4}$. As R departs from equality more correlation between x_n and y_n is obtained.

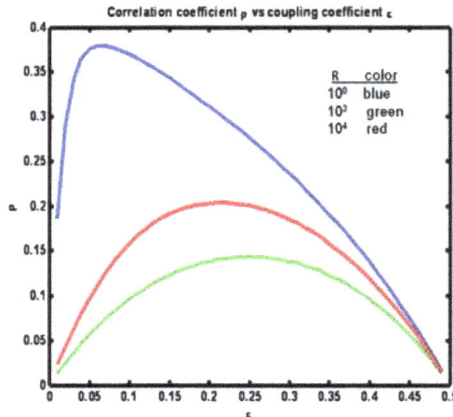

Figure 11. Example 2: Correlation coefficient ρ vs coupling coefficient ε for a set of R values (see figure legend).

The reason for this increase is that when the noise driving one process is greater in amplitude than the amplitude of the noise driving the other process, the first process becomes dominant over the other. This domination increases as the disparity between the process noise variances increases (R *versus* Q). Note also that as the disparity increases, the maximum correlation occurs at increasingly lower values of the coupling term ε. As the disparity increases at fixed $\varepsilon = \frac{1}{4}$ the correlation coefficient ρ increases. However, the variance in the denominator of ρ can be made smaller and thus ρ larger, if the variance of either of the two processes can be reduced. This can be accomplished by reducing ε.

The sensitivities of the transfer entropies to changes in coupling term ε are shown in Figure 12. Consistent with intuition, all entropies increase with increasing ε. Also, when R < Q (blue trace) we have $TE_{x\to y} > TE_{y\to x}$ and the reverse for R > Q. (red). For R = Q, $TE_{x\to y} = TE_{y\to x}$ (green).

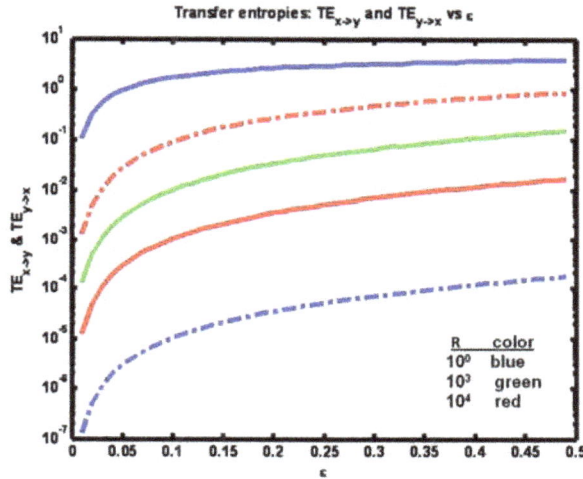

Figure 12. Example 2: Transfer entropies $TE_{x\to y}$ (solid lines) vs $TE_{y\to x}$ (dashed lines) vs coupling coefficient ε for a set of R values (see figure legend).

Finally, it is interesting to note that whenever we define three cases by fixing Q and varying the setting for R (one of R_1, R_2 and R_3 for each case) such that $R_1 < Q$, $R_2 = Q$ and $R_3 = Q^2/R_1$ (so that $R_{i+1} = QR_i/R_1$ for i = 1 and i = 2) we then obtain the symmetric relationships $TE_{x\text{-}>y}(R_1) = TE_{y\text{-}>x}(R_3)$ and $TE_{x\text{-}>y}(R_3) = TE_{y\text{-}>x}(R_1)$ for all ε in the interval $(1, \frac{1}{2})$. For these cases we also obtain $\rho(R_1) = \rho(R_3)$ on the same ε interval.

6. Conclusions

It has been shown how to compute transfer entropy values for Gaussian autoregressive processes for multiple timetags. The approach is based on the iterative computation of covariance matrices. Two examples were investigated: (1) a first-order filtered noise process whose state is measured with additive noise, and (2) two first-order symmetrically coupled processes each of which is driven by independent process noise. We found that, for the first example, increasing the first-order AR coefficient at a fixed correlation coefficient, transfer entropy increased since the entropy of the measured process was itself increased.

For the second example, it was discovered that the relationships between the coupling and correlation coefficients and the transfer entropies is more complicated. The minimum correlation coefficient occurs when the process noise variances match. It was seen that matching of these variances results in minimum information flow, expressed as the sum of both transfer entropies. Without a match, the transfer entropy is larger in the direction away from the process having the larger process noise. Fixing the process noise variances, transfer entropies in both directions increase with coupling strength ε.

Finally, it is worth noting that the method for computing covariance matrices for a variable number of timetags as presented here facilitates the calculation of many other information-theoretic quantities of interest. To this purpose, the authors have computed such quantities as crypticity [13] and normalized transfer entropy using the reported approach.

References

1. Schreiber, T. Measuring information transfer. *Phys. Rev. Lett.* **2000**, *85*, 461–464.
2. Barnett, L.; Barrett, A.B.; Seth, A.K. Granger causality and transfer entropy are equivalent for Gaussian variables. *Phys. Rev. Lett.* **2009**, *103*, 238701.
3. Ay, N.; Polani, D. Information Flows in Causal Networks. *Adv. Complex Syst.* **2008**, *11*, 17–41.
4. Lizier, J.T.; Prokopenko, M. Differentiating information transfer and causal effect. *Eur. Phys. J. B* **2010**, *73*, 605-615.
5. Chicharro, D.; Ledberg, A. When two become one: the limits of causality analysis of brain dynamics. *PLoS ONE* **2012**, *7*, e32466.
6. Hahs, D.W.; Pethel, S.D. Distinguishing anticipation from causality: anticipatory bias in the estimation of information flow. *Phys. Rev. Lett.* **2011**, *107*, 128701.
7. Gourevitch, B.; Eggermont, J.J. Evaluating information transfer between auditory cortical neurons. *J. Neurophysiol.* **2007**, *97*, 2533–2543.
8. Kaiser, A.; Schreiber, T. Information transfer in continuous processes. *Physica D* **2002**, *166*, 43–62.
9. Cover, T.M.; Thomas, J.A. *Elements of Information Theory*; Wiley Series in Telecommunications, Wiley: New York, NY, USA, 1991.
10. Kotz, S.; Balakrishnan, N.; Johnson, N.L. *Continuous Multivariate Distributions, Models and Applications*, 2nd ed.; John Wiley and Sons, Inc.: New York, NY, USA, 2000; Volume 1.
11. Lizier, J.T.; Prokopenko, M.; Zomaya, A.Y. Local information transfer as a spatiotemporal filter for complex systems. *Phys. Rev. E* **2008**, *77*, 026110.
12. Williams, P.L.; Beer, R.D. Nonnegative decomposition of multivariate information. 2010; arXiv:1004:2515.
13. Crutchfield, J.P; Ellison, C.J.; Mahoney, J.R. Time's barbed arrow: irreversibility, crypticity, and stored information. *Phys. Rev. Lett.* **2009**, *103*, 094101.

MDPI

Article

Assessing Coupling Dynamics from an Ensemble of Time Series

Germán Gómez-Herrero [1], Wei Wu [2], Kalle Rutanen [3], Miguel C. Soriano [4], Gordon Pipa [5] and Raul Vicente [6],*

[1] Netherlands Institute for Neuroscience, Meibergdreef 47, Amsterdam 1105 BA, The Netherlands;
E-Mail: german.gomezherrero@gmail.com

[2] Lab of Neurophysics and Neurophysiology, Hefei National Laboratory for Physical Sciences at the Microscale, University of Science and Technology of China, 96 JinZhai Rd., Hefei 230026, China;
E-Mail: wu@fias.uni-frankfurt.de

[3] Department of Mathematics, Tampere University of Technology, Korkeakoulunkatu 10, Tampere FI-33720, Finland; E-Mail: kalle.rutanen@tut.fi

[4] Instituto de Fisica Interdisciplinar y Sistemas Complejos (CSIC-UIB), Campus Universitat de les Illes Balears E-07122 Palma de Mallorca, Spain; E-Mail: miguel@ifisc.uib-csic.es

[5] Institut für Kognitionswissenschaft, University of Osnabrück, Albrechtstrasse 28, Osnabrück 49076 , Germany; E-Mail: gpipa@uos.de

[6] Institute of Computer Science, University of Tartu, J. Liivi 2, Tartu 50409, Estonia

* Author to whom correspondence should be addressed; E-Mail: raulvicente@gmail.com; Tel.: +372-737-5445; Fax: +372-737-5468.

Academic Editor: Deniz Gencaga

Received: 30 November 2014 / Accepted: 19 March 2015 / Published: 2 April 2015

Abstract: Finding interdependency relations between time series provides valuable knowledge about the processes that generated the signals. Information theory sets a natural framework for important classes of statistical dependencies. However, a reliable estimation from information-theoretic functionals is hampered when the dependency to be assessed is brief or evolves in time. Here, we show that these limitations can be partly alleviated when we have access to an ensemble of independent repetitions of the time series. In particular, we gear a data-efficient estimator of probability densities to make use of the full structure of trial-based measures. By doing so, we can obtain time-resolved estimates for a family of entropy combinations (including mutual information, transfer entropy and their conditional counterparts), which are more accurate than the simple average of individual estimates over trials. We show with simulated and real data generated by coupled electronic circuits that the proposed approach allows one to recover the time-resolved dynamics of the coupling between different subsystems.

Keywords: entropy; transfer entropy; estimator; ensemble; trial; time series

---

1. Introduction

An important problem is that of detecting interdependency relations between simultaneously measured time series. Finding an interdependency is the first step in elucidating how the subsystems underlying the time series interact. Fruitful applications of this approach abound in different fields, including neuroscience [1], ecology [2] or econometrics [3]. In these examples, the discovery of certain statistical interdependency is usually taken as an indicator that some interrelation exists between subsystems, such as different brain regions [4], animal populations or economical indexes.

Classical measures to unveil an interdependency include linear techniques, such as cross-correlation, coherence or Granger causality [6]. These measures quantify the strength of different linear relations and, thus, belong to the larger class of parametric measures, which assume a specific form for the interdependency between two or more processes. In particular, parametric techniques are often data-efficient, generalizable to multivariate settings and easy to interpret.

In general, statistical relationships between processes are more naturally and generally formulated within the probabilistic framework, which relaxes the need to assume explicit models on how variables relate to each other. For this reason, when a model of the underlying dynamics and of the assumed interaction is not available, a sound non-parametric approach can be stated in terms of information theory [7]. For example, mutual information is widely used to quantify the information statically shared between two random variables. Growing interest in interdependency measures that capture information flow rather than information sharing lead to the definition of transfer entropy [9]. In particular, transfer entropy quantifies how much the present and past of a random variable condition the future transitions of another. Thus, transfer entropy embodies an operational principle of causality first championed by Norbert Wiener [8], which was explicitly formulated for linear models by Clive Granger [5]. However, it is important to note that transfer entropy should not be understood as a quantifier of interaction strength nor interventional causality. See [10–12] for a detailed discussion on the relation between transfer entropy and different notions of causality and information transfer. See also [13] for a detailed account of how spatially- and temporally-local versions of information theoretic functionals, including transfer entropy, can be used to study the dynamics of computation in complex systems.

A practical pitfall is that without simplifying assumptions, a robust estimation of information theoretic functionals might require a large number of data samples. This requisite directly confronts situations in which the dependency to be analyzed evolves in time or is subjected to fast transients. When the non-stationarity is only due to a slow change of a parameter, over-embedding techniques can partially solve the problem by capturing the slow dynamics of the parameter as an additional variable [14]. It is also habitual to de-trend the time series or divide them into small windows within which the signals can be considered as approximately stationary. However, the above-mentioned procedures become unpractical when the relevant interactions change in a fast time scale. This is the common situation in brain responses and other complex systems where external stimuli elicit a rapid functional reorganization of information-processing pathways.

Fortunately, in several disciplines, the experiments leading to the multivariate time series can be systematically repeated. Thus, a typical experimental paradigm might render an ensemble of presumably independent repetitions or trials per experimental condition. In other cases, the processes under study display a natural cyclic variation and, thus, also render an ensemble of almost independent cycles or repetitions. This is often the case of seasonal time series that are common in economical and ecological studies and, more generally, of any cyclo-stationary process.

Here, we show how this multi-trial nature can be efficiently exploited to produce time-resolved estimates for a family of information-theoretic measures that we call entropy combinations. This family includes well-known functionals, such as mutual information, transfer entropy and their conditional counterparts: partial mutual information (PMI) [15,16] and partial transfer entropy (PTE) [17,18]. Heuristically, our approach can be motivated using the ergodic theorem. In other words, the time average of a measure converges to the space or ensemble average for an ergodic process. We can associate the conventional computation of entropies with a time average of log probabilities. Crucially, these should converge to the ensemble averages of the equivalent log probabilities, which we exploit with our (ensemble averaging) approach. In our case, the ensemble is constituted by multiple realizations of repeated trials. We use both simulations and experimental data to demonstrate that the proposed ensemble estimators of entropy combinations are more accurate than simple averaging of individual trial estimates.

2. Entropy Combinations

We consider three simultaneously measured time series generated from stochastic processes X, Y and Z, which can be approximated as stationary Markov processes [19] of finite order. The state space of X can then be reconstructed using the delay embedded vectors $x(n) = (x(n), ..., x(n - d_x + 1))$ for $n = 1, \ldots, N$, where n is a discrete time index and d_x is the corresponding Markov order. Similarly, we could construct $y(n)$ and $z(n)$ for processes Y and Z, respectively. Let $V = (V_1, ..., V_m)$ denote a random m-dimensional vector and $H(V)$ its Shannon entropy. Then, an entropy combination is defined by:

$$C(V_{\mathcal{L}_1}, ..., V_{\mathcal{L}_p}) = \sum_{i=1}^{p} s_i H(V_{\mathcal{L}_i}) - H(V) \tag{1}$$

where $\forall i \in [1, p] : \mathcal{L}_i \subset [1, m]$ and $s_i \in \{-1, 1\}$, such that $\sum_{i=1}^{p} s_i \chi_{\mathcal{L}_i} = \chi_{[1,m]}$, where χ_S is the indicator function of a set S (having the value one for elements in the set S and zero for elements not in S).

In particular, MI, TE, PMI and PTE all belong to the class of entropy combinations, since:

$$
\begin{aligned}
I_{X \leftrightarrow Y} &\equiv -H_{XY} + H_X + H_Y \\
T_{X \leftarrow Y} &\equiv -H_{WXY} + H_{WX} + H_{XY} - H_X \\
I_{X \leftrightarrow Y|Z} &\equiv -H_{XZY} + H_{XZ} + H_{ZY} - H_Z \\
T_{X \leftarrow Y|Z} &\equiv -H_{WXZY} + H_{WXZ} + H_{XZY} - H_{XZ}
\end{aligned}
$$

where random variable $W \equiv X^+ \equiv x(n+1)$, so that H_{WX} is the differential entropy of $p(x(n+1), x(n))$. The latter denotes the joint probability of finding X at states $x(n+1), x(n), ..., x(n - d_x + 1)$ during time instants $n + 1, n, n - 1, ..., n - d_x + 1$. Notice that, due to stationarity, $p(x(n+1), x(n))$ is invariant under variations of the time index n.

3. Ensemble Estimators for Entropy Combinations

A straightforward approach to the estimation of entropy combinations would be to add separate estimates of each of the multi-dimensional entropies appearing in combination. Popular estimators of differential entropy include plug-in estimators, as well as fixed and adaptive histogram or partition methods. However, other non-parametric techniques, such as kernel and nearest-neighbor estimators, have been shown to be extremely more data efficient [20,21]. An asymptotically unbiased estimator based on nearest-neighbor statistics is due to Kozachenko and Leonenko (KL) [22]. For N realizations $x[1], x[2], ..., x[N]$ of a d-dimensional random vector X, the KL estimator takes the form:

$$\hat{H}_X = -\psi(k) + \psi(N) + \log(v_d) + \frac{d}{N} \sum_{i=1}^{N} \log(\epsilon(i)) \tag{2}$$

where ψ is the digamma function, v_d is the volume of the d-dimensional unit ball and $\epsilon(i)$ is the distance from $x[i]$ to its k-th nearest neighbor in the set $\{x[j]\}_{\forall j \neq i}$. The KL estimator is based on the assumption that the density of the distribution of random vectors is constant within an ϵ-ball. The bias of the final entropy estimate depends on the validity of this assumption and, thus, on the values of $\epsilon(n)$. Since the size of the ϵ-balls depends directly on the dimensionality of the random vector, the biases of estimates for the differential entropies in Equation (1) will, in general, not cancel, leading to a poor estimator of the entropy combination. This problem can be partially overcome by noticing that Equation (2) holds for any value of k, so that we do not need to have a fixed k. Therefore, we can vary the value of k in each data point, so that the radius of the corresponding ϵ-balls would be approximately the same for the joint and the marginal spaces. This idea was originally proposed in [23] for estimating mutual information and was used in [16] to estimate PMI, and we generalize it here to the following estimator of entropy combinations:

$$\hat{C}(V_{\mathcal{L}_1}, ..., V_{\mathcal{L}_p}) = F(k) - \sum_{i=1}^{p} s_i \langle F(k_i(n)) \rangle_n \qquad (3)$$

where $F(k) = \psi(k) - \psi(N)$ and $\langle \cdots \rangle_n = \frac{1}{N} \sum_{n=1}^{N} (\cdots)$ denotes averaging with respect to the time index. The term $k_i(n)$ accounts for the number of neighbors of the n-th realization of the marginal vector $V_{\mathcal{L}_i}$ located at a distance strictly less than $\epsilon(n)$, where $\epsilon(n)$ denotes the radius of the ϵ-ball in the joint space. Note that the point itself is included in the counting neighbors in marginal spaces ($k_i(n)$), but not when selecting $\epsilon(n)$ from the k-th nearest neighbor in the full join space. Furthermore, note that estimator Equation (3) corresponds to extending "Algorithm 1" in [23] to entropy combinations. Extensions to conditional mutual information and conditional transfer entropy using "Algorithm 2" in [23] have been discussed recently [12].

A fundamental limitation of estimator Equation (3) is the assumption that the involved multidimensional distributions are stationary. However, this is hardly the case in many real applications, and time-adaptation becomes crucial in order to obtain meaningful estimates. A trivial solution is to use the following time-varying estimator of entropy combinations:

$$\hat{C}(\{V_{\mathcal{L}_1}, ..., V_{\mathcal{L}_p}\}, n) = F(k) - \sum_{i=1}^{p} s_i F(k_i(n)) \qquad (4)$$

This naive time-adaptive estimator is not useful in practice, due to its large variance, which stems from the fact that a single data point is used for producing the estimate at each time instant. More importantly, the neighbor searches in the former estimator run across the full time series and, thus, ignore possible non-stationary changes.

However, let us consider the case of an ensemble of r' repeated measurements (trials) from the dynamics of V. Let us also denote by $\left\{ v^{(r)}[n] \right\}_r$ the measured dynamics for those trials ($r = 1, 2, ...r'$). Similarly, we denote by $\{v_i^{(r)}[n]\}_r$ the measured dynamics for the marginal vector $V_{\mathcal{L}_i}$. A straightforward approach for integrating the information from different trials is to average together estimates obtained from individual trials:

$$\hat{C}^{\text{avg}}(\{V_{\mathcal{L}_1}, ..., V_{\mathcal{L}_p}\}, n) = \frac{1}{r'} \sum_{r=1}^{r'} \hat{C}^{(r)}(\{V_{\mathcal{L}_1}, ..., V_{\mathcal{L}_p}\}, n) \qquad (5)$$

where $\hat{C}^{(r)}(\{V_{\mathcal{L}_1}, ..., V_{\mathcal{L}_p}\}, n)$ is the estimate obtained from the r-th trial. However, this approach makes poor use of the available data and will typically produce useless estimates, as will be shown in the experimental section of this text.

A more effective procedure takes into account the multi-trial nature of our data by searching for neighbors across ensemble members, rather than from within each individual trial. This nearest ensemble neighbors [24] approach is illustrated in Figure 1 and leads to the following ensemble estimator of entropy combinations:

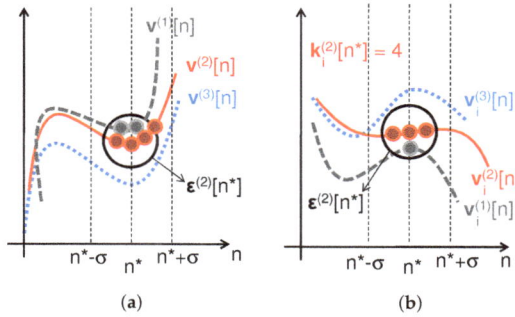

Figure 1. Nearest neighbor statistics across trials. (**a**): For each time instant $n = n^*$ and trial $r = r^*$, we compute the (maximum norm) distance $\epsilon^{(r^*)}(n^*)$ from $v^{(r^*)}[n^*]$ to its k-th nearest neighbor among all trials. Here, the procedure is illustrated for $k = 5$. (**b**): $k_i^{(r^*)}[n^*]$ counts how many neighbors of $v_i^{(r^*)}[n^*]$ are within a radius $\epsilon^{r^*}(n^*)$. The point itself (*i.e.*, $v_i^{(r^*)}[n^*]$) is also included in this count. These neighbor counts are obtained for all $i = 1, ...p$ marginal trajectories.

$$\hat{C}^{\text{en}}(\{V_{\mathcal{L}_1}, ..., V_{\mathcal{L}_p}\}, n) = F(k) - \frac{1}{r'}\sum_{r=1}^{r'}\sum_{i=1}^{p} s_i F\left(k_i^{(r)}(n)\right) \qquad (6)$$

where the counts of marginal neighbors $\{k_i^{(r)}(n)\}_{\forall i=1,...,p}^{\forall r=1,...,r'}$ are computed using overlapping time windows of size 2σ, as shown in Figure 1. For rapidly changing dynamics, small values of σ might be needed to increase the temporal resolution, thus, being able to track more volatile non-stationarities. On the other hand, larger values of σ will lead to lower estimator variance and are useful when non-stationarities develop over slow temporal scales.

4. Tests on Simulated and Experimental Data

To demonstrate that $\hat{C}^{\text{en}}$ can be used to characterize dynamic coupling patterns, we apply the ensemble estimator of PTE to multivariate time series from coupled processes.

In particular, we simulated three non-linearly-coupled autoregressive processes with a time-varying coupling factor:

$$
\begin{aligned}
x^r[n] &= 0.4x^r[n-1] + \eta_x, \\
y^r[n] &= 0.5y^r[n-1] + \kappa_{yx}[n]\sin\left(x^r[n-\tau_{yx}]\right) + \eta_y, \\
z^r[n] &= 0.5z^r[n-1] + \kappa_{zy}[n]\sin\left(y^r[n-\tau_{zy}]\right) + \eta_z.
\end{aligned}
$$

during 1,500 time steps and repeated R = 50 trials with new initial conditions. The terms η_x, η_y and η_z represent normally-distributed noise processes, which are mutually independent across trials and time instants. The coupling delays amount to $\tau_{yx} = 10$, $\tau_{zy} = 15$, while the dynamics of the coupling follows a sinusoidal variation:

$$
k_{yx}[n] = \begin{cases} \sin\left(\frac{2\pi n}{500}\right) & \text{for } 250 \le n < 750 \\ 0 & \text{otherwise} \end{cases}
$$

$$
k_{zy}[n] = \begin{cases} \cos\left(\frac{2\pi n}{500}\right) & \text{for } 750 \le n < 1250 \\ 0 & \text{otherwise.} \end{cases}
$$

Before PTE estimation, each time series was mapped via a delay embedding to its approximate state space. The dimension of the embedding was set using the Cao criterion [25], while the embedding delay time was set as the autocorrelation decay time. Other criteria to obtain embedding parameters, such as described in [19], provide similar results. Furthermore, each time series was time-delayed, so that they had maximal mutual information with the destination of the flow. That is, before computing some $T_{a \leftarrow b|c}(n)$, the time series b and c were delayed, so that they shared maximum information with the time series a, as suggested in [16]. For a rigorous and formal way to investigate the lag in the information flow between systems, we refer to [26,27].

To assess the statistical significance of the PTE values (at each time instant) we applied a permutation test with surrogate data generated by randomly shuffling trials [28]. Figure 2 shows the time-varying PTEs obtained for these data with the ensemble estimator of entropy combinations given in Equation (6). Indeed, the PTE analysis accurately describes the underlying interaction dynamics. In particular, it captures both the onset/offset and the oscillatory profile of the effective coupling across the three processes. On the other hand, the naive average estimator Equation (5) did not reveal any significant flow of information between the three time series.

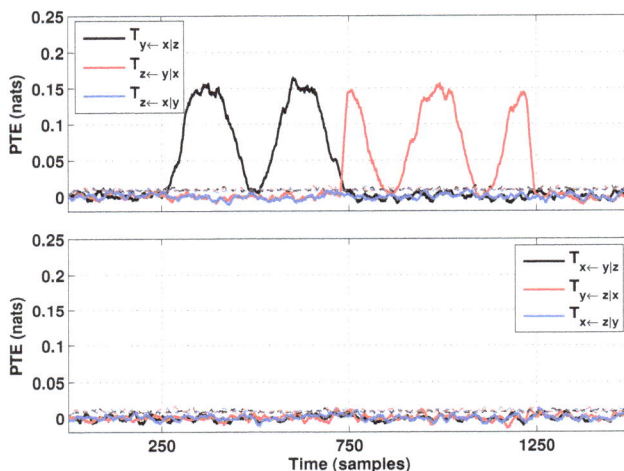

Figure 2. Partial transfer entropy between three non-linearly coupled Gaussian processes. The upper panel displays the partial transfer entropy (PTE) in directions compatible with the structural coupling of Gaussian processes (X to Y to Z). The lower panel displays the PTE values in directions non-compatible with the structural coupling. The solid lines represent PTE values, while the color-matched dashed lines denote corresponding $p = 0.05$ significance levels. $k = 20$. The time window for the search of neighbors is $2\sigma = 10$. The temporal variance of the PTE estimates was reduced with a post-processing moving average filter of order 20.

To evaluate the robustness and performance of the entropy combination estimator to real levels of noise and measurements variability, we also present a second example derived from experimental data on electronic circuits. The system consists of two nonlinear Mackey–Glass circuits unidirectionally coupled through their voltage variables. The master circuit is subject to a feedback loop responsible for generating high dimensional chaotic dynamics. A time-varying effective coupling is then induced by periodically modulating the strength of the coupling between circuits as controlled by an external CPU. Thus, the voltage variables of Circuits 1 and 2 are assumed to follow a stochastic dynamics of the type:

$$
\begin{aligned}
\frac{dx_1}{dt} &= \beta_1 \frac{x_{1\delta}}{1 + x_{1\delta}{}^n} - \gamma_1 x_1 + \eta_1, \\
\frac{dx_2}{dt} &= \beta_2 \frac{(1/2 + 1/4 \sin(\omega t)) x_{1\tau}}{1 + x_{1\tau}{}^n} - \gamma_2 x_2 + \eta_2,
\end{aligned}
\tag{7}
$$

where x_δ represents the value of the variable x at time $t - \delta$, γ, β and n are positive numbers and η represent noise sources. The feedback loop of the first circuit and time-varying coupling between the two circuits are represented by the first terms of each equation, respectively. We note that the former set of equations was not used to sample data. Instead, time series were directly obtained from the voltage variables of the electronic circuits. The equations above just serve to illustrate in mathematical terms the type of dynamics expected from the electronic circuits.

Thus, we applied transfer entropy between the voltage signals directly generated from the two electric circuits for 180 trials, each 1,000 sampling times long. Delay embedding and statistical significance analysis proceeded as in the previous example. Figure 3 shows the TE ensemble estimates between the master and slave circuit obtained with Equation (6) *versus* the temporal lag introduced between the two voltage signals (intended to scan the unknown coupling delay τ). Clearly, there is a directional flow of information time-locked at lag $\tau = 20$ samples, which is significant for all time instants ($p < 0.01$).

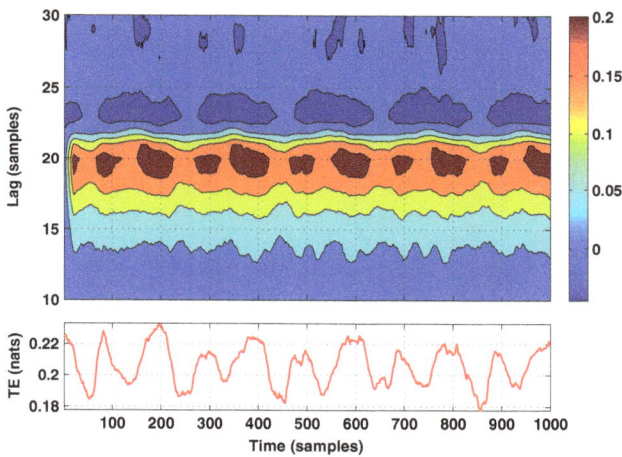

Figure 3. Transfer entropy from the first electronic circuit towards the second. The upper figure shows time-varying TE *versus* the lag introduced in the temporal activation of the first circuit. The lower figure shows that the temporal pattern of information flow for $\tau = 20$, *i.e.*, $T_{2\leftarrow1}(n, \tau = 20)$, which resembles a sinusoid with a period of roughly 100 data samples.

The results show that the TE ensemble estimates accurately capture the dynamics of the effect exerted by the master circuit on the slave circuit. On the other hand, the flow of information in the opposite direction was much smaller ($T_{1\leftarrow2} < 0.0795$ nats $\forall(t, \tau)$) and only reached significance ($p < 0.01$) for about 1% of the tuples (n, τ) Figure 4. Both the period of the coupling dynamics (100 samples) and the coupling delay (20 samples) can be accurately recovered from Figure 3.

Finally, we also performed numerical simulations to study the behavior of the bias and variance of the ensemble estimator with respect to the number of neighbors chosen and the sample size. In particular, we simulated two unidirectionally-coupled Gaussian linear autoregressive processes ($Y \to X$) for which the analytical values of TE can be known [29], so that we could compare the numerical and expected values. Then, we systematically varied the level of nominal TE (which was controlled by the chosen level of correlation coefficient between $X(t + 1)$ and $X(t)$), the number of neighbors chosen and the sample size and compute measures of bias and variance. Figure 5 and 6 display in a color-coded manner the quantities $-20 \times log_{10}(\textbf{\textit{bias}})$ and $-20 \times log_{10}(\textbf{\textit{var}})$, so large values of these quantities correspond to small bias and variances, respectively. In particular, Figure 5 shows the bias and variance of the estimator as a function of the number of samples and cross-correlation coefficient. As observed in the plot, the smaller the value of the underlying TE (smaller cross-correlation), the better its estimation (smaller bias and variance). For a given value of TE, the estimation improves as more samples are included, as is expected. Regarding the number of neighbors (Figure 6), we obtain that beyond a minimum number of samples, the accuracy obtained increased by either increasing the sample size or the number of neighbors.

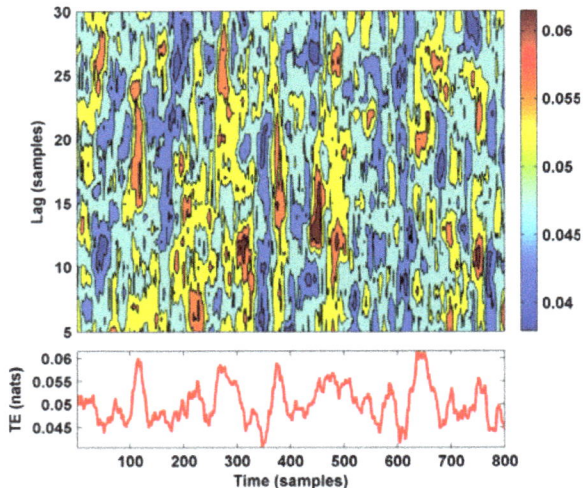

Figure 4. Transfer entropy from the second electronic circuit towards the first. The upper figure shows time-varying TE *versus* the lag introduced in the temporal activation of the first circuit. The lower figure shows that the temporal pattern of information flow for $\tau = 20$, *i.e.*, $T_{1\leftarrow 2}(n, \tau = 20)$.

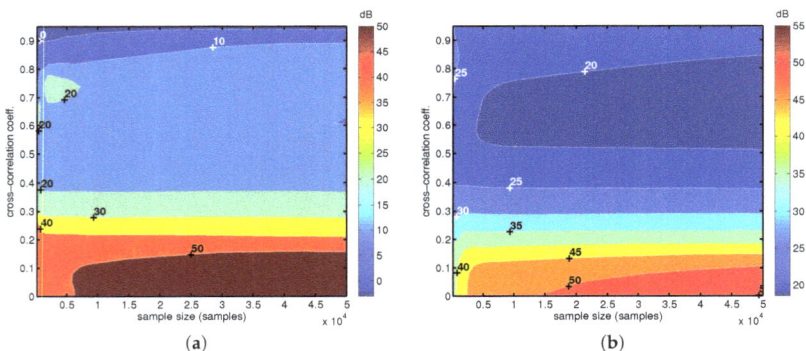

Figure 5. (**a**): $-20 \times log_{10}(\textbf{bias})$ of ensemble estimator TE($Y \rightarrow X$) as a function of the number of samples and cross-correlation coefficient for X (which controls the nominal TE value for ($Y \rightarrow X$)). (**b**): $-20 \times log_{10}(\textbf{variance})$ as a function of the number of samples and cross-correlation coefficient for X (which controls the nominal TE value for ($Y \rightarrow X$)).

Entropy **2015**, *17*, 1958–1970

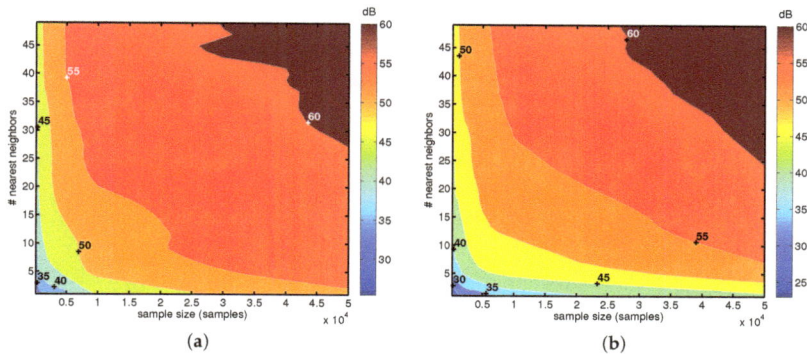

Figure 6. (**a**): $-20 \times log_{10}(\boldsymbol{bias})$ of ensemble estimator TE($Y \rightarrow X$) as a function of the number of samples and the number of nearest neighbors used in the estimator. (**b**): $-20 \times log_{10}(\boldsymbol{variance})$ as a function of the number of samples and the number of nearest neighbors used in the estimator.

5. Conclusions

In conclusion, we have introduced an ensemble estimator of entropy combinations that is able to detect time-varying information flow between dynamical systems, provided that an ensemble of repeated measurements is available for each system. The proposed approach allows one to construct time-adaptive estimators of MI, PMI, TE and PTE, which are the most common information-theoretic measures for dynamical coupling analyses. Using simulations and real physical measurements from electronic circuits, we showed that these new estimators can accurately describe multivariate coupling dynamics. However, strict causal interpretations of the transfer entropy analyses are discouraged [10].

It is also important to mention that intrinsic to our approach is the assumption that the evolution of the interdependencies to be detected are to some degree "locked" to the trial onset. In the setting of electrophysiology and the analysis of event-related potentials, the dispersion of the dynamics with respect to their onset is clearly an acute issue. Indeed, the key distinction between evoked and induced responses rests upon time-locking to a stimulus onset. In principle, one could apply the ensemble average entropic measures to induced responses as measured in terms of the power of the signals, even when they are not phase-locked to a stimulus. In general, the degree of locking determines the maximum temporal resolution achievable by the method (which is controlled via σ). Nevertheless, it is possible to use some alignment techniques [30] to reduce the possible jitter across trials and, thus, increase the resolution.

The methods presented here are general, but we anticipate that a potential application might be the analysis of the mechanisms underlying the generation of event-related brain responses and the seasonal variations of geophysical, ecological or economic variables. Efficient implementations of the ensemble estimators for several information-theoretic methods can be found in [31,32].

Acknowledgments: We are indebted to the anonymous referees for their constructive and valuable comments and discussions that helped to improve this manuscript. This work has been supported by the EU project GABA(FP6-2005-NEST-Path 043309), the Finnish Foundation for Technology Promotion, the Estonian Research Council through the personal research grants P.U.T.program (PUT438 grant), the Estonian Center of Excellence in Computer Science (EXCS) and a grant from the Estonian Ministry of Science and Education (SF0180008s12).

Author Contributions: Germán Gómez-Herrero and Raul Vicente conceptualized the problem and the technical framework. Germán Gómez-Herrero, Gordon Pipa, and Raul Vicente managed the project. Wei Wu, Kalle Rutanen, Germán Gómez-Herrero, and Raul Vicente developed and tested the algorithms. Miguel C. Soriano collected the experimental data. All authors wrote the manuscript. All authors have read and approved the final manuscript.

Conflicts of Interest: The authors declare no conflict of interest.

References

1. Gray, C.; Konig, P.; Engel, A.; Singer, W. Oscillatory responses in cat visual cortex exhibit inter-columnar synchronization which reflects global stimulus properties. *Nature* **1989**, *338*, 334.

2. Bjornstad, O.; Grenfell, B. Noisy clockwork: Time series analysis of population fluctuations in animals. *Science* **2001**, *293*, 638.

3. Granger, C.; Hatanaka, M. *Spectral Analysis of Economic Time Series*; Princeton University Press: Princeton, NJ, USA, 1964.

4. Valdes-Sosa, P.A.; Roebroeck, A.; Daunizeau, J.; Friston, K. Effective connectivity: Influence, causality and biophysical modeling. *Neuroimage* **2011**, *58*, 339.

5. Granger, C. Investigating causal relations by econometric models and cross-spectral methods. *Econometrica* **1969**, *37*, 424.

6. Pereda, E.; Quian Quiroga, R.; Bhattacharya, J. Nonlinear multivariate analysis of neurophysiological signals. *Prog. Neurobio.* **2005**, *77*, 1.

7. Cover, T.; Thomas, J. *Elements of Information Theory*; Wiley: Hoboken, NY, USA, 2006.

8. Wiener, N. The theory of prediction. In *Modern Mathematics for Engineers*; McGraw-Hill: New York, NY, USA, 1956.

9. Schreiber, T. Measuring information transfer. *Phys. Rev. Lett.* **2000**, *85*, 461.

10. Chicharro, D.; Ledberg, A. When two become one: the limits of causality analysis of brain dynamics. *PLoS One* **2012**, *7*, e32466.

11. Wibral, M.; Vicente, R.; Lizier, J.T. *Directed Information Measures in Neuroscience*; Springer: Berlin, Germany, 2014.

12. Wibral, M.; Vicente, R.; Lindner, M. Transfer entropy in Neuroscience. In *Directed Information Measures in Neuroscience*; Wibral, M., Vicente, R., Lizier, J.T., Eds.; Springer: Berlin, Germany, 2014.

13. Lizier, J.T. *The Local Information Dynamics of Distributed Computation in Complex Systems*; Springer: Berlin, Germany, 2013.

14. Kantz, H.; Schreiber, T. *Nonlinear Time Series Analysis*, 2nd ed.; Cambridge University Press: Cambridge, UK, 2004.

15. Wyner, A.D. A Definition of Conditional Mutual Information for Arbitrary Ensembles. *Inf. Control* **1978**, *38*, 51.

16. Frenzel, S.; Pompe, B. Partial mutual information for coupling analysis of multivariate time series. *Phys. Rev. Lett.* **2007**, *99*, 204101.

17. Verdes, P.F. Assessing causality from multivariate time series. *Phys. Rev. E* **2005**, *72*, 026222.

18. Gómez-Herrero, G. Ph.D. thesis, Department of Signal Processing, Tampere University of Technology, Finland, 2010.

19. Ragwitz, M.; Kantz, H. Markov models from data by simple nonlinear time series predictors in delay embedding spaces. *Phys. Rev. E* **2002**, *65*, 056201.

20. Victor, J.D. Binless strategies for estimation of information from neural data. *Phys. Rev. E* **2002**, *66*, 051903.

21. Vicente, R.; Wibral, M. Efficient estimation of information transfer. In *Directed Information Measures in Neuroscience*; Wibral, M., Vicente, R., Lizier, J.T., Eds.; Springer: Berlin, Germany, 2014.

22. Kozachenko, L.; Leonenko, N. Sample Estimate of the Entropy of a Random Vector. *Problemy Peredachi Informatsii* **1987**, *23*, 9.

23. Kraskov, A.; Stögbauer, H.; Grassberger, P. Estimating mutual information. *Phys. Rev. E* **2004**, *69*, 066138.

24. Kramer, M.A.; Edwards, E.; Soltani, M.; Berger, M.S.; Knight, R.T.; Szeri, A.J. Synchronization measures of bursting data: application to the electrocorticogram of an auditory event-related experiment. *Phys. Rev. E* **2004**, *70*, 011914.

25. Cao, L. Practical method for determining the minimum embedding dimension of a scalar time series. *Physica D* **1997**, *110*, 43.

26. Wibral, M.; Pampu, N.; Priesemann, V; Siebenhuhner; Seiwert; Lindner; Lizier; Vicente. R. Measuring information-transfer delays. *PLoS One* **2013**, *8*, e55809.

27. Wollstadt, P.; Martinez-Zarzuela, M.; Vicente, R.; Diaz-Pernas, F.J.; Wibral, M. Efficient transfer entropy analysis of non-stationary neural time series. *PLoS One* **2014**, *9*, e102833.

28. Pesarin, F. *Multivariate Permutation Tests*; John Wiley and Sons: Hoboken, NJ, USA, 2001.

29. Kaiser, A.; Schreiber, T. Information transfer in continuous processes. *Physica D* **2002**, *166*, 43.
30. Kantz, H.; Ragwitz, M. Phase space reconstruction and nonlinear predictions for stationary and nonstationary Markovian processes. *Int. J. Bifurc. Chaos* **2004**, *14*, 1935.
31. Rutanen, K. TIM 1.2.0. Available online: http://www.tut.fi/tim (accessed on 2 April 2015).
32. Lindner, M.; Vicente, R.; Priesemann, V.; Wibral, M. TRENTOOL: A Matlab open source toolbox to analyse information flow in time series data with transfer entropy. *BMC Neurosci.* **2011**, *12*, 119.

entropy

MDPI

Review

The Relation between Granger Causality and Directed Information Theory: A Review

Pierre-Olivier Amblard [1,2],* and Olivier J. J. Michel [1]

[1] GIPSAlab/CNRS UMR 5216/ BP46, 38402 Saint Martin d'Hères cedex, France;
E-Mail: olivier.michel@gipsa-lab.grenoble-inp.fr (O.J.J.M.)

[2] The University of Melbourne, Department of Mathematics and Statistics, Parkville, VIC, 3010, Australia

* Author to whom correspondence should be addressed; E-Mail: bidou.amblard@gipsa-lab.inpg.fr;
Tel.: +33-4-76826333; Fax: +33-4-76574790.

Received: 14 November 2012; in revised form: 19 December 2012 / Accepted: 19 December 2012 /
Published: 28 December 2012

Abstract: This report reviews the conceptual and theoretical links between Granger causality and directed information theory. We begin with a short historical tour of Granger causality, concentrating on its closeness to information theory. The definitions of Granger causality based on prediction are recalled, and the importance of the observation set is discussed. We present the definitions based on conditional independence. The notion of instantaneous coupling is included in the definitions. The concept of Granger causality graphs is discussed. We present directed information theory from the perspective of studies of causal influences between stochastic processes. Causal conditioning appears to be the cornerstone for the relation between information theory and Granger causality. In the bivariate case, the fundamental measure is the directed information, which decomposes as the sum of the transfer entropies and a term quantifying instantaneous coupling. We show the decomposition of the mutual information into the sums of the transfer entropies and the instantaneous coupling measure, a relation known for the linear Gaussian case. We study the multivariate case, showing that the useful decomposition is blurred by instantaneous coupling. The links are further developed by studying how measures based on directed information theory naturally emerge from Granger causality inference frameworks as hypothesis testing.

Keywords: granger causality; transfer entropy; information theory; causal conditioning; conditional independence

1. Introduction

This review deals with the analysis of influences that one system, be it physical, economical, biological or social, for example, can exert over another. In several scientific fields, the finding of the influence network between different systems is crucial. As examples, we can think of gene influence networks [1,2], relations between economical variables [3,4], communication between neurons or the flow of information between different brain regions [5], or the human influence on the Earth climate [6,7], and many others.

The context studied in this report is illustrated in Figure 1. For a given system, we have at disposal a number of different measurements. In neuroscience, these can be local field potentials recorded in the brain of an animal. In solar physics, these can be solar indices measured by sensors onboard some satellite. In the study of turbulent fluids, these can be the velocity measured at different scales in the fluid (or can be as in Figure 1, the wavelet analysis of the velocity at different scales). For these different examples, the aim is to find dependencies between the different measurements, and if possible, to give a direction to the dependence. In neuroscience, this will allow to understand how information flows between different areas of the brain. In solar physics, this will allow to understand the links between indices and their influence on the total solar irradiance received on Earth. In the study of turbulence, this can confirm the directional cascade of energy from large down to small scales.

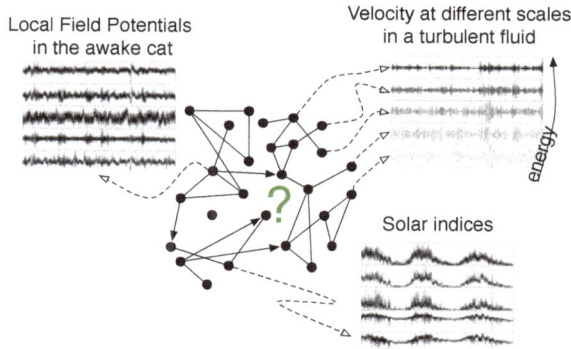

Figure 1. Illustration of the problem of information flow in networks of stochastic processes. Each node of the network is associated to a signal. Edges between nodes stand for dependence (shared information) between the signals. The dependence can be directed or not. This framework can be applied to different situations such as solar physics, neuroscience or the study of turbulence in fluids, as illustrated by the three examples depicted here.

In a graphical modeling approach, each signal is associated to a particular node of a graph, and dependencies are represented by edges, directed if a directional dependence exists. The questions addressed in this paper concern the assessment of directional dependence between signals, and thus concern the inference problem of estimating the edge set in the graph of signals considered.

Climatology and neuroscience were already given as examples by Norbert Wiener in 1956 [8], a paper which inspired econometrist Clive Granger to develop what is now termed Granger causality [9]. Wiener proposed in this paper that a signal x causes another time series y, if the past of x has a strictly positive influence on the quality of prediction of y. Let us quote Wiener [8]:

> "As an application of this, let us consider the case where $f_1(\alpha)$ represents the temperature at 9 A.M. in Boston and $f_2(\alpha)$ represents the temperature at the same time in Albany. We generally suppose that weather moves from west to east with the rotation of the earth; the two quantities $1 - C$ and its correlate in the other direction will enable us to make a precise statement containing some if this content and then verify whether this statement is true or not. Or again, in the study of brain waves we may be able to obtain electroencephalograms more or less corresponding to electrical activity in different part of the brain. Here the study of coefficients of causality running both ways and of their analogues for sets of more than two functions f may be useful in determining what part of the brain is driving what other part of the brain *in its normal activity*."

In a wide sense, Granger causality can be summed up as a theoretical framework based on conditional independence to assess directional dependencies between time series. It is interesting to note that Norbert Wiener influenced Granger causality, as well as another field dedicated to the analysis of dependencies: information theory. Information theory has led to the definition of quantities that measure the uncertainty of variables using probabilistic concepts. Furthermore, this has led to the definition of measures of dependence based on the decrease in uncertainty relating to one variable after observing another one. Usual information theory is, however, symmetrical. For example, the well-known mutual information rate between two stationary time series is symmetrical under an exchange of the two signals: the mutual information assesses the undirectional dependence. Directional dependence analysis viewed as an information-theoretic problem requires the breaking of the usual symmetry of information theory. This was realized in the 1960s and early 1970s by Hans Marko, a German professor of communication. He developed the bidirectional information theory in the Markov case [10]. This theory was later generalized by James Massey and Gerhard Kramer, to what we may now call directed information theory [11,12].

It is the aim of this report to review the conceptual and theoretical links between Granger causality and directed information theory.

Many information-theoretic tools have been designed for the practical implementation of Granger causality ideas. We will not show all of the different measures proposed, because they are almost always particular cases of the measures issued from directed information theory. Furthermore, some measures might have been proposed in different fields (and/or at different periods of time) and have received different names. We will only consider the well-accepted names. This is the case, for example, of "transfer entropy", as coined by Schreiber in 2000 [13], but which appeared earlier under different names, in different fields, and might be considered under slightly different hypotheses. Prior to developing a unified view of the links between Granger causality and information theory, we will provide a survey of the literature, concentrating on studies where information theory and Granger causality are jointly presented.

Furthermore, we will not review any practical aspects, nor any detailed applications. In this spirit, this report is different from [14], which concentrated on the estimation of information quantities, and where the review is restricted to transfer entropy. For reviews on the analysis of dependencies between systems and for applications of Granger causality in neuroscience, we refer to [15,16]. We will mention however some important practical points in our conclusions, where we will also discuss some current and future directions of research in the field.

1.1. What Is, and What Is Not, Granger Causality

We will not debate the meaning of causality or causation. We instead refer to [17]. However, we must emphasize that Granger causality actually measures a statistical dependence between the past of a process and the present of another. In this respect, the word causality in Granger causality takes on the usual meaning that a cause occurs *prior to* its effect. However, nothing in the definitions that we will recall precludes that signal x can simultaneously be Granger caused by y and be a cause of y! This lies in the very close connection between Granger causality and the feedback between times series.

Granger causality is based on the usual concept of conditioning in probability theory, whereas approaches developed for example in [17,18] relied on causal calculus and the concept of intervention. In this spirit, intervention is closer to experimental sciences, where we imagine that we can really, for example, freeze some system and measure the influence of this action on another process. It is now well-known that causality in the sense of between random variables can be inferred unambiguously only in restricted cases, such as directed acyclic graph models [17–20]. In the Granger causality context, there is no such ambiguity and restriction.

1.2. A Historical Viewpoint

In his Nobel Prize lecture in 2003, Clive W. Granger mentioned that in 1959, Denis Gabor pointed out the work of Wiener to him, as a hint to solve some of the difficulties he met in his work. Norbert Wiener's paper is about the theory of prediction [8]. At the end of his paper, Wiener proposed that prediction theory could be used to define causality between time series. Granger further developed this idea, and came up with a definition of causality and testing procedures [3,21].

In these studies, the essential stones were laid. Granger's causality states that a cause must occur before the effect, and that causality is relative to the knowledge that is available. This last statement deserves some comment. When testing for causality of one variable on another, it is assumed that the cause has information about the effect that is unique to it; *i.e.*, this information is unknown to any other variable. Obviously, this cannot be verified for variables that are not known. Therefore, the conclusion drawn in a causal testing procedure is relative to the set of measurements that are available. A conclusion reached based on a set of measurements can be altered if new measurements are taken into account.

Mention of information theory is also present in the studies of Granger. In the restricted case of two Gaussian signals, Granger already noted the link between what he called the "causality indices" and the mutual information (Equation 5.4 in [21]). Furthermore, he already foresaw the generalization to the multivariate case, as he wrote in the same paper:

> "In the case of q variables, similar equations exist if coherence is replaced by partial coherence, and a new concept of 'partial information' is introduced."

Granger's paper in 1969 does not contain much new information [3], but rather, it gives a refined presentation of the concepts.

During the 1970's, some studies, e.g., [4,22,23], appeared that generalized along some of the directions Granger's work, and related some of the applications to economics. In the early 1980's, several studies were published that established the now accepted definitions of Granger causality [24–27]. These are natural extensions of the ideas built upon prediction, and they rely on conditional independence. Finally, the recent studies of Dalhaus and Eichler allowed the definitions of Granger causality graphs [28–30]. These studies provide a counterpart of graphical models of multivariate random variables to multivariable stochastic processes.

In two studies published in 1982 and 1984 [31,32], Geweke, another econometrician, set up a full treatment of Granger causality, which included the idea of feedback and instantaneous coupling. In [31], the study was restricted to the link between two time series (possibly multidimensional). In this study, Geweke defined an index of causality from x to y; It is the logarithm of the *ratio* of the asymptotic mean square error when predicting y from its past only, to the asymptotic mean square error when predicting y from its past and from the past of x. Geweke also defined the same kind of index for instantaneous coupling. When the innovation sequence is Gaussian, the mutual information rate between x and y decomposes as the sum of the indices of causality from x to y and from y to x with the index of instantaneous coupling. This decomposition was shown in the Gaussian case, and it remains valid in any case when the indices of causality are replaced by transfer entropy rates, and the instantaneous coupling index is replaced by an instantaneous information exchange rate. This link between Granger causality and directed information theory was further supported by [33–35] (without mention of instantaneous coupling in [34,35]), and the generalization to the non-Gaussian case by [36] (see also [37] for related results). However, prior to these recent studies, the generalization of Geweke's idea to some general setting was reported in 1987, in econometry by Gouriéroux *et al.* [38], and in engineering by Rissannen&Wax [39]. Gouriéroux and his co-workers considered a joint Markovian representation of the signals, and worked in a decision-theoretic framework. They defined a sequence of nested hypotheses, whether causality was true or not, and whether instantaneous coupling was present or not. They then worked out the decision statistics using the Kullback approach to decision theory [40], in which discrepancies between hypotheses are measured according to the Kullback divergence between the probability measures under the hypotheses involved. In this setting, the decomposition obtained by Geweke in the Gaussian case was evidently generalised. In [39], the approach taken was closer to Geweke's study, and it relied on system identification, in which the complexity of the model was taken into account. The probability measures were parameterized, and an information measure that jointly assessed the estimation procedure and the complexity of the model was used when predicting a signal. This allowed Geweke's result to be extended to nonlinear modeling (and hence the non-Gaussian case), and provided an information-theoretic interpretation of the tests. Once again, the same kind of decomposition of dependence was obtained by these authors. We will see in Section 3 that the decomposition holds due to Kramers causal conditioning. These studies were limited to the bivariate case [38,39].

In the late 1990s, some studies began to develop in the physics community on influences between dynamical systems. A first route was taken that followed the ideas of dynamic system studies for the prediction of chaotic systems. To determine if one signal influenced another, the idea was to consider each of the signals as measured states of two different dynamic systems, and then to study the master-slave relationships between these two systems (for examples, see [41–43]). The dynamics of the systems was built using phase space reconstruction [44]. The influence of one system on another was then defined by making a prediction of the dynamics in the reconstructed phase space of one of the processes. To our knowledge, the setting was restricted to the bivariate case. A second route, which was also restricted to the bivariate case, was taken and relied on information-theoretic tools. The main contributions were from Paluš and Schreiber [13,45], with further developments appearing some years later [46–49]. In these studies, the influence of one process on the other was measured by the discrepancy between the probability measures under the hypotheses of influence or no influence. Naturally, the measures defined very much resembled the measures proposed by Gouriéroux *et. al* [38], and used the concept of conditional mutual information. The measure to assess whether one signal influences the other was termed *transfer entropy* by Schreiber. Its definition was proposed under a Markovian assumption, as was exactly done in [38]. The presentation by Paluš [45] was more direct and was not based on a decision-theoretic idea. The measure defined is, however, equivalent to the transfer entropy. Interestingly, Paluš noted in this 2001 paper the closeness of the approach to Granger causality, as per the quotation:

> "the latter measure can also be understood as an information theoretic formulation of the Granger causality concept."

Note that most of these studies considered bivariate analysis, with the notable exception of [46], in which the presence of side information (other measured time series) was explicitly considered.

In parallel with these studies, many others were dedicated to the implementation of Granger causality testing in fields as diverse as climatology (with applications to the controversial questions of global warming) and neuroscience; see [6,7,15,30,50–54], to cite but a few.

In a very different field, information theory, the problem of feedback has lead to many questions since the 1950s. We will not review or cite anything on the problem created by feedback in information theory as this is not within the scope of the present study, but some information can be found in [55]. Instead, we will concentrate on studies that are directly related to the subject of this review. A major breakthrough was achieved by James Massey in 1990 in a short conference paper [12]. Following the ideas of Marko on bidirectional information theory that were developed in the Markovian case [10], Massey re-examined the usual definition of what is called a discrete memoryless channel in information theory, and he showed that the usual definition based on some probabilistic assumptions prohibited the use of feedback. He then clarified the definition of memory and feedback in a communication channel. As a consequence, he showed that in a general channel used with feedback, the usual definition of capacity that relies on mutual information was not adequate. Instead, the right measure was shown to be *directed information*, an asymmetrical measure of the flow of information. These ideas were further examined by Kramer, who introduced the concept of causal conditioning, and who developed the first applications of directed information theory to communication in networks [11]. After some years, the importance of causal conditioning for the analysis of communication in systems with feedback was realized. Many studies were then dedicated to the analysis of the capacity of channels with feedback and the dual problem of rate-distortion theory [56–59]. Due to the rapid development in the study of networks (e.g., social networks, neural networks) and of the afferent connectivity problem, more recently many authors made connections between information theory and Granger causality [33,34,36,37,60–62]. Some of these studies were restricted to the Gaussian case, and to the bivariate case. Most of these studies did not tackle the problem of instantaneous coupling. Furthermore, several authors realized the importance of directed information theory to assess the circulation of information in networks [1,2,63,64].

Entropy **2013**, *15*, 113–143

1.3. Outline

Tools from directed information theory appear as natural measures to assess Granger causality. Although Granger causality can be considered as a powerful theoretical framework to study influences between signals mathematically, directed information theory provides the measures to test theoretical assertions practically. As already mentioned, these measures are transfer entropy (and its conditional versions), which assesses the dynamical part of Granger causality, and instantaneous information exchange (and its conditional versions), which assesses instantaneous coupling.

This review is structured here as follows. We will first give an overview of the definitions of Granger causality. These are presented in a multivariate setting. We go gradually from weak definitions based on prediction, to strong definitions based on conditional independence. The problem of instantaneous coupling is then discussed, and we show that there are two possible definitions for it. Causality graphs (after Eichler [28]) provide particular reasons to prefer one of these definitions. Section 3 introduces an analysis of Granger causality from an information-theoretic perspective. We insist on the concept of causal conditioning, which is at the root of the relationship studied. Section 4 then highlights the links. Here, we first restate the definitions of Granger causality using concepts from directed information theory. Then from a different point of view, we show how conceptual inference approaches lead to the measures defined in directed information theory. The review then closes with a discussion of some of the aspects that we do not present here intentionally, and on some lines for further research.

1.4. Notations

All of the random variables, vectors and signals considered here are defined in a common probability space $(\Omega, \mathcal{B}, P)$. They take values either in $\mathbb{R}$ or $\mathbb{R}^d$, d being some strictly positive integer, or they can even take discrete values. As we concentrate on conceptual aspects rather than technical aspects, we assume that the variables considered are "well behaved". In particular, we assume finiteness of moments of sufficient order. We assume that continuously valued variables have a measure that is absolutely continuous with respect to the Lebesgue measure of the space considered. Hence, the existence of probability density functions is assumed. Limits are supposed to exist when needed. All of the processes considered in this report are assumed to be stationary.

We work with discrete time. A signal will generically be denoted as $x(k)$. This notation stands also for the value of the signal at time k. The collection of successive samples of the signal, $x_k, x_{k+1}, \ldots, x_{k+n}$ will be denoted as x_k^{k+n}. Often, an initial time will be assumed. This can be $0, 1$, or $-\infty$. In any case, if we collect all of the sample of the signals from the initial time up to time n, we will suppress the lower index and write this collection as x^n.

When dealing with multivariate signals, we use a graph-theoretic notation. This will simplify some connections with graphical modeling. Let V be an index set of finite cardinality $|V|$. $x_V = \{x_V(k), k \in \mathbb{Z}\}$ is a d-dimensional discrete time stationary multivariate process for the probability space considered. For $a \in V$, x_a is the corresponding component of x_V. Likewise, for any subset $A \subset V$, x_A is the corresponding multivariate process $(x_{a_1}, \ldots, x_{|A|})$. We say that subsets A, B, C form a partition of V if they are disjoint and if $A \cup B \cup C = V$. The information obtained by observing x_A up to time k is given by the filtration generated by $\{x_A(l), \forall l \leq k\}$. This is denoted as x_A^k. Furthermore, we will often identify x_A with A in the discussion.

The probability density functions (p.d.f.) or probability mass functions (p.m.f) will be denoted by the same notation as $p(x_A^n)$. The conditional p.d.f. and p.m.f. are written as $p(x_A^n|x_B^m)$. The expected value is denoted as $E[.], E_x[.]$ or $E_p[.]$ if we want to specify which variable is averaged, or under which probability measure the expected value is evaluated.

Independence between random variables and vectors x and y will be denoted as $x \perp\!\!\!\perp y$, while conditional independence given z will be written as $x \perp\!\!\!\perp y \mid z$.

2. Granger's Causality

The early definitions followed the ideas of Wiener: A signal x causes a signal y if the past of x helps in the prediction of y. Implementing this idea requires the performing of the prediction and the quantification of its quality. This leads to a weak, but operational, form of the definitions of Granger causality. The idea of improving a prediction is generalized by encoding it into conditional dependence or independence.

2.1. From Prediction-Based Definitions...

Consider a cost function $g : \mathbb{R}^k \longrightarrow \mathbb{R}$ (k is some appropriate dimension), and the associated risk $E[g(e)]$, where e stands for an error term. Let a predictor of $x_B(n)$ be defined formally as $\widehat{x_B}(n+1) = f(x_A^n)$, where A and B are subsets of V, and f is a function between appropriate spaces, chosen to minimize the risk with $e(n) := x_B(n+1) - \widehat{x_B}(n+1)$. Solvability may be granted if f is restricted to an element of a given class of functions, such as the set of linear functions. Let $\mathcal{F}$ be such a function class. Define:

$$R_{\mathcal{F}}\big(B(n+1)\big|A^n\big) = \inf_{f \in \mathcal{F}} E\big[g\big(x_B(n+1) - f(x_A^n)\big)\big] \tag{1}$$

$R_{\mathcal{F}}\big(B(n+1)\big|A^n\big)$ is therefore the optimal risk when making a one-step-ahead prediction of the multivariate signal x_B from the past samples of the multivariate signal x_A. We are now ready to measure the influence of the past of a process on the prediction of another. To be relatively general and to prepare comments on the structure of the graph, this can be done for subsets of V. We thus choose A and B to be two disjoint subsets of V, and we define $C := V \backslash (A \cup B)$ (we use $\backslash$ to mean subtraction of a set). We study causality from x_A to x_B by measuring the decrease in the quality of the prediction of $x_B(n)$ when excluding the past of x_A.

Let $R_{\mathcal{F}}\big(B(n+1)\big|V^n\big)$ be the optimal risk obtained for the prediction of x_B from the past of all of the signals grouped in x_V. This risk is compared with $R_{\mathcal{F}}\big(B(n+1)\big|(V \backslash A)^n\big)$, where the past of x_A is omitted. Then, for the usual costs functions, we have necessarily:

$$R_{\mathcal{F}}\big(B(n+1)\big|V^n\big) \le R_{\mathcal{F}}\big(B(n+1)\big|(V \backslash A)^n\big) \tag{2}$$

A natural first definition for Granger causality is:

Definition 1. x_A Granger does not cause x_B relative to V if and only if $R_{\mathcal{F}}\big(B(n+1)\big|V^n\big) = R_{\mathcal{F}}\big(B(n+1)\big|(V \backslash A)^n\big)$

This definition of Granger causality depends on the cost g chosen as well as on the class $\mathcal{F}$ of the functions considered. Usually, a quadratic cost function is chosen, for its simplicity and for its evident physical interpretation (a measure of the power of the error). The choice of the class of functions $\mathcal{F}$ is crucial. The result of the causality test in definition 1 can change when the class is changed. Consider the very simple example of $x_{n+1} = \alpha x_n + \beta y_n^2 + \varepsilon_{n+1}$, where y_n and ε_n are zero-mean Gaussian independent and identically distributed (i.i.d.) sequences that are independent of each other. The covariance between x_{n+1} and y_n is zero, and using the quadratic loss and the class of linear functions, we conclude that y does not Granger cause x, because using a linear function of x_n, y_n to predict x would lead to the same minimal risk as using a linear function of x_n only. However, y_n obviously causes x_n, but in a nonlinear setting.

The definition is given using the negative of the proposition. If by using the positive way, *i.e.*, $R_{\mathcal{F}}\big(B(n+1)\big|V^n\big) < R_{\mathcal{F}}\big(B(n+1)\big|(V \backslash A)^n\big)$, Granger proposes to say that x_A is a *prima facie* cause of x_B relative to V, *prima facie* can be translated as "at a first glance". This is used to insist that if V is enlarged by including other measurements, then the conclusion might be changed. This can be seen as redundant with the mention of the relativity to the observation set V, and we therefore do not use this terminology. However, a mention of the relativity to V must be used, as modification of this set can alter the conclusion. A very simple example of this situation is the chain $x_n \to y_n \to z_n$, where, for example, x_n is an i.i.d. sequence, $y_{n+1} = x_n + \varepsilon_{n+1}$, $z_{n+1} = y_n + \eta_{n+1}$, ε_n, η_n being independent i.i.d. sequences. Relative to $V = \{x, z\}$, x causes z if we use the quadratic loss and linear functions of the past samples of x (note here that the predictor z_{n+1} must be a function of not only x_n, but also of x_{n-1}). However, if we include the past samples of y and $V = \{x, y, z\}$, then the quality of the prediction of z does not deteriorate if we do not use past samples of x. Therefore, x does not cause z relative to $V = \{x, y, z\}$.

The advantage of the prediction-based definition is that it leads to operational tests. If the quadratic loss is chosen, working in a parameterized class of functions, such as linear filters or Volterra filters, or even working in reproducing kernel Hilbert spaces, allows the implementation of the definition [65–67]. In such cases, the test can be evaluated efficiently from the data. From a theoretical point of view, the quadratic loss can be used to find the optimal function in a much wider class of functions: the measurable functions. In this class, the optimal function for the quadratic loss is widely known to be the conditional expectation [68]. When predicting x_B from the whole observation set V, the optimal predictor is written as $\widehat{x_B}(n+1) = E\left[x_B(n+1)\big|x_V^n\right]$. Likewise, elimination of A from V to study its influence on B leads to the predictor $\widehat{x_B}(n+1) = E\left[x_B(n+1)\big|x_B^n, x_C^n\right]$, where $V = C \cup A \cup B$. These estimators are of little use, because they are too difficult, or even impossible, to compute. However, they highlight the importance of conditional distributions $p(x_B(n+1)|x_V^n)$ and $p(x_B(n+1)|x_B^n, x_C^n)$ in the problem of testing whether x_A Granger causes x_B relative to V or not.

2.2. ... To a Probabilistic Definition

The optimal predictors studied above are equal if the conditional probability distributions $p(x_B(n+1)|x_V^n)$ and $p(x_B(n+1)|x_B^n, x_C^n)$ are equal. These distributions are identical if and only if $x_B(n+1)$ and x_A^n are independent conditionally to x_B^n, x_C^n. A natural extension of definition 1 relies on the use of conditional independence. Once again, let $A \cup B \cup C$ be a partition of V.

Definition 2. x_A does not Granger cause x_B relative to V if and only if $x_B(n+1) \perp\!\!\!\perp x_A^n \mid x_B^n, x_C^n, \quad \forall n \in \mathbb{Z}$

This definition means that conditionally to the past of x_C, the past of x_A does not bring more information about $x_B(n+1)$ than is contained in the past of x_B.

Definition 2 is far more general than definition 1. If x_A does not Granger cause x_B relatively to V in the sense of definition 1, it also does not in the sense of definition 2. Then, definition 2 does not rely on any function class and on any cost function. However, it lacks an inherent operational character: the tools to evaluate conditional independence remain to be defined. The assessment of conditional independence can be achieved using measures of conditional independence, and some of these measures will be the cornerstone to link directed information theory and Granger causality.

Note also that the concept of causality in this definition is again a relative concept, and that adding or deleting data from the observation set V might modify the conclusions.

2.3. Instantaneous Coupling

The definitions given so far concern the influence of the past of one process on the present of another. This is one reason that justifies the use of the term "causality", when the definitions are actually based on statistical dependence. For an extensive discussion on the differences between causality and statistical dependence, we refer to [17].

There is another influence between the processes that is not taken into account by definitions 1 and 2. This influence is referred to as "instantaneous causality" [21,27]. However, we will prefer the term "instantaneous coupling", specifically to insist that it is not equivalent to a causal link *per se*, but actually a statistical dependence relationship. The term "contemporaneous conditional independence" that is used in [28] could also be chosen.

Instantaneous coupling measures the common information between $x_A(n+1)$ and $x_B(n+1)$ that is not shared with their past. A definition of instantaneous coupling might then be that $x_A(n+1)$ and

$x_B(n+1)$ are not instantaneously coupled if $x_A(n+1) \perp\!\!\!\perp x_B(n+1) \mid x_A^n, x_B^n, \quad \forall n$. This definition makes perfect sense if the observation set is reduced to A and B, a situation we refer to as the bivariate case. However, in general, there is also side information C, and the definition must include this knowledge. However, this presence of side information then leads to two possible definitions of instantaneous coupling.

Definition 3. x_A and x_B are not conditionally instantaneously coupled relative to V if and only if $x_A(n+1) \perp\!\!\!\perp x_B(n+1) \mid x_A^n, x_B^n, x_C^{n+1}, \quad \forall n \in \mathbb{Z}$, where $A \cup B \cup C$ is a partition of V.

The second possibility is the following:

Definition 4. x_A and x_B are not instantaneously coupled relative to V if and only if $x_A(n+1) \perp\!\!\!\perp x_B(n+1) \mid x_A^n, x_B^n, x_C^n, \quad \forall n \in \mathbb{Z}$

Note that definitions 3 and 4 are symmetrical in A and B (the application of Bayes theorem). The difference between definitions 3 and 4 resides in the conditioning on x_C^{n+1} instead of x_C^n. If the side information up to time n is considered only as in definition 4, the instantaneous dependence or independence is not conditional on the presence of the remaining nodes in C. Thus, this coupling is a bivariate instantaneous coupling: it does measure instantaneous dependence (or independence between A and B) without considering the possible instantaneous coupling between either A and C or B and C. Thus, instantaneous coupling found with definition 4 between A and B does not preclude the possibility that the coupling is actually due to couplings between A and C and/or B and C.

Inclusion of all of the information up to time $n+1$ in the conditioning variables allows the dependence or independence to be tested between $x_A(n+1)$ and $x_B(n+1)$ *conditionally* to $x_C(n+1)$.

We end up here with the same differences as those between correlation and partial correlation, or dependence and conditional independence for random variables. In graphical modeling, the usual graphs are based on conditional independence between variables [19,20]. These conditional independence graphs are preferred to independence graphs because of their geometrical properties (e.g., d-separation [17]), which match the Markov properties possibly present in the multivariate distribution they represent. From a physical point of view, conditional independence might be preferable, specifically to eliminate "false" coupling due to third parties. In this respect, conditional independence is not the panacea, as independent variables can be conditionally dependent. The well-known example is the conditional coupling of independent x and y by their addition. Indeed, even if independent, x and y are conditionally dependent to $z = x + y$.

2.4. More on Graphs

Granger causality graphs were defined and studied in [28]. A causality graph is a mixed graph (V, E_d, E_u) that encodes Granger causality relationships between the components of x_V. The vertex set V stores the indices of the components of x_V. E_d is a set of directed edges between vertices. A directed edge from a to b is equivalent to "x_a Granger causes x_b relatively to V". E_u is a set of undirected edges. An undirected edge between x_a and x_b is equivalent to "x_a and x_b are (conditionally if def.4 adopted) instantaneously coupled". Interestingly, a Granger causality graph may have Markov properties (as in usual graphical models) reflecting a particular (spatial) structure of the joint probability distribution of the whole process $\{x_V^t\}$ [28]. A taxonomy of Markov properties (local, global, block recursive) is studied in [28], and equivalence between these properties is put forward. More interestingly, these properties are linked with topological properties of the graph. Therefore, structural properties of the graphs are equivalent to a particular factorization of the joint probability of the multivariate process. We will not continue on this subject here, but this must be known since it paves the way to more efficient inference methods for Granger graphical modeling of multivariate processes (see first steps in this direction in [69,70]).

3. Directed Information Theory and Directional Dependence

Directed information theory is a recent extension of information theory, even if its roots go back to the 1960s and 1970s and the studies of Marko [10]. The developments began in the late 1990s, after the *impetus* given by James Massey in 1990 [12]. The basic theory was then extended by Gerhard Kramer [11], and then further developed by many authors [56–59,71] to cite a few. We provide here a short review of the essentials of directed information theory. We will, moreover, adopt a presentation close to the spirit of Granger causality to highlight the links between Granger causality and information theory. We begin by recalling some basics from information theory. Then, we describe the information-theoretic approach to study directional dependence between stochastic processes, first in the bivariate case, and then, from Section 3.5, for networks, *i.e.*, the multivariate case.

3.1. Notation and Basics

Let $H(x_A^n) = -E[\log p(x_A^n)]$ be the entropy of a random vector x_A^n, the density of which is p. Let the conditional entropy be defined as $H(x_A^n|x_B^n) = -E[\log p(x_A^n|x_B^n)]$. The mutual information $I(x_A^n; y_B^n)$ between x_A^n and x_B^n is defined as [55]:

$$\begin{aligned} I(x_A^n; x_B^n) &= H(x_B^n) - H(x_B^n|x_A^n) \\ &= D_{KL}\left(p(x_A^n, x_B^n)\|p(x_A^n)p(x_B^n)\right) \end{aligned} \tag{3}$$

where $D_{KL}(p\|q) = E_p[\log p(x)/q(x)]$ is the Kulback–Leibler divergence. $D_{KL}(p\|q)$ is 0 if and only if $p = q$, and it is positive otherwise. The mutual information effectively measures independence since it is 0 if and only if x_A^n and x_B^n are independent random vectors. As $I(x_A^n; x_B^n) = I(x_B^n; x_A^n)$, mutual information cannot handle directional dependence.

Let x_C^n be a third time series. It might be a multivariate process that accounts for side information (all of the available observations, but x_A^n and x_B^n). To account for x_C^n, the conditional mutual information is introduced:

$$\begin{aligned} I(x_A^n; x_B^n|x_C^n) &= E\left[D_{KL}\left(p(x_A^n, x_B^n|x_C^n)\|p(x_A^n|x_C^n)p(x_B^m|x_C^n)\right)\right] \tag{4} \\ &= D_{KL}\left(p(x_A^n, x_B^n, x_C^n)\|p(x_A^n|x_C^n)p(x_B^n|x_C^n)p(x_C^n)\right) \tag{5} \end{aligned}$$

$I(x_A^n; x_B^n|x_C^n)$ is zero if and only if x_A^n and x_B^n are independent *conditionally* to x_C^n. Stated differently, conditional mutual information measures the divergence between the actual observations and those that would be observed under the Markov assumption ($x \to z \to y$). Arrows can be misleading here, as by reversibility of Markov chains, the equality above holds also for ($y \to z \to x$). This emphasizes how mutual information cannot provide answers to the information flow directivity problem.

3.2. Directional Dependence between Stochastic Processes; Causal Conditioning

The dependence between the components of the stochastic process x_V is encoded in the full generality by the joint probability distributions $p(x_V^n)$. If V is partitioned into subsets A, B, C, studying dependencies between A and B then requires that $p(x_V^n)$ is factorized into terms where x_A and x_B appear. For example, as $p(x_V^n) = p(x_A^n, x_B^n, x_C^n)$, we can factorize the probability distribution as $p(x_B^n|x_A^n, x_C^n)p(x_A^n, x_C^n)$, which appears to emphasize a link from A to B. Two problems appear, however: first, the presence of C perturbs the analysis (more than this, A and C have a symmetrical role here); secondly, the factorization does not take into account the arrow of time, as the conditioning is considered over the whole observations up to time n.

Marginalizing x_C out makes it possible to work directly on $p(x_A^n, x_B^n)$. However, this eliminates all of the dependence between A and B that might exist *via* C, and therefore this might lead to an incorrect assessment of the dependence. As for Granger causality, this means that dependence analysis is relative to the observation set. Restricting the study to A and B is what we referred to as the bivariate case, and this allows the basic ideas to be studied. We will therefore present directed information first in the bivariate case, and then turn to the full multivariate case.

The second problem is at the root of the measure of directional dependence between stochastic processes. Assuming that $x_A(n)$ and $x_B(n)$ are linked by some physical (e.g., biological, economical) system, it is natural to postulate that their dependence is constrained by causality: if $A \to B$, then an event occurring at some time in A will influence B later on. Let us come back to the simple factorization above for the bivariate case. We have $p(x_A^n, x_B^n) = p(x_B^n | x_A^n) p(x_A^n)$, and furthermore (We implicitly choose 1 here as the initial time):

$$p(x_B^n | x_A^n) \quad = \quad \prod_{i=1}^{n} p\big(x_B(i) \big| x_B^{i-1}, x_A^n\big) \tag{6}$$

where for $i = 1$, the first term is $p(x_B(1) | x_A(1))$. The conditional distribution quantifies a directional dependence from A to B, but it lacks the causality property mentioned above, as $p\big(x_B(i) | x_B^{i-1}, x_A^n\big)$ quantifies the influence of the whole observation x_A^n (past and future of i) on the present $x_B(i)$ knowing its past x_B^{i-1}. The causality principle would require the restriction of the *prior* time i to the past of A only. Kramer defined "causal conditioning" precisely in this sense [11]. Modifying Equation (6) accordingly, we end up with the definition of the causal conditional probability distribution:

$$p(x_B^n \| x_A^n) \quad := \quad \prod_{i=1}^{n} p\big(x_B(i) \big| x_B^{i-1}, x_A^i\big) \tag{7}$$

Remarkably this provides an alternative factorization of the joint probability. As noted by Massey [12], $p(x_A^n, y_B^n)$ can then be factorized as (x_B^{n-1} stands for the delayed collections of samples of x_B. If the time origin is finite, 0 or 1, the first element of the list x_B^{n-1} should be understood as a wild card $\varnothing$, which does not influence the conditioning.):

$$p(x_A^n, x_B^n) \quad = \quad p(x_B^n \| x_A^n) p(x_A^n \| x_B^{n-1}) \tag{8}$$

Assuming that x_A is the input of a system that creates x_B, $p(x_A^n \| x_B^{n-1}) = \prod_i p(x_A(i) | x_A^{i-1}, x_B^{i-1})$ characterizes the feedback in the system: each of the factors controls the probability of the input x_A at time i conditionally to its past and to the past values of the output x_B. Likewise, the term $p(x_B^n \| x_A^n) = \prod_i p(x_B(i) | x_B^{i-1}, x_A^i)$ characterizes the direct (or feedforward) link in the system.

Several interesting simple cases occur:

- In the absence of feedback in the link from A to B, there is the following:

$$p(x_A(i) | x_A^{i-1}, x_B^{i-1}) = p(x_A(i) | x_A^{i-1}), \ \forall i \geq 2 \tag{9}$$

  or equivalently, in terms of entropies,

$$H(x_A(i) | x_A^{i-1}, x_B^{i-1}) = H(x_A(i) | x_A^{i-1}), \ \forall i \geq 2 \tag{10}$$

  and as a consequence:

$$p(x_A^n \| x_B^{n-1}) = p(x_A^n) \tag{11}$$

- Likewise, if there is only a feedback term, then $p(x_B(i) | x_B^{i-1}, x_A^i) = p(x_B(i) | x_B^{i-1})$ and then:

$$p(x_B^n \| x_A^n) = p(x_B^n) \tag{12}$$

- If the link is memoryless, *i.e.*, the output x_B does not depend on the past, then:

$$p(x_B(i) | x_A^i, y_B^{i-1}) = p(x_B(i) | x_A(i)) \ \forall i \geq 1 \tag{13}$$

These results allow the question of whether x_A influences x_B to be addressed. If it does, then the joint distribution has the factorization of Equation (8). However, if x_A does not influence x_B, then $p(x_B^n \| x_A^n) = p(x_B^n)$, and the factorization of the joint probability distribution simplifies to $p(x_A^n \| x_B^{n-1}) p(x_B^n)$. Kullback divergence between the probability distributions for each case generalizes the definition of mutual information to the directional mutual information:

$$I(x_A^n \to x_B^n) = D_{KL}\left(p(x_A^n, x_B^n) \| p(x_A^n \| x_B^{n-1}) p(x_B^n)\right) \tag{14}$$

This quantity measures the loss of information when it is incorrectly assumed that x_A does not influence x_B. This was called *directed information* by Massey [12]. Expanding the Kullback divergence allows different forms for the directed information to be obtained:

$$I(x_A^n \to x_B^n) = \sum_{i=1}^{n} I\left(x_A^i; x_B(i) \big| x_B^{i-1}\right) \tag{15}$$

$$= H(x_B^n) - H(x_B^n \| x_A^n) \tag{16}$$

where we define the "causal conditional entropy":

$$H(x_B^n \| x_A^n) = -E\left[\log p(x_B^n \| x_A^n)\right] \tag{17}$$

$$= \sum_{i=1}^{n} H\left(x_B(i) \big| x_B^{i-1}, x_A^i\right) \tag{18}$$

Note that causal conditioning might involve more than one process. This leads to define the causal conditional directed information as:

$$I(x_A^n \to x_B^n \| x_C^n) := H(x_B^n \| x_C^n) - H(x_B^n \| x_A^n, x_C^n)$$

$$= \sum_{i=1}^{n} I\left(x_A^i; x_B(i) \big| x_B^{i-1}, x_C^i\right) \tag{19}$$

The basic properties of the directed information were studied by Massey and Kramer [11,12,72], and some are recalled below. As a Kullback divergence, the directed information is always positive or zero. Then, simple algebraic manipulation allows the decomposition to be obtained:

$$I(x_A^n \to x_B^n) + I(x_B^{n-1} \to x_A^n) = I(x_A^n; x_B^n) \tag{20}$$

Equation (20) is fundamental, as it shows how mutual information splits into the sum of a feedforward information flow $I(x_A^n \to x_B^n)$ and a feedback information flow $I(x_B^{n-1} \to x_A^n)$. In the absence of feedback, $p(x_A^n \| x_B^{n-1}) = p(x_A^n)$ and $I(x_A^n; x_B^n) = I(x_A^n \to x_B^n)$. Equation (20) allows the conclusion that the mutual information is always greater than the directed information, as $I(x_B^{n-1} \to x_A^n)$ is always positive or zero (as directed information). It is zero if and only if:

$$I(x_A(i); x_B^{i-1} | x_A^{i-1}) = 0 \ \forall i = 2, \ldots, n \tag{21}$$

or equivalently:

$$H(x_A(i) | x_A^{i-1}, x_B^{i-1}) = H(x_A(i) | x_A^{i-1}) \ \forall i = 2, \ldots, n \tag{22}$$

This situation corresponds to the absence of feedback in the link $A \to B$, whence the fundamental result that the directed information and the mutual information are equal if the channel is free of feedback. This result implies that mutual information over-estimates the directed information between two processes in the presence of feedback. This was thoroughly studied in [11,57–59], in a communication-theoretic framework.

The decomposition of Equation (20) is surprising, as it shows that the mutual information is not the sum of the directed information flowing in both directions. Instead, the following decomposition holds:

$$I(x_A^n \to x_B^n) + I(x_B^n \to x_A^n) \;=\; I(x_A^n; x_B^n) + I(x_A^n \to x_B^n \| x_A^{n-1}) \tag{23}$$

where:

$$
\begin{aligned}
I(x_A^n \to x_B^n \| x_A^{n-1}) &= \sum_i I(x_A^i; x_B(i) | x_B^{i-1}, x_A^{i-1}) \\
&= \sum_i I(x_A(i); x_B(i) | x_B^{i-1}, x_A^{i-1})
\end{aligned}
\tag{24}
$$

This demonstrates that $I(x_A^n \to x_B^n) + I(x_B^n \to x_A^n)$ is symmetrical, but is in general not equal to the mutual information, except if and only if $I(x_A(i); x_B(i) | x_B^{i-1}, x_A^{i-1}) = 0, \forall i = 1, \ldots, n$. As the term in the sum is the mutual information between the present samples of the two processes conditioned on their joint past values, this measure is a measure of instantaneous dependence. It is indeed symmetrical in A and B. The term $I(x_A^n \to x_B^n \| x_A^{n-1}) = I(x_B^n \to x_A^n \| x_B^{n-1})$ will thus be named the *instantaneous information exchange* between x_A and x_B, and will hereafter be denoted as $I(x_A^n \leftrightarrow x_B^n)$. Like directed information, conditional forms of the instantaneous information exchange can be defined, as for example:

$$I(x_A^n \leftrightarrow x_B^n \| x_C^n) := I(x_A^n \to x_B^n \| x_A^{n-1}, x_C^n) \tag{25}$$

which quantifies an instantaneous information exchange between A and B causally conditionally to C.

3.3. Directed Information Rates

Entropy and mutual information in general increase linearly with the length n of the recorded time series. Shannon's information rate for stochastic processes compensates for the linear growth by considering $A_\infty(x) = \lim_{n \to +\infty} A(x^n)/n$ (if the limit exists), where $A(x^n)$ denotes any information measure on the sample x^n of length n. For the important class of stationary processes (see e.g., [55]), the entropy rate turns out to be the limit of the conditional entropy:

$$\lim_{n \to +\infty} \frac{1}{n} H(x_A^n) = \lim_{n \to +\infty} H(x_A(n) | x_A^{n-1}) \tag{26}$$

Kramer generalized this result for causal conditional entropies [11], thus defining the directed information rate for stationary processes as:

$$
\begin{aligned}
I_\infty(x_A \to x_B) &= \lim_{n \to +\infty} \frac{1}{n} \sum_{i=1}^{n} I(x_A^i; x_B(i) | x_B^{i-1}) \\
&= \lim_{n \to +\infty} I(x_A^n; x_B(n) | x_B^{n-1})
\end{aligned}
\tag{27}
$$

This result holds also for the instantaneous information exchange rate. Note that the proof of the result relies on the positivity of the entropy for discrete valued stochastic processes. For continuously valued processes, for which the entropy can be negative, the proof is more involved and requires the methods developed in [73–75], and see also [58].

3.4. Transfer Entropy and Instantaneous Information Exchange

As introduced by Schreiber in [13,47], *transfer entropy* evaluates the deviation of the observed data from a model, assuming the following joint Markov property:

$$p(x_B(n) | x_{B\,n-k+1}^{n-1}, x_{A\,n-l+1}^{n-1}) = p(x_B(n) | x_{B\,n-k+1}^{n-1}) \tag{28}$$

This leads to the following definition:

$$T(x_{A\,n-l+1}^{\quad n-1} \to x_{B\,n-k+1}^{\quad n}) = E\left[\log \frac{p(x_B(n)|x_{B\,n-k+1}^{\quad n-1}, x_{A\,n-l+1}^{\quad n-1})}{p(x_B(n)|x_{B\,n-k+1}^{\quad n-1})}\right] \tag{29}$$

Then $T(x_{A\,n-l+1}^{\quad n-1} \to x_{B\,n-k+1}^{\quad n}) = 0$ if and only if Equation (28) is satisfied. Although in the original definition, the past of x in the conditioning might begin at a different time $m \neq n$, for practical reasons $m = n$ is considered. Actually, no *a priori* information is available about possible delays, and setting $m = n$ allows the transfer entropy to be compared with the directed information.

By expressing the transfer entropy as a difference of conditional entropies, we get:

$$\begin{aligned}
T(x_{A\,n-l+1}^{\quad n-1} \to x_{B\,n-k+1}^{\quad n}) &= H(x_B(n)|x_{B\,n-k+1}^{\quad n-1}) - H(x_B(n)|x_{B\,n-k+1}^{\quad n-1}, x_{A\,n-l+1}^{\quad n-1}) \\
&= I(x_{A\,n-l+1}^{\quad n-1}; x_B(n)|x_{B\,n-k+1}^{\quad n-1})
\end{aligned} \tag{30}$$

For $l = n = k$ and choosing 1 as the time origin, the identity $I(x,y;z|w) = I(x;z|w) + I(y;z|x,w)$ leads to:

$$\begin{aligned}
I(x_A^n; x_B(n)|x_B^{n-1}) &= I(x_A^{n-1}; x_B(n)|x_B^{n-1}) + I(x_A(n); x_B(n)|x_A^{n-1}, x_B^{n-1}) \\
&= T(x_A^{n-1} \to x_B^n) + I(x_A(n); x_B(n)|x_A^{n-1}, x_B^{n-1})
\end{aligned} \tag{31}$$

For stationary processes, letting $n \to \infty$ and provided the limits exist, for the rates, we obtain:

$$I_\infty(x_A \to x_B) = T_\infty(x_A \to x_B) + I_\infty(x_A \leftrightarrow x_B) \tag{32}$$

Transfer entropy is the part of the directed information that measures the influence of the past of x_A on the present of x_B. However it does not take into account the possible instantaneous dependence of one time series on another, which is handled by directed information.

Moreover, as defined by Schreiber in [13,47], only $I(x_A^{i-1}; x_B(i)|x_B^{i-1})$ is considered in T, instead of its sum over i in the directed information. Thus stationarity is implicitly assumed and the transfer entropy has the same meaning as a rate. A sum over delays was considered by Paluš as a means of reducing errors when estimating the measure [48]. Summing over n in Equation (31), the following decomposition of the directed information is obtained:

$$I(x_A^n \to x_B^n) = I(x_A^{n-1} \to x_B^n) + I(x_A^n \leftrightarrow x_B^n) \tag{33}$$

Equation (33) establishes that the influence of one process on another can be decomposed into two terms that account for the past and for the instantaneous contributions. Moreover, this explains the presence of the term $I(x_A^n \leftrightarrow x_B^n)$ in the r.h.s. of Equation (23): Instantaneous information exchange is counted twice in the l.h.s. terms $I(x_A^n \to x_B^n) + I(x_B^n \to x_A^n)$, but only once in the mutual information $I(x_A^n; x_B^n)$. This allows Equation (23) to be written in a slightly different form, as:

$$I(x_A^{n-1} \to x_B^n) + I(x_B^{n-1} \to x_A^n) + I(x_A^n \leftrightarrow x_B^n) = I(x_A^n; x_B^n) \tag{34}$$

which is very appealing, as it shows how dependence as measured by mutual information decomposes as the sum of the measures of directional dependences and the measure of instantaneous coupling.

3.5. Accounting for Side Information

The preceding developments aimed at the proposing of definitions of the information flow between x_A and x_B; however, whenever A and B are connected to other parts of the network, the flow of information between A and B might be mediated by other members of the network. Time series observed on nodes other than A and B are hereafter referred to as side information. The available side information at time n is denoted as x_C^n, with A, B, C forming a partition of V. Then, depending on the type of conditioning (usual or causal) two approaches are possible. Usual conditioning considers directed information from A to B that is conditioned on the whole observation x_C^n. However, this leads to the consideration of causal flows from A to B that possibly include a flow that goes from A to B *via* C in the future! Thus, an alternate definition for conditioning is required. This is given by the definition of Equation (19) of the causal conditional directed information:

$$I(x_A^n \to x_B^n \| x_C^n) \quad := \quad H(x_B^n \| x_C^n) - H(x_B^n \| x_A^n, x_C^n)$$

$$= \quad \sum_{i=1}^n I(x_A^i; x_B(i) | x_B^{i-1}, x_C^i) \tag{35}$$

Does the causal conditional directed information decompose as the sum of a causal conditional transfer entropy and a causal conditional instantaneous information exchange, as it does in the bivariate case? Applying twice the chain rule for conditional mutual information, we obtain:

$$I(x_A^n \to x_B^n \| x_C^n) = I(x_A^{n-1} \to x_B^n \| x_C^{n-1}) + I(x_A^n \leftrightarrow x_B^n \| x_C^n) + \Delta I(x_C^n \leftrightarrow x_B^n) \tag{36}$$

In this equation, $I(x_A^{n-1} \to x_B^n \| x_C^{n-1})$ is termed the "causal conditional transfer entropy". This measures the flow of information from A to B by taking into account a possible route *via* C. If the flow of information from A to B is entirely relayed by C, the "causal conditional transfer entropy" is zero. In this situation, the usual transfer entropy is not zero, indicating the existence of a flow from A to B. Conditioning on C allows the examination of whether the route goes through C. The term:

$$I(x_A^n \leftrightarrow x_B^n \| x_C^n) \quad := \quad I(x_A^n \to x_B^n \| x_A^{n-1}, x_C^n) \tag{37}$$

$$= \quad \sum_{i=1}^n I(x_A(i); x_B(i) | x_B^{i-1}, x_A^{i-1}, x_C^i) \tag{38}$$

is the "causal conditional information exchange". It measures the conditional instantaneous coupling between A and B. The term $\Delta I(x_C^n \leftrightarrow x_B^n)$ emphasizes the difference between the bivariate and the multivariate cases. This extra term measures an instantaneous coupling and is defined by:

$$\Delta I(x_C^n \leftrightarrow x_B^n) = I(x_C^n \leftrightarrow x_B^n \| x_A^{n-1}) - I(x_C^n \leftrightarrow x_B^n) \tag{39}$$

An alternate decomposition to Equation (36) is:

$$I(x_A^n \to x_B^n \| x_C^n) = I(x_A^{n-1} \to x_B^n \| x_C^n) + I(x_A^n \leftrightarrow x_B^n \| x_C^n) \tag{40}$$

which emphasizes that the extra term comes from:

$$I(x_A^{n-1} \to x_B^n \| x_C^n) = I(x_A^{n-1} \to x_B^n \| x_C^{n-1}) + \Delta I(x_C^n \leftrightarrow x_B^n) \tag{41}$$

This demonstrates that the definition of the conditional transfer entropy requires conditioning on the past of C. If not, the extra term appears and accounts for instantaneous information exchanges between C and B, due to the addition of the term $x_C(i)$ in the conditioning. This extra term highlights the difference between the two different natures of instantaneous coupling. The first term,

$$I(x_C^n \leftrightarrow x_B^n \| x_A^{n-1}) = \sum_i I(x_C(i); x_B(i) | x_A^{i-1}, x_B^{i-1}, x_C^{i-1}) \tag{42}$$

describes the intrinsic coupling in the sense that it does not depend on parties other than C and B. The second coupling term,

$$I(x_C^n \leftrightarrow x_B^n) = \sum_i I(x_C(i); x_B(i) | x_B^{i-1}, x_C^{i-1})$$

is relative to the extrinsic coupling, as it measures the instantaneous coupling at time i that is created by variables other than B and C.

As discussed in Section 2.3, the second definition for instantaneous coupling considers conditioning on the past of the side information *only*. Causally conditioning on x_C^{n-1} does not modify the results of the bivariate case. In particular, we still get the elegant decomposition:

$$I(x_A^n \to x_B^n \| x_C^{n-1}) = I(x_A^{n-1} \to x_B^n \| x_C^{n-1}) + I(x_A^n \leftrightarrow x_B^n \| x_C^{n-1}) \tag{43}$$

and therefore, the decomposition of Equation (34) is generalized to:

$$I(x_A^{n-1} \to x_B^n \| x_C^{n-1}) + I(x_B^{n-1} \to x_A^n \| x_C^{n-1}) + I(x_A^n \leftrightarrow x_B^n \| x_C^{n-1}) = I(x_A^n; x_B^n \| x_C^{n-1}) \tag{44}$$

where:

$$I(x_A^n; x_B^n \| x_C^{n-1}) = \sum_i I\left(x_A^n; x_B(i) \big| x_B^{i-1}, x_C^{i-1}\right) \tag{45}$$

is the causally conditioned mutual information.

Finally, let us consider that for jointly stationary time series, the causal directed information rate is defined similarly to the bivariate case as:

$$I_\infty(x_A \to x_B \| x_C) = \lim_{n \to +\infty} \frac{1}{n} \sum_{i=1}^n I\left(x_A^i; x_B(i) | x_B^{i-1}, x_C^i\right) \tag{46}$$

$$= \lim_{n \to +\infty} I\left(x_A^n; x_B(n) | x_B^{n-1}, x_C^n\right) \tag{47}$$

In this section we have emphasized on Kramer's causal conditioning, both for the definition of directed information and for taking into account side information. We have also shown that Schreiber's transfer entropy is the part of the directed information that is dedicated to the strict sense of causal information flow (not accounting for simultaneous coupling). The next section more explicitly revisits the links between Granger causality and directed information theory.

4. Inferring Granger Causality and Instantaneous Coupling

Granger causality in its probabilistic form is not operational. In practical situations, for assessing Granger causality between time series, we cannot use the definition directly. We have to define dedicated tools to assess the conditional independence. We use this inference framework to show the links between information theory and Granger causality. We begin by re-expressing Granger causality definitions in terms of some measures that arise from directed information theory. Therefore, in an inference problem, these measures can be used as tools for inference. However, we show in the following sections that these measures naturally emerge from the more usual statistical inference strategies. In the following, and as above, we use the same partitioning of V into the union of disjoint subsets of A, B and C.

4.1. Information-theoretic Measures and Granger Causality

As anticipated in the presentation of directed information, there are profound links between Granger causality and directed information measures. Granger causality relies on conditional independence, and it can also be defined using measures of conditional independence. Information-theoretic measures appear as natural candidates. Recall that two random elements are independent if and only if their mutual information is zero. Moreover, two random elements are independent conditionally to a third one if and only if the conditional mutual information is zero. We can reconsider definitions 2, 3 and 4 and recast them in terms of information-theoretic measures.

Definition 2 stated that x_A does not Granger cause x_B relative to V if and only if $x_B(r+1) \perp\!\!\!\perp x_A^n \mid x_B^n, x_C^n, \;\; \forall n \geq 1$. This can be alternatively rephrased into:

Definition 5. x_A does not Granger cause x_B relative to V if and only if $I(x_A^{n-1} \to x_B^n \| x_C^{n-1}) = 0 \;\; \forall n \geq 1$

since $x_B(i) \perp\!\!\!\perp x_A^i \mid x_A^{i-1}, x_C^{i-1}, \;\; \forall 1 \leq i \leq n$ is equivalent to $I(x_B(i); x_A^i \mid x_A^{i-1}, x_C^{i-1}) = 0 \;\; \forall 1 \leq i \leq n$.

Otherwise stated, the transfer entropy from A to B causally conditioned on C is zero if and only if A does not Granger cause B relative to V. This shows that causal conditional transfer entropy can be used to assess Granger causality.

Likewise, we can give alternative definitions of instantaneous coupling.

Definition 6. x_A and x_B are not conditionally instantaneously coupled relative to V if and only if $I(x_A^n \leftrightarrow x_B^n \| x_C^n) \forall n \geq 1$,

or if and only if the instantaneous information exchange causally conditioned on C is zero. The second possible definition of instantaneous coupling is equivalent to:

Definition 7. x_A and x_B are not instantaneously coupled relative to V if and only if $I(x_A^n \leftrightarrow x_B^n \| x_C^{n-1}) \forall n \geq 1$,

or if and only if the instantaneous information exchange causally conditioned on the past of C is zero.

Note that in the bivariate case only (when C is not taken into account), the directed information $I(x_A^n \to x_B^n)$ summarizes both the Granger causality and the coupling, as it decomposes as the sum of the transfer entropy $I(x_A^{n-1} \to x_B^n)$ and the instantaneous information exchange $I(x_A^{n-1} \leftrightarrow x_B^n)$.

4.2. Granger Causality Inference

We consider the practical problem of inferring the graph of dependence between the components of a multivariate process. Let us assume that we have measured a multivariate process $x_V(n)$ for $n \leq T$. We want to study the dependence between each pair of components (Granger causality and instantaneous coupling between any pair of components relative to V).

We can use the result of the preceding section to evaluate the directed information measures on the data. When studying the influence from any subset A to any subset B, if the measures are zero, then there is no causality (or no coupling); if they are strictly positive, then A Granger causes B relative to V (or A and B are coupled relative to V). This point of view has been adopted in many of the studies that we have already referred to (e.g., [14,16,37,47,76]), and it relies on estimating the measures from the data. We will not review the estimation problem here.

However, it is interesting to examine more traditional frameworks for testing Granger causality, and to examine how directed information theory naturally emerges from these frameworks. To begin with, we show how the measures defined emerge from a binary hypothesis-testing view of Granger causality inference. We then turn to prediction and model-based approaches. We will review how Geweke's measures of Granger causality in the Gaussian case are equivalent to directed information measures. We will then present a more general case adopted by [37–39,77–79] and based on a model of the data.

4.2.1. Directed Information Emerges from a Hypotheses-testing Framework

In the inference problem, we want to determine whether or not x_A Granger causes (is coupled with) x_B relative to V. This can be formulated as a binary hypothesis testing problem. For inferring dependencies between A and B relative to V, we can state the problem as follows.

Assume we observe $x_V(n), \forall n \leq T$. Then, we want to test: "x_A does not Granger cause x_B", against "x_A causes x_B"; and "x_A and x_B are instantaneously coupled" against "x_A are x_B not instantaneously coupled". We will refer to the first test as the Granger causality test, and to the second one, as the instantaneous coupling test.

In the bivariate case, for which the Granger causality test indicates:

$$\begin{cases} H_0 & : \quad p_0(x_B(i) \mid x_A^{i-1}, x_B^{i-1}) \quad = \quad p(x_B(i) \mid x_B^{i-1}), \forall i \leq T \\ H_1 & : \quad p_1(x_B(i) \mid x_A^{i-1}, x_B^{i-1}) \quad = \quad p(x_B(i) \mid x_A^{i-1}, x_B^{i-1}), \forall i \leq T \end{cases} \tag{48}$$

this leads to the testing of different functional forms of the conditional densities of $x_B(i)$ given the past of x_A. The likelihood of the observation under H_1 is the full joint probability $p(x_A^T, x_B^T) = p(x_A^T \| x_B^T) p(x_B^T \| x_A^{T-1})$. Under H_0 we have $p(x_B^T \| x_A^{T-1}) = p(x_B^T)$ and the likelihood reduces to $p(x_A^T \| x_B^T) p(x_B^T \| x_A^{T-1}) = p(x_A^T \| x_B^T) p(x_B^T)$. The log likelihood *ratio* for the test is:

$$l(x_A^T, x_B^T) \quad := \quad \log \frac{p(x_A^T, x_B^T \mid H_1)}{p(x_A^T, x_B^T \mid H_0)} = \log \frac{p(x_B^T \| x_A^{T-1})}{p(x_B^T)} \tag{49}$$

$$= \quad \sum_{i=1}^{T} \log \frac{p(x_B(i) \mid x_A^{i-1}, x_B^{i-1})}{p(x_B(i) \mid x_B^{i-1})} \tag{50}$$

For example, in the case where the multivariate process is a positive Harris recurrent Markov chain [80], the law of large numbers applies and we have under hypothesis H_1:

$$\frac{1}{T} l(x_A^T, x_B^T) \xrightarrow{T \to +\infty} T_\infty(x_A \to x_B) \text{ a.s.} \tag{51}$$

where $T_\infty(x_A \to x_B)$ is the transfer entropy rate. Thus from a practical point of view, as the amount of data increases, we expect the log likelihood *ratio* to be close to the transfer entropy rate (under H_1). Turning the point of view, this can justify the use of an estimated transfer entropy to assess Granger causality. Under H_0, $\frac{1}{T} l(x_A^T, x_B^T)$ converges to $\lim_{T \to +\infty}(1/T) D_{KL}\big(p(x_A^T \| x_B^T) p(x_B^T) \| p(x_A^T \| x_B^T) p(x_B^T \| x_A^{T-1})\big)$, which can be termed "the Lautum transfer entropy rate" that extends the "Lautum directed information" defined in [71]. Directed information can be viewed as a measure of the loss of information when assuming x_A does not causally influence x_B when it actually does. Likewise, "Lautum directed information" measures the loss of information when assuming x_A does causally influence x_B, when actually it does not.

For testing instantaneous coupling, we will use the following:

$$\begin{cases} H_0 & : \quad p_0(x_A(i), x_B(i) \mid x_A^{i-1}, x_B^{i-1}) \quad = \quad p(x_A(i) \mid x_A^{i-1}, x_B^{i-1}) p(x_B(i) \mid x_A^{i-1}, x_B^{i-1}), \forall i \leq T \\ H_1 & : \quad p_1(x_A(i), x_B(i) \mid x_A^{i-1}, x_B^{i-1}) \quad = \quad p(x_A(i), x_B(i) \mid x_A^{i-1}, x_B^{i-1}), \forall i \leq T \end{cases} \tag{52}$$

where under H_0, there is no coupling. Then, under H_1 and some hypothesis on the data, the likelihood ratio converges almost surely to the information exchange rate $I_\infty(x_A \leftrightarrow x_B)$.

A related encouraging result due to [71] is the emergence of the directed information in the false-alarm probability error rate. Merging the two tests Equations (48) and (52), *i.e.*, testing both for causality and coupling, or neither, the test is written as:

$$\begin{cases} H_0 & : \quad p_0(x_B(i) \mid x_A^i, x_B^{i-1}) \quad = \quad p(x_B(i) \mid x_B^{i-1}), \forall i \leq T \\ H_1 & : \quad p_1(x_B(i) \mid x_A^i, x_B^{i-1}) \quad = \quad p(x_B(i) \mid x_A^i, x_B^{i-1}), \forall i \leq T \end{cases} \tag{53}$$

Among the tests with a probability of miss P_M that is lower than some positive value $\varepsilon > 0$, the best probability of false alarm P_{FA} follows $\exp\big(-T I(x_A \to x_B)\big)$ when T is large. For the case studied here, this is the so-called Stein lemma [55]. In the multivariate case, there is no such result in the literature. An extension is proposed here. However, this is restricted to the case of instantaneously *uncoupled* time series. Thus, we assume for the end of this subsection that:

$$p(x_A(i), x_B(i), x_C(i) \mid x_A^{i-1}, x_B^{i-1}, x_C^{i-1}) = \prod_{\alpha = A,B,C} p(x_\alpha(i) \mid x_A^{i-1}, x_B^{i-1}, x_C^{i-1}), \; \forall i \leq T \tag{54}$$

which means that there is no instantaneous exchange of information between the three subsets that form a partition of V. This assumption has held in most of the recent studies that have applied Granger causality tests. It is, however, unrealistic in applications where the dynamics of the processes involved are faster than the sampling period adopted (see [27] for a discussion in econometry). Consider now the problem of testing Granger causality of A on B relative to V. The binary hypothesis test is given by:

$$\begin{cases} H_0 & : \quad p_0(x_B(i) \mid x_A^{i-1}, x_B^{i-1}, x_C^{i-1}) \quad = \quad p(x_B(i) \mid x_B^{i-1}, x_C^{i-1}), \forall i \leq T \\ H_1 & : \quad p_1(x_B(i) \mid x_A^{i-1}, x_B^{i-1}, x_C^{i-1}) \quad = \quad p(x_B(i) \mid x_A^{i-1}, x_B^{i-1}, x_C^{i-1}), \forall i \leq T \end{cases} \tag{55}$$

The log likelihood *ratio* reads as:

$$l(x_A^T, x_B^T, x_C^T) \quad = \quad \sum_{i=1}^{T} \log \frac{p(x_B(i) \mid x_A^{i-1}, x_B^{i-1}, x_C^{i-1})}{p(x_B(i) \mid x_B^{i-1}, x_C^{i-1})} \tag{56}$$

Again, by assuming that the law of large numbers applies, we can conclude that under H_1

$$\frac{1}{T} l(x_A^T, x_B^T, x_C^T) \xrightarrow{T \to +\infty} T_\infty(x_A \to x_B \| x_C) \text{ a.s.} \tag{57}$$

This means that the causal conditional transfer entropy rate is the limit of the log likelihood *ratio* as the amount of data increases.

4.2.2. Linear Prediction based Approach and the Gaussian Case

Following definition 1 and focusing on linear models and the quadratic risk $R(e) = E[e^2]$, Geweke introduced the following indices for the study of stationary processes [31,32]:

$$F_{x_A \leftrightarrow x_B} \quad = \quad \lim_{n \to +\infty} \frac{R(x_B(n) | x_B^{n-1}, x_A^{n-1})}{R(x_B(n) | x_B^{n-1}, x_A^n)} \tag{58}$$

$$F_{x_A \leftrightarrow x_B \| x_C} \quad = \quad \lim_{n \to +\infty} \frac{R(x_B(n) | x_B^{n-1}, x_A^{n-1}, x_C^n)}{R(x_B(n) | x_B^{n-1}, x_A^n, x_C^n)} \tag{59}$$

$$F_{x_A \to x_B} \quad = \quad \lim_{n \to +\infty} \frac{R(x_B(n) | x_B^{n-1})}{R(x_B(n) | x_B^{n-1}, x_A^{n-1})} \tag{60}$$

$$F_{x_A \to x_B \| x_C} \quad = \quad \lim_{n \to +\infty} \frac{R(x_B(n) | x_B^{n-1}, x_C^{n-1})}{R(x_B(n) | x_B^{n-1}, x_A^{n-1}, x_C^{n-1})} \tag{61}$$

Geweke demonstrated the efficiency of these indices for testing Granger causality and instantaneous coupling (bivariate and multivariate cases). In the particular Gaussian and bivariate case, he gave explicit results for the statistics of the tests, and furthermore he showed that:

$$F_{x_A \to x_B} + F_{x_B \to x_A} + F_{x_A \leftrightarrow x_B} = I_\infty(x_A; x_B) \tag{62}$$

where $I_\infty(x_A; x_B)$ is the mutual information rate. This relationship, which was already sketched out in [21], is nothing but Equation (34). Indeed, in the Gaussian case, $F_{x_A \leftrightarrow x_B} = I_\infty(x_A \leftrightarrow x_B)$ and $F_{x_A \to x_B} = I_\infty(x_A \to x_B)$ stem from the knowledge that the entropy rate of a Gaussian stationary process is the logarithm of the asymptotic power of the one-step-ahead prediction [55]. Likewise, we can show that $F_{x_A \leftrightarrow x_B \| x_C} = I_\infty(x_A \leftrightarrow x_B \| x_C)$ and $F_{x_A \to x_B \| x_C} = I_\infty(x_A \to x_B \| x_C)$ holds.

In the multivariate case, conditioning on the past of the side information, *i.e.*, x_C^{n-1}, in the definition of $F_{x_A \leftrightarrow x_B \| x_C}$, a decomposition analogous to Equation (62) holds and is exactly that given by Equation (44).

4.2.3. The Model-based Approach

In a more general framework, we examine how a model-based approach can be used to test for Granger causality, and how directed information comes into play.

Let us consider a rather general model in which $x_V(t)$ is a multivariate Markovian process that statisfies:

$$x_V(t) = f_\theta\left(x_{Vt-k}^{t-1}\right) + w_V(t) \tag{63}$$

where $f_\theta : \mathbb{R}^{k|V|} \longrightarrow \mathbb{R}^{|V|}$ is a function belonging to some functional class $\mathcal{F}$, and where w_V is a multivariate i.i.d. sequence, the components of which are not necessarily mutually independent. Function f_θ might (or might not) depend on θ, a multidimensional parameter. This general model considers each signal as an AR model (linear or not) with exogenous inputs; f_θ can also stand for a function belonging to some reproducing kernel Hilbert space, which can be estimated from the data [66,67,81]. Using the partition A, B, C, this model can be written equivalently as:

$$\begin{cases} x_A(t) &= f_{A,\theta_A}\left(x_{At-k}^{t-1}, x_{Bt-k}^{t-1}, x_{Ct-k}^{t-1}\right) + w_A(t) \\ x_B(t) &= f_{B,\theta_B}\left(x_{At-k}^{t-1}, x_{Bt-k}^{t-1}, x_{Ct-k}^{t-1}\right) + w_B(t) \\ x_C(t) &= f_{C,\theta_C}\left(x_{At-k}^{t-1}, x_{Bt-k}^{t-1}, x_{Ct-k}^{t-1}\right) + w_C(t) \end{cases} \tag{64}$$

where the functions $f_{.,\theta}$ are the corresponding components of f_θ. This relation can be used for inference in a parametric setting: the functional form is assumed to be known and the determination of the function is replaced by the estimation of the parameters $\theta_{A,B,C}$. This can also be used in a nonparametric setting, in which case the function f is searched for in an appropriate functional space, such as an rkHs associated to a kernel [81].

In any case, for studying the influence of x_A to x_B relative to V, two models are required for x_B: one in which x_B explicitly depends on x_A, and the other one in which x_B does not depend on x_A. In the parametric setting, the two models can be merged into a single model, in such a way that some components of the parameter θ_B are, or not, zero, which depends on whether A causes B or not. The procedure then consists of testing nullity (or not) of these components. In the linear Gaussian case, this leads to the Geweke indices discussed above. In the nonlinear (non-Gaussian) case, the Geweke indices can be used to evaluate the prediction in some classes of nonlinear models (in the minimum mean square error sense). In this latter case, the decomposition of the mutual information, Equation (62), has no reason to remain valid.

Another approach base relies on directly modeling the probability measures. This approach has been used recently to model spiking neurons and to infer Granger causality between several neurons working in the class of generalized linear models [37,79]. Interestingly, the approach has been used either to estimate the directed information [37,77] or to design a likelihood ratio test [38,79]. Suppose we wish to test whether "x_A Granger causes x_B relative to V" as a binary hypothesis problem (as in Section 4.2.1). Forgetting the problem of instantaneous coupling, the problem is then to choose between the hypotheses:

$$\begin{cases} H_0 &: \quad p_0(x_B(i) \mid x_V^{i-1}) &= \quad p(x_B(i) \mid x_V^{i-1}; \theta_0), \forall i \le T \\ H_1 &: \quad p_1(x_B(i) \mid x_V^{i-1}) &= \quad p(x_B(i) \mid x_V^{i-1}; \theta_1), \forall i \le T \end{cases} \tag{65}$$

where the existence of causality is entirely reflected into the parameter θ. To be more precise, θ_0 should be seen as a restriction of θ_1 when its components linked to x_A are set to zero. As a simple example using the model approach discussed above, consider the simple linear Gaussian model

$$x_B(t) = \sum_{i>0} \theta_A(i) x_A(t-i) + \sum_{i>0} \theta_B(i) x_B(t-i) + \sum_{i>0} \theta_C(i) x_C(t-i) + w_B(t) \tag{66}$$

where $w_B(t)$ is an i.i.d. Gaussian sequence, and $\theta_A, \theta_B, \theta_C$ are multivariate impulse responses of appropriate dimensions. Define $\theta_1 = (\theta_A, \theta_B, \theta_C)$ and $\theta_0 = (0, \theta_B, \theta_C)$. Testing for Granger causality is then equivalent to testing $\theta = \theta_1$; furthermore, the likelihood *ratio* can be implemented due to the Gaussian assumption. The example developed in [37,79] assumes that the probability that neuron b $(b \cup A \cup C = V)$ sends a message at time t $(x_b(t) = 1)$ to its connected neighbors is given by the conditional probability

$$\Pr\big(x_b(t) = 1\big|x_V^t; \theta\big) = U\Big(\sum_{i>0} \theta_A(i) x_A(t-i) + \sum_{i>0} \theta_b(i) x_b(t-i) + \sum_{i>0} \theta_{Eb}(i) x_{Eb}(t-i) + w_b(t) \Big)$$

where U is some decision function, the output of which belongs to $[0;1]$, A represents the subset of neurons that can send information to b, and Eb represents external inputs to b. Defining this probability for all $b \in V$ completely specifies the behavior of the neural network V.

The problem is a composite hypothesis testing problem, in which parameters defining the likelihoods have to be estimated. It is known that there is no definitive answer to this problem [82]. An approach that relies on an estimation of the parameters using maximum likelihood can be used. Letting Ω be the space where parameter θ is searched for and Ω_0 the subspace where θ_0 lives, then the generalized log likelihood *ratio* test reads:

$$l(x_A^T, x_B^T) \quad := \quad \log \frac{\sup_{\theta \in \Omega} p(x_V^T; \theta)}{\sup_{\theta \in \Omega_0} p(x_V^T; \theta)} = \log \frac{p(x_V^T; \widehat{\theta_1^T})}{p(x_V^T; \widehat{\theta_0^T})} \tag{67}$$

where $\widehat{\theta_i^T}$ denotes the maximum likelihood estimator of θ under hypothesis i. In the linear Gaussian case, we will recover exactly the measures developed by Geweke. In a more general case, and as illustrated in Section 4.2.1, as the maximum likelihood estimates are efficient, we can conjecture that the generalized log likelihood *ratio* will converge to the causal conditional transfer entropy rate if sufficiently relevant conditions are imposed on the models (e.g., Markov processes with recurrent properties). This approach was described in [38] in the bivariate case.

5. Discussion and Extensions

Granger causality was developed originally in econometrics, and it is now transdisciplinary, with the literature on the subject being widely dispersed. We have tried here to sum up the profound links that exist between Granger causality and directed information theory. The key ingredients to build these links are conditional independence and the recently introduced causal conditioning.

We have eluded the important question of the practical use of the definitions and measures presented here. Some of the measures can be used and implemented easily, especially in the linear Gaussian case. In a more general case, different approaches can be taken. The information-theoretic measures can be estimated, or the prediction can be explicitly carried out and the residuals used to assess causality.

Many studies have been carried out over the last 20 years on the problem of estimation of information-theoretic measures. We refer to [83–87] for information on the different ways to estimate information measures. Recent studies into the estimation of entropy and/or information measures are [88–90]. The recent report by [76] extensively details and applies transfer entropy in neuroscience using k-nearest neighbors type of estimators. Concerning the applications, important reviews include [14,16], where some of the ideas discussed here are also mentioned, and where practicalities such as the use of surrogate data, for example, are extensively discussed. Applications for neuroscience are discussed in [15,30,50,51,79].

Information-theoretic measures of conditional independence based on Kullback divergence were chosen here to illustrate the links between Granger causality and (usual) directed information theory. Other type of divergence could have been chosen (see e.g., [91,92]); metrics in probability space could also be useful in the assessing of conditional independence. As an illustration, we refer to the study of Fukumizu and co-workers [93], where conditional independence was evaluated using the Hilbert–Schmidt norm of an operator between reproducing kernel Hilbert spaces. The operator generalizes the partial covariance between two random vectors given a third one, and is called the conditional covariance operator. Furthermore, the Hilbert–Schmidt norm of conditional covariance operator can be efficiently estimated from data. A related approach is also detailed in [94].

Many important directions can be followed. An issue is in the time horizon over which the side information is considered in definition 2. As done for instantaneous coupling, we could have chosen to condition by x_C^{n+1} instead of x_C^n. This proposition made recently in [35,95] allows in certain circumstances to eliminate the effect of common inputs to A, B and C. It is denoted as partial Granger causality. As noted in [35] this is particularly useful when the common inputs are very powerful and distributed equally likely among all the nodes. If this definition is adopted, then according to Equation (40), the directed information $I(x_A^n \to x_B^n \| x_C^n)$ decomposes as the sum of instantaneous information exchange $I(x_A^n \leftrightarrow x_B^n \| x_C^n)$ with the adequate formulation of the transfer entropy for this definition $I(x_A^{n-1} \to x_B^n \| x_C^n)$. Despite this nice result, a definitive interpretation remains unclear within the probabilistic description presented here. Even in the usual linear setting as developed in [32] this definition leads to some difficulties. Indeed, Geweke's analysis relies on the possibility to invert the Wold decomposition of the time series, representing the times series as a possibly infinite autoregression with the innovation sequence as input. All the existing dynamical structure (finite order autoregression and moving average input representing exogenous inputs) is then captured by Geweke's approach. The analysis in [35,95] assumes that residuals may not be white, and identifiability issues may then arise in this case. Other important issues are the following. Causality between nonstationary processes has rarely been considered (see however [76] for an *ad-hoc* approach in neuroscience). A very promising methodology is to adopt a graphical modeling way of thinking. The result of [28] on the structural properties of Markov–Granger causality graphs can be used to identify such graphs from real datasets. First steps in this direction were proposed by [69,70]. Assuming that the network under study is a network of sparsely connected nodes and that some Markov properties hold, efficient estimation procedures can be designed, as is the case in usual graphical modeling.

Acknowledgments: P.O.A. is supported by a Marie Curie International Outgoing Fellowship from the European Community. We thank the anonymous reviewers for pointing out to interesting references that were missing in the early draft of the paper.

References

1. Rao, A.; Hero, A.O.; States, D.J.; Engel, J.D. Inference of Biologically Relevant Gene Influence Networks Using the Directed Information Criterion. In Proceedings of the ICASSP, Toulouse, France, 15–19 May, 2006.
2. Rao, A.; Hero, A.O.; States, D.J.; Engel, J.D. Motif discovery in tissue-specific regulatory sequences using directed information. *EURASIP J. Bioinf. Syst. Biol.* **2007**, 13853.
3. Granger, C.W.J. Investigating causal relations by econometrics models and cross-spectral methods. *Econometrica* **1969**, *37*, 424–438.
4. Sims, C.A. Money, income and causality. *Am. Econ. Rev.* **1972**, *62*, 540–552.
5. Sporns, O. *The Networks of the Brain*; MIT Press: Cambridge, MA, USA, 2010.
6. Kaufmann, R.K.; Stern, D.I. Evidence for human influence on climat from hemispheric temperature relations. *Nature* **1997**, *388*, 39–44.
7. Triacca, U. On the use of Granger causality to investigate the human influence on climate. *Theor. Appl. Clim.* **2001**, *69*, 137–138.
8. Wiener, N. The theory of prediction. In *Modern Mathematics for the Engineer*; MacGrawHill: New York, NY, USA, 1956; pp. 165–190.
9. Granger, C.W.J. Times Series Anlaysis, Cointegration and Applications. In *Nobel Lecture*; Stockholm, Sweden, 2003.
10. Marko, H. The bidirectional communication theory–a generalization of information theory. *IEEE Trans. Commun.* **1973**, *21*, 1345–1351.

11. Kramer, G. Directed Information for Channels with Feedback. PhD thesis, Swiss Federal Institute of Technology Zurich, 1998.

12. Massey, J.L. Causality, Feedback and Directed Information. In Proceedings of the International Symposium on Information Theory and its Applications, Waikiki, HI, USA, November 1990.

13. Schreiber, T. Measuring information transfer. *Phys. Rev. Lett.* **2000**, *85*, 461–465.

14. Hlavackova-Schindler, K.; Palus, M.; Vejmelka, M.; Bhattacharya, J. Causality detection based on information-theoretic approaches in time series analysis. *Phys. Rep.* **2007**, *441*, 1–46.

15. Gourévitch, B.; Bouquin-Jeannès, R.L.; Faucon, G. Linear and nonlinear causality between signals: Methods, example and neurophysiological applications. *Biol. Cybern.* **2006**, *95*, 349–369.

16. Palus, M. From nonlinearity to causality: Statistical testing and inference of physical mechanisms underlying complex dynamics. *Contemp. Phys.* **2007**, *48*, 307–348.

17. Pearl, J. *Causality: Models, Reasoning and Inference*; Cambridge University Press: Cambridge, UK, 2000.

18. Lauritzen, S. Chapter 2: Causal inference from graphical models. In *Complex Stochastic Systems*; Barndroff-Nielsen, O., Cox, D.R., Kluppelberg, C., Eds.; Chapman and Hall: London, UK, 2001; pp. 63–108.

19. Lauritzen, S. *Graphical Models*; Oxford University Press: Oxford, UK, 1996.

20. Whittaker, J. *Graphical Models in Applied Multivariate Statistics*; Wiley& Sons: Weinheim, Germany, 1989.

21. Granger, C.W.J. Economic processes involving feedback. *Inf. Control* **1963**, *6*, 28–48.

22. Caines, P.E.; Chan, C.W. Feedback between stationary stochastic processes. *IEEE Trans. Autom. Control* **1975**, *20*, 498–508.

23. Hosoya, Y. On the granger condition for non-causality. *Econometrica* **1977**, *45*, 1735–1736.

24. Chamberlain, G. The general equivalence of granger and sims causality. *Econometrica* **1982**, *50*, 569–581.

25. Florens, J.P.; Mouchart, M. A note on noncausality. *Econometrica* **1982**, *50*, 583–591.

26. Granger, C.W.J. Some recent developments in a concept of causality. *J. Econ.* **1988**, *39*, 199–211.

27. Granger, C.W.J. Testing for causality: A personal viewpoint. *J. Econ. Dyn. Control* **1980**, *2*, 329–352.

28. Eichler, M. Graphical modeling of multivariate time series. *Proba. Theory Relat. Fields* **2011**, doi: 10.1007/s00440-011-0345-8.

29. Dahlaus, R.; Eichler, M. Causality and graphical models in time series analysis. In *Highly Structured Stochastic Systems*; Green, P., Hjort, N., Richardson, S., Eds.; Oxford University Press: Oxford, UK, 2003.

30. Eichler, M. On the evaluation of information flow in multivariate systems by the directed transfer function. *Biol. Cybern.* **2006**, *94*, 469–482.

31. Geweke, J. Measurement of linear dependence and feedback between multiple time series. *J. Am. Stat. Assoc.* **1982**, *77*, 304–313.

32. Geweke, J. Measures of conditional linear dependence and feedback between times series. *J. Am. Stat. Assoc.* **1984**, *79*, 907–915.

33. Amblard, P.O.; Michel, O.J.J. Sur Différentes Mesures de Dépendance Causales Entre Signaux Alé Atoires (On Different Measures of Causal Dependencies between Random Signals). In Proceedings of the GRETSI, Dijon, France, September 8-11, 2009.

34. Barnett, L.; Barrett, A.B.; Seth, A.K. Granger causality and transfer entropy are equivalent for Gaussian variables. *Phys. Rev. Lett.* **2009**, *103*, 238707.

35. Barrett, A.B.; Barnett, L.; Seth, A.K. Multivariate Granger causality and generalized variance. *Phys. Rev. E* **2010**, *81*, 041907.

36. Amblard, P.O.; Michel, O.J.J. On directed information theory and Granger causality graphs. *J. Comput. Neurosci.* **2011**, *30*, 7–16.

37. Quinn, C.J.; Coleman, T.P.; Kiyavash, N.; Hastopoulos, N.G. Estimating the directed information to infer causal relationships in ensemble neural spike train recordings. *J. Comput. Neurosci.* **2011**, *30*, 17–44.

38. Gouriéroux, C.; Monfort, A.; Renault, E. Kullback causality measures. *Ann. Econ. Stat.* **1987**, *(6–7)*, 369–410.

39. Rissanen, J.; Wax, M. Measures of mutual and causal dependence between two time series. *IEEE Trans. Inf. Theory* **1987**, *33*, 598–601.

40. Kullback, S. *Information Theory and Statistics*; Dover: NY, USA, 1968.

41. Quian Quiroga, R.; Arnhold, J.; Grassberger, P. Learning driver-response relashionship from synchronisation patterns. *Phys. Rev. E* **2000**, *61*, 5142–5148.

42. Lashermes, B.; Michel, O.J.J.; Abry, P. Measuring Directional Dependences of Information Flow between Signal and Systems. In Proceedings of the PSIP'03, Grenoble, France, 29–31 January, 2003.

43. Le Van Quyen, M.; Martinerie, J.; Adam, C.; Varela, F. Nonlinear analyses of interictal eeg map the brain interdependences in human focal epilepsy. *Physica D* **1999**, *127*, 250–266.

44. Kantz, H.; Schreiber, T. *Nonlinear Time Series Analysis*, 2nd ed.; Cambridge University Press: Cambridge, UK, 2004.

45. Palus, M.; Komarek, V.; Hrncir, Z.; Sterbova, K. Synchronisation as adjustment of information rates: Detection from bivariate time series. *Phys. Rev. E* **2001**, *63*, 046211:1–6.

46. Frenzel, S.; Pompe, B. Partial mutual information for coupling analysis of multivariate time series. *Phys. Rev. Lett.* **2007**, *99*, 204101.

47. Kaiser, A.; Schreiber, T. Information transfer in continuous processes. *Physica D* **2002**, *166*, 43–62.

48. Palus, M.; Vejmelka, M. Directionality of coupling from bivariate time series: How to avoid false causalities and missed connections. *Phys. Rev. E* **2007**, *75*, 056211:2–056211:14.

49. Vejmelka, M.; Palus, M. Inferring the directionality of coupling with conditional mutual information. *Phys. Rev. E* **2008**, *77*, 026214.

50. Eichler, M. A graphical approach for evaluating effective connectivity in neural systems. *Phil. Trans. R. Soc. B* **2005**, *360*, 953–967.

51. Kaminski, M.; Ding, M.; Truccolo, W.; Bressler, S. Evaluating causal relations in neural systems: Granger causality, directed transfer functions and statistical assessment of significance. *Biol. Cybern.* **2001**, *85*, 145–157.

52. Lungarella, M.; Sporns, O. Mapping information flow in sensorimotor networks. *PLOS Comput. Biol.* **2006**, *2*, 1301–1312.

53. Mosedale, T.J.; Stephenson, D.B.; Collins, M.; Mills, T.C. Granger causality of coupled climate processes: Ocean feedback on the north Atlantic oscillation. *J. Clim.* **2006**, *19*, 1182–1194.

54. Saito, Y.; Harashima, H. *Recent Advances in EEG and EMG Data Processing*; chapter Tracking of information within multichannel EEG record-causal analysis in EEG. Elsevier: Amsterdam, The Netherlands, 1981; pp. 133–146.

55. Cover, J.; Thomas, B. *Elements of Information Theory*, 2nd ed.; Wiley: Weinheim, Germany, 2006.

56. Kim, Y.H. A coding theorem for a class of stationary channel with feedback. *IEEE Trans. Inf. Theory* **2008**, *54*, 1488–1499.

57. Tatikonda, S.C. *Control Under Communication Constraints*; PhD thesis, MIT: Cambridge, MA, USA, 2000.

58. Tatikonda, S.; Mitter, S. The capacity of channels with feedback. *IEEE Trans. Inf. Theory* **2009**, *55*, 323–349.

59. Venkataramanan, R.; Pradhan, S.S. Source coding with feed-forward: Rate-distortion theorems and error exponents for a general source. *IEEE Trans Inf. Theory* **2007**, *53*, 2154–2179.

60. Amblard, P.O.; Michel, O.J.J. Information Flow through Scales. In Proceedings of the IMA Conference on Maths and Signal processing, Cirencester, UK, 16–18 December, 2008; p. 78

61. Amblard, P.O.; Michel, O.J.J. Measuring information flow in networks of stochastic processes. **2009**, arXiv:0911.2873.

62. Solo, V. On Causality and Mutual Information. In Proceedings of the 47th IEEE conference on Decision and Control, Cancun, Mexico, 9–11 December, 2008.

63. Kamitake, T.; Harashima, H.; Miyakawa, H.; Saito, Y. A time-series analysis method based on the directed transinformation. *Electron. Commun. Jpn.* **1984**, *67*, 1–9.

64. Al-Khassaweneh, M.; Aviyente, S. The relashionship between two directed information measures. *IEEE Sig. Proc. Lett.* **2008**, *15*, 801–804.

65. Amblard, P.-O.; Michel, O.J.J.; Richard, C.; Honeine, P. A Gaussian Process Regression Approach for Testing Granger Causality between Time Series Data. In Proceedings of the ICASSP, Osaka, Japan, 25–30 March, 2012.

66. Amblard, P.O.; Vincent, R.; Michel, O.J.J.; Richard, C. Kernelizing Geweke's Measure of Granger Causality. In Proceedings of the IEEE Workshop on MLSP, Santander, Spain, 23–26 September, 2012.

67. Marinazzo, D.; Pellicoro, M.; Stramaglia, S. Kernel-Granger causality and the analysis of dynamical networks. *Phys. Rev. E* **2008**, *77*, 056215.

68. Lehmann, E.L.; Casella, G. *Theory of Point Estimation*, 2nd ed.; Springer: Berlin/Heidelberg, Germany, 1998.

69. Quinn, C.J.; Kiyavas, N.; Coleman, T.P. Equivalence between Minimal Generative Model Graphs and Directed Information Graph. In Proceeding of the ISIT, St. Petersburg, Russia, 31 July–5 August, 2011.

70. Runge, J.; Heitzig, J.; Petoukhov, V.; Kurths, J. Escping the curse of dimensionality in estimating multivariate transfer entropy. *Phys. Rev. Lett.* **2012**, *108*, 258701.

71. Permuter, H.H.; Kim, Y.-H.; Weissman, T. Interpretations of directed information in portfolio theory, data compression, and hypothesis testing. *IEEE Trans. Inf. Theory* **2011**, *57*, 3248–3259.

72. Massey, J.L.; Massey, P.C. Conservation of Mutual and Directed Information. In Proceedings of the International Symposium on Information Theory and its Applications, Adelalaïde, Australia, 4–7 September, 2005.

73. Gray, R.M.; Kieffer, J.C. Mutual information rate, distorsion and quantization in metric spaces. *IEEE Trans. Inf. Theory* **1980**, *26*, 412–422.

74. Gray, R.M. *Entropy and Information Theory*; Springer-Verlag: Berlin/Heidelberg, Germany, 1990.

75. Pinsker, M.S. *Information and Information Stability of Random Variables*; Holden Day: San Francisco, USA, 1964.

76. Vicente, R.; Wibral, M.; Lindner, M.; Pipa, G. Transfer entropy–a model-free measure of effective connectivity for the neurosciences. *J. Comput. Neurosci.* **2011**, *30*, 45–67.

77. Barnett, L.; Bossomaier, T. Transfer entropy as log-likelihood ratio. *Phys. Rev. Lett.* **2012**, *109*, 138105.

78. Kim , S.; Brown, E.N. A General Statistical Framework for Assessing Granger Causality. In Proceedings of IEEE Icassp, Prague, Czech Republic, 22–27 May, 2010; pp. 2222–2225.

79. Kim, S.; Putrino, D.; Ghosh, S.; Brown, E.N. A Granger causality measure for point process models of ensembled neural spiking activity. *PLOS Comput. Biol.* **2011**, doi:10.1371/journal.pcbi.1001110.

80. Meyn, S.; Tweedie, R.L. *Markov Chains and Stochastic Stability*, 2nd ed.; Cambridge University Press: Cambridge, UK, 2009.

81. Schölkopf, B.; Smola, A.J. *Learning with Kernels*; MIT Press: Cambridge, MA, USA, 2002.

82. Lehmann, E.L.; Romano, J.P. *Testing Statistical Hypotheses*, 3rd ed.; Springer: Berlin/Heidelberg, Germany, 2005.

83. Beirlant, J.; Dudewicz, E.J.; Gyorfi, L.; van Der Meulen, E.C. Nonparametric entropy estimation: An overview. *Int. J. Math. Stat. Sci.* **1997**, *6*, 17–39.

84. Goria, M.N.; Leonenko, N.N.; Mergell, V.V.; Novi Invardi, P.L. A new class of random vector entropy estimators and its applications in testing statistical hypotheses. *J. Nonparam. Stat.* **2005**, *17*, 277–297.

85. Kraskov, A.; Stogbauer, H.; Grassberger, P. Estimating mutual information. *Phys. Rev. E* **2004**, *69*, 066138.

86. Kozachenko, L.F.; Leonenko, N.N. Sample estimate of the entropy of a random vector. *Problems Inf. Trans.* **1987**, *23*, 95–101.

87. Paninski, L. Estimation of entropy and mutual information. *Neural Comput.* **2003**, *15*, 1191–1253.

88. Leonenko, N.N.; Pronzato, L.; Savani, V. A class of Rényi information estimators for multidimensional densities. *Ann. Stat.* **2008**, *36*, 2153–2182.

89. Sricharan, K.; Raich, R.; Hero, A.O. Estimation of non-linear functionals of densities with confidence. *IEEE Trans. Inf. Theory* **2012**, *58*, 4135–4159.

90. Wang, Q.; Kulkarni, S.; Verdu, S. Divergence estimation for multidimensional densities via-nearest-neighbor distances. *IEEE Trans. Inf. Theory* **2009**, *55*, 2392–2405.

91. Basseville, M. Divergence measures for statistical data processing—an annotated bibliography. *Signal Process.* **2012**, in press.

92. Bercher, J.F. Escort entropies and divergences and related canonical distribution. *Phys. Lett. A* **2011**, *375*, 2969–2973.

93. FukumizuU, K.; Gretton, A.; Sun, X.; Scholkopf, B. Kernel Measures of Conditional Dependence. In *NIPS*; Vancouver: Canada, 3–8 December, 2007.

94. Seth, S.; Príncipe, J.C. Assessing Granger non-causality using nonparametric measure of conditional independence. *IEEE Trans. Neural Netw. Learn. Syst.* **2012**, *23*, 47–59.

95. Guo, S.; Seth, A.K.; Kendrick, K.M.; Zhou, C.; Feng, J. Partial Granger causality–eliminating exogeneous inputs and latent variables. *J. Neurosci. Methods* **2008**, *172*, 79–93.

Article

Moving Frames of Reference, Relativity and Invariance in Transfer Entropy and Information Dynamics

Joseph T. Lizier [1,2,3,]* and John R. Mahoney [4,5,]*

1   CSIRO Information and Communications Technology Centre, PO Box 76, Epping, NSW 1710, Australia
2   School of Information Technologies, The University of Sydney, NSW 2006, Australia
3   Max Planck Institute for Mathematics in the Sciences, Inselstrasse 22, D-04103 Leipzig, Germany
4   Physics Department, University of California, Davis, CA 95616, USA
5   School of Natural Sciences, University of California, Merced, CA 95344, USA

*   Author to whom correspondence should be addressed; E-Mail: joseph.lizier@csiro.au (J.T.L.); jmahoney3@ucmerced.edu (J.R.M.); Tel.: +61 2 9372 4711; Fax: +61 2 9372 4161.

Received: 16 November 2012 / Accepted: 31 December 2012 / Published: 10 January 2013

Abstract: We present a new interpretation of a local framework for information dynamics, including the transfer entropy, by defining a moving frame of reference for the observer of dynamics in lattice systems. This formulation is inspired by the idea of investigating "relativistic" effects on observing the dynamics of information—in particular, we investigate a Galilean transformation of the lattice system data. In applying this interpretation to elementary cellular automata, we demonstrate that using a moving frame of reference certainly alters the observed spatiotemporal measurements of information dynamics, yet still returns meaningful results in this context. We find that, as expected, an observer will report coherent spatiotemporal structures that are moving in their frame as information transfer, and structures that are stationary in their frame as information storage. Crucially, the extent to which the shifted frame of reference alters the results depends on whether the shift of frame retains, adds or removes relevant information regarding the source-destination interaction.

Keywords: information theory; information transfer; information storage; transfer entropy; information dynamics; cellular automata; complex systems

PACS: 89.75.Fb; 89.75.Kd; 89.70.+c; 05.65.+b

1. Introduction

Einstein's theory of relativity postulates that the laws of physics are the same for observers in all *moving frames of reference* (no frame is preferred) and that the speed of light is the same in all frames [1]. These postulates can be used to quantitatively describe the differences in measurements of the same events made by observers in different frames of reference.

Information-theoretic measures are always computed with reference to some *observer*. They are highly dependent on how the observer measures the data, the subtleties of how an observer asks a question of the data, how the observer attempts to interpret information from the data, and what the observer already knows [2,3]. We aim to take inspiration from the theory of relativity to explore the effect of a *moving observer* on information-theoretic measures here. To make such an investigation however, we need not only an observer for the information measures but specifically:

(1)   a *space-time interpretation* for the relevant variables in the system; and

(2)   some *frame of reference* for the observer, which can be moving in space-time in the system while the measures are computed.

A candidate for such investigations is a recently introduced framework for information dynamics [4–8], which measures information storage, transfer and modification at each local point in a spatiotemporal system. This framework has had success in various domains, particularly in application to cellular automata (CAs), a simple but theoretically important class of discrete dynamical system that is set on a regular space-time lattice. In application to CAs, the framework has provided quantitative evidence for long-held conjectures that the *moving* coherent structures known as particles are the dominant information transfer entities and that collisions between them are information modification events. In considering the dynamics of information, the framework examines the state updates of each variable in the system with respect to the past state of that variable. For example, in examining the information transfer into a destination variable using the *transfer entropy* [9], we consider how much information was contributed from some source, *in the context of* the past state of that destination. This past state can be seen as akin to a stationary frame of reference for the measurement. As such, we have the possibility to use this framework to explore "relativistic" effects on information; *i.e.*, as applied to a spatiotemporal system such as a CA, with a spatiotemporally moving frame of reference. We begin our paper by introducing CAs in Section 2, basic information-theoretic quantities in Section 3, and the measures for information dynamics in Section 4.

Our primary concern in this paper then lies in exploring a new interpretation of this framework for information dynamics by defining and incorporating a moving frame of reference for the observer (Section 5). The type of relativity presented for application to these lattice systems is akin to an *ether* relativity, where there is a preferred stationary frame in which information transfer is limited by the speed of light. (We note the existence of a discretized *special* relativity for certain CAs by Smith [10]. For special relativity to be applicable, the CA laws must obey the same rules in all frames of reference. Smith notes the difficulty to find any non-trivial CA rules that meet this requirement, and indeed uses only a simple diffusion process as an example. While in principle we could apply our measures within moving frames of reference in that particular discretization, and intend to do so in future work, we examine only an ether-type of relativity in this study, as this is more naturally applicable to lattice systems.) We also mathematically investigate the shift of frame to demonstrate the *invariance* of certain information properties. That is, while the total information required to predict a given variable's value remains the same, shifting the frame of reference redistributes that information amongst the measurements of information storage and transfer by the observer. The nature of that redistribution will depend on whether the shift of frame retains, adds or removes relevant information regarding the source-destination interactions.

We perform experiments on elementary cellular automata (ECAs) using the new perspective on information dynamics with shifted frames of reference in Section 6, comparing the results to those found in the stationary frame. We find that, as expected, the use of a moving frame of reference has a dramatic effect on the measurements of information storage and transfer, though the results are well-interpretable in the context of the shifted frame. In particular, particles only appear as information transfer in frames in which they are moving, otherwise they appear as information storage.

2. Dynamics of Computation in Cellular Automata

Cellular automata (CAs) have been a particular focus for experimentation with the framework for the information dynamics measures that we use here. This is because CAs have been used to model a wide variety of real-world phenomena (see [11]), and have attracted much discussion regarding the nature of computation in their dynamics.

CAs are discrete dynamical systems consisting of an array of cells that each synchronously update their state as a function of the states of a fixed number of spatially neighboring cells using a uniform rule. We focus on *Elementary CAs*, or *ECAs*, a simple variety of 1D CAs using binary states, deterministic rules and one neighbor on either side (*i.e.*, cell range $r = 1$). An example evolution of an ECA may be seen in Figure 1(a). For more complete definitions, including that of the Wolfram rule number convention for describing update rules (used here), see [12].

Studies of information dynamics in CAs have focused on their emergent structure: *particles, gliders, blinkers* and *domains*. A domain is a set of background configurations in a CA, any of which will update to another configuration in the set in the absence of any disturbance. Domains are formally defined by computational mechanics as spatial process languages in the CA [13]. Particles are considered to be dynamic elements of coherent spatiotemporal structure, as disturbances or in contrast to the background domain. Gliders are regular particles, and blinkers are stationary gliders. Formally, particles are defined by computational mechanics as a boundary between two domains [13]; as such, they can be referred to as *domain walls*, though this term is usually reserved for irregular particles. Several techniques exist to *filter* particles from background domains (e.g., [5–7,13–20]). As a visual example, see Figure 1(a) and Figure 1(b) – the horizontally moving gliders in Figure 1(a) are filtered using negative values of the measure in Figure 1(b) (which will be introduced in Section 4.1), while the domains (in the background) and the blinkers (the stationary large triangular structures) in Figure 1(a) are filtered using positive values of the measure in Figure 1(b).

These emergent structures have been quite important to studies of computation in CAs, for example in the design or identification of universal computation in CAs (see [11]), and in the analyses of the dynamics of intrinsic or other specific computation ([13,21,22]). This is because these studies typically discuss the computation in terms of the three primitive functions of computation and their apparent analogues in CA dynamics [11,21]:

- blinkers as the basis of information storage, since they periodically repeat at a fixed location;
- particles as the basis of information transfer, since they communicate information about the dynamics of one spatial part of the CA to another part; and
- collisions between these structures as information modification, since collision events combine and modify the local dynamical structures.

Previous to recent work however [4–7] (as discussed in Section 4), these analogies remained conjecture only.

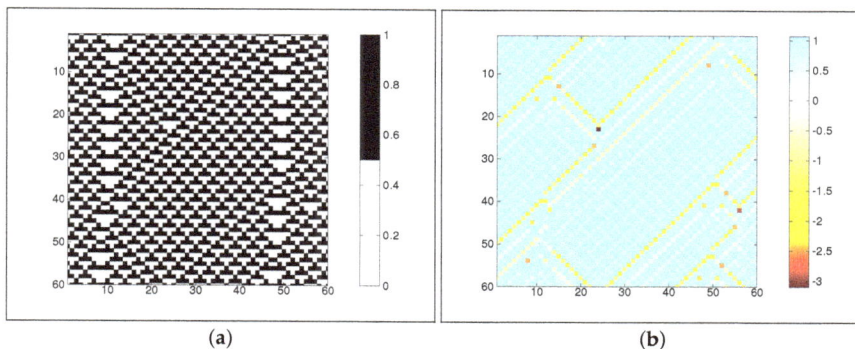

(a)   (b)

Figure 1. *Cont.*

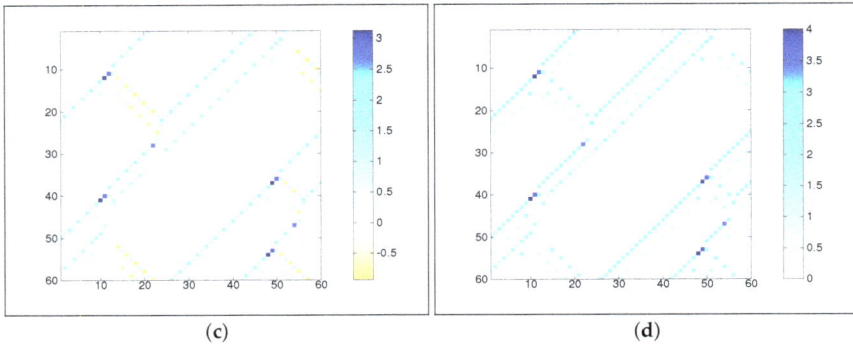

Figure 1. Measures of information dynamics applied to ECA Rule 54 with a stationary frame of reference (all units in (b)–(d) are in bits). Time increases down the page for all plots. (a) Raw CA; (b) Local active information storage $a(i, n, k = 16)$; (c) Local apparent transfer entropy $t(i, j = -1, n, k = 16)$; (d) Local complete transfer entropy $t^c(i, j = -1, n, k = 16)$.

3. Information-theoretic Quantities

To quantify these dynamic functions of computation, we look to information theory (e.g., see [2,3]) which has proven to be a useful framework for the design and analysis of complex self-organized systems, e.g., [23–27]. In this section, we give a brief overview of the fundamental quantities which will be built on in the following sections.

The *Shannon entropy* represents the uncertainty associated with any measurement x of a random variable X (logarithms are in base 2, giving units in bits): $H(X) = -\sum_x p(x) \log p(x)$. The *joint entropy* of two random variables X and Y is a generalization to quantify the uncertainty of their joint distribution: $H(X, Y) = -\sum_{x,y} p(x, y) \log p(x, y)$. The *conditional entropy* of X given Y is the average uncertainty that remains about x when y is known: $H(X|Y) = -\sum_{x,y} p(x, y) \log p(x|y)$. The *mutual information* between X and Y measures the average reduction in uncertainty about x that results from learning the value of y, or vice versa: $I(X; Y) = H(X) - H(X|Y)$. The *conditional mutual information* between X and Y given Z is the mutual information between X and Y when Z is known: $I(X; Y|Z) = H(X|Z) - H(X|Y, Z)$.

Moving to *dynamic* measures of information in time-series processes X, the *entropy rate* is the limiting value of the average entropy of the next realizations x_{n+1} of X conditioned on the realizations $x_n^{(k)} = \{x_{n-k+1}, \dots, x_{n-1}, x_n\}$ of the previous k values $X^{(k)}$ of X (up to and including time step n):

$$H_{\mu X} = \lim_{k \to \infty} H\left[X|X^{(k)}\right] = \lim_{k \to \infty} H_{\mu X}(k) \tag{1}$$

Finally, the *effective measure complexity* [28] or *excess entropy* [23] quantifies the total amount of structure or memory in a system, and is computed in terms of the slowness of the approach of the entropy rate estimates to their limiting value (see [23]). For our purposes, it is best formulated as the mutual information between the semi-infinite past and semi-infinite future of the process:

$$E_X = \lim_{k \to \infty} I\left[X^{(k)}; X^{(k^+)}\right] \tag{2}$$

where $X^{(k^+)}$ refers to the next k states with realizations $x^{(k^+)} = \{x_{n+1}, x_{n+2}, \dots, x_{n+k}\}$. This interpretation is known as the *predictive information* [29], as it highlights that the excess entropy captures the information in a system's past that can also be found in its future.

4. Framework for Information Dynamics

A local framework for information dynamics has recently been introduced in [4–8]. This framework examines the information composition of the next value x_{n+1} of a destination variable, in terms of how much of that information came from the past state of that variable (*information storage*), how much came from respective source variables (*information transfer*), and how those information sources were combined (*information modification*). The measures of the framework provide *information profiles* quantifying each element of computation at each spatiotemporal point in a complex system.

In this section, we describe the information storage and transfer components of the framework (the information modification component is not studied here; it may be seen in [6]). We also review example profiles of these information dynamics in ECA rule 54 (see raw states in Figure 1(a)). ECA rule 54 is considered a class IV complex rule, contains simple glider structures and collisions, and is therefore quite useful in illustrating the concepts around information dynamics.

4.1. Information Storage

We define *information storage* as the amount of information from the past of a process that is relevant to or will be used at some point in its future. The *statistical complexity* [30] measures the amount of information in the past of a process that is *relevant* to the prediction of its future states. It is known that the statistical complexity $C_{\mu X}$ provides an upper bound to the excess entropy [31]; *i.e.*, $E_X \leq C_{\mu X}$. This can be interpreted in that the statistical complexity measures *all* information stored by the system that *may be used* in the future, whereas the excess entropy only measures the information that *is used* by the system *at some point* in the future. Of course, this means that the excess entropy measures information storage that will possibly but not necessarily be used at the next time step $n + 1$. When focusing on the *dynamics* of information processing, we are particularly interested in how much of the stored information is actually *in use* at the next time step, so as to be examined in conjunction with information transfer.

As such, the *active information storage* A_X was introduced [7] to explicitly measure how much of the information from the past of the process is observed to be *in use* in computing its next state. The active information storage is the average mutual information between realizations $x_n^{(k)}$ of the past state $X^{(k)}$ (as $k \to \infty$) and the corresponding realizations x_{n+1} of the *next value* X' of a given time series X:

$$A_X = \lim_{k \to \infty} A_X(k) \tag{3}$$

$$A_X(k) = I\left[X^{(k)}; X'\right] \tag{4}$$

We note that the limit $k \to \infty$ is required in general, unless the next value x_{n+1} is conditionally independent of the far past values $x_{n-k}^{(\infty)}$ given $x_n^{(k)}$.

We can then extract the *local* active information storage $a_X(n + 1)$ [7] as the amount of information storage attributed to the specific configuration or realization $(x_n^{(k)}, x_{n+1})$ at time step $n + 1$; *i.e.*, the amount of information storage in use by the process at the particular time-step $n + 1$: (Descriptions of the manner in which local information-theoretical measures are obtained from averaged measures may be found in [5,31].)

$$A_X = \langle a_X(n + 1) \rangle_n \tag{5}$$

$$a_X(n + 1) = \lim_{k \to \infty} a_X(n + 1, k) \tag{6}$$

$$A_X(k) = \langle a_X(n + 1, k) \rangle_n \tag{7}$$

$$a_X(n + 1, k) = \log_2 \frac{p(x_n^{(k)}, x_{n+1})}{p(x_n^{(k)}) p(x_{n+1})} \tag{8}$$

$$= i(x_n^{(k)}; x_{n+1}) \tag{9}$$

By convention, we use lower case labels for the local values of information-theoretic quantities. Note that $A_X(k)$ and $a(i, n + 1, k)$ represent finite k estimates.

Where the process of interest exists for cells on a lattice structure, we include the index i to identify the variable of interest. This gives the following notation for local active information storage $a(i, n + 1)$ in a spatiotemporal system:

$$a(i, n + 1) = \lim_{k \to \infty} a(i, n + 1, k) \tag{10}$$

$$a(i, n + 1, k) = \log_2 \frac{p(x_{i,n}^{(k)}, x_{i,n+1})}{p(x_{i,n}^{(k)}) p(x_{i,n+1})} \tag{11}$$

We note that the local active information storage is defined for every spatiotemporal point (i, n) in the lattice system. We have $A(i, k) = \langle a(i, n, k) \rangle_n$ as the average for variable i. For stationary systems of homogeneous variables where the probability distribution functions are estimated over all variables, it is appropriate to average over all variables also, giving:

$$A(k) = \langle a(i, n, k) \rangle_{i,n} \tag{12}$$

Figure 2(a) shows the local active information as this mutual information between the destination cell and its past history. Importantly, $a(i, n, k)$ may be positive or negative, meaning the past history of the cell can either positively inform us or actually *misinform* us about its next state. An observer is misinformed where, conditioned on the past history, the observed outcome was *relatively* unlikely as compared with the unconditioned probability of that outcome (*i.e.*, $p(x_{n+1}|x_n^{(k)}) < p(x_{n+1})$). In deterministic systems (e.g., CAs), negative local active information storage means that there must be strong information transfer from other causal sources.

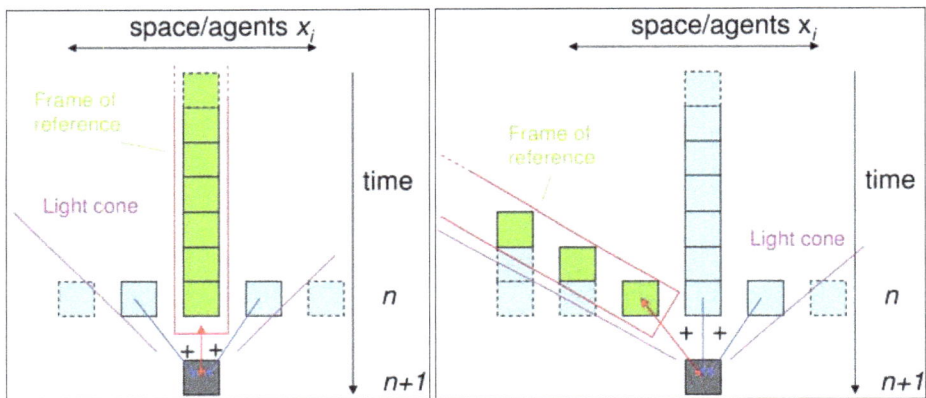

Figure 2. Local information dynamics for a lattice system with speed of light $c = 1$ unit per time step: (**a**) (left) with stationary frame of reference ($f = 0$); (**b**) (right) with moving frame of reference $f = 1$ (*i.e.*, at one cell to the right per unit time step). Red double-headed arrow represents active information storage $a(i, n + 1, f)$ from the frame of reference; the blue single-headed arrow represent transfer entropy $t(i, j, n + 1, f)$ from each source orthogonal to the frame of reference. Note that the frame of reference in the figures is the path of the moving observer through space-time.

As reported in [7], and shown in the sample application to rule 54 in Figure 1(b), when applied to CAs the local active information storage identifies strong positive values in the domain and in blinkers (vertical gliders). For each of these entities, the next state is effectively predictable from the destination's past. This was the first direct quantitative evidence that blinkers and domains were the dominant information storage entities in CAs. Interestingly for rule 54, the amount of predictability from the past (*i.e.*, the active information storage) is roughly the same for both the blinkers and the background domain (see further discussion in [7]). Furthermore, negative values are typically measured at (the leading edge of) traveling gliders, because the past of the destination (being in the regular domain) would predict domain continuation, which is misinformative when the glider is encountered.

4.2. Information Transfer

Information transfer is defined as the amount of information that a source provides about a destination's next state that was not contained in the destination's past. This definition pertains to Schreiber's transfer entropy measure [9] (which we will call the *apparent* transfer entropy, as discussed later). The transfer entropy captures the average mutual information from realizations $y_n^{(l)}$ of the state $Y^{(l)}$ of a source Y to the corresponding realizations x_{n+1} of the next value X' of the destination X, conditioned on realizations $x_n^{(k)}$ of the previous state $X^{(k)}$:

$$T_{Y \to X}(k, l) = I\left[Y^{(l)}; X' \mid X^{(k)}\right] \tag{13}$$

Schreiber emphasized that, unlike the (unconditioned) time-differenced mutual information, the transfer entropy was a properly directed, dynamic measure of information transfer rather than shared information.

In general, one should take the limit as $k \to \infty$ in order to properly represent the previous state $X^{(k)}$ as relevant to the relationship between the next value X' and the source Y [5]. Note that k can be limited here where the next value x_{n+1} is conditionally independent of the far past values $x_{n-k}^{(\infty)}$ given $(x_n^{(k)}, y_n)$. One than then interpret the transfer entropy as properly representing information transfer [5,32]. Empirically of course one is restricted to finite-k estimates $T_{Y \to X}(k, l)$. Furthermore, where only the previous value y_n of Y is a direct causal contributor to x_{n+1}, it is appropriate to use $l = 1$ [5,32]. So for our purposes, we write:

$$T_{Y \to X} = \lim_{k \to \infty} T_{Y \to X}(k) \tag{14}$$

$$T_{Y \to X}(k) = I\left[Y; X' \mid X^{(k)}\right] \tag{15}$$

We can then extract the *local transfer entropy* $t_{Y \to X}(n + 1)$ [5] as the amount of information transfer attributed to the specific configuration or realization $(x_{n+1}, x_n^{(k)}, y_n)$ at time step $n + 1$; *i.e.*, the amount of information transfered from Y to X at time step $n + 1$:

$$T_{Y \to X} = \langle t_{Y \to X}(n + 1) \rangle \tag{16}$$

$$t_{Y \to X}(n + 1) = \lim_{k \to \infty} t_{Y \to X}(n + 1, k) \tag{17}$$

$$T_{Y \to X}(k) = \langle t_{Y \to X}(n + 1, k) \rangle \tag{18}$$

$$t_{Y \to X}(n + 1, k) = \log_2 \frac{p(x_{n+1} \mid x_n^{(k)}, y_n)}{p(x_{n+1} \mid x_n^{(k)})} \tag{19}$$

$$= i(y_n; x_{n+1} \mid x_n^{(k)}) \tag{20}$$

Again, where the processes Y and X exist on cells on a lattice system, we denote i as the index of the destination variable X_i and $i - j$ as the source variable X_{i-j}, such that we consider the local transfer entropy *across* j cells in:

$$t(i, j, n + 1) = \lim_{k \to \infty} t(i, j, n + 1, k) \tag{21}$$

$$t(i, j, n + 1, k) = \log \frac{p(x_{i,n+1} | x_{i,n}^{(k)}, x_{i-j,n})}{p(x_{i,n+1} | x_{i,n}^{(k)})} \tag{22}$$

The local transfer entropy is defined for every channel j for the given destination i, but for proper interpretation as information transfer j is constrained among causal information contributors to the destination [32] (*i.e.*, within the past *light cone* [33]). For CAs for example we have $|j| \leq r$, being $|j| \leq 1$ for ECAs as shown in Figure 2(a).

We have $T(i, j, k) = \langle t(i, j, n, k) \rangle_n$ as the average transfer from variable $i - j$ to variable i. For systems of homogeneous variables where the probability distribution functions for transfer across j cells are estimated over all variables, it is appropriate to average over all variables also, giving:

$$T(j, k) = \langle t(i, j, n, k) \rangle_{i,n} \tag{23}$$

Importantly, the information conditioned on by the transfer entropy (*i.e.*, that contained in the destination's past about its next state) is that provided by the local active information storage. (Note however that a conditional mutual information may be either larger or smaller than the corresponding unconditioned mutual information [3]; the conditioning removes information redundantly held by the source and the conditioned variable, but also includes synergistic information that can only be decoded with knowledge of both the source and conditioned variable [34].)

Also, the *local* transfer entropy may also be positive or negative. As reported in [5], when applied to CAs it is typically strongly positive when measured at a glider in the same direction j as the macroscopic motion of the glider (see the sample application to rule 54 in Figure 1(c)). Negative values imply that the source *misinforms* an observer about the next state of the destination in the context of the destination's past. Negative values are typically only found at gliders for measurements in the orthogonal direction to macroscopic glider motion (see the right moving gliders in Figure 1(c)); at these points, the source (still part of the domain) would suggest that the domain pattern in the destination's past would continue, which is misinformative. Small positive non-zero values are also often measured in the domain and in the orthogonal direction to glider motion (see Figure 1(c)). These correctly indicate non-trivial information transfer in these regions (e.g., indicating the *absence* of a glider), though they are dominated by the positive transfer in the direction of glider motion. These results for local transfer entropy provided the first quantitative evidence for the long-held conjecture that particles are the information transfer agents in CAs.

We note that the transfer entropy can also be conditioned on other possible causal contributors Z in order to account for their effects on the destination. We introduced the *conditional* transfer entropy for this purpose [5,6]:

$$T_{Y \to X|Z} = \lim_{k \to \infty} T_{Y \to X|Z}(k) \tag{24}$$

$$T_{Y \to X|Z}(k) = I \left[Y; X' \mid X^{(k)}, Z \right] \tag{25}$$

$$T_{Y \to X|Z}(k) = \left\langle t_{Y \to X|Z}(n + 1, k) \right\rangle \tag{26}$$

$$t_{Y \to X|Z}(n + 1, k) = \log_2 \frac{p(x_{n+1} \mid x_n^{(k)}, y_n, z_n)}{p(x_{n+1} \mid x_n^{(k)}, z_n)} \tag{27}$$

$$= i(y_n; x_{n+1} \mid x_n^{(k)}, z_n) \tag{28}$$

This extra conditioning can exclude the (redundant) influence of a common drive Z from being attributed to Y, and can also include the synergistic contribution when the source Y acts in conjunction with another source Z (e.g., where X is the outcome of an XOR operation on Y and Z).

We specifically refer to the conditional transfer entropy as the *complete transfer entropy* (with notation $T^c_{Y \to X}(k)$ and $t^c_{Y \to X}(n+1, k)$ for example) when it conditions on all other causal sources Z to the destination X [5]. For CAs, this means conditioning on the only other causal contributor to the destination. For example, for the $j = 1$ channel, we can write

$$t^c(i, j = 1, n + 1) = \lim_{k \to \infty} t^c(i, j = 1, n + 1, k) \tag{29}$$

$$t^c(i, j = 1, n + 1, k) = \log \frac{p(x_{i,n+1} | x^{(k)}_{i,n}, x_{i-1,n}, x_{i+1,n})}{p(x_{i,n+1} | x^{(k)}_{i,n}, x_{i+1,n})} \tag{30}$$

with $T^c(j, k)$ for the spatiotemporal average in homogeneous, stationary systems. To differentiate the conditional and complete transfer entropies from the original measure, we often refer to $T_{Y \to X}$ simply as the *apparent* transfer entropy [5]—this nomenclature conveys that the result is the information transfer that is apparent without accounting for other sources.

In application to CAs, we note that the results for $t^c(i, j, n + 1, k)$ are largely the same as for $t(i, j, n + 1, k)$ (e.g., compare Figure 1(d) with Figure 1(c) for rule 54), with some subtle differences. These results are discussed in detail in [5]. First, in deterministic systems such as CAs, $t^c(i, j, n + 1, k)$ cannot be negative since by accounting for all causal sources (and without noise) there is no way that our source can misinform us about the next state of the destination. Also, the strong transfer measured in gliders moving in the macroscopic direction of the measured channel j is slightly stronger with $t^c(i, j, n + 1, k)$. This is because, by accounting for the other causal source, we can be sure that there is no other incoming glider to disturb this one, and thus attribute more influence to the source of the ongoing glider here. Other scenarios regarding synergistic interactions in other rules are discussed in [5].

5. Information Dynamics for a Moving Observer

In this section, we consider how these measures of information dynamics would change for a moving observer. First, we consider the meaning of the past state $x^{(k)}_n$ in these measures, and how it can be interpreted as a frame of reference. We then provide a formulation to interpret these measures for an observer with a moving frame of reference. We consider what aspects of the dynamics would remain invariant, and finally consider what differences we may expect to see from measures of information dynamics by moving observers.

5.1. Meaning of the Use of the Past State

Realizations $x^{(k)}_n$ of the past state $X^{(k)}$ of the destination variable X play a very important role in the measures of information dynamics presented above. We see that the active information storage directly considers the amount of information contained in $x^{(k)}_n$ about the next value x_{n+1} of X, while the transfer entropy considers how much information the source variable adds to this next value conditioned on $x^{(k)}_n$.

The role of the past state $x^{(k)}_n$ can be understood from three complementary perspectives here:

(1) To *separate* information storage and transfer. As described above, we know that $x^{(k)}_n$ provides information storage for use in computation of the next value x_{n+1}. The conditioning on the past state in the transfer entropy ensures that none of that information storage is counted as information transfer (where the source and past hold some information redundantly) [5,6].

(2) To capture the *state transition* of the destination variable. We note that Schreiber's original description of the transfer entropy [9] can be rephrased as the information provided by the

source about the state transition in the destination. That $x_n^{(k)} \to x_{n+1}$ (or including redundant information $x_n^{(k)} \to x_{n+1}^{(k)}$) is a *state transition* is underlined in that the $x_n^{(k)}$ are *embedding vectors* [35], which capture the underlying *state* of the process.

(3)   To examine the information composition of the next value x_{n+1} of the destination *in the context of* the past state $x_n^{(k)}$ of the destination. With regard to the transfer entropy, we often describe the conditional mutual information as "conditioning out" the information contained in $x_n^{(k)}$, but this nomenclature can be slightly misleading. This is because, as pointed out in Section 4.2, a conditional mutual information can be larger or smaller than the corresponding unconditioned form, since the conditioning both removes information redundantly held by the source variable and the conditioned variable (e.g., if the source is a copy of the conditioned variable) and adds information synergistically provided by the source and conditioned variables together (e.g., if the destination is an XOR-operation of these variables). As such, it is perhaps more useful to describe the conditioned variable as providing context to the measure, rather than "conditioning out" information. Here then, we can consider the past state $x_n^{(k)}$ as providing context to our analysis of the information composition of the next value x_{n+1}.

Note that we need $k \to \infty$ to properly capture each perspective here (see discussion in Section 4.1 and Section 4.2 regarding conditions where finite-k is satisfactory).

Importantly, we note that the final perspective of $x_n^{(k)}$ as providing context to our analysis of the information composition of the computation of the next state can also be viewed as a "frame of reference" for the analysis.

5.2. Information Dynamics with a Moving Frame of Reference

Having established the perspective of $x_n^{(k)}$ as providing a frame of reference for our analysis, we now examine how the measures of our framework are altered if we consider a moving frame of reference for our observer in lattice systems.

It is relatively straightforward to define a frame of reference for an observer moving at f cells per unit time towards the destination cell $x_{i,n+1}$. Our measures consider the set of k cells backwards in time from $x_{i,n+1}$ at $-f$ cells per time step:

$$x_{i-f,n}^{(k,f)} = \{x_{i-(q+1)f,n-q}|0 \le q < k\} \tag{31}$$

$$= \{x_{i-kf,n-k+1}, ..., x_{i-2f,n-1}, x_{i-f,n}\} \tag{32}$$

Notice that $x_{i,n}^{(k)} = x_{i-0,n}^{(k,0)}$ with $f = 0$, as it should.

We can then define measures for each of the information dynamics in this new frame of reference f. As shown with the double headed arrow in Figure 2(b), the local active information in this frame becomes the local mutual information between the observer's frame of reference $x_{i-f,n}^{(k,f)}$ and the next state of the destination cell $x_{i,n+1}$; mathematically this is represented by:

$$a(i, n+1, f) = \lim_{k \to \infty} a(i, n+1, k, f) \tag{33}$$

$$a(i, n+1, k, f) = \log \frac{p(x_{i-f,n}^{(k,f)}, x_{i,n+1})}{p(x_{i-f,n}^{(k,f)})p(x_{i,n+1})} \tag{34}$$

Crucially, $a(i, n+1, k, f)$ *is still a measure of local information storage for the moving observer*: it measures how much information is contained in the past of their frame of reference about the next state that appears in their frame. The observer, as well as the shifted measure itself, is oblivious to the fact that these observations are in fact taken over different variables. Finally, we write $A(k, f) =$

$\langle a(i, n + 1, k, f) \rangle_{i,n}$ as the average of finite-k estimates over all space-time points (i, n) in the lattice, for stationary homogeneous systems.

As shown by directed arrows in Figure 2(b), the local transfer entropy becomes the local conditional mutual information between the source cell $x_{i-j,n}$ and the destination $x_{i,n+1}$, conditioned on the moving frame of reference $x_{i-f,n}^{(k,f)}$:

$$t(i, j, n + 1, f) = \lim_{k \to \infty} t(i, j, n + 1, k, f) \tag{35}$$

$$t(i, j, n + 1, k, f) = \log \frac{p(x_{i,n+1} | x_{i-f,n}^{(k,f)}, x_{i-j,n})}{p(x_{i,n+1} | x_{i-f,n}^{(k,f)})} \tag{36}$$

The set of sensible values to use for j remains those within the light-cone (*i.e.*, those that represent causal information sources to the destination variable); otherwise we only measure correlations rather than information transfer. That said, we also do not consider the transfer entropy for the channel $j = f$ here, since this source is accounted for by the local active information. Of course, we can now also consider $j = 0$ for moving frames $f \neq 0$. Writing the local complete transfer entropy $t^c(i, j, n + 1, k, f)$ for the moving frame trivially involves adding conditioning on the remaining causal source (that which is not the source $x_{i-j,n}$ itself, nor the source $x_{i-f,n}$ in the frame) to Equation (36).

Again, $t(i, j, n + 1, f)$ *is still interpretable as a measure of local information transfer for the moving observer*: it measures how much information was provided by the source cell about the state transition of the observer's frame of reference. The observer is oblivious to the fact that the states in its frame of reference are composed of observations taken over different variables.

Also, note that while $t(i, j, n + 1, f)$ describes the transfer across j cells in a stationary frame as observed in a frame moving at speed f, we could equally express it as the transfer observed across $j - f$ cells in the frame f.

Finally, we write $T(j, k, f) = \langle t(i, j, n + 1, k, f) \rangle_{i,n}$ as the average of finite-k estimates over all space-time points (i, n) in the lattice, for stationary homogeneous systems.

In the next two subsections, we describe what aspects of the information dynamics remain invariant, and how we can expect the measures to change, with a moving frame of reference.

5.3. Invariance

This formulation suggests the question of why we consider the same set of information sources j in the moving and stationary frames (*i.e.*, those within the light-cone), rather than say a symmetric set of sources around the frame of reference (as per a stationary frame). To examine this, consider the local (single-site) entropy $h(i, n + 1) = \log p(x_{i,n+1})$ as a sum of incrementally conditioned mutual information terms as presented in [6]. For ECAs (a deterministic system), in the stationary frame of reference, this sum is written as:

$$h(i, n + 1) = i(x_{i,n}^{(k)}; x_{i,n+1}) + i(x_{i-j,n}; x_{i,n+1} | x_{i,n}^{(k)})$$
$$+ i(x_{i+j,n}; x_{i,n+1} | x_{i,n}^{(k)}, x_{i-j,n}) \tag{37}$$
$$h(i, n + 1) = a(i, n + 1, k) + t(i, j, n + 1, k)$$
$$+ t^c(i, -j, n + 1, k) \tag{38}$$

with either $j = 1$ or $j = -1$. Since $h(i, n + 1)$ represents the information required to predict the state at site $(i, n + 1)$, Equation (37) shows that one can obtain this by considering the information contained in the past of the destination, then the information contributed through channel j that was not in this past, then that contributed through channel $-j$ which was not in this past or the channel j. The first term here is the active information storage, the first local conditional mutual information term here is a transfer entropy, the second is a complete transfer entropy. Considering any sources in addition to or instead of these will only return correlations to the information provided by these entities.

Note that there is no need to take the limit $k \rightarrow \infty$ for the correctness of Equation (37) (unless one wishes to properly interpret the terms as information storage and transfer). In fact, the sum of incrementally conditional mutual information terms in Equation (37) is *invariant* as long as all terms use the same context. We can also consider a moving *frame of reference* as this context and so construct this sum for a moving frame of reference f. Note that the choice of f determines which values to use for j, so we write an example with $f = 1$:

$$
\begin{aligned}
h(i, n+1) = \quad & a(i, n+1, k, f=1) + t(i, j=0, n+1, k, f=1) \\
& + t^c(i, j=-1, n+1, k, f=1)
\end{aligned}
\tag{39}
$$

Obviously this is true because the set of causal information contributors is invariant, and we are merely considering the same causal sources but in a different context. Equation (39) demonstrates that prediction of the next state for a given cell in a moving frame of reference depends on the same causal information contributors. Considering the local transfer entropy from sources outside the light cone instead may be insufficient to predict the next state [32].

Choosing the frame of reference here merely sets the context for the information measures, and redistributes the attribution of the invariant amount of information in the next value $x_{i,n+1}$ between the various storage and transfer sources. This could be understood in terms of the different context redistributing the information atoms in a *partial information diagram* (see [34]) of the sources to the destination.

Note that we examine a type of *ether* relativity for local information dynamics. That is to say, there is a preferred stationary frame of reference $f = 0$ in which the velocity for information is bounded by the speed of light c. The stationary frame of reference is preferred because it is the *only* frame that has an even distribution of causal information sources on either side, while other frames observe an asymmetric distribution of causal information sources. It is also the only frame of reference that truly represents the information storage in the causal variables. As pointed out in Section 1, we do not consider a type of relativity where the rules of physics (*i.e.*, CA rules) are invariant, remaining observationally symmetric around the frame of reference.

5.4. Hypotheses and Expectations

In general, we expect the measures $a(i, n, k, f)$ and $t(i, j, n, k, f)$ to be different from the corresponding measurements in a stationary frame of reference. Obviously, this is because the frames of reference $x_{i-f,n}^{(k,f)}$ provide in general different contexts for the measurements. As exceptional cases however, the measurements would not change if:

- The two contexts or frames of reference in fact provide the same information redundantly about the next state (and in conjunction with the sources for transfer entropy measurements).
- Neither context provides any relevant information about the next state at all.

Despite such differences to the standard measurements, as described in Section 5.2 the measurements in a moving frame of reference are still interpretable as information storage and transfer for the moving observer, and still provide relevant insights into the dynamics of the system.

In the next section, we will examine spatiotemporal information profiles of CAs, as measured by a moving observer. We hypothesize that in a moving frame of reference f, we shall observe:

- Regular background domains appearing as information storage regardless of movement of the frame of reference, since their spatiotemporal structure renders them predictable in both moving and stationary frames. In this case, both the stationary and moving frames would *retain* the same information redundantly regarding how their spatiotemporal pattern evolves to give the next value of the destination in the domain;

- Gliders moving at the speed of the frame appearing as information storage in the frame, since the observer will find a large amount of information in their past observations that predict the next state observed. In this case, the shift of frame incorporates different information into the new frame of reference, making that *added* information appear as information storage;

- Gliders that were stationary in the stationary frame appearing as information transfer in the channel $j = 0$ when viewed in moving frames, since the $j = 0$ source will add a large amount of information for the observer regarding the next state they observe. In this case, the shift of frame of reference *removes* relevant information from the new frame of reference, allowing scope for the $j = 0$ source to add information about the next observed state.

6. Results and Discussion

To investigate the local information dynamics in a moving frame of reference, we study ECA rule 54 here with a frame of reference moving at $f = 1$ (*i.e.*, one step to the right per unit time). Our experiments used 10,000 cells initialized in random states, with 600 time steps captured for estimation of the probability distribution functions (similar settings used in introducing the local information dynamics in [5–7]). We fixed $k = 16$ for our measures (since the periodic background domain for ECA rule 54 has a period of 4, this captures an adequate amount of history to properly separate information storage and transfer as discussed in [5]). We measure the local information dynamics measures in both the stationary frame of reference (Figure 1) and the moving frame of reference $f = 1$ (Figure 3). The results were produced using the "Java Information Dynamics Toolkit" [36], and can be reproduced using the Matlab/Octave script `movingFrame.m` in the `demos/octave/CellularAutomata` example distributed with this toolkit.

We first observe that the background domain is captured as a strong information storage process irrespective of whether the frame of reference is moving (with $a(i, n, k = 16, f = 1)$, Figure 3(b)) or stationary (with $a(i, n, k = 16, f = 0)$, Figure 1(b)). That is to say that the frame of reference is strongly predictive of the next state in the domain, regardless of whether the observer is stationary or moving. This is as expected, because the background domain is not only temporally periodic, but *spatiotemporally* periodic, and the moving frame provides much redundant information with the stationary frame about the next observed state.

While it is not clear from the local profiles however, the average active information storage is significantly lower for the moving frame than the stationary frame ($A(i, n, k = 16, f = 1) = 0.468$ bits versus $A(i, n, k = 16, f = 0) = 0.721$ bits). At first glance, this seems strange since the background domain is dominated by information storage, and the observer in both frames should be able to adequately detect the periodic domain process. On closer inspection though, we can see that the storage process in the domain is significantly more disturbed by glider incidence in the moving frame, with a larger number and magnitude of negative local values encountered, and more time for the local values to recover to their usual levels in the domain. This suggests that the information in the moving frame is not fully redundant with the stationary frame, which could be explained in that the stationary frame (being centred in the light cone) is better able to retain information about the surrounding dynamics that could influence the next state. The moving frame (moving at the speed of light itself) is not able to contain any information regarding incoming dynamics from neighboring cells. Thus, in the moving frame, more of the (invariant) information in the next observed state is distributed amongst the transfer sources.

As expected also, we note that gliders that are moving at the same speed as the frame of reference $f = 1$ are now considered as information storage in that frame. That is, the right moving gliders previously visible as misinformative storage in Figure 1(b) now blend in with the background information storage process in the moving frame in Figure 3(b). As previously discussed, this is because the moving frame brings new information for the observer about these gliders into the frame of reference.

Figure 3(b) also shows that it is only gliders moving in orthogonal directions to the frame $f = 1$ (including blinkers, which were formerly considered stationary) that contain negative local active information storage, and are therefore information transfer processes in this frame. Again, this is as expected, since these gliders contribute new information about the observed state in the context of the frame of reference. For gliders that now become moving in the moving frame of reference, this is because the information about those gliders is no longer in the observer's frame of reference but can now be contributed to the observer by the neighboring sources. To understand these processes in more detail however, we consider the various sources of that transfer via the transfer entropy measurements in Figure 3(c)–Figure 3(f).

First, we focus on the vertical gliders that were stationary in the stationary frame of reference (*i.e.*, the blinkers): we had expected that these entities would be captured as information transfer processes in the $j = 0$ (vertical) channel in the $j = 1$ moving frame. This expectation is upheld, but the dynamics are more complicated than the foreseen in our hypothesis. Here, we see that the apparent transfer entropy from the $j = 0$ source alone does not dominate the dynamics for this vertical glider (Figure 3(c)). Instead, the information transfer required to explain the vertical gliders is generally a combination of both apparent and complete transfer entropy measures, requiring the $j = -1$ source for interpretation as well. The full information may be accounted for by either taking Figure 3(c) plus Figure 3(f) or Figure 3(e) plus Figure 3(d) (as per the two different orders of considering sources to sum the invariant information in Equation (38)). Further, we note that some of the points within the glider are even considered as strong information storage processes - note how there are positive storage points amongst the negative points (skewed by the moving frame) for this glider in Figure 3(b). These vertical gliders are thus observed in this frame of reference to be a complex structure consisting of some information storage, as well as information transfer requiring both other sources for interpretation. This is a perfectly valid result, demonstrating that switching frames of reference does not lead to the simple one-to-one correspondence between individual information dynamics that one may naively expect.

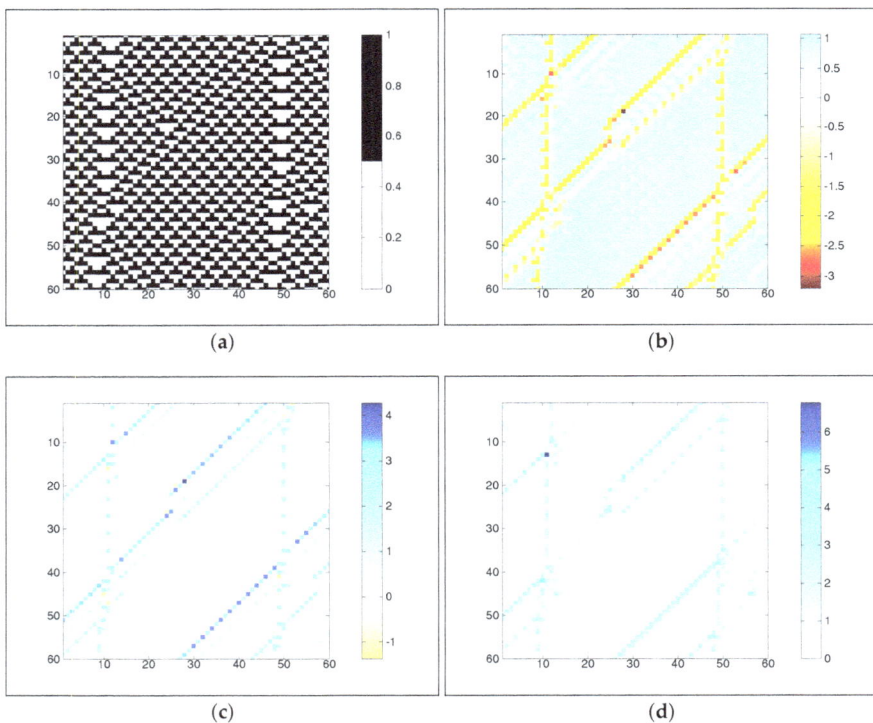

(a)

(b)

(c)

(d)

Figure 3. *Cont.*

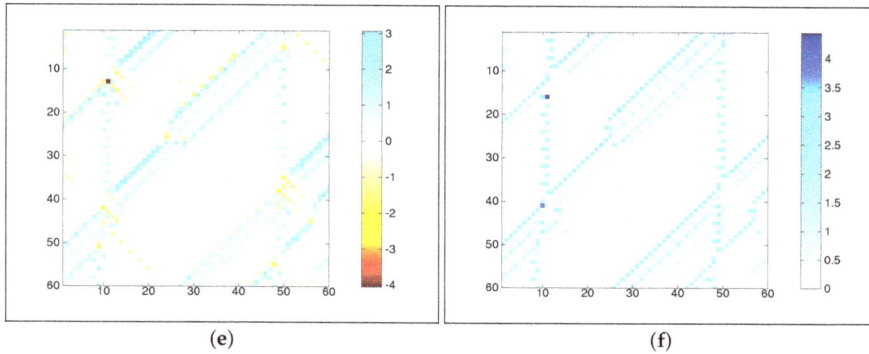

(e)   (f)

Figure 3. Measures of local information dynamics applied to ECA rule 54, computed in frame of reference $f = 1$, *i.e.*, moving 1 cell to the right per unit time (all units in (b)–(f) are in bits). Note that raw states are the same as in Figure 1. (**a**) Raw CA; (**b**) Local active information storage $a(i, n, k = 16, f = 1)$; (**c**) Local apparent transfer entropy $t(i, j = 0, n, k = 16, f = 1)$; (**d**) Local complete transfer entropy $t^c(i, j = 0, n, k = 16, f = 1)$; (**e**) Local apparent transfer entropy $t(i, j = -1, n, k = 16, f = 1)$; (**f**) Local complete transfer entropy $t^c(i, j = -1, n, k = 16, f = 1)$.

We note a similar result for the left-moving gliders in the $j = -1$ channel, which are considering moving both in the stationary and $j = 1$ frames of reference: here we see that the complete transfer entropy from the $j = 0$ source (Figure 3(d)) is required to completely explain some of these gliders. What is interesting is that the (extra) complete transfer entropy from the $j = 0$ source orthogonal to the glider is a greater proportion here than for orthogonal sources in the stationary frame (see the complete transfer entropy for the right moving gliders in Figure 1(d)). This suggests that there was less information pertinent to these gliders in the moving frame of reference than there was in the stationary frame. Clearly, a change of frame of reference can lead to complicated interplays between the information dynamics in each frame, with changes in both the *magnitude* and *source attribution* of the information.

Finally, note that while one can easily write down the differences between the measures in each frame (e.g., subtracting Equation (11) from Equation (34)), there do not appear to be any clear general principals regarding how the information will be redistributed between storage and transfers for an observer, since this depends on the common information between each frame of reference.

7. Conclusions

In this paper, we have presented a new interpretation of a framework for local information dynamics (including transfer entropy), which incorporates a moving *frame of reference* for the observer. This interpretation was inspired by the idea of investigating relativistic effects on information dynamics, and indeed contributes some interesting perspectives to this field.

We reported the results from investigations to explore this perspective applied to cellular automata, showing that moving elements of coherent spatiotemporal structure (particles or gliders) are identified as information transfer in frames in which they are moving and as information storage in frames where they are stationary, as expected. Crucially, the extent to which the shifted frame of reference alters the results depends on whether the shift of frame *retains*, *adds* or *removes* relevant information regarding the source-destination interaction. We showed examples illustrating each of these scenarios, and it is important to note that we showed all three to occur at different local points in the same coupled system (*i.e.*, these differences are not mutually exclusive).

Future work may include exploring mathematically formalizing transformation laws between individual information dynamics under shifts of frames of reference, as well as time-reversibility, and the use of different frames of reference as a classification tool.

Acknowledgments: This work was partially supported by the Santa Fe Institute through NSF Grant No. 0200500 entitled "A Broad Research Program in the Sciences of Complexity." JL acknowledges travel support that contributed to this work from the ARC Complex Open Systems Research Network (COSNet) and the CSIRO Complex Systems Science Theme.

References

1. Halliday, D.; Resnick, R.; Walker, J. *Fundamentals of Physics*, 4th ed.; John Wiley & Sons: New York, NY, USA, 1993.
2. Cover, T.M.; Thomas, J.A. *Elements of Information Theory*; John Wiley & Sons: New York, NY, USA, 1991.
3. MacKay, D.J. *Information Theory, Inference, and Learning Algorithms*; Cambridge University Press: Cambridge, UK, 2003.
4. Lizier, J.T.; Prokopenko, M.; Zomaya, A.Y. Detecting Non-trivial Computation in Complex Dynamics. In Proceedings of the 9th European Conference on Artificial Life (ECAL 2007), Lisbon, Portugal, 10–14 September 2007; Almeidae Costa, F., Rocha, L.M., Costa, E., Harvey, I., Coutinho, A., Eds.; Springer: Berlin / Heidelberg, 2007; Vol. 4648, *Lecture Notes in Computer Science*, pp. 895–904.
5. Lizier, J.T.; Prokopenko, M.; Zomaya, A.Y. Local information transfer as a spatiotemporal filter for complex systems. *Phys. Rev. E* **2008**, *77*, 026110+.
6. Lizier, J.T.; Prokopenko, M.; Zomaya, A.Y. Information modification and particle collisions in distributed computation. *Chaos* **2010**, *20*, 037109+.
7. Lizier, J.T.; Prokopenko, M.; Zomaya, A.Y. Local measures of information storage in complex distributed computation. *Inform. Sciences* **2012**, *208*, 39–54.
8. Lizier, J.T. *The local information dynamics of distributed computation in complex systems*; Springer Theses, Springer: Berlin / Heidelberg, Germany, 2013.
9. Schreiber, T. Measuring Information Transfer. *Phys. Rev. Lett.* **2000**, *85*, 461–464.
10. Smith, M.A. Cellular automata methods in mathematical physics. PhD thesis, Massachusetts Institute of Technology, 1994.
11. Mitchell, M. Computation in Cellular Automata: A Selected Review. In *Non-Standard Computation*; Gramss, T., Bornholdt, S., Gross, M., Mitchell, M.; Pellizzari, T., Eds.; VCH Verlagsgesellschaft: Weinheim, Germany, 1998; pp. 95–140.
12. Wolfram, S. *A New Kind of Science*; Wolfram Media: Champaign, IL, USA, 2002.
13. Hanson, J.E.; Crutchfield, J.P. The Attractor-Basin Portait of a Cellular Automaton. *J. Stat. Phys.* **1992**, *66*, 1415–1462.
14. Grassberger, P. New mechanism for deterministic diffusion. *Phys. Rev. A* **1983**, *28*, 3666.
15. Grassberger, P. Long-range effects in an elementary cellular automaton. *J. Stat. Phys.* **1986**, *45*, 27–39.
16. Hanson, J.E.; Crutchfield, J.P. Computational mechanics of cellular automata: An example. *Physica D* **1997**, *103*, 169–189.
17. Wuensche, A. Classifying cellular automata automatically: Finding gliders, filtering, and relating space-time patterns, attractor basins, and the Z parameter. *Complexity* **1999**, *4*, 47–66.
18. Helvik, T.; Lindgren, K.; Nordahl, M.G. Local information in one-dimensional cellular automata. Proceedings of the International Conference on Cellular Automata for Research and Industry, Amsterdam; Sloot, P.M., Chopard, B., Hoekstra, A.G., Eds.; Springer: Berlin/Heidelberg, Germany, 2004; Vol. 3305, *Lecture Notes in Computer Science*, pp. 121–130.
19. Helvik, T.; Lindgren, K.; Nordahl, M.G. Continuity of Information Transport in Surjective Cellular Automata. *Commun. Math. Phys.* **2007**, *272*, 53–74.
20. Shalizi, C.R.; Haslinger, R.; Rouquier, J.B.; Klinkner, K.L.; Moore, C. Automatic filters for the detection of coherent structure in spatiotemporal systems. *Phys. Rev. E* **2006**, *73*, 036104.
21. Langton, C.G. Computation at the edge of chaos: phase transitions and emergent computation. *Physica D* **1990**, *42*, 12–37.

22. Mitchell, M.; Crutchfield, J.P.; Hraber, P.T. Evolving Cellular Automata to Perform Computations: Mechanisms and Impediments. *Physica D* **1994**, *75*, 361–391.

23. Crutchfield, J.P.; Feldman, D.P. Regularities unseen, randomness observed: Levels of entropy convergence. *Chaos* **2003**, *13*, 25–54.

24. Prokopenko, M.; Gerasimov, V.; Tanev, I. Evolving Spatiotemporal Coordination in a Modular Robotic System. In *Ninth International Conference on the Simulation of Adaptive Behavior (SAB'06)*; Nolfi, S., Baldassarre, G., Calabretta, R., Hallam, J., Marocco, D., Meyer, J.A., Parisi, D., Eds.; Springer Verlag: Rome, Italy, 2006; Volume 4095, *Lecture Notes in Artificial Intelligence*, pp. 548–559.

25. Prokopenko, M.; Lizier, J.T.; Obst, O.; Wang, X.R. Relating Fisher information to order parameters. *Phys. Rev. E* **2011**, *84*, 041116+.

26. Mahoney, J.R.; Ellison, C.J.; James, R.G.; Crutchfield, J.P. How hidden are hidden processes? A primer on crypticity and entropy convergence. *Chaos* **2011**, *21*, 037112+.

27. Crutchfield, J.P.; Ellison, C.J.; Mahoney, J.R. Time's Barbed Arrow: Irreversibility, Crypticity, and Stored Information. *Phys. Rev. Lett.* **2009**, *103*, 094101.

28. Grassberger, P. Toward a quantitative theory of self-generated complexity. *Int. J. Theor. Phys.* **1986**, *25*, 907–938.

29. Bialek, W.; Nemenman, I.; Tishby, N. Complexity through nonextensivity. *Physica A* **2001**, *302*, 89–99.

30. Crutchfield, J.P.; Young, K. Inferring statistical complexity. *Phys. Rev. Lett.* **1989**, *63*, 105.

31. Shalizi, C.R. Causal Architecture, Complexity and Self-Organization in Time Series and Cellular Automata. PhD thesis, University of Wisconsin-Madison, 2001.

32. Lizier, J.T.; Prokopenko, M. Differentiating information transfer and causal effect. *Eur. Phys. J. B* **2010**, *73*, 605–615.

33. Shalizi, C.R.; Shalizi, K.L.; Haslinger, R. Quantifying Self-Organization with Optimal Predictors. *Phys. Rev. Lett.* **2004**, *93*, 118701.

34. Williams, P.L.; Beer, R.D. Nonnegative Decomposition of Multivariate Information **2010**. arXiv:1004.2515.

35. Takens, F. Detecting strange attractors in turbulence. In *Dynamical Systems and Turbulence, Warwick 1980*; Rand, D.; Young, L.S., Eds.; Springer: Berlin / Heidelberg, Germany, 1981; Vol. 898, *Lecture Notes in Mathematics*, chapter 21, pp. 366–381.

36. Lizier, J.T. JIDT: An information-theoretic toolkit for studying the dynamics of complex systems. Available online: https://code.google.com/p/information-dynamics-toolkit/ (accessed on 15 November 2012).

entropy

MDPI

Article

On Thermodynamic Interpretation of Transfer Entropy

Mikhail Prokopenko [1,2,]*, Joseph T. Lizier [1] and Don C. Price [3]

[1] CSIRO Information and Communications Technologies Centre, PO Box 76, Epping, NSW 1710, Australia;
     E-Mail: joseph.lizier@csiro.au
[2] School of Physics A28, University of Sydney, NSW 2006, Australia
[3] CSIRO Materials Science and Engineering, Bradfield Road, West Lindfield, NSW 2070, Australia;
     E-Mail: don.price@csiro.au

* Author to whom correspondence should be addressed; E-Mail: mikhail.prokopenko@csiro.au;
   Tel.: +61-29372-4716; Fax: +61-29372-4161.

Received: 16 November 2012; in revised form: 16 January 2013 / Accepted: 28 January 2013 /
Published: 1 February 2013

Abstract: We propose a thermodynamic interpretation of transfer entropy near equilibrium, using a specialised Boltzmann's principle. The approach relates conditional probabilities to the probabilities of the corresponding state transitions. This in turn characterises transfer entropy as a difference of two entropy rates: the rate for a resultant transition and another rate for a possibly irreversible transition within the system affected by an additional source. We then show that this difference, the local transfer entropy, is proportional to the external entropy production, possibly due to irreversibility. Near equilibrium, transfer entropy is also interpreted as the difference in equilibrium stabilities with respect to two scenarios: a default case and the case with an additional source. Finally, we demonstrated that such a thermodynamic treatment is not applicable to information flow, a measure of causal effect.

Keywords: transfer entropy; information transfer; entropy production; irreversibility; Kullback–Leibler divergence; thermodynamic equilibrium; Boltzmann's principle; causal effect

1. Introduction

Transfer entropy has been introduced as an information-theoretic measure that quantifies the statistical coherence between systems evolving in time [1]. Moreover, it was designed to detect asymmetry in the interaction of subsystems by distinguishing between "driving" and "responding" elements. In constructing the measure, Schreiber considered several candidates as measures of directional information transfer, including symmetric mutual information, time-delayed mutual information, as well as asymmetric conditional information. All these alternatives were argued to be inadequate for determining the direction of information transfer between two, possibly coupled, processes.

In particular, defining information transfer simply as the dependence of the next state of the receiver on the previous state of the source [2] is incomplete according to Schreiber's criteria requiring the definition to be both *directional* and *dynamic*. Instead, the (predictive) information transfer is defined as the average information contained in the source about the next state of the destination in the context of what was already contained in the destination's past.

Following the seminal work of Schreiber [1] numerous applications of transfer entropy have been successfully developed, by capturing information transfer within complex systems, e.g., the stock market [3], food webs [4], EEG signals [5], biochemicals [6], cellular automata and distributed computation in general [7–10], modular robotics [11], random and small-world Boolean networks [12,13], inter-regional interactions within a brain [14], swarm dynamics [15], cascading failures in power grids [16], *etc*. Also, several studies further capitalised on transition probabilities used in the measure, highlighting fundamental connections of the measure to entropy rate and Kullback–Leibler divergence noted by Kaiser and Schreiber [17], as well as causal flows [18]. At the same time there are several recent studies investigating ties between information theory and thermodynamics [19–23]. This is primarily through Landauer's principle [24], which states that irreversible destruction of one bit of information results in dissipation of at least $kT \ln 2$ J of energy (*T* is the absolute temperature and *k* is Boltzmann's constant.) into the environment (*i.e.*, an entropy increase in the environment by this amount). (Maroney [25] argues that while a logically irreversible transformation of information does generate this amount of heat, it can in fact be accomplished by a *thermodynamically reversible* mechanism.)

Nevertheless, transfer entropy *per se* has not been precisely interpreted thermodynamically. Of course, as a measure of directed information transfer, it does not need to have an explicit thermodynamic meaning. Yet, one may still put forward several questions attempting to cast the measure in terms more familiar to a physicist rather than an information theorist or a computer scientist: Is transfer entropy a measure of some entropy transferred between subsystems or coupled processes? Is it instead an entropy of some transfer happening within the system under consideration (and what is then the nature of such transfer)? If it is simply a difference between some entropy rates, as can be seen from the definition itself, one may still inquire about the thermodynamic nature of the underlying processes.

Obviously, once the subject relating entropy definitions from information theory and thermodynamics is touched, one may expect vigorous debates that have been ongoing since Shannon introduced the term entropy itself. While this paper will attempt to produce a thermodynamic interpretation of transfer entropy, it is out of scope to comment here on rich connections between Boltzmann entropy and Shannon entropy, or provide a review of quite involved discussions on the topic. It suffices to point out prominent works of Jaynes [26,27] who convincingly demonstrated that information theory can be applied to the problem of justification of statistical mechanics, producing predictions of equilibrium thermodynamic properties. The statistical definition of entropy is widely considered more general and fundamental than the original thermodynamic definition, sometimes allowing for extensions to the situations where the system is not in thermal equilibrium [23,28]. In this study, however, we treat the problem of finding a thermodynamic interpretation of transfer entropy somewhat separately from the body of work relating Boltzmann and Shannon entropies—and the reason for this is mainly that, even staying within Jaynes' framework, one still needs to provide a possible thermodynamic treatment for transfer entropy *per se*. As will become clear, this task is not trivial, and needs to be approached carefully.

Another contribution of this paper is a clarification that similar thermodynamic treatment is not applicable to information flow—a measure introduced by Ay and Polani [18] in order to capture causal effect. That correlation is not causation is well-understood. Yet while authors increasingly consider the notions of information transfer and information flow and how they fit with our understanding of correlation and causality [1,18,29–34], several questions nag. Is information transfer, captured by transfer entropy, akin to causal effect? If not, what is the distinction between them? When examining the "effect" of one variable on another (e.g., between brain regions), should one seek to measure information transfer or causal effect?

Unfortunately, these concepts have become somewhat tangled in discussions of information transfer. Measures for both predictive transfer [1] and causal effect [18] have been inferred to capture information transfer in general, and measures of predictive transfer have been used to infer causality [33,35–37] with the two sometimes (problematically) directly equated (e.g., [29,32,34,38–40]). The study of Lizier and Prokopenko [41] clarified the relationship between these concepts and described the manner in which they should be considered separately. Here, in addition, we demonstrate that a thermodynamic interpretation of transfer entropy is not applicable to causal effect (information flow), and clarify the reasons behind this.

This paper is organised as follows. We begin with Section 2 that introduces relevant information-theoretic measures both in average and local terms. Section 3 defines the system and the range of applicability of our approach. In providing a thermodynamic interpretation for transfer entropy in Section 4 we relate conditional probabilities to the probabilities of the corresponding state transitions, and use a specialised Boltzmann's principle. This allows us to define components of transfer entropy with the entropy rate of (i) the resultant transition and (ii) the internal entropy production. Sub-section 4.3 presents an interpretation of transfer entropy near equilibrium. The following Section 5 discusses the challenges for supplying a similar interpretation to causal effect (information flow). A brief discussion in Section 6 concludes the paper.

2. Definitions

In the following sections we describe relevant background on transfer entropy and causal effect (information flow), along some technical preliminaries.

2.1. Transfer Entropy

Mutual information $I_{Y;X}$ has been something of a de facto measure for information transfer between Y and X in complex systems science in the past (e.g., [42–44]). A major problem however is that mutual information contains no inherent *directionality*. Attempts to address this include using the previous state of the "source" variable Y and the next state of the "destination" variable X' (known as *time-lagged mutual information* $I_{Y;X'}$). However, Schreiber [1] points out that this ignores the more fundamental problem that mutual information measures the *statically* shared information between the two elements. (The same criticism applies to equivalent non-information-theoretic definitions such as that in [2].)

To address these inadequacies Schreiber introduced *transfer entropy* [1] (TE), the deviation from independence (in bits) of the state transition (from the previous state to the next state) of an information destination X from the previous state of an information source Y:

$$T_{Y \to X}(k,l) = \sum_{x_{n+1}, x_n^{(k)}, y^{(l)}} p(x_{n+1}, x_n^{(k)}, y_n^{(l)}) \log_2 \frac{p(x_{n+1} \mid x_n^{(k)}, y_n^{(l)})}{p(x_{n+1} \mid x_n^{(k)})} \quad (1)$$

Here n is a time index, $x_n^{(k)}$ and $y_n^{(l)}$ represent past states of X and Y (i.e., the k and l past values of X and Y up to and including time n). Schreiber points out that this formulation is a truly *directional*, *dynamic* measure of information transfer, and is a generalisation of the entropy rate to more than one element to form a mutual information *rate*. That is, transfer entropy may be seen as the difference between two entropy rates:

$$T_{Y \to X}(k,l) = h_X - h_{X,Y} \quad (2)$$

where h_X is the entropy rate:

$$h_X = -\sum p(x_{n+1}, x_n^{(k)}) \log_2 p(x_{n+1} \mid x_n^{(k)}) \quad (3)$$

and $h_{X,Y}$ is a generalised entropy rate conditioning on the source state as well:

$$h_{X,Y} = -\sum p(x_{n+1}, x_n^{(k)}, y_n^{(l)}) \log_2 p(x_{n+1} \mid x_n^{(k)}, y_n^{(l)}) \quad (4)$$

The entropy rate h_X accounts for the average number of bits needed to encode one additional state of the system if all previous states are known [1], while the entropy rate $h_{X,Y}$ is the entropy rate capturing the average number of bits required to represent the value of the next destination's state if source states are included in addition. Since one can always write

$$h_X = -\sum p(x_{n+1}, x_n^{(k)}) \log_2 p(x_{n+1} \mid x_n^{(k)}) = -\sum p(x_{n+1}, x_n^{(k)}, y_n^{(l)}) \log_2 p(x_{n+1} \mid x_n^{(k)}) \qquad (5)$$

it is easy to see that the entropy rate h_X is equivalent to the rate $h_{X,Y}$ when the next state of destination is independent of the source [1]:

$$p(x_{n+1} \mid x_n^{(k)}) = p(x_{n+1} \mid x_n^{(k)}, y_n^{(l)}) \qquad (6)$$

Thus, in this case the transfer entropy reduces to zero.

Similarly, the TE can be viewed as a *conditional* mutual information $I(Y^{(l)}; X' \mid X^{(k)})$ [17], that is as the average information contained in the source about the next state X' of the destination that was not already contained in the destination's past $X^{(k)}$:

$$T_{Y \to X}(k, l) = I_{Y^{(l)}; X' \mid X^{(k)}} = H_{X' \mid X^{(k)}} - H_{X' \mid X^{(k)}, Y^{(l)}} \qquad (7)$$

This could be interpreted (following [44,45]) as the diversity of state transitions in the destination minus assortative noise between those state transitions and the state of the source.

Furthermore, we note that Schreiber's original description can be rephrased as the information provided by the source about the state transition in the destination. That $x_n^{(k)} \to x_{n+1}$ (or including redundant information $x_n^{(k)} \to x_{n+1}^{(k)}$) is a *state transition* is underlined in that the $x_n^{(k)}$ are *embedding vectors* [46], which capture the underlying *state* of the process. Indeed, since all of the above mathematics for the transfer entropy is equivalent if we consider the next source *state* $x_{n+1}^{(k)}$ instead of the next source value x_{n+1}, we shall adjust our notation from here onwards to consider the next source state $x_{n+1}^{(k)}$, so that we are always speaking about interactions between source states $\mathbf{y_n}$ and destination state transitions $\mathbf{x_n} \to \mathbf{x_{n+1}}$ (with embedding lengths l and k implied).

Importantly, the TE remains a measure of observed (conditional) *correlation* rather than direct effect. In fact, the TE is a non-linear extension of a concept known as the "Granger causality" [47], the nomenclature for which may have added to the confusion associating information transfer and causal effect. Importantly, as an information-theoretic measure based on observational probabilities, the TE is applicable to both deterministic and stochastic systems.

2.2. Local Transfer Entropy

Information-theoretic variables are generally defined and used as an *average* uncertainty or information. We are interested in considering *local* information-theoretic values, *i.e.*, the uncertainty or information associated with a *particular observation* of the variables rather than the average over all observations. *Local* information-theoretic measures are sometimes called *point-wise* measures [48,49]. Local measures within a global average are known to provide important insights into the *dynamics* of non-linear systems [50].

Using the technique originally described in [7], we observe that the TE is an average (or *expectation value*) of a *local transfer entropy* at each observation n, *i.e.*,:

$$T_{Y \to X} = \langle t_{Y \to X}(n+1) \rangle \qquad (8)$$

$$t_{Y \to X}(n+1) = \log_2 \frac{p(\mathbf{x_{n+1}} \mid \mathbf{x_n}, \mathbf{y_n})}{p(\mathbf{x_{n+1}} \mid \mathbf{x_n})} \qquad (9)$$

with embedding lengths l and k implied as described above. The local transfer entropy quantifies the information contained in the source state $\mathbf{y_n}$ about the next state of the destination $\mathbf{x_{n+1}}$ at time step $n+1$, in the context of what was already contained in the past state of the destination $\mathbf{x_n}$. The measure is *local* in that it is defined at each time n for each destination X in the system and each causal information source Y of the destination.

The local TE may also be expressed as a local conditional mutual information, or a difference between local conditional entropies:

$$t_{Y \to X}(n+1) = i(\mathbf{y_n}; \mathbf{x_{n+1}} \mid \mathbf{x_n}) = h(\mathbf{x_{n+1}} \mid \mathbf{x_n}) - h(\mathbf{x_{n+1}} \mid \mathbf{x_n}, \mathbf{y_n}) \tag{10}$$

where local conditional mutual information is given by

$$i(\mathbf{y_n}; \mathbf{x_{n+1}} \mid \mathbf{x_n}) = \log_2 \frac{p(\mathbf{x_{n+1}} \mid \mathbf{x_n}, \mathbf{y_n})}{p(\mathbf{x_{n+1}} \mid \mathbf{x_n})} \tag{11}$$

and local conditional entropies are defined analogously:

$$h(\mathbf{x_{n+1}} \mid \mathbf{x_n}) = -\log_2 p(\mathbf{x_{n+1}} \mid \mathbf{x_n}) \tag{12}$$

$$h(\mathbf{x_{n+1}} \mid \mathbf{x_n}, \mathbf{y_n}) = -\log_2 p(\mathbf{x_{n+1}} \mid \mathbf{x_n}, \mathbf{y_n}) \tag{13}$$

The average transfer entropy $T_{Y \to X}(k)$ is always positive but is bounded above by the information capacity of a single observation of the destination. For a discrete system with b possible observations this is $\log_2 b$ bits. As a conditional mutual information, it can be either larger *or* smaller than the corresponding mutual information [51]. The *local* TE however is not constrained so long as it averages into this range: it can be greater than $\log_2 b$ for a large local information transfer, and can also in fact be measured to be negative. Local transfer entropy is negative where (in the context of the history of the destination) the probability of observing the actual next state of the destination given the source state $p(\mathbf{x_{n+1}} \mid \mathbf{x_n}, \mathbf{y_n})$, is lower than that of observing that actual next state independently of the source $p(\mathbf{x_{n+1}} \mid \mathbf{x_n})$. In this case, the source variable is actually *misinformative* or misleading about the state transition of the destination. It is possible for the source to be misleading where other causal information sources influence the destination, or in a stochastic system. Full examples are described by Lizier *et al.* [7].

2.3. Causal Effect as Information Flow

As noted earlier, predictive information transfer refers to the amount of information that a source variable adds to the next state of a destination variable; *i.e.*, "if I know the state of the source, how much does that help to predict the state of the destination?". Causal effect, on the contrary, refers to the extent to which the source variable has a direct influence or drive on the next state of a destination variable, *i.e.*, "if I change the state of the source, to what extent does that alter the state of the destination?". Information from causal effect can be seen to *flow* through the system, like injecting dye into a river [18].

It is well-recognised that measurement of causal effect necessitates some type of *perturbation* or *intervention* of the source so as to detect the effect of the intervention on the destination (e.g., see [52]). Attempting to infer causality without doing so leaves one measuring correlations of observations, regardless of how directional they may be [18]. In this section, we adopt the measure information flow for this purpose, and describe a method introduced by Lizier and Prokopenko [41] for applying it on a local scale.

Following Pearl's probabilistic formulation of causal Bayesian networks [52], Ay and Polani [18] consider how to measure causal information flow via *interventional conditional probability distribution functions*. For instance, an interventional conditional PDF $p(y \mid \hat{s})$ considers the distribution of y resulting from *imposing* the value of $\hat{s}$. *Imposing* means intervening in the system to *set* the value of the imposed variable, and is at the essence of the definition of causal information flow. As an illustration of the difference between interventional and standard conditional PDFs, consider two correlated variables S and Y: their correlation alters $p(y \mid s)$ in general from $p(y)$. If both variables are solely caused by another variable G however, then even where they remain correlated we have $p(y \mid \hat{s}) = p(y)$ because imposing a value $\hat{s}$ has no effect on the value of y.

In a similar fashion to the definition of transfer entropy as the deviation of a destination from *stochastic* independence on the source in the content of the destination's past, Ay and Polani propose the measure *information flow* as the deviation of the destination X from *causal* independence on the source Y *imposing* another set of nodes **S**. Mathematically, this is written as:

$$I_p(Y \to X \mid \hat{\mathbf{S}}) = \sum_{\mathbf{s}} p(\mathbf{s}) \sum_{y} p(y \mid \hat{\mathbf{s}}) \sum_{x} p(x \mid \hat{y}, \hat{\mathbf{s}}) \log_2 \frac{p(x \mid \hat{y}, \hat{\mathbf{s}})}{\sum_{y'} p(y' \mid \hat{\mathbf{s}}) p(x \mid \hat{y'}, \hat{\mathbf{s}})} \qquad (14)$$

The value of the measure is dependent on the choice of the set of nodes **S**. It is possible to obtain a measure of apparent causal information flow $I_p(Y \to X)$ from Y to X without any **S** (*i.e.*, $\mathbf{S} = \varnothing$), yet this can be misleading. In particular, it ignores causal information flow arising from interactions of the source with another source variable. For example, if $x = y$ XOR s and $p(y, s) = 0.25$ for each combination of binary y and s, then $I_p(Y \to X) = 0$ despite the clear causal effect of Y, while $I_p(Y \to X \mid \hat{S}) = 1$ bit. Also, we may have $I_p(Y \to X) > 0$ only because Y effects **S** which in turn effects X; where we are interested in *direct* causal information flow from Y to X only $I_p(Y \to X \mid \hat{\mathbf{S}})$ validly infers no direct causal effect.

Here we are interested in measuring the *direct* causal information flow from Y to X, so we must either include all possible other sources in **S** or at least include enough sources to "block" (A set of nodes U *blocks* a path of causal links where there is a node v on the path such that either:

- $v \in U$ and the causal links through v on the path are not both into v, or
- the causal links through v on the path are both into v, and v and all its causal descendants are not in U.)

all non-immediate directed paths from Y to X [18]. The minimum to satisfy this is the set of all direct causal sources of X excluding Y, including any past states of X that are direct causal sources. That is, in alignment with transfer entropy **S** would include $X^{(k)}$.

The major task in computing $I_p(Y \to X \mid \hat{\mathbf{S}})$ is the determination of the underlying interventional conditional PDFs in Equation (14). By definition these may be gleaned by observing the results of intervening in the system, however this is not possible in many cases.

One alternative is to use detailed knowledge of the dynamics, in particular the structure of the causal links and possibly the underlying rules of the causal interactions. This also is often not available in many cases, and indeed is often the very goal for which one turned to such analysis in the first place. Regardless, where such knowledge is available it may allow one to make direct inferences.

Under certain constrained circumstances, one can construct these values from observational probabilities only [18], e.g., with the "back-door adjustment" [52]. A particularly important constraint on using the back-door adjustment here is that *all* $\{\mathbf{s}, y\}$ combinations must be observed.

2.4. Local Information Flow

A *local information flow* can be defined following the argument that was used to define local information transfer:

$$f(y \to x \mid \hat{\mathbf{s}}) = \log_2 \frac{p(x \mid \hat{y}, \hat{\mathbf{s}})}{\sum_{y'} p(y' \mid \hat{\mathbf{s}}) p(x \mid \hat{y'}, \hat{\mathbf{s}})} \qquad (15)$$

The meaning of the local information flow is slightly different however. Certainly, it is an *attribution* of local causal effect of y on x were $\hat{s}$ imposed at the given observation $(y, x, \mathbf{s})$. However, one must be aware that $I_p(Y \to X \mid \hat{\mathbf{S}})$ is not the *average* of the local values $f(y \to x \mid \hat{\mathbf{s}})$ in exactly the same manner as the local values derived for information transfer. Unlike standard information-theoretical measures, the information flow is averaged over a product of *interventional* conditional probabilities

$(p(\mathbf{s})p(y \mid \hat{\mathbf{s}})p(x \mid \hat{y}, \hat{\mathbf{s}})$, see Equation (14) which in general does not reduce down to the probability of the given observation $p(\mathbf{s}, y, x) = p(\mathbf{s})p(y \mid \mathbf{s})p(x \mid y, \mathbf{s})$. For instance, it is possible that not all of the tuples $\{y, x, \mathbf{s}\}$ will actually be observed, so averaging over observations would ignore the important contribution that any unobserved tuples provide to the determination of information flow. Again, the local information flow is specifically tied not to the *given observation* at time step n but to the *general configuration* $(y, x, \mathbf{s})$, and only *attributed* to the associated observation of this configuration at time n.

3. Preliminaries

3.1. System Definition

Let us consider the non-equilibrium thermodynamics of a physical system close to equilibrium. At any given moment in time, n, the thermodynamic state of the physical system X is given by a vector $\mathbf{x} \in R^d$, comprising d variables, for instance the (local) pressure, temperature, chemical concentrations and so on. A state vector completely describes the physical macrostate as far as predictions of the outcomes of all possible measurements performed on the system are concerned [53]. The state space of the system is the set of all possible states of the system.

The thermodynamic state is generally considered as a fluctuating entity so that transition probabilities like $p(\mathbf{x_{n+1}}|\mathbf{x_n})$ are clearly defined and can be related to a sampling procedure. Each macrostate can be realised by a number of different microstates consistent with the given thermodynamic variables. Importantly, in the theory of non-equilibrium thermodynamics close to equilibrium, the microstates belonging to one macrostate $\mathbf{x}$ are equally probable.

3.2. Entropy Definitions

The thermodynamic entropy was originally defined by Clausius as a state function S that satisfies

$$S_B - S_A = \int_A^B \mathrm{d}q_{rev}/T \tag{16}$$

where q_{rev} is the heat transferred to an equilibrium thermodynamic system during a reversible process from state A to state B. Note that this *path integral* is the same for all reversible paths between the past and next states.

It was shown by Jaynes that thermodynamic entropy could be interpreted, from the perspective of statistical mechanics, as a measure of the amount of information about the microstate of a system that an observer lacks if they know only the macrostate of the system [53].

This is encapsulated in the famous Boltzmann's equation $S = k \log W$, where k is Boltzmann's constant and W is the number of microstates corresponding to a given macrostate (an integer greater than or equal to one). While it is not a mathematical probability between zero and one, it is sometimes called "thermodynamic probability", noting that W can be normalized to a probability $p = W/N$, where N is the number of possible microstates for all macrostates.

The Shannon entropy that corresponds to the Boltzmann entropy $S = k \log W$ is the uncertainty in the microstate that has produced the given macrostate. That is, given the number W of microscopic configurations that correspond to the given macrostate, we have $p_i = 1/W$ for each equiprobable microstate i. As such, we can compute the local entropy for each of these W microstates as $-\log_2 1/W = \log_2 W$ bits. Note that the average entropy across all of these equiprobable microstates is $\log_2 W$ bits also. This is equivalent to the Boltzmann entropy up to Boltzmann's constant k and the base of the logarithms (see [54,55] for more details).

3.3. Transition Probabilities

A specialisation of Boltzmann's principle by Einstein [56], for two states with entropies S and S_0 and "relative probability" W_r (the ratio of numbers W and W_0 that account for the numbers of microstates in the macrostates with S and S_0 respectively), is given by:

$$S - S_0 = k \log W_r \tag{17}$$

The expression in these relative terms is important, as pointed out by Norton [57], because the probability W_r *is the probability of the transition between the two states under the system's normal time evolution.*

In the example considered by Einstein [56,57], S_0 is the entropy of an (equilibrium) state, e.g., "a volume V_0 of space containing n non-interacting, moving points, whose dynamics are such as to favor no portion of the space over any other", while S is the entropy of the (non-equilibrium) state with the "same system of points, but now confined to a sub-volume V of V_0". Specifically, Einstein defined the transition probability $W_r = (V/V_0)^n$, yielding

$$S - S_0 = kn \log(V/V_0) \tag{18}$$

Since dynamics favour no portion of the space over any other, all the microstates are equiprobable.

3.4. Entropy Production

In general, the variation of entropy of a system ΔS is equal to the sum of the internal entropy production σ inside the system and the entropy change due to the interactions with the surroundings ΔS_{ext}:

$$\Delta S = \sigma + \Delta S_{ext} \tag{19}$$

In the case of a closed system, ΔS_{ext} is given by the expression

$$\Delta S_{ext} = \int dq/T \tag{20}$$

where q represents the heat flow received by the system from the exterior and T is the temperature of the system. This expression is often written as

$$\sigma = \Delta S - \Delta S_{ext} = (S - S_0) - \Delta S_{ext} \tag{21}$$

so that when the transition from the initial state S_0 to the final state S is irreversible, the entropy production $\sigma > 0$, while for reversible processes $\sigma = 0$, that is

$$S - S_0 = \int dq_{rev}/T \tag{22}$$

We shall consider another state vector, **y**, describing a state of a part Y of the exterior possibly coupled to the system represented by X. In other words, X and Y may or may not be dependent. In general, we shall say that σ_y is the internal entropy production *in the context of* some source Y, while ΔS_{ext} is the entropy production *attributed to* Y.

Alternatively, one may consider two scenarios for such a general physical system. In the first scenario, the entropy changes only due to reversible transitions, amounting to $S - S_0$. In the second scenario, the entropy changes partly irreversibly due to the interactions with the external environment affected by **y**, but still achieves the same total change $S - S_0$.

3.5. Range of Applicability

In an attempt to provide a thermodynamic interpretation of transfer entropy we make two important assumptions, defining the range of applicability for such an interpretation. The first one relates the transition probability W_{r_1} of the system's reversible state change to the conditional probability $p(\mathbf{x_{n+1}} \mid \mathbf{x_n})$, obtained by sampling the process X:

$$p(\mathbf{x_{n+1}} \mid \mathbf{x_n}) = \frac{1}{Z_1} W_{r_1} \tag{23}$$

where Z_1 is a normalisation factor that depends on $\mathbf{x_n}$. According to the expression for transition probability (17), under this assumption the conditional probability of the system's transition from state $\mathbf{x_n}$ to state $\mathbf{x_{n+1}}$ corresponds to some number W_{r_1}, such that $S(\mathbf{x_{n+1}}) - S(\mathbf{x_n}) = k \log W_{r_1}$, and hence

$$p(\mathbf{x_{n+1}} \mid \mathbf{x_n}) = \frac{1}{Z_1} e^{(S(\mathbf{x_{n+1}}) - S(\mathbf{x_n}))/k} \tag{24}$$

The second assumption relates the transition probability W_{r_2} of the system's possibly irreversible internal state change, due to the interactions with the external surroundings represented in the state vector $\mathbf{y}$, to the conditional probability $p(\mathbf{x_{n+1}} \mid \mathbf{x_n}, \mathbf{y_n})$, obtained by sampling the systems X and Y:

$$p(\mathbf{x_{n+1}} \mid \mathbf{x_n}, \mathbf{y_n}) = \frac{1}{Z_2} W_{r_2} \tag{25}$$

Under this assumption the conditional probability of the system's (irreversible) transition from state $\mathbf{x_n}$ to state $\mathbf{x_{n+1}}$ in the context of $\mathbf{y_n}$, corresponds to some number W_{r_2}, such that $\sigma_y = k \log W_{r_2}$, where σ_y is the system's internal entropy production in the context of $\mathbf{y}$, and thus

$$p(\mathbf{x_{n+1}} \mid \mathbf{x_n}, \mathbf{y_n}) = \frac{1}{Z_2} e^{\sigma_y/k} \tag{26}$$

where Z_2 is a normalisation factor that depends on $\mathbf{x_n}$.

3.6. An Example: Random Fluctuation Near Equilibrium

Let us consider the above-defined stochastic process X for a small random fluctuation around equilibrium:

$$\mathbf{x_{n+1}} = \Lambda \mathbf{x_n} + \xi \tag{27}$$

where ξ is a multi-variate Gaussian noise process, with covariance matrix Σ_ξ, uncorrelated in time. Starting at time n with state $\mathbf{x_n}$ having entropy $S(\mathbf{x_n})$, the state develops into $\mathbf{x_{n+1}}$, with entropy $S(\mathbf{x_{n+1}})$.

From the probability distribution function of the above multi-variate Gaussian process, we obtain

$$p(\mathbf{x_{n+1}} \mid \mathbf{x_n}) = \frac{1}{Z} e^{-\frac{1}{2}(\mathbf{x_{n+1}} - \Lambda \mathbf{x_n})^T \Sigma_\xi^{-1}(\mathbf{x_{n+1}} - \Lambda \mathbf{x_n})} \tag{28}$$

We now demonstrate that this expression concurs with the corresponding expression obtained under assumption (24). To do so we expand the entropies around $\mathbf{x} = 0$ with entropy $S(0)$:

$$S(\mathbf{x_n}) = S(0) - k\frac{1}{2}\mathbf{x_n}^T \Sigma_x^{-1} \mathbf{x_n} \tag{29}$$

where Σ_x is the covariance matrix of the process X.

Then, according to the assumption (24)

$$p(\mathbf{x_{n+1}} \mid \mathbf{x_n}) = \frac{1}{Z_1} e^{(S(\mathbf{x_{n+1}}) - S(\mathbf{x_n}))/k} = \frac{1}{Z_1} e^{-\frac{1}{2}\left(\mathbf{x_{n+1}}^T \Sigma_x^{-1} \mathbf{x_{n+1}} - \mathbf{x_n}^T \Sigma_x^{-1} \mathbf{x_n}\right)} = \frac{1}{\tilde{Z}_1} e^{-\frac{1}{2}\mathbf{x_{n+1}}^T \Sigma_x^{-1} \mathbf{x_{n+1}}} \tag{30}$$

where the term $e^{\frac{1}{2}\mathbf{x_n}^T \Sigma^{-1} \mathbf{x_n}}$ is absorbed into the normalisation factor being only dependent on $\mathbf{x_n}$. In general [58,59], we have

$$\Sigma_x = \sum_{j=0}^{\infty} \Lambda^j \, \Sigma_\varsigma \, \Lambda^{j^T} \tag{31}$$

Given the quasistationarity of the relaxation process, assumed near an equilibrium, $\Lambda \to 0$, and hence $\Sigma_x \to \Sigma_\varsigma$. Then the Equation (30) reduces to

$$p(\mathbf{x_{n+1}} \mid \mathbf{x_n}) = \frac{1}{\tilde{Z}_1} e^{-\frac{1}{2}\left(\mathbf{x_{n+1}}^T \Sigma_\varsigma^{-1} \mathbf{x_{n+1}}\right)} \tag{32}$$

The last expression concurs with Equation (28) when $\Lambda \to 0$.

4. Transfer Entropy: Thermodynamic Interpretation

4.1. Transitions Near Equilibrium

Supported by this background, we proceed to interpret transfer entropy via transitions between states. In doing so, we shall operate with local information theoretic measures (such as the local transfer entropy (9)), as we are dealing with (transitions between) *specific* states $\mathbf{y_n}, \mathbf{x_n}, \mathbf{x_{n+1}}$, etc. and not with all possible state-spaces X, Y, etc. containing all realizations of specific states.

Transfer entropy is a difference not between entropies, but rather between entropy rates or conditional entropies, specified on average by Equations (2) or (7), or for local values by Equation (10):

$$t_{Y \to X}(n+1) = h(\mathbf{x_{n+1}} \mid \mathbf{x_n}) - h(\mathbf{x_{n+1}} \mid \mathbf{x_n}, \mathbf{y_n}) \tag{33}$$

As mentioned above, the first assumption (23), taken to define the range of applicability for our interpretation, entails (24). It then follows that the first component of Equation (33), $h(\mathbf{x_{n+1}} \mid \mathbf{x_n})$, accounts for $S(\mathbf{x_{n+1}}) - S(\mathbf{x_n})$:

$$h(\mathbf{x_{n+1}} \mid \mathbf{x_n}) = -\log_2 p(\mathbf{x_{n+1}} \mid \mathbf{x_n}) = -\log_2 \frac{1}{Z_1} e^{(S(\mathbf{x_{n+1}}) - S(\mathbf{x_n}))/k} \tag{34}$$

$$= \log_2 Z_1 - \frac{1}{k \log 2} \left(S(\mathbf{x_{n+1}}) - S(\mathbf{x_n}) \right) \tag{35}$$

That is, the local conditional entropy $h(\mathbf{x_{n+1}} \mid \mathbf{x_n})$ corresponds to resultant entropy change of the transition from the past state $\mathbf{x_n}$ to the next state $\mathbf{x_{n+1}}$.

Now we need to interpret the second component of Equation (33): the local conditional entropy $h(\mathbf{x_{n+1}} \mid \mathbf{x_n}, \mathbf{y_n})$ in presence of some other factor or extra source, $\mathbf{y_n}$. Importantly, we must keep both the past state $\mathbf{x_n}$ and the next state $\mathbf{x_{n+1}}$ the same—only then we can characterise the internal entropy change, offset by some contribution of the source $\mathbf{y_n}$.

Our second constraint on the system (25) entails (26), and so

$$h(\mathbf{x_{n+1}} \mid \mathbf{x_n}, \mathbf{y_n}) = -\log_2 p(\mathbf{x_{n+1}} \mid \mathbf{x_n}, \mathbf{y_n}) = -\log_2 \frac{1}{Z_2} e^{\sigma_y/k} = \log_2 Z_2 - \frac{1}{k \log 2} (\sigma_y) \tag{36}$$

4.2. Transfer Entropy as Entropy Production

At this stage we can bring two right-hand side components of transfer entropy (33), represented by Equations (35) and (36), together:

$$t_{Y \to X}(n+1) = \log_2 \frac{Z_1}{Z_2} + \frac{1}{k \log 2} \left(-(S(\mathbf{x_{n+1}}) - S(\mathbf{x_n})) + \sigma_y \right) \tag{37}$$

When one considers a small fluctuation near an equilibrium, $Z_1 \approx Z_2$, as the number of microstates does not change much in the relevant macrostates. This removes the additive constant. Then, using the expression for entropy production (21), we obtain

$$t_{Y \to X}(n+1) = -\frac{\Delta S_{ext}}{k \log 2} \tag{38}$$

If $Z_1 \neq Z_2$, the relationship includes some additive constant $\log_2 \frac{Z_1}{Z_2}$.

That is, the transfer entropy is proportional to the external entropy production, brought about by the source of irreversibility Y. It captures the difference between the entropy rates that correspond to two scenarios: the reversible process and the irreversible process affected by another source Y. It is neither a transfer of entropy, nor an entropy of some transfer—it is formally a difference between two entropy rates. The opposite sign reflects the different direction of entropy production attributed to the source Y: when $\Delta S_{ext} > 0$, *i.e.*, the entropy increased during the transition in X more than the entropy produced internally, then the local transfer entropy is negative, and the source misinforms about the macroscopic state transition. When, on the other hand, $\Delta S_{ext} < 0$, *i.e.*, some of the internal entropy produced during the transition in X dissipated to the exterior, then the local transfer entropy is positive, and better predictions can be made about the macroscopic state transitions in X if source Y is measured.

As mentioned earlier, while transfer entropy is non-negative on average, some local transfer entropies can be negative when (in the context of the history of the destination) the source variable is misinformative or misleading about the state transition. This, obviously, concurs with the fact that, while a statistical ensemble average of time averages of the entropy change is always non-negative, at certain times entropy change can be negative. This follows from the fluctuation theorem [60], the Second law inequality [61], and can be illustrated with other examples of backward transformations and local violations of the second law [62,63].

Another observation follows from our assumptions (24) and (26) and the representation (37) when $Z_1 \approx Z_2$. If the local conditional entropy $h(\mathbf{x_{n+1}} \mid \mathbf{x_n})$, corresponding to the resultant entropy change of the transition, is different from the local conditional entropy $h(\mathbf{x_{n+1}} \mid \mathbf{x_n}, \mathbf{y_n})$ capturing the internal entropy production in context of the external source Y, then X and Y are dependent. Conversely, whenever these two conditional entropies are equal to each other, X and Y are independent.

4.3. Transfer Entropy as a Measure of Equilibrium's Stability

There is another possible interpretation that considers a fluctuation near the equilibrium. Using Kullback–Leibler divergence between discrete probability distributions p and q:

$$D_{KL}(p\|q) = \sum_i p(i) \log \frac{p(i)}{q(i)} \tag{39}$$

and its local counterpart:

$$d_{KL}(p\|q) = \log \frac{p(i)}{q(i)} \tag{40}$$

we may also express the local conditional entropy as follows:

$$h(\mathbf{x_{n+1}} \mid \mathbf{x_n}) = h(\mathbf{x_{n+1}}, \mathbf{x_n}) - h(\mathbf{x_n}) = d_{KL}\left(p(\mathbf{x_{n+1}}, \mathbf{x_n})\|p(\mathbf{x_n})\right) \tag{41}$$

It is known in macroscopic thermodynamics that stability of an equilibrium can be measured with Kullback–Leibler divergence between the initial (past) state, e.g., $\mathbf{x_n}$, and the state brought about by some fluctuation (a new observation), e.g., $\mathbf{x_{n+1}}$ [64]. That is, we can also interpret the local conditional entropy $h(\mathbf{x_{n+1}} \mid \mathbf{x_n})$ as the entropy change (or entropy rate) of the fluctuation near the equilibrium.

Analogously, the entropy change in another scenario, where an additional source **y** contributes to the fluctuation around the equilibrium, corresponds now to Kullback–Leibler divergence

$$h(\mathbf{x_{n+1}} \mid \mathbf{x_n, y_n}) = h(\mathbf{x_{n+1}, x_n, y_n}) - h(\mathbf{x_n, y_n}) = d_{\mathrm{KL}}\left(p(\mathbf{x_{n+1}, x_n, y_n}) \| p(\mathbf{x_n, y_n})\right) \qquad (42)$$

and can be seen as a measure of stability with respect to the fluctuation that is now affected by the extra source **y**.

Contrasting both these fluctuations around the same equilibrium, we obtain in terms of Kullback–Leibler divergences:

$$t_{Y \to X}(n+1) = d_{\mathrm{KL}}\left(p(\mathbf{x_{n+1}, x_n}) \| p(\mathbf{x_n})\right) - d_{\mathrm{KL}}\left(p(\mathbf{x_{n+1}, x_n, y_n}) \| p(\mathbf{x_n, y_n})\right) \qquad (43)$$

In these terms, transfer entropy contrasts stability of the equilibrium between two scenarios: the first one corresponds to the original system, and the second one disturbs the system by the source Y. If, for instance, the source Y is such that the system X is independent of it, then there is no difference in the extents of disturbances to the equilibrium, and the transfer entropy is zero.

4.4. Heat Transfer

It is possible to provide a similar thermodynamic interpretation relating directly to the Clausius definition of entropy. However, in this case we need to make assumptions stronger than Equations (23) and (25). Specifically, we assume Equations (24) and (26) which do not necessarily entail Equations (23) and (25) respectively. For example, setting the conditional probability $p(\mathbf{x_{n+1}} \mid \mathbf{x_n}) = \frac{1}{Z_1} e^{(S-S_0)/k}$ does not mean that $W_1 = e^{(S-S_0)/k}$ is the transition probability.

Under the new stronger assumptions, the conditional entropies can be related to the heat transferred in the transition, per temperature. Specifically, assumption (24) entails

$$h(\mathbf{x_{n+1}} \mid \mathbf{x_n}) = \log_2 Z_1 - \frac{1}{k \log 2}\left(S(\mathbf{x_{n+1}}) - S(\mathbf{x_n})\right) = \log_2 Z_1 - \frac{1}{k \log 2}\int_{\mathbf{x_n}}^{\mathbf{x_{n+1}}} dq_{rev}/T \qquad (44)$$

where the last step used the definition of Clausius entropy (16). As per (16), this quantity is the same for all reversible paths between the past and next states. An example illustrating the transition $(\mathbf{x_n} \to \mathbf{x_{n+1}})$ can be given by a simple thermal system $\mathbf{x_n}$ that is connected to a heat bath—that is, to a system in contact with a source of energy at temperature T. When the system X reaches a (new) equilibrium, e.g., the state $\mathbf{x_{n+1}}$, due to its connection to the heat bath, the local conditional entropy $h(\mathbf{x_{n+1}} \mid \mathbf{x_n})$ of the transition undergone by system X represents the heat transferred in the transition, per temperature.

Similarly, assumption (26) leads to

$$h(\mathbf{x_{n+1}} \mid \mathbf{x_n, y_n}) = \log_2 Z_2 - \frac{1}{k \log 2}\left(\sigma_y\right) = \log_2 Z_2 - \frac{1}{k \log 2}\int_{\mathbf{x_n}\xrightarrow{\mathbf{y_n}}\mathbf{x_{n+1}}} dq/T \qquad (45)$$

where $\mathbf{x_n} \xrightarrow{\mathbf{y_n}} \mathbf{x_{n+1}}$ is the new path between $\mathbf{x_n}$ and $\mathbf{x_{n+1}}$ brought about by $\mathbf{y_n}$, and the entropy produced along this path is σ_y. That is, the first and the last points of the path over which we integrate heat transfers per temperature are unchanged but the path is affected by the source **y**. This can be illustrated by a modified thermal system, still at temperature T but with heat flowing through some thermal resistance Y, while the system X repeats its transition from $\mathbf{x_n}$ to $\mathbf{x_{n+1}}$.

Transfer entropy captures the difference between expressions (44) and (45), *i.e.*, between the relevant amounts of heat transferred to the system X, per temperature.

$$t_{Y \to X}(n+1) = \log_2 \frac{Z_1}{Z_2} + \frac{1}{k \log 2}\left(\int_{\mathbf{x_n}\xrightarrow{\mathbf{y_n}}\mathbf{x_{n+1}}} dq/T - \int_{\mathbf{x_n}}^{\mathbf{x_{n+1}}} dq_{rev}/T\right) \qquad (46)$$

Assuming that $Z_1 \approx Z_2$ is realistic, e.g., for quasistatic processes, then the additive constant disappears as well.

It is clear that if the new path is still reversible (e.g., when the thermal resistance is zero) then the source **y** has not affected the resultant entropy change and we must have

$$\int_{x_n}^{x_{n+1}} dq_{rev}/T = \int_{x_n \xrightarrow{y_n} x_{n+1}} dq/T \qquad (47)$$

and $t_{Y \to X}(n+1) = 0$. This obviously occurs if and only if the source Y satisfies the independence condition (6), making the transfer entropy (46) equal to zero. In other words, we may again observe that if the local conditional entropy $h(x_{n+1} \mid x_n)$ corresponds to the resultant entropy change of the transition, then X and Y are dependent only when the external source Y, captured in the local conditional entropy $h(x_{n+1} \mid x_n, y_n)$, brings about an irreversible internal change. If, however, the source Y changed the path in such a way that the process became irreversible, then $t_{Y \to X}(n+1) \neq 0$.

Finally, according to Equations (19) and (20), the difference between the relevant heats transferred is $\int dq/T$, where q represents the heat flow received by the system from the exterior via the source Y, and hence

$$t_{Y \to X}(n+1) = \log_2 \frac{Z_1}{Z_2} - \frac{1}{k \log 2} \int dq/T \qquad (48)$$

In other words, local transfer entropy is proportional to the heat received or dissipated by the system from/to the exterior.

5. Causal Effect: Thermodynamic Interpretation?

In this section we shall demonstrate that a similar treatment is not possible in general for causal effect. Again, we begin by considering local causal effect (15) of the source y_n on destination x_{n+1}, while selecting **s** as the destination's past state x_n:

$$f(y_n \to x_{n+1} \mid \hat{x}_n) = \log_2 \frac{p(x_{n+1} \mid \hat{y}_n, \hat{x}_n)}{\sum_{y'_n} p(y'_n \mid \hat{x}_n) p(x_{n+1} \mid \hat{y'_n}, \hat{x}_n)} \qquad (49)$$

Let us first consider conditions under which this representation reduces to the local transfer entropy. As pointed out by Lizier and Prokopenko [41], there are several conditions for such a reduction.

Firstly, y_n and x_n must be the only causal contributors to x_{n+1}. In a thermodynamic setting, this means that there are no other sources affecting the transition from x_n to x_{n+1}, apart from y_n.

Whenever this condition is met, and in addition, the combination (y_n, x_n) is observed, it follows that

$$p(x_{n+1} \mid \hat{y}_n, \hat{x}_n) = p(x_{n+1} \mid y_n, x_n) \qquad (50)$$

simplifying the numerator of Equation (49).

Furthermore, there is another condition:

$$p(y_n \mid \hat{x}_n) \equiv p(y_n \mid x_n) \qquad (51)$$

For example, it is met when the source y_n is both causally and conditionally independent of the destination's past x_n. Specifically, causal independence means $p(y_n) \equiv p(y_n \mid \hat{x}_n)$, while conditional independence is simply $p(y_n) \equiv p(y_n \mid x_n)$. Intuitively, the situation of causal and conditional independence means that inner workings of the system X under consideration do not interfere with the source Y. Alternatively, if X is the only causal influence on Y, the condition (51) also holds, as Y is perfectly "explained" by X, whether X is observed or imposed on. In general, though, the condition (51) means that the probability of y_n if we impose a value $\hat{x}_n$ is the same as if we had simply observed the value $x_n = \hat{x}_n$ without imposing in the system X.

Under the conditions (50) and (51), the denominator of Equation (49) reduces to $p(\mathbf{x_{n+1}} \mid \mathbf{x_n})$, yielding the equivalence between local causal effect and local transfer entropy

$$f(\mathbf{y_n} \to \mathbf{x_{n+1}} \mid \hat{\mathbf{x}}_n) = t_{Y \to X}(n+1) \tag{52}$$

In this case, the thermodynamic interpretation of transfer entropy would be applicable to causal effect as well.

Whenever one of these conditions is not met, however, the reduction fails. Consider, for instance, the case when the condition (51) is satisfied, but the condition (50) is violated. For example, we may assume that there is some hidden source affecting the transition to $\mathbf{x_{n+1}}$. In this case, the denominator of Equation (49) does not simplify much, and the component that may have corresponded to the entropy rate of the transition between $\mathbf{x_n}$ and $\mathbf{x_{n+1}}$ becomes

$$\log_2 \sum_{\mathbf{y}'_n} p(\mathbf{y}'_n \mid \mathbf{x_n}) p(\mathbf{x_{n+1}} \mid \hat{\mathbf{y}}'_n, \hat{\mathbf{x}}_n) \tag{53}$$

The interpretation of this irreducible component is important: the presence of the imposed term $\hat{\mathbf{y}}'_n$ means that one should estimate individual contributions of all possible states $\mathbf{y}$ of the source Y, while varying (*i.e.*, imposing on) the state $\mathbf{x_n}$. This procedure becomes necessary because, in order to estimate the causal effect of source $\mathbf{y}$, *in presence of some other hidden source*, one needs to check all possible impositions on the source state $\mathbf{y}$. The terms of the sum under the logarithm in Equation (53) inevitably vary in their specific contribution, and so the sum cannot be analytically expressed as a single product under the logarithm. This means that we cannot construct a direct thermodynamic interpretation of causal effect in the same way that we did for the transfer entropy.

6. Discussion and Conclusions

In this paper we proposed a thermodynamic interpretation of transfer entropy: an information-theoretic measure introduced by Schreiber [1] as the average information contained in the source about the next state of the destination in the context of what was already contained in the destination's past. In doing so we used a specialised Boltzmann's principle. This in turn produced a representation of transfer entropy $t_{Y \to X}(n+1)$ as a difference of two entropy rates: one rate for a resultant transition within the system of interest X and another rate for a possibly irreversible transition within the system affected by an addition source Y. This difference was further shown to be proportional to the external entropy production, Δ_{ext}, attributed to the source of irreversibility Y.

At this stage we would like to point out a difference between our main result, $t_{Y \to X}(n+1) \propto -\Delta_{ext}$, and a representation for entropy production discussed by Parrondo *et al.* [22]. The latter work characterised the entropy production in the total device, in terms of relative entropy, the Kullback–Leibler divergence between the probability density ρ in phase space of some forward process and the probability density $\tilde{\rho}$ of the corresponding and suitably defined time-reversed process. The consideration of Parrondo *et al.* [22] does not involve any additional sources Y, and so transfer entropy is outside of the scope of their study. Their main result characterised entropy production as $k\, d_{KL}\,(\rho \| \tilde{\rho})$, which is equal to the total entropy change in the total device. In contrast, in our study we consider the system of interest X specifically, and characterise various entropy rates of X, but in doing so compare how these entropy rates are affected by some source of irreversibility Y. In short, transfer entropy is shown to concur with the entropy produced/dissipated by the system attributed to the external source Y.

We also briefly considered a case of fluctuations in the system X near an equilibrium, relating transfer entropy to the difference in stabilities of the equilibrium, with respect to two scenarios: a default case and the case with an additional source Y. This comparison was carried out with Kullback–Leibler divergences of the corresponding transition probabilities.

Entropy **2013**, *15*, 524–543

Finally, we demonstrated that such a thermodynamic treatment is not applicable to information flow, a measure introduced by Ay and Polani [18] in order to capture a causal effect. We argue that the main reason is the interventional approach adopted in the definition of causal effect. We identified several conditions ensuring certain dependencies between the involved variables, and showed that the causal effect may also be interpreted thermodynamically—but in this case it reduces to transfer entropy anyway. The highlighted difference once more shows a fundamental difference between transfer entropy and causal effect: the former has a thermodynamic interpretation relating to the source of irreversibility Y, while the latter is a construct that in general assumes an observer intervening in the system in a particular way.

We hope that the proposed interpretation will further advance studies relating information theory and thermodynamics, both in equilibrium and non-equilibrium settings, reversible and irreversible scenarios, average and local scopes, *etc.*

Acknowledgments: The Authors are thankful to Ralf Der (Max Planck Institute for Mathematics in the Sciences, Leipzig) who suggested and co-developed the example in subsection 3.6, and anonymous reviewers whose suggestions significantly improved the paper.

References

1. Schreiber, T. Measuring information transfer. *Phys. Rev. Lett.* **2000**, *85*, 461–464.
2. Jakubowski, M.H.; Steiglitz, K.; Squier, R. Information transfer between solitary waves in the saturable Schrödinger equation. *Phys. Rev. E* **1997**, *56*, 7267.
3. Baek, S.K.; Jung, W.S.; Kwon, O.; Moon, H.T. Transfer Entropy Analysis of the Stock Market, **2005**, arXiv:physics/0509014v2.
4. Moniz, L.J.; Cooch, E.G.; Ellner, S.P.; Nichols, J.D.; Nichols, J.M. Application of information theory methods to food web reconstruction. *Ecol. Model.* **2007**, *208*, 145–158.
5. Chávez, M.; Martinerie, J.; Le Van Quyen, M. Statistical assessment of nonlinear causality: Application to epileptic EEG signals. *J. Neurosci. Methods* **2003**, *124*, 113–128.
6. Pahle, J.; Green, A.K.; Dixon, C.J.; Kummer, U. Information transfer in signaling pathways: A study using coupled simulated and experimental data. *BMC Bioinforma.* **2008**, *9*, 139.
7. Lizier, J.T.; Prokopenko, M.; Zomaya, A.Y. Local information transfer as a spatiotemporal filter for complex systems. *Phys. Rev. E* **2008**, *77*, 026110.
8. Lizier, J.T.; Prokopenko, M.; Zomaya, A.Y. Information modification and particle collisions in distributed computation. *Chaos* **2010**, *20*, 037109.
9. Lizier, J.T.; Prokopenko, M.; Zomaya, A.Y. Coherent information structure in complex computation. *Theory Biosci.* **2012**, *131*, 193–203.
10. Lizier, J.T.; Prokopenko, M.; Zomaya, A.Y. Local measures of information storage in complex distributed computation. *Inf. Sci.* **2012**, *208*, 39–54.
11. Lizier, J.T.; Prokopenko, M.; Tanev, I.; Zomaya, A.Y. Emergence of Glider-like Structures in a Modular Robotic System. In Proceedings of the Eleventh International Conference on the Simulation and Synthesis of Living Systems (ALife XI), Winchester, UK, 5–8 August 2008; Bullock, S., Noble, J., Watson, R., Bedau, M.A., Eds.; MIT Press: Cambridge, MA, USA, 2008; pp. 366–373.
12. Lizier, J.T.; Prokopenko, M.; Zomaya, A.Y. The Information Dynamics of Phase Transitions in Random Boolean Networks. In Proceedings of the Eleventh International Conference on the Simulation and Synthesis of Living Systems (ALife XI), Winchester, UK, 5–8 August 2008; Bullock, S., Noble, J., Watson, R., Bedau, M.A., Eds.; MIT Press: Cambridge, MA, USA, 2008; pp. 374–381.
13. Lizier, J.T.; Pritam, S.; Prokopenko, M. Information dynamics in small-world Boolean networks. *Artif. Life* **2011**, *17*, 293–314.
14. Lizier, J.T.; Heinzle, J.; Horstmann, A.; Haynes, J.D.; Prokopenko, M. Multivariate information-theoretic measures reveal directed information structure and task relevant changes in fMRI connectivity. *J. Comput. Neurosci.* **2011**, *30*, 85–107.
15. Wang, X.R.; Miller, J.M.; Lizier, J.T.; Prokopenko, M.; Rossi, L.F. Quantifying and tracing information cascades in swarms. *PLoS One* **2012**, *7*, e40084.

16. Lizier, J.T.; Prokopenko, M.; Cornforth, D.J. The Information Dynamics of Cascading Failures in Energy Networks. In Proceedings of the European Conference on Complex Systems (ECCS), Warwick, UK, 21–25 October 2009; p. 54. ISBN: 978-0-9554123-1-8.

17. Kaiser, A.; Schreiber, T. Information transfer in continuous processes. *Physica D* **2002**, *166*, 43–62.

18. Ay, N.; Polani, D. Information flows in causal networks. *Adv. Complex Syst.* **2008**, *11*, 17–41.

19. Bennett, C.H. Notes on Landauer's principle, reversible computation, and Maxwell's Demon. *Stud. History Philos. Sci. Part B* **2003**, *34*, 501–510.

20. Piechocinska, B. Information erasure. *Phys. Rev. A* **2000**, *61*, 062314.

21. Lloyd, S. *Programming the Universe*; Vintage Books: New York, NY, USA, 2006.

22. Parrondo, J.M.R.; den Broeck, C.V.; Kawai, R. Entropy production and the arrow of time. *New J. Phys.* **2009**, *11*, 073008.

23. Prokopenko, M.; Lizier, J.T.; Obst, O.; Wang, X.R. Relating Fisher information to order parameters. *Phys. Rev. E* **2011**, *84*, 041116.

24. Landauer, R. Irreversibility and heat generation in the computing process. *IBM J. Res. Dev.* **1961**, *5*, 183–191.

25. Maroney, O.J.E. Generalizing Landauer's principle. *Phys. Rev. E* **2009**, *79*, 031105.

26. Jaynes, E.T. Information theory and statistical mechanics. *Phys. Rev.* **1957**, *106*, 620–630.

27. Jaynes, E.T. Information theory and statistical mechanics. II. *Phys. Rev.* **1957**, *108*, 171–190.

28. Crooks, G. Measuring thermodynamic length. *Phys. Rev. Lett.* **2007**, *99*, 100602+.

29. Liang, X.S. Information flow within stochastic dynamical systems. *Phys. Rev. E* **2008**, *78*, 031113.

30. Lüdtke, N.; Panzeri, S.; Brown, M.; Broomhead, D.S.; Knowles, J.; Montemurro, M.A.; Kell, D.B. Information-theoretic sensitivity analysis: A general method for credit assignment in complex networks. *J. R. Soc. Interface* **2008**, *5*, 223–235.

31. Auletta, G.; Ellis, G.F.R.; Jaeger, L. Top-down causation by information control: From a philosophical problem to a scientific research programme. *J. R. Soc. Interface* **2008**, *5*, 1159–1172.

32. Hlaváčková-Schindler, K.; Paluš, M.; Vejmelka, M.; Bhattacharya, J. Causality detection based on information-theoretic approaches in time series analysis. *Phys. Rep.* **2007**, *441*, 1–46.

33. Lungarella, M.; Ishiguro, K.; Kuniyoshi, Y.; Otsu, N. Methods for quantifying the causal structure of bivariate time series. *Int. J. Bifurc. Chaos* **2007**, *17*, 903–921.

34. Ishiguro, K.; Otsu, N.; Lungarella, M.; Kuniyoshi, Y. Detecting direction of causal interactions between dynamically coupled signals. *Phys. Rev. E* **2008**, *77*, 026216.

35. Sumioka, H.; Yoshikawa, Y.; Asada, M. Causality Detected by Transfer Entropy Leads Acquisition of Joint Attention. In Proceedings of the 6th IEEE International Conference on Development and Learning (ICDL 2007), London, UK, 11–13 July 2007; pp. 264–269.

36. Vejmelka, M.; Palus, M. Inferring the directionality of coupling with conditional mutual information. *Phys. Rev. E* **2008**, *77*, 026214.

37. Verdes, P.F. Assessing causality from multivariate time series. *Phys. Rev. E* **2005**, *72*, 026222:1–026222:9.

38. Tung, T.Q.; Ryu, T.; Lee, K.H.; Lee, D. Inferring Gene Regulatory Networks from Microarray Time Series Data Using Transfer Entropy. In Proceedings of the Twentieth IEEE International Symposium on Computer-Based Medical Systems (CBMS '07), Maribor, Slovenia, 20–22 June 2007; Kokol, P., Podgorelec, V., Mičetič-Turk, D., Zorman, M., Verlič, M., Eds.; IEEE: Los Alamitos, CA, USA, 2007; pp. 383–388.

39. Van Dijck, G.; van Vaerenbergh, J.; van Hulle, M.M. Information Theoretic Derivations for Causality Detection: Application to Human Gait. In Proceedings of the International Conference on Artificial Neural Networks (ICANN 2007), Porto, Portugal, 9–13 September 2007; Sá, J.M.d., Alexandre, L.A., Duch, W., Mandic, D., Eds.; Springer-Verlag: Berlin/Heidelberg, Germany, 2007; Volume 4669, *Lecture Notes in Computer Science*, pp. 159–168.

40. Hung, Y.C.; Hu, C.K. Chaotic communication via temporal transfer entropy. *Phys. Rev. Lett.* **2008**, *101*, 244102.

41. Lizier, J.T.; Prokopenko, M. Differentiating information transfer and causal effect. *Eur. Phys. J. B* **2010**, *73*, 605–615.

42. Wuensche, A. Classifying cellular automata automatically: Finding gliders, filtering, and relating space-time patterns, attractor basins, and the Z parameter. *Complexity* **1999**, *4*, 47–66.

43. Solé, R.V.; Valverde, S. Information transfer and phase transitions in a model of internet traffic. *Physica A* **2001**, *289*, 595–605.

44. Solé, R.V.; Valverde, S. Information Theory of Complex Networks: On Evolution and Architectural Constraints. In *Complex Networks*; Ben-Naim, E., Frauenfelder, H., Toroczkai, Z., Eds.; Springer: Berlin/Heidelberg, Germany, 2004; Volume 650, *Lecture Notes in Physics*, pp. 189–207.

45. Prokopenko, M.; Boschietti, F.; Ryan, A.J. An information-theoretic primer on complexity, self-organization, and emergence. *Complexity* **2009**, *15*, 11–28.

46. Takens, F. Detecting Strange Attractors in Turbulence. In *Dynamical Systems and Turbulence, Warwick 1980*; Rand, D., Young, L.S., Eds.; Springer: Berlin/Heidelberg, Germany, 1981; pp. 366–381.

47. Granger, C.W.J. Investigating causal relations by econometric models and cross-spectral methods. *Econometrica* **1969**, *37*, 424–438.

48. Fano, R. *Transmission of Information: A Statistical Theory of Communications*; The MIT Press: Cambridge, MA, USA, 1961.

49. Manning, C.D.; Schütze, H. *Foundations of Statistical Natural Language Processing*; The MIT Press: Cambridge, MA, USA, 1999.

50. Dasan, J.; Ramamohan, T.R.; Singh, A.; Nott, P.R. Stress fluctuations in sheared Stokesian suspensions. *Phys. Rev. E* **2002**, *66*, 021409.

51. MacKay, D.J. *Information Theory, Inference, and Learning Algorithms*; Cambridge University Press: Cambridge, MA, USA, 2003.

52. Pearl, J. *Causality: Models, Reasoning, and Inference*; Cambridge University Press: Cambridge, MA, USA, 2000.

53. Goyal, P. Information physics–towards a new conception of physical reality. *Information* **2012**, *3*, 567–594.

54. Sethna, J.P. *Statistical Mechanics: Entropy, Order Parameters, and Complexity*; Oxford University Press: Oxford, UK, 2006.

55. Seife, C. *Decoding the Universe*; Penguin Group: New York, NY, USA, 2006.

56. Einstein, A. Über einen die Erzeugung und Verwandlung des Lichtes betreffenden heuristischen Gesichtspunkt. *Ann. Phys.* **1905**, *322*, 132–148.

57. Norton, J.D. Atoms, entropy, quanta: Einstein's miraculous argument of 1905. *Stud. History Philos. Mod. Phys.* **2006**, *37*, 71–100.

58. Barnett, L.; Buckley, C.L.; Bullock, S. Neural complexity and structural connectivity. *Phys. Rev. E* **2009**, *79*, 051914.

59. Ay, N.; Bernigau, H.; Der, R.; Prokopenko, M. Information-driven self-organization: The dynamical system approach to autonomous robot behavior. *Theory Biosci.* **2012**, *131*, 161–179.

60. Evans, D.J.; Cohen, E.G.D.; Morriss, G.P. Probability of second law violations in shearing steady states. *Phys. Rev. Lett.* **1993**, *71*, 2401–2404.

61. Searles, D.J.; Evans, D.J. Fluctuations relations for nonequilibrium systems. *Aus. J. Chem.* **2004**, *57*, 1129–1123.

62. Crooks, G.E. Entropy production fluctuation theorem and the nonequilibrium work relation for free energy differences. *Phys. Rev. E* **1999**, *60*, 2721–2726.

63. Jarzynski, C. Nonequilibrium work relations: Foundations and applications. *Eur. Phys. J. B-Condens. Matter Complex Syst.* **2008**, *64*, 331–340.

64. Schlögl, F. Information Measures and Thermodynamic Criteria for Motion. In Structural Stability in Physics: Proceedings of Two International Symposia on Applications of Catastrophe Theory and Topological Concepts in Physics, Tübingen, Germany, 2–6 May and 11–14 December 1978; Güttinger, W., Eikemeier, H., Eds.; Springer: Berlin/Heidelberg, Germany, 1979; Volume 4, *Springer series in synergetics*, pp. 199–209.

Entropy **2013**, *15*, 2635–2661; doi:10.3390/e15072635

Article

Simulation Study of Direct Causality Measures in Multivariate Time Series

Angeliki Papana [1,*], Catherine Kyrtsou [1,2], Dimitris Kugiumtzis [3] and Cees Diks [4]

[1] Department of Economics, University of Macedonia, Egnatias 156, 54006, Thessaloniki, Greece; E-Mail: ckyrtsou@uom.gr

[2] University of Strasbourg, BETA, University of Paris 10, Economix, ISC-Paris, Ile-de-France

[3] Faculty of Engineering, Aristotle University of Thessaloniki, University Campus, 54124, Thessaloniki, Greece; E-Mail: dkugiu@gen.auth.gr

[4] Faculty of Economics, Department of Economics and Econometrics, University of Amsterdam, Valckenierstraat 65-67, 1018 XE, Amsterdam, The Netherlands; E-Mail: c.g.h.diks@uva.nl

* Author to whom correspondence should be addressed; E-Mail: angeliki.papana@gmail.com; Tel.: +30-2310891764.

Received: 28 March 2013; in revised form: 5 June 2013 / Accepted: 27 June 2013 / Published: 4 July 2013

Abstract: Measures of the direction and strength of the interdependence among time series from multivariate systems are evaluated based on their statistical significance and discrimination ability. The best-known measures estimating direct causal effects, both linear and nonlinear, are considered, *i.e.*, conditional Granger causality index (CGCI), partial Granger causality index (PGCI), partial directed coherence (PDC), partial transfer entropy (PTE), partial symbolic transfer entropy (PSTE) and partial mutual information on mixed embedding (PMIME). The performance of the multivariate coupling measures is assessed on stochastic and chaotic simulated uncoupled and coupled dynamical systems for different settings of embedding dimension and time series length. The CGCI, PGCI and PDC seem to outperform the other causality measures in the case of the linearly coupled systems, while the PGCI is the most effective one when latent and exogenous variables are present. The PMIME outweighs all others in the case of nonlinear simulation systems.

Keywords: direct Granger causality; multivariate time series; information measures

PACS: 05.45.Tp; 05.45.-a; 02.70.-c

1. Introduction

The quantification of the causal effects among simultaneously observed systems from the analysis of time series recordings is essential in many scientific fields, ranging from economics to neurophysiology. Estimating the inter-dependence among the observed variables provides valuable knowledge about the processes that generate the time series. Granger causality has been the leading concept for the identification of directional interactions among variables from their time series, and it has been widely used in economics [1]. However, the last few years, it has become popular also in many different fields, e.g., for the analysis of electroencephalograms.

The mathematical formulation of linear Granger causality is based on linear regression modeling of stochastic processes. Many modifications and extensions of the Granger causality test have been developed; see e.g., [2–7]. Most of the non-causality tests, built on the Granger causality concept and applied in economics, are therefore based on the modeling of the multivariate time series. Despite the success of these strategies, the model-based methods may suffer from the shortcomings of model mis-specification.

Entropy **2013**, *15*, 2635–2661

The majority of the measures determining the interrelationships among variables that have been developed so far are for bivariate data, e.g., state-space based techniques [8,9], information measures [10–12] and techniques based on the concept of synchronization [13,14].

Bivariate causality tests may erroneously detect couplings when two variables are conditionally independent. To address this, techniques accounting for the effect of the confounding variables have been introduced, termed direct causality measures, which are more appropriate when dealing with multivariate time series [15–17]. Direct causality methods emerged as extensions of bivariate Granger causality. For example, the Granger causality index (GCI), implementing the initial idea for two variables in the time domain, has been extended to the conditional and partial Granger causality index (CGCI and PGCI) [2,18]. Directed coherence (DC) was introduced in the frequency domain, and being a bivariate measure, it cannot discriminate between direct and indirect coupling. The direct transfer function (DTF) is similarly defined as DC [19]. The partial directed coherence (PDC) is an extension of DC to multivariate time series measuring only the direct influences among the variables [20]. Similarly, direct Directed Transfer Function (dDTF) modified DTF to detect only direct information transfer [21].

Information theory sets a natural framework for non-parametric methodologies of several classes of statistical dependencies. Several techniques from information theory have been used in the last few years for the identification of causal relationships in multivariate systems, and the best known is transfer entropy (TE) [11]. Test for causality using the TE has also been suggested [22]. However, the TE is, again, bivariate and its natural extension to account for the presence of confounding variables has been recently introduced, namely, the partial TE (PTE), under different estimating schemes, using bins [23], correlation sums [24] and nearest neighbors [25]. The TE and PTE are actually expressions of conditional mutual information, and with this respect, an improved version of TE making use of a properly restricted non-uniform state space reconstruction was recently developed, termed mutual information on mixed embedding (MIME) [26]. Later, a similar approach to TE/PTE was implemented, which takes into consideration the conditional entropy [27]. Recently MIME was extended for multivariate time series to the partial MIME (PMIME) [28]. Other coupling methods have also been suggested, such as Renyi's information transfer [29]. In a different approach, the TE has been defined on rank vectors instead of sample vectors, called the symbolic transfer entropy (STE) [30], and, respectively, to the multivariate case termed partial STE (PSTE) (for a correction of STE and PSTE, see, respectively, [31,32]).

Most comparative works on the effectiveness of causality measures concentrate on bivariate tests, e.g., [33–36], while some works evaluating multivariate methodologies include only model-based tests, see, e.g., [37–39], or compare direct and indirect causality measures, e.g., [36,40].

In this work, we compare model-based methods, both in the time and frequency domain, and information theoretic multivariate causality measures that are able to distinguish between direct and indirect causal effects. We include in the study most of the known direct causality measures of these classes, *i.e.*, CGCI and PGCI (linear in time domain), PDC (linear in frequency domain), PTE, PSTE and PMIME (from information theory). The statistical significance of the test statistics is assessed with resampling methods, bootstraps or randomization tests using appropriate surrogates, whenever it is not theoretically known.

The structure of the paper is as follows. The multivariate causality measures considered in this study are presented in Section 2. The statistical significance of the coupling measures is assessed on simulated systems. The simulation systems and the setup of the simulation study are presented in Section 3, while the results of this study and the performance of the causality measures are discussed in Section 4. Finally, the conclusions are drawn in Section 5.

2. Direct Causality Measures

Let $\{x_{1,t}, \ldots, x_{K,t}\}$, $t = 1, \ldots, n$, denote a K-variate time series, consisting of K simultaneously observed variables, $X_1, \ldots, X_K$, belonging to a dynamical system or representing respective subsystems of a global system. The reconstructed vectors of each X_i are formed as

175

$\mathbf{x}_{i,t} = (x_{i,t}, x_{i,t-\tau_i}, \ldots, x_{i,t-(m_i-1)\tau_i})'$, where $t = 1, \ldots, n'$, $n' = n - \max_i\{(m_i - 1)\tau_i\}$, and m_i and τ_i are, respectively, the reconstruction parameters of embedding dimension and time delay for X_i. The notation, $X_2 \rightarrow X_1$, denotes the Granger causality from X_2 to X_1, while $X_2 \rightarrow X_1|Z$ denotes the direct Granger causality from X_2 to X_1, accounting for the presence of the other (confounding) variables, *i.e.*, $Z = \{X_3, \ldots, X_K\}$. The notation of Granger causality for other pairs of variables is analogous.

Almost all of the causality measures require the time series be stationary, *i.e.*, their mean and variance do not change over time. If the time series are non-stationary, then the data should be pre-processed, e.g., for time series that are non-stationary in mean, one can apply the measures on the first or higher order differences. Different transformations are needed in case of non-stationary data in variance or co-integrated time series.

2.1. Conditional Granger Causality Index

Granger causality is based on the concept that if the value of a time series, X_1, to be predicted is improved by using the values of X_2, then we say that X_2 is driving X_1. A vector autoregressive model (VAR) in two variables and of order P, fitted to the time series, $\{x_{1,t}\}$, is:

$$x_{1,t+1} = \sum_{j=0}^{P-1} a_{1,j} x_{1,t-j} + \sum_{j=0}^{P-1} b_{1,j} x_{2,t-j} + \epsilon_{1,t+1} \tag{1}$$

where $a_{1,j}, b_{1,j}$ are the coefficients of the model and ϵ_1 the residuals from fitting the model with variance s_{1U}^2. The model in Equation (1) is referred to as the unrestricted model, while the restricted model is obtained by omitting the terms regarding the driving variable [the second sum in Equation (1)] and has residual variance, s_{1R}^2. According to the concept of Granger causality, the variable, X_2, Granger causes X_1 if $s_{1R}^2 > s_{1U}^2$ [1]. The magnitude of the effect of X_2 on X_1 is given by the Granger Causality Index (GCI), defined as:

$$\text{GCI}_{X_2 \rightarrow X_1} = \ln(s_{1R}^2 / s_{1U}^2) \tag{2}$$

Considering all K variables, the unrestricted model for X_1 is a VAR model in K variables and involves the P lags of all K variables [K sum terms instead of two in Equation (1)]; the restricted model will have all but the P lags of the driving variable, X_2. Likewise, the conditional Granger causality index (CGCI) is:

$$\text{CGCI}_{X_2 \rightarrow X_1|Z} = \ln(s_{1R}^2 / s_{1U}^2) \tag{3}$$

where s_{1U}^2 and s_{1R}^2 are the residual variances for the unrestricted and restricted model defined for all K variables.

A parametric significance test for GCI and CGCI can be conducted for the null hypothesis that variable X_2 is not driving X_1, making use of the F-significance test for all P coefficients, $b_{1,j}$ [41]. When we want to assess collectively the causal effects among all pairs of the K variables, a correction for multiple testing should be performed, e.g., by means of the false discovery rate [42].

The order, P, of the VAR model is usually chosen using an information criterion, such as the Akaike Information Criterion (AIC) [43] and the Bayesian Information Criterion (BIC) [44]. The estimation of the coefficients of the VAR models and the residual variances of the models are described analytically in [45].

2.2. Partial Granger Causality Index

The partial Granger causality index (PGCI) is associated with the concept of Granger causality and partial correlation [18]. The PGCI addresses the problem of exogenous inputs and latent variables. The intuition is that the influence of exogenous and/or latent variables on a system will be reflected by correlations among the residuals of a VAR model of the measured variables. Thus, in the PGCI, one makes use of the residual covariance matrix of the VAR unrestricted and restricted model, denoted

Σ and ρ, respectively, and not only of the residual variance of the response variable, X_1, s_{1U}^2 and s_{1R}^2, subsequently. For example, for $X_2 \rightarrow X_1 | X_3$, denoting the components of Σ as Σ_{ij}, $i, j = 1, 2, 3$ and the components of ρ as ρ_{ij}, $i, j = 1, 2$, the PGCI is given as:

$$\text{PGCI}_{X_2 \rightarrow X_1 | X_3} = \ln \frac{\rho_{11} - \rho_{12} \rho_{22}^{-1} \rho_{21}}{\Sigma_{11} - \Sigma_{13} \Sigma_{33}^{-1} \Sigma_{31}} \tag{4}$$

Note that $\Sigma_{11} = s_{1U}^2$ and $\rho_{11} = s_{1R}^2$. The PGCI constitutes an improved estimation of the direct Granger causality as compared to the CGCI when the residuals of the VAR models are correlated; otherwise, it is identical to the CGCI. The estimation procedure for the PGCI is described analytically in [18].

2.3. Partial Directed Coherence

The partial directed coherence (PDC) is related to the same VAR model as the CGCI, but is defined in the frequency domain [20]. Denoting the $K \times K$ matrix of the Fourier transform of the coefficients of the VAR model in K variables and order P by $A(f)$, the PDC from X_2 to X_1 at a frequency f is given by [20]:

$$\text{PDC}_{X_2 \rightarrow X_1 | Z}(f) = \frac{|A_{1,2}(f)|}{\sqrt{\sum_{k=1}^{K} |A_{k,2}(f)|^2}} \tag{5}$$

where $A_{i,j}(f)$ is the component at the position, (i, j), in the matrix, $A(f)$. $\text{PDC}_{X_2 \rightarrow X_1 | Z}(f)$ provides a measure for the directed linear influence of X_2 on X_1 at frequency, f, conditioned on the other $K - 2$ variables in Z and takes values in the interval, $[0, 1]$. The $\text{PDC}_{X_2 \rightarrow X_1}(f)$ is computed at each frequency, f, within an appropriate range of frequencies. Parametric inference and significance tests for PDC have been studied in [46,47].

2.4. Partial Transfer Entropy

The transfer entropy (TE) is a nonlinear measure that quantifies the amount of information explained in X_1 at h steps ahead from the state of X_2, accounting for the concurrent state of X_1 [11]. The TE is given here in terms of entropies. For a discrete variable, X (scalar or vector), the Shannon entropy is $H(X) = -\sum p(x_i) \ln p(x_i)$, where $p(x_i)$ is the probability mass function of variable, X, at the value, x_i. Further, the TE is expressed as:

$$\begin{aligned} \text{TE}_{X_2 \rightarrow X_1} &= I(x_{1,t+h}; x_{2,t} | \mathbf{x}_{1,t}) = H(x_{1,t+h} | \mathbf{x}_{1,t}) - H(x_{1,t+h} | \mathbf{x}_{2,t}, \mathbf{x}_{1,t}) \\ &= H(\mathbf{x}_{2,t}, \mathbf{x}_{1,t}) - H(x_{1,t+h}, \mathbf{x}_{2,t}, \mathbf{x}_{1,t}) + H(x_{1,t+h}, \mathbf{x}_{1,t}) - H(\mathbf{x}_{1,t}) \end{aligned} \tag{6}$$

The first equality is inserted to show that the TE is equivalent to the conditional mutual information (CMI), where $I(X, Y) = H(X) + H(Y) - H(X, Y)$ is the mutual information (MI) of two variables, X and Y. The time horizon, h, is introduced here instead of the single time step, originally used in the definition of TE.

The partial transfer entropy (PTE) is the extension of the TE designed for the direct causality of X_2 to X_1 conditioning on the remaining variables in Z

$$\text{PTE}_{X_2 \rightarrow X_1 | Z} = H(x_{1,t+h} | \mathbf{x}_{1,t}, \mathbf{z}_t) - H(x_{1,t+h} | \mathbf{x}_{2,t}, \mathbf{x}_{1,t}, \mathbf{z}_t) \tag{7}$$

The entropy terms of PTE are estimated here using the k-nearest neighbors method [48].

2.5. Symbolic Transfer Entropy

The symbolic transfer entropy (STE) is the continuation of the TE estimated on rank-points formed by the reconstructed vectors of the variables [30]. For each vector, $\mathbf{x}_{2,t}$, the ranks of its components assign a rank-point, $\hat{\mathbf{x}}_{2,t} = [r_1, r_2, \ldots, r_{m_2}]$, where $r_j \in \{1, 2, \ldots, m_2\}$ for $j = 1, \ldots, m_2$. Following this sample-point to rank-point conversion, the sample, $x_{1,t+h}$, in Equation (7) is taken as the rank point at time, $t + h$, $\hat{\mathbf{x}}_{1,t+h}$, and STE is defined as:

$$\text{STE}_{X_2 \to X_1} = H(\hat{\mathbf{x}}_{1,t+h}|\hat{\mathbf{x}}_{1,t}) - H(\hat{\mathbf{x}}_{1,t+h}|\hat{\mathbf{x}}_{2,t}, \hat{\mathbf{x}}_{1,t}) \tag{8}$$

where the entropies are computed based on the rank-points.

In complete analogy to the derivation of the PTE from the TE, the partial symbolic transfer entropy (PSTE) extends the STE for multivariate time series and is expressed as:

$$\text{PSTE}_{X_2 \to X_1|Z} = H(\hat{\mathbf{x}}_{1,t+h}|\hat{\mathbf{x}}_{1,t}, \hat{\mathbf{z}}_t) - H(\hat{\mathbf{x}}_{1,t+h}|\hat{\mathbf{x}}_{2,t}, \hat{\mathbf{x}}_{1,t}, \hat{\mathbf{z}}_t) \tag{9}$$

where the rank vector, $\hat{\mathbf{z}}_t$, is the concatenation of the rank vectors for each of the embedding vectors of the variables in Z.

2.6. Partial Mutual Information on Mixed Embedding

The mutual information on mixed embedding (MIME) is derived directly from a mixed embedding scheme based on the conditional mutual information (CMI) criterion [26]. In the bivariate case and for the driving of X_2 on X_1, the scheme gives a mixed embedding of varying delays from the variables, X_1 and X_2, that explains best the future of X_1, defined as $\mathbf{x}_{1,t}^h = [x_{1,t+1}, \dots, x_{1,t+h}]$. The mixed embedding vector, $\mathbf{w}_t$, may contain lagged components of X_1, forming the subset, $\mathbf{w}_t^{X_1}$, and of X_2, forming $\mathbf{w}_t^{X_2}$, where $\mathbf{w}_t = [\mathbf{w}_t^{X_1}, \mathbf{w}_t^{X_2}]$. The MIME is then estimated as:

$$\text{MIME}_{X_2 \to X_1} = \frac{I(\mathbf{x}_{1,t}^h; \mathbf{w}_t^{X_2}|\mathbf{w}_t^{X_1})}{I(\mathbf{x}_{1,t}^h; \mathbf{w}_t)} \tag{10}$$

The numerator in Equation (10) is the CMI as for the TE in Equation (7), but for non-uniform embedding vectors of X_1 and X_2. Therefore, the MIME can be considered as a normalized version of the TE for optimized non-uniform embedding of X_1 and X_2 [26].

For multivariate time series, the partial mutual information on mixed embedding (PMIME) has been developed in analogy to the MIME [28]. The mixed embedding vector that best describes the future of X_1, $\mathbf{x}_{1,t}^h$, is now formed potentially by all K lagged variables, *i.e.*, X_1, X_2 and the other $K-2$ variables in Z, and it can be decomposed to the three respective subsets as $\mathbf{w}_t = (\mathbf{w}_t^{X_1}, \mathbf{w}_t^{X_2}, \mathbf{w}_t^Z)$. The PMIME is then estimated as:

$$\text{PMIME}_{X_2 \to X_1|Z} = \frac{I(\mathbf{x}_{1,t}^h; \mathbf{w}_t^{X_2}|\mathbf{w}_t^{X_1}, \mathbf{w}_t^Z)}{I(\mathbf{x}_{1,t}^h; \mathbf{w}_t)} \tag{11}$$

Similarly to the MIME, the PMIME can be considered as a normalized version of the PTE for optimized non-uniform embedding of all K variables. Thus, the PMIME takes values between zero and one, where zero indicates the absence of components of X_2 in the mixed embedding vector and, consequently, no direct Granger causality from X_2 to X_1.

A maximum lag to search for components in the mixed embedding vector is set for each variable, here being the same maximum lag, L_{max}, for all variables. L_{max} can be set equal to a sufficiently large number without affecting the performance of the measure; however, the larger it is, the higher the computational cost is. For the estimation of the MI and the CMI, the k-nearest neighbors method is used [48].

3. Simulation Study

The multivariate causality measures are evaluated in a simulation study. All the considered direct coupling measures are computed on 100 realizations of multivariate uncoupled and coupled systems, for increasing coupling strengths and for all directions. The simulation systems that have been used in this study are the following.

- System 1: A vector autoregressive process of order one [VAR(1)] in three variables with $X_1 \to X_2$ and $X_2 \to X_3$

$$
\begin{aligned}
x_{1,t} &= \theta_t \\
x_{2,t} &= x_{1,t-1} + \eta_t \\
x_{3,t} &= 0.5x_{3,t-1} + x_{2,t-1} + \epsilon_t
\end{aligned}
$$

where θ_t, η_t and ϵ_t are independent to each other Gaussian white noise processes, with standard deviations one, 0.2 and 0.3, respectively.

- System 2: A VAR(5) process in four variables with $X_1 \to X_3$, $X_2 \to X_1$, $X_2 \to X_3$ and $X_4 \to X_2$ (Equation 12 in [49])

$$
\begin{aligned}
x_{1,t} &= 0.8x_{1,t-1} + 0.65x_{2,t-4} + \epsilon_{1,t} \\
x_{2,t} &= 0.6x_{2,t-1} + 0.6x_{4,t-5} + \epsilon_{2,t} \\
x_{3,t} &= 0.5x_{3,t-3} - 0.6x_{1,t-1} + 0.4x_{2,t-4} + \epsilon_{3,t} \\
x_{4,t} &= 1.2x_{4,t-1} - 0.7x_{4,t-2} + \epsilon_{4,t}
\end{aligned}
$$

where $\epsilon_{i,t}$, $i = 1, \dots, 4$ are independent to each other Gaussian white noise processes with unit standard deviation.

- System 3: A VAR(4) process in five variables with $X_1 \to X_2$, $X_1 \to X_4$, $X_2 \to X_4$, $X_4 \to X_5$, $X_5 \to X_1$, $X_5 \to X_2$ and $X_5 \to X_3$ [46]

$$
\begin{aligned}
x_{1,t} &= 0.4x_{1,t-1} - 0.5x_{1,t-2} + 0.4x_{5,t-1} + \epsilon_{1,t} \\
x_{2,t} &= 0.4x_{2,t-1} - 0.3x_{1,t-4} + 0.4x_{5,t-2} + \epsilon_{2,t} \\
x_{3,t} &= 0.5x_{3,t-1} - 0.7x_{3,t-2} - 0.3x_{5,t-3} + \epsilon_{3,t} \\
x_{4,t} &= 0.8x_{4,t-3} + 0.4x_{1,t-2} + 0.3x_{2,t-3} + \epsilon_{4,t} \\
x_{5,t} &= 0.7x_{5,t-1} - 0.5x_{5,t-2} - 0.4x_{4,t-1} + \epsilon_{5,t}
\end{aligned}
$$

and $\epsilon_{i,t}$, $i = 1, \dots, 5$, as above.

- System 4: A coupled system of three variables with linear ($X_2 \to X_3$) and nonlinear causal effects ($X_1 \to X_2$ and $X_1 \to X_3$) (Model 7 in [50])

$$
\begin{aligned}
x_{1,t} &= 3.4x_{1,t-1}(1 - x_{1,t-1})^2 \exp\left(-x_{1,t-1}^2\right) + 0.4\epsilon_{1,t} \\
x_{2,t} &= 3.4x_{2,t-1}(1 - x_{2,t-1})^2 \exp\left(-x_{2,t-1}^2\right) + 0.5x_{1,t-1}x_{2,t-1} + 0.4\epsilon_{2,t} \\
x_{3,t} &= 3.4x_{3,t-1}(1 - x_{3,t-1})^2 \exp\left(-x_{3,t-1}^2\right) + 0.3x_{2,t-1} + 0.5x_{1,t-1}^2 + 0.4\epsilon_{3,t}
\end{aligned}
$$

and $\epsilon_{i,t}$, $i = 1, \dots, 3$, as above.

- System 5: Three coupled Hénon maps with nonlinear couplings, $X_1 \to X_2$ and $X_2 \to X_3$

$$
\begin{aligned}
x_{1,t} &= 1.4 - x_{1,t-1}^2 + 0.3x_{1,t-2} \\
x_{2,t} &= 1.4 - cx_{1,t-1}x_{2,t-1} - (1-c)x_{2,t-1}^2 + 0.3x_{2,t-2} \\
x_{3,t} &= 1.4 - cx_{2,t-1}x_{3,t-1} - (1-c)x_{3,t-1}^2 + 0.3x_{3,t-2}
\end{aligned}
$$

with equal coupling strengths, c, and $c = 0, 0.05, 0.3, 0.5$.

The time series of this system become completely synchronized for coupling strengths, $c \geq 0.7$. In order to investigate the effect of noise on the causality measures, we also consider the case of addition of Gaussian white noise to each variable of System 5, with standard deviation 0.2 times their standard deviation.

- System 6: Three coupled Lorenz systems with nonlinear couplings, $X_1 \to X_2$ and $X_2 \to X_3$

$$
\begin{aligned}
\dot{x}_1 &= 10(y_1 - x_1) & \dot{x}_2 &= 10(y_2 - x_2) + c(x_1 - x_2) & \dot{x}_3 &= 10(y_3 - x_3) + c(x_2 - x_3) \\
\dot{y}_1 &= 28x_1 - y_1 - x_1 z_1 \; , & \dot{y}_2 &= 28x_2 - y_2 - x_2 z_2 & \; , \; \dot{y}_3 &= 28x_3 - y_3 - x_3 z_3 \\
\dot{z}_1 &= x_1 y_1 - 8/3 z_1 & \dot{z}_2 &= x_2 y_2 - 8/3 z_2 & \dot{z}_3 &= x_3 y_3 - 8/3 z_3
\end{aligned}
$$

The first variables of the three interacting systems are observed at a sampling time of 0.05 units. The couplings, $X_1 \to X_2$ and $X_2 \to X_3$, have the same strength, c, and $c = 0, 1, 3, 5$. The time series of the system become completely synchronized for coupling strengths, $c \geq 8$. For a more detailed description of the synchronization of the coupled Systems 5 and 6, see [51].

- System 7: A linear coupled system in five variables with $X_1 \to X_2$, $X_1 \to X_3$, $X_1 \to X_4$, $X_4 \leftrightarrow X_5$ with latent end exogenous variables [18]

$$
\begin{aligned}
x_{1,t} &= 0.95\sqrt{2} x_{1,t-1} - 0.9025 x_{1,t-2} + \epsilon_{1,t} + a_1 \epsilon_{6,t} + b_1 \epsilon_{7,t-1} + c_1 \epsilon_{7,t-2} \\
x_{2,t} &= 0.5 x_{1,t-2} + \epsilon_{2,t} + a_2 \epsilon_{6,t} + b_2 \epsilon_{7,t-1} + c_2 \epsilon_{7,t-2} \\
x_{3,t} &= -0.4 x_{1,t-3} + \epsilon_{3,t} + a_3 \epsilon_{6,t} + b_3 \epsilon_{7,t-1} + c_3 \epsilon_{7,t-2} \\
x_{4,t} &= -0.5 x_{1,t-2} + 0.25\sqrt{2} x_{4,t-1} + 0.25\sqrt{2} x_{5,t-1} + \epsilon_{4,t} + a_4 \epsilon_{6,t} + b_4 \epsilon_{7,t-1} + c_4 \epsilon_{7,t-2} \\
x_{5,t} &= -0.25\sqrt{2} x_{4,t-1} + 0.25\sqrt{2} x_{5,t-1} + \epsilon_{5,t} + a_5 \epsilon_{6,t} + b_5 \epsilon_{7,t-1} + c_5 \epsilon_{7,t-2}
\end{aligned}
$$

where $\epsilon_{i,t}$ are zero mean uncorrelated processes with variances $0.8, 0.6, 1, 1.2, 1, 0.9, 1$, respectively, $a_1 = 5$, $a_2 = a_3 = a_4 = a_5 = 1$ and $b_i = 2$, $c_i = 5$, $i = 1, \dots, 5$.

The time series lengths considered in the simulation study are $n = 512$ and $n = 2048$. Regarding the CGCI, the PGCI and the PDC, the order, P, of the VAR model is selected by combining the Akaike Information Criterion (AIC), the Bayesian Information Criterion (BIC), as well as our knowledge for the degrees of freedom of each coupled system, as follows. The range of model orders, for which the AIC and the BIC are calculated, is selected to be at the level of the 'true' model order based on the equations of each system. Specifically, for Systems 1, 4, 5, 6 and 7, we considered the range of model orders, $[1, 5]$, for the calculation of AIC and BIC, and for Systems 2 and 3, we considered the range, $[1, 10]$. Further, we estimate the PDC for a range of frequencies determined by the power spectrum of the variables of each system. We specify this range by selecting those frequencies that display the highest values in the auto-spectra of the variables [52]. The p-values from a non-parametric test for the PDC are estimated for the selected range of frequencies (using bootstraps [53]), and in order to decide whether a coupling is significant, at least 80% of the p-values from this range of frequencies should be significant.

The embedding dimension, m, for the PTE and the PSTE and the maximum lag, L_{max}, for the PMIME are set equal to P and $\tau = 1$ for the PTE and the PSTE. Note that this choice of L_{max} may be very restrictive, and the PMIME may not be optimal; but, we adopt it here to make the choice for VAR order and embedding uniform. The time step ahead, h, for the estimation of the PTE, PSTE and PMIME is set to one for the first five systems and System 7 (the common choice for discrete-time systems), while for the continuous-time system (System 6), h is set to be equal to m. The number of nearest neighbors for the estimation of the PTE and the PMIME is set to $k = 10$. We note that the k-nearest neighbors methods for the estimation of the measures is found to be stable and not significantly affected by the choice of k [48]. The threshold for the stopping criterion for the mixed embedding scheme for the PMIME is set to $A = 0.95$ (for details, see [26]).

3.1. Statistical Significance of the Causality Measures

In the simulation study, we assess the statistical significance of the causality measures by means of parametric tests, when applicable, and nonparametric (resampling) tests, otherwise, in the way these have been suggested in the literature for each measure. The correction for multiple testing regarding the significance of a measure on all possible variable pairs is not considered here, as the interest is in comparing the performance of the different direct causality measures rather than providing rigorous statistical evidence for the significance of each coupling.

Starting with the CGCI, it bears a parametric significance test, and this is the F-test for the null hypothesis that the coefficients of the lagged driving variables in the unrestricted VAR model are all zero [41]. If P_1 and P_2 are the numbers of variables in the restricted and the unrestricted autoregressive model, respectively, $(P_2 > P_1)$, and n is the length of the time series, then the test statistic is $F = ((RSS_1 - RSS_2)/(P_1 - P_2))/(RSS_2/(n - P_2))$, where RSS_i is the residual sum of squares of model, i. Under the null hypothesis that the unrestricted model does not provide a significantly better fit than the restricted model, the F-statistic follows the Fisher-Snedecor, or F, distribution with $(P_2 - P_1, n - P_2)$ degrees of freedom. The null hypothesis is rejected if the F-statistic calculated on the data is greater than the critical value of the F-distribution for some desired false-rejection probability (here $\alpha = 0.05$).

The statistical significance of the PGCI is assessed by means of confidence intervals formed by bootstrapping [53], since the null distribution is unknown. The empirical distribution of any statistic using bootstrapping is formed from the values of the statistic computed on a number of new samples obtained by random sampling with replacement from the observed data. In the context of vector autoregressive models, this can be realized by subdividing the data matrix (of the predictor and response jointly) into a number of windows, which are repeatedly sampled with replacement to generate bootstrap data matrices. By this procedure, the causal relationships within each window are not affected. The PGCI is computed for each bootstrapped data matrix. The confidence interval of the PGCI is formed by the lower and upper empirical quantiles of the bootstrap distribution of the PGCI for the significance level, $\alpha = 0.05$. The bootstrap confidence interval for the PGCI can be considered as a significance test, where the test decision depends on whether zero is included in the confidence interval. The details for the estimation of the bootstrap confidence intervals of the PGCI can be found in [18].

The statistical significance of the PDC can be determined using both parametric testing [46,47], and randomization (surrogate) testing [54]. Here, we choose the parametric approach. The statistical significance of a nonzero value, $\text{PDC}_{X_2 \to X_1}(f)$, is investigated by means of a critical value, c_{PDC}. Under the null hypothesis that there is no Granger causality, $X_2 \to X_1$, it holds $|A_{12}(f)| = 0$, and c_{PDC} can be derived from theoretical considerations for each frequency, f, at a given α-significance level by:

$$c_{PDC}(f) = ((\hat{C}_{ij}(f)\chi^2_{1,1-a})/(N\sum_k |\hat{A}_{kj}(f)|^2))^{1/2}$$

The term, $\chi^2_{1,1-a}$, denotes the $(1 - \alpha)$-quantile of the χ^2 distribution with one degree of freedom, and $\hat{C}_{ij}(f)$ is an estimate of the expression:

$$C_{ij}(f) = \Sigma_{ii}(\sum_{k,l=1}^{P} \Sigma_{jj}^{-1}[\cos(kf)\cos(lf) + \sin(kf)\sin(lf)])$$

where Σ_{jj}^{-1} denotes the entries of the inverse of the covariance matrix, Σ, of the VAR process [47].

The statistical significance of the PTE and the PSTE is evaluated assuming a randomization test with appropriate surrogate time series, as their null distribution is not known (for the PSTE, in [32], analytic approximations were built, but found to be inferior to approximations using surrogates). We create M surrogate time series consistent with the non-causality null hypotheses, H_0, *i.e.*, X_2 does not Granger causes X_1. To destroy any causal effect of X_2 on X_1 without changing the dynamics in each time series, we randomly choose a number, d, less than the time series length, n, and the d-first values of the time series of X_2 are moved to the end, while the other series remain unchanged. The random number, d, for the time-shifted surrogates is an integer within the range, $[0.05n, 0.95n]$, where n is the time series length. This scheme for generating surrogate time series is termed time-shifted surrogates [55]. We estimate the causality measure (PTE or PSTE) from the original multivariate time series, let us denote it q_0, and for each of the M multivariate surrogate time series, let us denote them $q_1, q_2, \ldots, q_M$. If q_0 is at the tail of the empirical null distribution formed by $q_1, q_2, \ldots, q_M$, then H_0 is rejected. For the two-sided test, if r_0 is the rank of q_0 in the ordered list of $q_0, q_1, \ldots, q_M$, the p-value for the test is $2(r_0 - 0.326)/(M + 1 + 0.348)$ if $r_0 < (M+1)/2$ and $2[1 - (r_0 - 0.326)/(M + 1 + 0.348)]$ if $r_0 \geq (M+1)/2$, by applying the correction for the empirical cumulative density function in [56].

Finally, PMIME does not rely on any significance test, as it gives zero values in the uncoupled case and positive values, otherwise. This was confirmed using time-shifted surrogates also for the PMIME in the simulation study, and the PMIME values of the surrogate time series were all zero.

For the estimation of CGCI and PGCI and their statistical significance, we used the 'Causal Connectivity Analysis' toolbox [57]. The programs for the computations of the remaining causality measures have been implemented in Matlab.

4. Evaluation of Causality Measures

In order to evaluate the multivariate causality measures, the percentage of rejection of the null hypothesis of no causal effects (H_0) in 100 realizations of the system is calculated for each possible pair of variables and for different time series lengths and free parameters of the measures. The focus when presenting the results is on the sensitivity of the measure or, respectively, the power of the significance test (the percentage of rejection at the significance level 5% or $\alpha = 0.05$ when there is true direct causality), as well as the specificity of the measure or size of the test (the percentage of rejection at $\alpha = 0.05$ when there is no direct causality) and how these properties depend on the time series length and the measure-specific parameter.

4.1. Results for System 1

For the estimation of the linear measures, the order of the model, P, is set to one, as indicated from the Bayesian Information Criterion (BIC) and the Akaike Information Criterion (AIC), while for the estimation of PTE, m is also set to one. The PDC is estimated for the range of frequencies $[0, 0.5]$, since the auto-spectra of the variables do not suggest a narrower range. Indeed, the p-values of PDC are all significant in $[0, 0.5]$, when there is direct causality, and not significant, when there is no direct causality. The CGCI, PGCI, PDC and PTE correctly detect the direct causal effect for both time series lengths, $n = 512$ and 2048. All the aforementioned measures indicate 100% rejection of H_0 for the true couplings, $X_1 \rightarrow X_2$ and $X_2 \rightarrow X_3$, and low percentages for all other couplings. Their performance is not affected by the time series length, for the time series lengths considered. The estimated percentages are displayed for both n in Table 1.

Table 1. Percentage of statistically significant values of the causality measures for System 1, $P = m = L_{max} = 1$ [$m = 2$ for partial symbolic transfer entropy (PSTE)]. The directions of direct causal effects are pointed out in bold. When the same percentage has been found for both n, a single number is displayed in the cell.

$n = 512/2048$	CGCI	PGCI	PDC	PTE	PSTE	PMIME
$\mathbf{X_1 \rightarrow X_2}$	100	100	100	100	100	1 / 0
$X_2 \rightarrow X_1$	4 / 3	2 / 1	2 / 3	8 / 5	58 / 100	2 / 7
$\mathbf{X_2 \rightarrow X_3}$	100	100	100	100	100	100
$X_3 \rightarrow X_2$	7 / 6	8 / 1	3	3 / 5	7 / 25	0
$X_1 \rightarrow X_3$	3 / 5	0 / 2	2 / 3	5 / 7	93 / 100	0
$X_3 \rightarrow X_1$	2 / 7	0	3	3 / 2	14 / 43	7 / 7

The PSTE can only be estimated for $m \geq 2$, and therefore, results are obtained for $m = 2$. The PSTE correctly detects the direct causalities for $m = 2$; however, it also indicates the indirect effect, $X_1 \rightarrow X_3$, and the spurious causal effect, $X_2 \rightarrow X_1$.

Only for this system, the PMIME with the threshold of $A = 0.95$ failed to detect the true direct effects, and the randomization test gave partial improvement (detection of one of the two true direct effects, $X_2 \rightarrow X_3$). This is merely a problem of using the fixed threshold, $A = 0.95$, in this system, and following the adapted threshold proposed in [28], the two true direct effects could be detected for all realizations with $n = 512$ and $n = 2048$ with the largest rate of false rejection being 8%.

4.2. Results for System 2

For the second simulation system, the model order is set to $P = 5$, as indicated both by BIC and AIC. The embedding dimension, m, and L_{max} are also set to five. The PDC is estimated for the range of frequencies, $[0, 0.4]$, since the auto-spectra of all the variables are higher in the range, $[0, 0.2]$, while variable, X_3, exhibits a peak in the range, $[0.2, 0.4]$. Indicatively, the p-values from one realization of the system for the range of frequencies, $[0, 0.5]$, is displayed in Figure 1a.

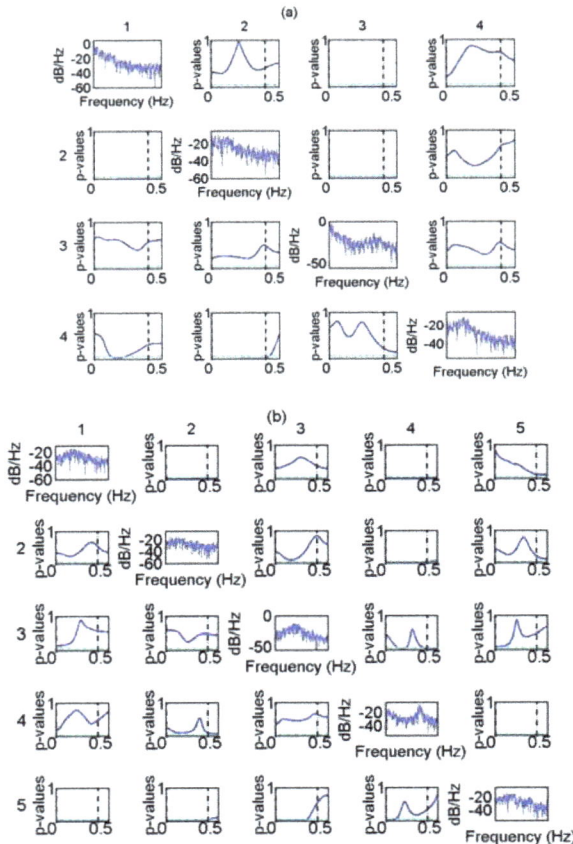

Figure 1. Graph summarizing the causal influences for one realization of System 2 (rows → columns) in (**a**) and System 3 in (**b**). The p-values from the partial directed coherence (PDC) are displayed for the range of frequencies, $[0, 0.5]$, while the dotted vertical lines indicate the frequency, 0.4. The horizontal cyan lines indicate the 5%-significance level. The auto-spectra are shown on the diagonal.

The CGCI, PGCI, PDC and PMIME correctly detect the direct couplings ($X_2 \to X_1$, $X_1 \to X_3$, $X_2 \to X_3$, $X_4 \to X_2$), as shown in Table 2. The performance of the CGCI and PDC is not affected by the time series length. The PGCI is also not affected by n, except for the causal effect, $X_2 \to X_4$, where the PGCI falsely indicates causality for $n = 2048$ (20%). The PMIME indicates lower power of the test compared to the linear measures only for $X_2 \to X_3$ and $n = 512$.

Table 2. Percentage of statistically significant values of the causality measures for System 2, $P = m = L_{max} = 5$. The directions of direct causal effects are displayed in bold face. CGCI, conditional Granger causality index; PGCI, partial Granger causality index; PDC, partial directed coherence; PTE, partial transfer entropy; PSTE, partial symbolic transfer entropy; PMIME, partial mutual information on mixed embedding.

$n = 512/2048$	CGCI	PGCI	PDC	PTE	PSTE	PMIME
$X_1 \to X_2$	6 / 2	1	0	6 / 11	7 / 20	0
$\mathbf{X_2 \to X_1}$	100	100	100	100	5 / 11	100
$\mathbf{X_1 \to X_3}$	100	100	100	100	7 / 15	100
$X_3 \to X_1$	5 / 3	0	0	6 / 14	6 / 9	0
$X_1 \to X_4$	6 / 7	0	0	7 / 50	2 / 24	3 / 0
$X_4 \to X_1$	3 / 2	0	0	2 / 5	5	0
$\mathbf{X_2 \to X_3}$	100	98 / 100	94 / 100	14 / 39	9 / 18	64 / 99
$X_3 \to X_2$	8 / 5	0	0	4 / 16	5 / 3	1 / 0
$X_2 \to X_4$	7 / 6	3 / 20	0	5 / 8	1 / 20	2 / 0
$\mathbf{X_4 \to X_2}$	100	100	100	100	7 / 2	100
$X_3 \to X_4$	4 / 3	0 / 2	0	8 / 29	2 / 8	3 / 0
$X_4 \to X_3$	4 / 5	0	0	7	5 / 6	0

Table 3. Mean PTE values from 100 realizations of System 2, for $P = 5$ and $n = 512, 2048$. The values of the true direct couplings are highlighted.

mean	$X_1 \to X_2$	$X_2 \to X_1$	$X_1 \to X_3$	$X_3 \to X_1$	$X_1 \to X_4$	$X_4 \to X_1$
$n = 512$	0.0042	**0.0920**	**0.0772**	0.0034	0.0067	0.0043
$n = 2048$	0.0029	**0.1221**	**0.0965**	0.0016	0.0034	0.0020
	$\mathbf{X_2 \to X_3}$	$X_3 \to X_2$	$X_2 \to X_4$	$\mathbf{X_4 \to X_2}$	$X_3 \to X_4$	$X_4 \to X_3$
$n = 512$	**0.0060**	0.0052	0.0095	**0.0998**	0.0071	0.0033
$n = 2048$	**0.0059**	0.0032	0.0061	**0.1355**	0.0042	0.0013

The PTE detects the direct causal relationships, apart from the coupling, $X_2 \to X_3$, although the percentage of rejection in this direction increases with n (from 14% for $n = 512$ to 39% for $n = 2048$). Further, the erroneous relationships, $X_1 \to X_4$ (50%) and $X_3 \to X_4$ (29%), are observed for $n = 2048$. Focusing on the PTE values, it can be observed that they are much higher for the directions of direct couplings than for the remaining directions. Moreover, the percentages of significant PTE values increase with n for the directions with direct couplings and decrease with n for all other couplings (see Table 3).

We note that the standard deviation of the estimated PTE values from the 100 realizations are low (on the order of 10^{-2}). Thus, the result of having falsely statistically significant PTE values for $X_1 \to X_4$ and $X_3 \to X_4$ is likely due to insufficiency of the randomization test.

PSTE fails to detect the causal effects for the second coupled system for both time series lengths, giving rejections at a rate between 1% and 24% for all directions. The failure of PSTE may be due to the the the high dimensionality of the rank vectors (the joint rank vector has dimension 21).

Entropy **2013**, *15*, 2635–2661

4.3. Results for System 3

The CGCI, PGCI, PDC, PTE and PMIME correctly detect all direct causal effects ($X_1 \to X_2$, $X_1 \to X_4$, $X_5 \to X_1$, $X_2 \to X_4$, $X_5 \to X_2$, $X_5 \to X_3$, $X_4 \to X_5$) for $P = m = L_{max} = 4$ (based on BIC and AIC), as shown in Table 4.

Table 4. Percentages of statistically significant values of causality measures for System 3, for $P = m = L_{max} = 4$.

$n = 512/2048$	CGCI	PGCI	PDC	PTE	PSTE	PMIME
$X_1 \to X_2$	100	86 / 100	92 / 100	81 / 100	8 / 7	100
$X_2 \to X_1$	7 / 3	2 / 0	0	7 / 2	3 / 8	18 / 0
$X_1 \to X_3$	6 / 5	2 / 0	0	3 / 4	4 / 10	2 / 0
$X_3 \to X_1$	1 / 2	0	0	5 / 4	2 / 4	14 / 0
$X_1 \to X_4$	100	100	100	52 / 100	2 / 9	100
$X_4 \to X_1$	7 / 4	1 / 0	0	6 / 4	4 / 10	12 / 0
$X_1 \to X_5$	5	2	0	6 / 10	7 / 12	16 / 0
$X_5 \to X_1$	100	98 / 100	99 / 100	100	2 / 16	100
$X_2 \to X_3$	4	2 / 0	0	4 / 6	9 / 5	2 / 0
$X_3 \to X_2$	6 / 2	1 / 0	0	4 / 2	5 / 4	9 / 1
$X_2 \to X_4$	100	99 / 100	96 / 100	18 / 77	7	94 / 100
$X_4 \to X_2$	5 / 9	0	0	5 / 4	2 / 4	9 / 0
$X_2 \to X_5$	6 / 4	2 / 3	0	5 / 4	4 / 6	22 / 0
$X_5 \to X_2$	100	96 / 100	99 / 100	99 / 100	3 / 8	100
$X_3 \to X_4$	3 / 7	1 / 0	0	3 / 4	5 / 8	0
$X_4 \to X_3$	4	1 / 0	0	3 / 6	4 / 5	4 / 0
$X_3 \to X_5$	4 / 3	1	0	7 / 4	6	14 / 0
$X_5 \to X_3$	100	87 / 100	84 / 100	49 / 97	3 / 15	100
$X_4 \to X_5$	100	100	100	100	6 / 4	100
$X_5 \to X_4$	5 / 2	0	0	17 / 37	6 / 14	1 / 0

The PDC is again estimated in the range of frequencies, $[0, 0.4]$, (see in Figure 1b the auto-spectra of the variables and the p-values from the parametric test of PDC from one realization of the system). The CGCI, PGCI and PDC perform similarly for the two time series lengths. The PTE indicates 100% significant values for $n = 2048$ when direct causality exists. However, the PTE also indicates the spurious causality, $X_5 \to X_4$, for $n = 2048$ (37%). The specificity of the PMIME is improved by the increase of n, and the percentage of positive PMIME values in case of no direct causal effects varies from 0% to 22% for $n = 512$, while for $n = 2048$, it varies from 0% to 1%. The PSTE again fails to detect the causal effects, giving very low percentage of rejection at all directions (2% to 16%).

Since the linear causality measures CGCI, PGCI and PDC have been developed for the detection of direct causality in linear coupled systems, it was expected that these methods would be successfully applied to all linear systems. The nonlinear measures PMIME and PTE also seem to be able to capture the direct linear couplings in most cases, with PMIME following close the linear measures both in specificity and sensitivity.

In the following systems, we investigate the ability of the causality measures to correctly detect direct causal effects when nonlinearities are present.

4.4. Results for System 4

For the fourth coupled system, the BIC and AIC suggest to set $P = 1, 2$ and 3. The performance of the linear measures does not seem to be affected by the choice of P. The PDC is estimated for frequencies in $[0.1, 0.4]$. The auto-spectra of the three variables do not display any peaks or any upward/downward trends. No significant differences in the results are observed if a wider or narrower range of frequencies is considered. The linear measures, CGCI, PGCI and PDC, capture only the linear direct causal effect,

$X_2 \rightarrow X_3$, while they fail to detect the nonlinear relationships, $X_1 \rightarrow X_2$ and $X_1 \rightarrow X_3$, for both time series lengths.

The PTE and the PMIME correctly detect all the direct couplings for the fourth coupled system for $m = L_{max} = 1, 2$ and 3. The percentage of significant values of the causality measures are displayed in Table 5. The PTE gives equivalent results for $m = 1$ and $m = 2$. The PTE correctly detects the causalities for $m = 3$, but at a smaller power of the significance test for $n = 512$ (63% for $X_1 \rightarrow X_2$, 46% for $X_2 \rightarrow X_3$ and 43% for $X_1 \rightarrow X_3$). The percentage of significant PMIME values is 100% for the directions of direct couplings, and falls between 0% and 6% for all other couplings, and this holds for any $L_{max} = 1, 2$ or 3 and for both n.

Table 5. Percentage of statistically significant values of the causality measures for System 4, $P = m = L_{max} = 2$.

$n = 512/2048$	CGCI	PGCI	PDC	PTE	PSTE	PMIME
$X_1 \rightarrow X_2$	12 / 7	1	2	97 / 100	10 / 69	100
$X_2 \rightarrow X_1$	2 / 7	0 / 1	1 / 0	8 / 9	4 / 8	3 / 0
$X_2 \rightarrow X_3$	100	73 / 100	100	76 / 100	69 / 100	100
$X_3 \rightarrow X_2$	7 / 4	3 / 1	1	4	3 / 9	4 / 0
$X_1 \rightarrow X_3$	7	1 / 0	0 / 1	86 / 100	1 / 7	100
$X_3 \rightarrow X_1$	4 / 5	0	0	4 / 6	8 / 21	0

The PSTE indicates the link $X_2 \rightarrow X_3$ for both time series lengths, while $X_1 \rightarrow X_2$ is detected only for $n = 2048$. The PSTE fails to point out the causality, $X_1 \rightarrow X_3$. The results for $m = 2$ and 3 are equivalent. In order to investigate whether the failure of PSTE to show $X_1 \rightarrow X_3$ is due to finite sample data, we estimate the PSTE also for $n = 4096$. For $m = 2$, it indicates the same results as for $n = 2048$. For $m = 3$, the PSTE detects all the direct causal effects, $X_1 \rightarrow X_2$ (99%), $X_2 \rightarrow X_3$ (100%), $X_1 \rightarrow X_3$ (86%), but $X_3 \rightarrow X_1$ (62%) is also erroneously detected.

4.5. Results for System 5

For the fifth coupled simulation system, we set the model order, $P = 2$, based on the complexity of the system, and $P = 3, 4$ and 5 using the AIC and BIC. The auto-spectra of the variables display peaks in $[0.1, 0.2]$ and $[0.4, 0.5]$. The PDC is estimated for different ranges of frequencies to check its sensitivity with respect to the selection of the frequency range. When small frequencies are considered, the PDC seems to indicate larger percentages of spurious couplings; however, also, the percentages of significant PDC values at the directions of true causal effects are smaller. The results are presented for System 5 considering the range of frequencies, $[0.4, 0.5]$.

The CGCI seems to be sensitive to the selection of the model order P, indicating some spurious couplings for the different P. The best performance for CGCI is achieved for $P = 3$; therefore, only results for $P = 3$ are shown. On the other hand, the PGCI turns out to be less dependent on P, giving similar results for $P = 2, 3, 4$ and 5. The PTE is not substantially affected by the selection of the embedding dimension, m (at least for the examined coupling strengths); therefore, only results for $m = 2$ are discussed. The PSTE is sensitive to the selection of m, performing best for $m = 2$ and 3, while for $m = 4$ and 5, it indicates spurious and indirect causal effects. The PMIME does not seem to depend on L_{max}. Results are displayed for $L_{max} = 5$. The percentage of significant values for each measure are displayed in Figure 2, for all directions, for increasing coupling strength and for both n.

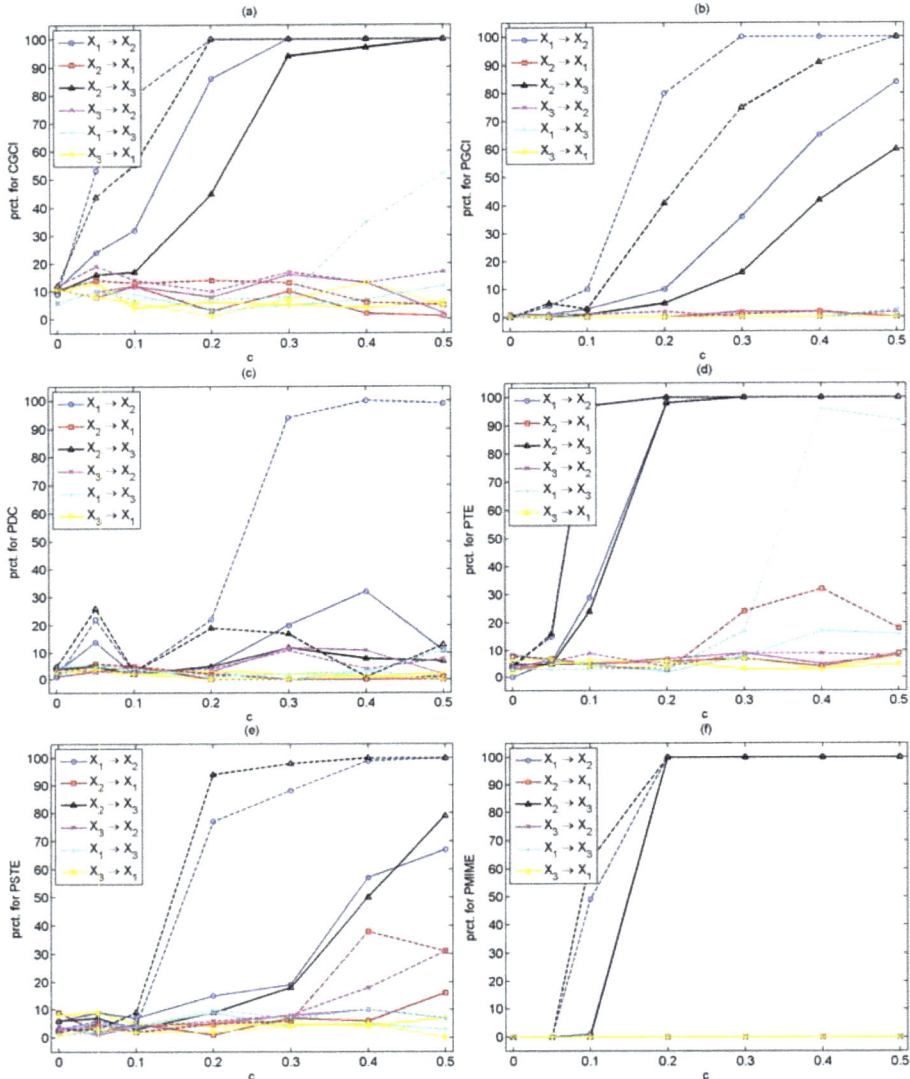

Figure 2. Percentage of significant (**a**) CGCI ($P = 3$); (**b**) PGCI ($P = 3$); (**c**) PDC ($P = 3$); (**d**) PTE ($m = 2$); (**e**) PSTE ($m = 2$); and (**f**) PMIME ($L_{max} = 5$) values, for System 5, for increasing coupling strengths, c, at all directions and for both n (for $n = 512$, solid lines, for $n = 2048$, dotted lines).

Most of the measures show good specificity, and the percentage of rejection for all pairs of the variables of the uncoupled system ($c = 0$) is at the significance level, $\alpha = 0.05$, with only CGCI scoring a somehow larger percentage of rejection up to 17%.

For the weak coupling strengths, $c = 0.05$ and 0.1, the causality measures cannot effectively detect the causal relationships or have a low sensitivity. The CGCI and the PTE seem to have the best performance, while the PMIME seems to be effective only for $n = 2048$ and $c = 0.1$.

As the coupling strength increases, the sensitivity of the causality measures is improved. For $c = 0.2$, the CGCI, PTE and PMIME correctly indicate the true couplings for both n, while the PGCI and the PSTE do this only for $n = 2048$. The PDC has low power, even for $n = 2048$. For $c = 0.3$, nearly all measures correctly point out the direct causal effects (see Table 6). The best results are obtained with the PMIME, while the CGCI and PTE display similar performance. The PGCI and the PSTE are sensitive to the time series length and have a high power only for $n = 2048$. The PDC performs poorly, giving low percentage of significant PDC values, even for $n = 2048$. All measures have good specificity, with CGCI and PTE giving rejections well above the nominal level for some non-existing couplings.

Considering larger coupling strengths, the causality measures correctly indicate the true couplings, but also some spurious ones. The PMIME outperforms the other measures giving 100% positive values for both n for $X_1 \rightarrow X_2$ and $X_2 \rightarrow X_3$ and 0% at the remaining directions for $c \geq 0.2$. Indicative results for all measures are displayed for the strong coupling strength $c = 0.5$ in Table 7.

In order to investigate the effect of noise on each measure, we consider the coupled Hénon map (System 5) with the addition of Gaussian white noise with standard deviation 0.2 times the standard deviation of the original time series. Each measure is estimated again from 100 realizations from the noisy system for the same free parameters as considered in the noise-free case.

The CGCI is not significantly affected by the addition of noise, giving equivalent results for $P = 3$ as for the noise-free system. The CGCI detects the true causal effects even for weak coupling strength ($c \geq 0.05$). For different P values ($P = 2, 4$ or 5), some spurious and/or indirect causal effects are observed for $c > 0.3$.

Table 6. Percentage of statistically significant values of the causality measures for System 5 for $c = 0.3$, where $P = 3$, $m = 2$ and $L_{max} = 5$.

$n = 512/2048$	CGCI	PGCI	PDC	PTE	PSTE	PMIME
$X_1 \rightarrow X_2$	100	36 / 100	20 / 94	100	19 / 88	100
$X_2 \rightarrow X_1$	10 / 13	0 / 1	0 / 2	7 / 24	7 / 6	0
$X_2 \rightarrow X_3$	94 / 100	16 / 75	12 / 19	100	18 / 98	100
$X_3 \rightarrow X_2$	16 / 17	2 / 0	12 / 4	9	8	0
$X_1 \rightarrow X_3$	5 / 8	0	0 / 1	8 / 17	4 / 7	0
$X_3 \rightarrow X_1$	5 / 7	0	2 / 0	3 / 7	5 / 4	0

Table 7. Percentages of statistically significant values of the causality measures for System 5 for $c = 0.5$, where $P = 3$, $m = 2$ and $L_{max} = 5$.

$n = 512/2048$	CGCI	PGCI	PDC	PTE	PSTE	PMIME
$X_1 \rightarrow X_2$	100	84 / 100	11 / 99	100	67 / 100	100
$X_2 \rightarrow X_1$	1 / 5	0	0 / 1	9 / 18	16 / 31	0
$X_2 \rightarrow X_3$	100	60 / 100	7 / 13	100	79 / 100	100
$X_3 \rightarrow X_2$	2 / 17	0 / 2	2 / 8	8	7 / 31	0
$X_1 \rightarrow X_3$	12 / 52	0 / 3	1 / 11	16 / 92	3 / 7	0
$X_3 \rightarrow X_1$	6 / 5	0	2 / 0	8 / 5	7 / 0	0

The PGCI is also not considerably affected by the addition of noise. The causal effects are detected only for coupling strengths, $c \geq 0.3$, for $n = 512$, and for $c \geq 0.2$, for $n = 2048$, while the power of the test increases with c and with n (see Figure 3a).

The PDC fails in the case of the noisy coupled Hénon maps, detecting only the coupling $X_1 \rightarrow X_2$, for coupling strengths, $c \geq 0.2$ and $n = 2048$ (see Figure 3b).

The PTE seems to be significantly affected by the addition of noise, falsely detecting the coupling, $X_2 \rightarrow X_1$, $X_3 \rightarrow X_2$, and the indirect coupling, $X_1 \rightarrow X_3$, for strong coupling strengths. The performance of PTE is not significantly influenced by the choice of m. Indicative results are presented in Table 8 for $m = 2$.

Noise addition does not seem to affect the performance of PSTE. Results for $m = 2$ are equivalent to the results obtained for the noise-free case. The power of the significance test increases with c and n.

The PSTE is sensitive to the selection of m; as m increases, the percentage of significant PSTE values in the directions of no causal effects also increases.

The PMIME outperforms the other measures also for the noisy coupled Hénon maps, detecting the true couplings for $c \geq 0.2$ for $n = 512$ (100%) and for $c \geq 0.1$ for $n = 2048$ (for coupling strength $c = 0.1$ the percentages are 22% and 23% for $X_1 \rightarrow X_2$, $X_2 \rightarrow X_3$, respectively, and for $c \geq 0.2$, the percentages are 100%, for both couplings).

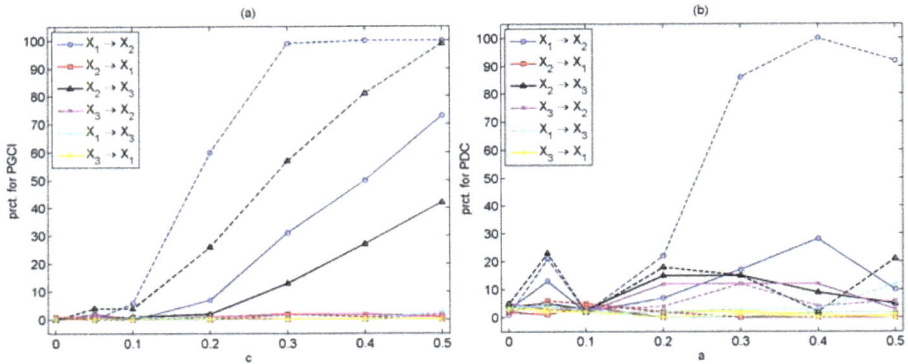

Figure 3. Percentage of significant (**a**) PGCI ($P = 3$) and (**b**) PDC ($P = 3$) values, for System 5 with addition of noise (solid lines for $n = 512$, dotted lines for for $n = 2048$).

Table 8. Percentages of statistically significant PTE ($m = 2$) values for System 5 with the addition of noise.

$n = 512/2048$	$X_1 \rightarrow X_2$	$X_2 \rightarrow X_1$	$X_2 \rightarrow X_3$	$X_3 \rightarrow X_2$	$X_1 \rightarrow X_3$	$X_3 \rightarrow X_1$
$c = 0$	**2 / 5**	5 / 6	**4**	5 / 7	10 / 1	7 / 6
$c = 0.05$	**6 / 17**	3	**5 / 20**	4 / 5	3 / 6	7 / 1
$c = 0.1$	**22 / 98**	6 / 2	**22 / 98**	3 / 7	4 / 5	5 / 8
$c = 0.2$	**100**	6 / 11	**99 / 100**	4 / 7	1 / 5	4 / 2
$c = 0.3$	**100**	10 / 52	**100**	8 / 22	12 / 27	4 / 8
$c = 0.4$	**100**	9 / 79	**100**	6 / 50	24 / 97	7 / 10
$c = 0.5$	**100**	23 / 95	**100**	7 / 48	39 / 100	8 / 13

4.6. Results for System 6

For System 6, we set $P = 3$ based on the complexity of the system and $P = 5$ regarding the AIC and BIC. The PTE, PSTE and PMIME are estimated for four different combinations of the free parameters, h and m (L_{max} for PMIME), i.e., for $h = 1$ and $m = 3$, for $h = 3$ and $m = 3$, for $h = 1$ and $m = 5$ and for $h = 5$ and $m = 5$. The PDC is computed for the range of frequencies, $[0, 0.2]$, based on the auto-spectra of the variables. As this system is a nonlinear flow, the detection of causal effects is more challenging compared to stochastic systems and nonlinear coupled maps. Indicative results for all causality measures are displayed for increasing coupling strengths in Figure 4.

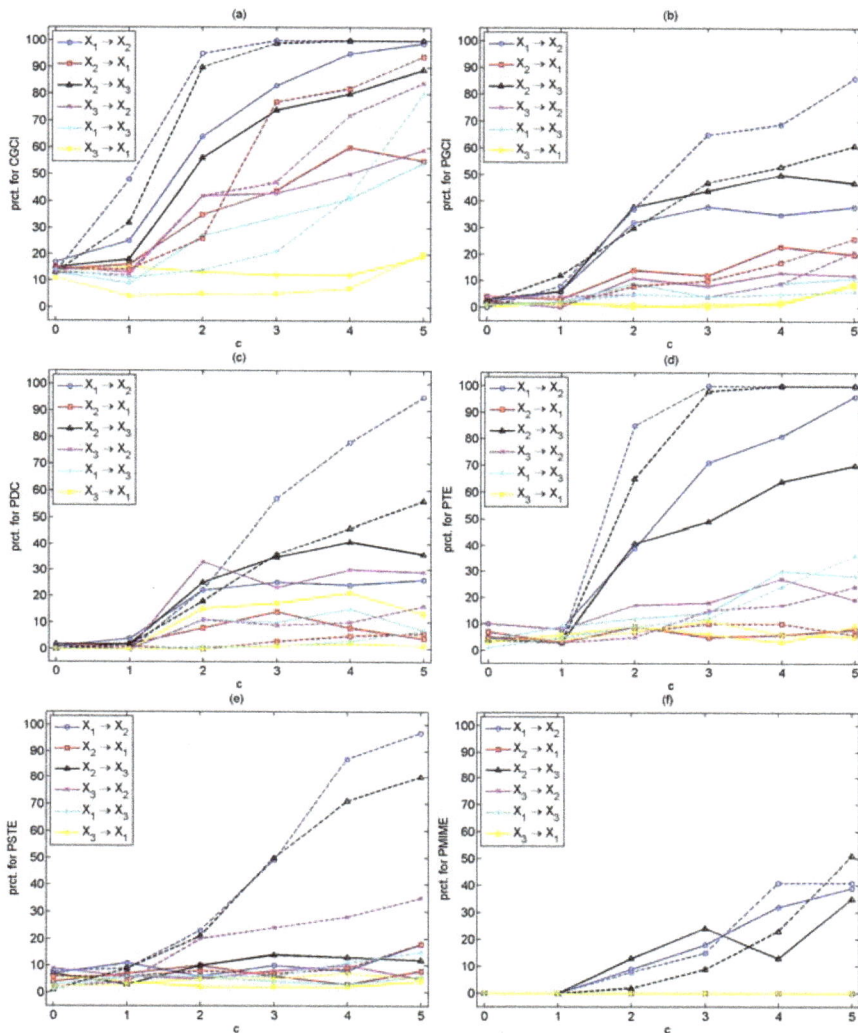

Figure 4. Percentage of significant (**a**) CGCI ($P = 5$); (**b**) PGCI ($P = 5$); (**c**) PDC ($P = 5$); (**d**) PTE ($h = 1$, $m = 5$); (**e**) PSTE ($h = 3$, $m = 3$); and (**f**) PMIME ($h = 1$, $L_{max} = 5$) values, for System 6, for increasing coupling strengths, c, at all directions and for both n (solid lines for $n = 512$, dotted lines for $n = 2048$).

The CGCI has poor performance, indicating many spurious causalities. The PGCI improves the specificity of the CGCI, but still, the percentages of statistically significant PGCI values increase with c for non-existing direct couplings (less for larger n). Similar results are obtained for $P = 3$ and 5. On the other hand, the PDC is sensitive to the selection of P, indicating spurious causal effects for all P. As P increases, the percentage of significant PDC values at the directions of no causal effects is reduced. However, the power of the test is also reduced.

The PTE is sensitive to the embedding dimension m and the number of steps ahead h, performing best for $h = 1$ and $m = 5$. It fails to detect the causal effects for small c; however, for $c > 1$, it effectively indicates the true couplings. The size of the test increases with c (up to 36% for $c = 5$ and $n = 2048$), while the power of the test increases with n.

The PSTE is also affected by its free parameters, performing best for $h = 3$ and $m = 3$. It is unable to detect the true causal effects for weak coupling strengths ($c \leq 1$) and for small time series lengths. The PSTE is effective only for $c > 2$ and $n = 2048$. Spurious couplings are observed for strong coupling strengths c and $n = 2048$.

The PMIME is also influenced by the choice of h and L_{max}, indicating low sensitivity when setting $h = 1$, but no spurious couplings, while for $h = L_{max}$, the percentage of significant PMIME values for $X_1 \rightarrow X_2$ and $X_2 \rightarrow X_3$ is higher, but the indirect coupling $X_1 \rightarrow X_3$ is detected for strong coupling strengths. The PMIME has a poor performance for weak coupling strength ($c < 2$).

As c increases, the percentages of significant values of almost all the causality measures increase, but not only at the true directions, $X_1 \rightarrow X_2$ and $X_2 \rightarrow X_3$. Indicative results are presented in Table 9 for strongly coupled systems ($c = 5$). The CGCI gives high percentages of rejection of H_0 for all couplings (very low specificity). This also holds for the PGCI, but at a lower significance level. The PTE correctly detects the two true direct causal effects for $h = 1$ and $m = 5$, but at some significant degree, also the indirect coupling, $X_1 \rightarrow X_3$, and the non-existing coupling, $X_3 \rightarrow X_2$. The PSTE does not detect the direct couplings for $n = 512$, but it does when $n = 2048$ (97% for $X_1 \rightarrow X_2$ and 80% for $X_2 \rightarrow X_3$), but then it detects also spurious couplings, most notably $X_3 \rightarrow X_2$ (35%). The PMIME points out only the direct causal effects, giving, however, a lower percentage than the other measures for $h = 1$, $L_{max} = 5$. Its performance seems to be affected by the selection of h and L_{max}. The nonlinear measures turn out to be more sensitive to their free parameters.

Table 9. Percentage of statistically significant values of the causality measures for System 6 with $c = 5$, where $P = 5$, $h = 1$ and $m = 5$ for PTE, $h = 3$ and $m = 3$ for PSTE and $h = 1$ and $L_{max} = 5$ for PMIME.

$n = 512/2048$	CGCI	PGCI	PDC	PTE	PSTE	PMIME
$X_1 \rightarrow X_2$	99 / 100	38 / 86	26 / 95	96 / 100	18 / 97	39 / 41
$X_2 \rightarrow X_1$	55 / 94	20 / 26	4 / 6	8 / 6	8 / 18	0
$X_2 \rightarrow X_3$	89 / 100	47 / 61	36 / 56	70 / 100	12 / 80	35 / 51
$X_3 \rightarrow X_2$	59 / 84	12 / 21	29 / 16	19 / 24	5 / 35	0
$X_1 \rightarrow X_3$	54 / 80	11 / 6	7 / 7	28 / 36	6 / 15	0
$X_3 \rightarrow X_1$	19 / 20	9 / 8	13 / 1	9 / 5	4 / 5	0

4.7. Results for System 7

For the last coupled simulation system, we set the model order $P = 2, 3, 4$ based on AIC and BIC, while the PDC is estimated in the range of frequencies, $[0.1, 0.2]$. The embedding dimension, m, for the estimation of PTE and PSTE, as well as L_{max} for the estimation of PMIME, are set equal to P. Results for all causality measures are displayed in Table 10.

The CGCI (for $P = 5$) correctly indicates the causal effects for $n = 512$, giving 100% percentage of significant values at the direction $X_1 \rightarrow X_2$ and $X_1 \rightarrow X_4$, but lower percentage at the directions $X_1 \rightarrow X_3$ (63%), $X_4 \rightarrow X_5$ (37%) and $X_5 \rightarrow X_4$ (42%). The power of the test increases with n, but spurious couplings are also detected for $n = 2048$. For $P = 2, 3$ and 4, the CGCI indicates more spurious couplings than for $P = 5$.

Table 10. Percentage of statistically significant values of the causality measures for System 7, where $P = m = L_{max} = 3$.

$n = 512/2048$	CGCI	PGCI	PDC	PTE	PSTE	PMIME
$X_1 \rightarrow X_2$	100	100	47 / 100	100	47 / 31	100
$X_2 \rightarrow X_1$	8 / 26	0	7 / 23	6 / 9	86 / 100	0
$X_1 \rightarrow X_3$	63 / 100	100	57 / 100	36 / 91	100	100
$X_3 \rightarrow X_1$	6 / 8	1 / 8	1 / 2	98 / 100	32 / 26	100
$X_1 \rightarrow X_4$	100	100	73 / 100	100	31 / 84	100
$X_4 \rightarrow X_1$	6	0	2 / 0	54 / 96	73 / 100	87 / 23
$X_1 \rightarrow X_5$	13 / 75	3 / 8	2 / 11	0 / 1	100	20 / 9
$X_5 \rightarrow X_1$	9 / 5	0	1 / 2	5 / 16	100	1 / 0
$X_2 \rightarrow X_3$	14 / 36	0	7 / 30	49 / 75	40 / 24	3 / 41
$X_3 \rightarrow X_2$	9 / 12	0	4 / 3	40 / 66	100	1 / 0
$X_2 \rightarrow X_4$	11 / 38	3 / 0	7 / 29	36 / 81	91 / 92	0
$X_4 \rightarrow X_2$	7 / 10	18 / 22	2 / 1	0 / 2	100	0
$X_2 \rightarrow X_5$	10 / 37	0	5 / 29	47 / 69	100	86 / 94
$X_5 \rightarrow X_2$	5 / 7	0	1 / 3	20 / 49	72 / 96	1 / 0
$X_3 \rightarrow X_4$	12	0	5 / 3	88 / 100	100	0
$X_4 \rightarrow X_3$	5 / 10	0	2 / 2	14 / 59	98 / 100	15 / 0
$X_3 \rightarrow X_5$	8 / 13	0	4 / 1	1 / 2	100	0
$X_5 \rightarrow X_3$	7 / 11	0	1 / 3	14 / 43	100	5 / 0
$X_4 \rightarrow X_5$	37 / 94	83 / 100	23 / 88	6 / 7	94 / 100	100
$X_5 \rightarrow X_4$	42 / 100	6 / 4	20 / 76	100	78 / 75	100

System 7 favors the PGCI, as it has been specifically defined for systems with latent and exogenous variables. The PGCI denotes the couplings, $X_1 \rightarrow X_2$, $X_1 \rightarrow X_3$, $X_1 \rightarrow X_4$ and $X_4 \rightarrow X_5$, even for $n = 512$ with a high percentage; however, it fails to detect the coupling, $X_5 \rightarrow X_4$, for both n.

The PDC detects the true couplings at low percentage for $n = 512$. The percentage increases with $n = 2048$ at the directions of the true couplings. However, there are also false indications of directed couplings.

The PTE does not seem to be effective in this setting for any of the considered m values, since it indicates many spurious causal effects, while it fails to detect $X_4 \rightarrow X_5$. The PSTE completely fails in this case, suggesting significant couplings at all directions. The true causal effects are indicated by the PMIME, but here, as well, many spurious causal effects are also observed.

5. Discussion

In this paper, we have presented six multivariate causality measures that are able to detect the direct causal effects among simultaneously measured time series. The multivariate direct coupling measures are tested on simulated data from coupled and uncoupled systems of different complexity, linear and nonlinear, maps and flows. The linear causality measures and the PMIME can be used in complex systems with a large number of observed variables, but the PTE and the PSTE fail, because they involve estimation of probability distributions of high dimensional variables.

The simulation results suggest that for real world data, it is crucial to investigate the presence of nonlinearities and confirm the existence of causal effects by estimating more than one causality measure, sensitive to linear, as well as nonlinear causalities. Concerning the specificity of the coupling measures (in absence of direct causality), the PMIME outperforms the other measures, but for weak coupling, it is generally less sensitive than the PTE. In general, the PMIME indicated fewer spurious causal effects. Here, we considered only systems of a few variables, and for larger systems, the PMIME was found to outperform the PTE and, also, the CGCI [28].

Regarding the three first linear coupled systems in the simulation study, the CGCI, PGCI and PDC are superior to the nonlinear causality measures, both in sensitivity and specificity. The PMIME correctly indicates the coupling among the variables, but tends to have smaller sensitivity than the linear tests. The PTE cannot detect the true direct causality in all the examined linear systems and gives also some spurious results. The PSTE was the less effective one. For the last simulation system with the exogenous and latent variables (System 7), all but PGCI measures had low specificity, indicating only the true direct causal effects.

Concerning the nonlinear coupled system, the PTE and the PMIME outperform the other methods. The linear measures (CGCI, PGCI and PDC) fail to consistently detect the true direct causal effects (low sensitivity). The failure of the linear measures may not only be due to the fact that the system is nonlinear, but also due to the small time series lengths and the low model order. For example, the PDC correctly indicated the causal effects on simulated data from the coupled Rössler system for $n = 50,000$ and model order, $P = 200$ (see [49]). The PSTE requires large data sets to have a good power, while it gives spurious couplings at many cases. Though the PSTE performed overall worst in the simulation study, there are other settings in which it can be useful, e.g., in the presence of outliers or non-stationarity in mean, as slow drifts do not have a direct effect on the ranks. The addition of noise does not seem to affect the causality measures, CGCI, PGCI, PSTE and PMIME.

The free parameters were not optimized separately for each measure. For all systems, the parameters of model order, P, embedding dimension, m, and maximum lag, L_{max}, were treated as one free parameter, the values of which were selected according to the complexity of each system and the standard criteria of AIC and BIC. The linear measures tend to be less sensitive to changes on this free parameter than the nonlinear ones. The PTE gave more consistent results than the PSTE for varying m, whereas the PMIME was not dependent on L_{max}. For the nonlinear measures and the continuous-time system (three coupled Lorenz systems), we considered also the causalities at more than one step ahead, and the PTE, PSTE and PMIME were found to be sensitive to the selection of the steps ahead.

A point of concern regarding all direct causality measures, but the PMIME, is that the size of the significance test was high in many settings. This was observed for both types of spurious direct causal effect, *i.e.*, when there is indirect coupling or when there is no causal effect. In many cases of non-existing direct causalities, although the observed test size was large, the estimated values of the measure were low compared to those in the presence of direct causalities. This raises also the question of the validity of the significance tests. The randomization test used time-shifted surrogates. Although it is simple and straightforward to implement, it may not always be sufficient, and further investigation for other randomization techniques is due for future work.

In conclusion, we considered six of the best-known measures of direct causality and studied their performance for different systems, time series lengths and free parameters. The worst performance was observed for the PSTE, since it completely failed in the case of the linear coupled systems, while for nonlinear systems, it required large data sets. The other measures scored differently in terms of sensitivity and specificity in the different settings. The CGCI, PGCI and PDC outperformed the nonlinear ones in the case of the linear coupled simulation systems, while in the presence of exogenous and latent variables, the PGCI seems to be the most effective one. The PMIME seems to have the best performance for nonlinear and noisy systems, while always obtaining the highest specificity, indicating no spurious effects. It is the intention of the authors to pursue the comparative study on selected real applications.

Acknowledgments: The research project is implemented within the framework of the Action "Supporting Postdoctoral Researchers" of the Operational Program, "Education and Lifelong Learning" (Action's Beneficiary: General Secretariat for Research and Technology), and is co-financed by the European Social Fund (ESF) and the Greek State.

Conflicts of Interest: The authors declare no conflict of interest.

References

1. Granger, C.W.J. Investigating causal relations by econometric models and cross-spectral methods. *Econometrica* **1969**, *37*, 424–438.
2. Geweke, J. Measurement of linear dependence and feedback between multiple time series. *J. Am. Stat. Assoc.* **1982**, *77*, 304–313.
3. Baek, E.; Brock, W. A general test for nonlinear Granger causality: Bivariate model, 1992. Working paper, University of Wisconsin, Madison, WI, USA.

4.   Hiemstra, C.; Jones, J.D.   Testing for linear and nonlinear Granger causality in the stock Price-Volume Relation. *J. Financ.* **1994**, *49*, 1639–1664.

5.   Aparicio, F.; Escribano, A. Information-theoretic analysis of serial correlation and cointegration. *Stud. Nonlinear Dyn. Econom.* **1998**, *3*, 119–140.

6.   Freiwald, V.A.; Valdes, P.; Bosch, J.; Biscay, R.; Jimenez, J.; Rodriguez, L. Testing non-linearity and directedness of interactions between neural groups in the macaque inferotemporal cortex. *J. Neurosci. Methods* **1999**, *94*, 105–119.

7.   Marinazzo, D.; Pellicoro, M.; Stramaglia, S. Kernel method for nonlinear Granger causality. *Phys. Rev. Lett.* **2008**, *100*, 144103.

8.   Arnhold, J.; Grassberger, P.; Lehnertz, K.; Elger, C.E. A robust method for detecting interdependences: Application to intracranially recorded EEG. *Physica D: Nolinear Phenom.* **1999**, *134*, 419–430.

9.   Romano, M.; Thiel, M.; Kurths, J.; Grebogi, C. Estimation of the direction of the coupling by conditional probabilities of recurrence. *Phys. Rev. E* **2007**, *76*, 036211.

10.   Marko, H. The bidirectional communication theory-a generalization of information theory. *IEEE Trans. Commun.* **1973**, *21*, 1345–1351.

11.   Schreiber, T. Measuring information transfer. *Phys. Rev. Lett.* **2000**, *85*, 461–464.

12.   Paluš, M. Coarse-grained entropy rates for characterization of complex time series. *Physica D: Nolinear Phenom.* **1996**, *93*, 64–77.

13.   Rosenblum, M.G.; Pikovsky, A.S. Detecting direction of coupling in interacting oscillators. *Phys. Rev. E* **2001**, *64*, 045202.

14.   Nolte, G.; Ziehe, A.; Nikulin, V.; Schlogl, A.; Kramer, N.; Brismar, T.; Muller, K.R. Robustly estimating the flow direction of information in complex physical systems. *Phys. Rev. Lett.* **2008**, *100*, 234101.

15.   Blinowska, K.J.; Kus, R.; Kaminski, M. Granger causality and information flow in multivariate processes. *Phys. Rev. E* **2004**, *70*, 050902.

16.   Kus, R.; Kaminski, M.; Blinowska, K.J. Determination of EEG activity propagation: Pair-wise *versus* multichannel estimate. *IEEE Trans. Biomed. Eng.* **2004**, *51*, 1501–1510.

17.   Eichler, M. Graphical modelling of multivariate time series. *Probab. Theory Relat. Fields* **2012**, *153*, 233–268.

18.   Guo, S.; Seth, A.K.; Kendrick, K.M.; Zhou, C.; Feng, J. Partial granger causality–eliminating exogenous inputs and latent variables. *J. Neurosci. Methods* **2008**, *172*, 79–93.

19.   Kaminski, M.J.; Blinowska, K.J. A new method of the description of the information flow in the brain structures. *Biol. Cybern.* **1991**, *65*, 203–210.

20.   Baccala, L.A.; Sameshima, K. Partial directed coherence: A new concept in neural structure determination. *Biol. Cybern.* **2001**, *84*, 463–474.

21.   Korzeniewska, A.; Manczak, M.; Kaminski, M.; Blinowska, K.; Kasicki, S. Determination of information flow direction between brain structures by a modified directed transfer function (dDTF) method. *J. Neurosci. Methods* **2003**, *125*, 195–207.

22.   Chavez, M.; Martinerie, J.; Quyen, M. Statistical assessment of non-linear causality: Application to epileptic EEG signals. *J. Neurosci. Methods* **2003**, *124*, 113–128.

23.   Verdes, P. Assessing causality from multivariate time series. *Phys. Rev. E* **2005**, *72*, 026222.

24.   Vakorin, V.A.; Krakovska, O.A.; McIntosh, A.R. Confounding effects of indirect connections on causality estimation. *J. Neurosci. Methods* **2009**, *184*, 152–160.

25.   Papana, A.; Kugiumtzis, D.; Larsson, P.G. Detection of direct causal effects and application in the analysis of electroencephalograms from patients with epilepsy.   *Int. J. Bifurc. Chaos* **2012**, *22*, doi: 10.1142/S0218127412502227.

26.   Vlachos, I.; Kugiumtzis, D. Non-uniform state space reconstruction and coupling detection. *Phys. Rev. E* **2010**, *82*, 016207.

27.   Faes, L.; Nollo, G.; Porta, A. Information-based detection of nonlinear Granger causality in multivariate processes via a nonuniform embedding technique. *Phys. Rev. E* **2011**, *83*, 051112.

28.   Kugiumtzis, D. Direct coupling information measure from non-uniform embedding. *Phys. Rev. E* **2013**, *87*, 062918.

29.   Jizba, P.; Kleinert, H.; Shefaat, M. Rényi's information transfer between financial time series. *Physica A: Stat. Mech. Appl.* **2012**, *391*, 2971–2989.

30.   Staniek, M.; Lehnertz, K. Symbolic transfer entropy. *Phys. Rev. Lett.* **2008**, *100*, 158101.

31. Kugiumtzis, D. Transfer entropy on rank vectors. *J. Nonlinear Syst. Appl.* **2012**, *3*, 73–81.
32. Kugiumtzis, D. Partial transfer entropy on rank vectors. *Eur. Phys. J. Spec. Top.* **2013**, *222*, 401–420.
33. Smirnov, D.A.; Andrzejak, R.G. Detection of weak directional coupling: Phase-dynamics approach *versus* state-space approach. *Phys. Rev. E* **2005**, *71*, 036207.
34. Faes, L.; Porta, A.; Nollo, G. Mutual nonlinear prediction as a tool to evaluate coupling strength and directionality in bivariate time series: Comparison among different strategies based on k nearest neighbors. *Phys. Rev. E* **2004**, *78*, 026201.
35. Papana, A.; Kugiumtzis, D.; Larsson, P.G. Reducing the bias of causality measures. *Phys. Rev. E* **2011**, *83*, 036207.
36. Silfverhuth, M.J.; Hintsala, H.; Kortelainen, J.; Seppanen, T. Experimental comparison of connectivity measures with simulated EEG signals. *Med. Biol. Eng. Comput.* **2012**, *50*, 683–688.
37. Dufour, J.M.; Taamouti, A. Short and long run causality measures: Theory and inference. *J. Econ.* **2010**, *154*, 42–58.
38. Florin, E.; Gross, J.; Pfeifer, J.; Fink, G.; Timmermann, L. Reliability of multivariate causality measures for neural data. *J. Neurosci. Methods* **2011**, *198*, 344–358.
39. Wu, M.-H.; Frye, R.E.; Zouridakis, G. A comparison of multivariate causality based measures of effective connectivity. *Comput. Biol. Med.* **2011**, *21*, 1132–1141.
40. Blinowska, K.J. Review of the methods of determination of directed connectivity from multichannel data. *Med. Biol. Eng. Comput.* **2011**, *49*, 521–529.
41. Brandt, P.T.; Williams, J.T. *Multiple Time Series Models*; Sage Publications, Inc.: Thousand Oaks, CA, USA, 2007; chapter 2, pp. 32–34.
42. Benjamini, Y.; Hochberg, Y. Controlling the false discovery rate: A practical and powerful approach to multiple testing. *J. R. Stat. Soc. Ser. B* **1995**, *57*, 289–300.
43. Akaike, H. A new look at the statistical model identification. *IEEE Trans. Autom. Control* **1974**, *19*, 716–723.
44. Schwartz, G. Estimating the dimension of a model. *Ann. Stat.* **1978**, *5*, 461–464.
45. Ding, M.; Chen, Y.; Bressler, S.L. Granger Causality: Basic Theory and Applications to Neuroscience. In *Handbook of Time Series Analysis: Recent Theoretical Developments and Applications*; Schelter, B., Winterhalder, M., Timmer, J., Eds.; Wiley-VCH Verlag: Berlin, Germany, 2006; Chapter 17, pp. 437–460.
46. Schelter, B.; Winterhalder, M.; Eichler, M.; Peifer, M.; Hellwig, B.; Guschlbauer, B.; Lucking, C.H.; Dahlhaus, R.; Timmer, J. Testing for directed influences in neuroscience using partial directed coherence. *J. Neurosci. Methods* **2006**, *152*, 210–219.
47. Takahashi, D.Y.; Baccala, L.A.; Sameshima, K. Connectivity inference between neural structures via partial directed coherence. *J. Appl. Stat.* **2007**, *34*, 1255–1269.
48. Kraskov, A.; Stögbauer, H.; Grassberger, P. Estimating mutual information. *Phys. Rev. E* **2004**, *69*, 066138.
49. Winterhalder, M.; Schelter, B.; Hesse, W.; Schwab, K.; Leistritz, L.; Klan, D.; Bauer, R.; Timmer, J.; Witte, H. Comparison of linear signal processing techniques to infer directed interactions in multivariate neural systems. *Signal Process.* **2005**, *85*, 2137–2160.
50. Gourévitch, B.; Le Bouquin-Jeannés, R.; Faucon, G. Linear and nonlinear causality between signals: methods, examples and neurophysiological applications. *Biol. Cybern.* **2006**, *95*, 349–369.
51. Stefanski, A. *Determining Thresholds of Complete Synchronization, and Application*; World Scientific: Singapore, Singapore, 2009.
52. Wilke, C.; van Drongelen, W.; Kohrman, M.; Hea, B. Identification of epileptogenic foci from causal analysis of ECoG interictal spike activity. *Clin. Neurophys.* **2009**, *120*, 1449–1456.
53. Efron, B.; Tibshirani, R. *An Introduction to the Bootstrap*; Chapman and Hall: New York, NY, USA, 1994.
54. Faes, L.; Porta, A.; Nollo, G. Testing frequency-domain causality in multivariate time series. *IEEE Trans. Biomed. Eng.* **2010**, *57*, 1897–1906.
55. Quian Quiroga, R.; Kraskov, A.; Kreuz, T.; Grassberger, P. Performance of different synchronization measures in real data: A case study on electroencephalographic signals. *Phys. Rev. E* **2002**, *65*, 041903.
56. Yu, G.H.; Huang, C.C. A distribution free plotting position. *Stoch. Environ. Res. Risk Assess.* **2001**, *15*, 462–476.
57. Seth, A.K. A MATLAB toolbox for Granger causal connectivity analysis. *J. Neurosci. Methods* **2010**, *186*, 262–273.

entropy

MDPI

Article

Compensated Transfer Entropy as a Tool for Reliably Estimating Information Transfer in Physiological Time Series

Luca Faes [1],*, Giandomenico Nollo [1] and Alberto Porta [2]

[1] Department of Physics and BIOtech Center, University of Trento, Via delle Regole 101, 38123 Mattarello, Trento, Italy; nollo@science.unitn.it

[2] Department of Biomedical Sciences for Health, Galeazzi Orthopaedic Institute, University of Milan, Via R. Galeazzi 4, 20161 Milano, Italy; alberto.porta@unimi.it

* Author to whom correspondence should be addressed; luca.faes@unitn.it; Tel.: +39-0461-282773; Fax: +39-0461-283091.

Received: 29 October 2012; in revised form: 21 December 2012; Accepted: 5 January 2013; Published: 11 January 2013

Abstract: We present a framework for the estimation of transfer entropy (TE) under the conditions typical of physiological system analysis, featuring short multivariate time series and the presence of instantaneous causality (IC). The framework is based on recognizing that TE can be interpreted as the difference between two conditional entropy (CE) terms, and builds on an efficient CE estimator that compensates for the bias occurring for high dimensional conditioning vectors and follows a sequential embedding procedure whereby the conditioning vectors are formed progressively according to a criterion for CE minimization. The issue of IC is faced accounting for zero-lag interactions according to two alternative empirical strategies: if IC is deemed as physiologically meaningful, zero-lag effects are assimilated to lagged effects to make them causally relevant; if not, zero-lag effects are incorporated in both CE terms to obtain a compensation. The resulting compensated TE (cTE) estimator is tested on simulated time series, showing that its utilization improves sensitivity (from 61% to 96%) and specificity (from 5/6 to 0/6 false positives) in the detection of information transfer respectively when instantaneous effect are causally meaningful and non-meaningful. Then, it is evaluated on examples of cardiovascular and neurological time series, supporting the feasibility of the proposed framework for the investigation of physiological mechanisms.

Keywords: cardiovascular variability; conditional entropy; instantaneous causality; magnetoencephalography; time delay embedding

PACS: 05.45.Tp; 02.50.Sk; 87.19.lo; 87.19.le; 87.19.Hh

1. Introduction

Since its first introduction by Schreiber [1], transfer entropy (TE) has been recognized as a powerful tool for detecting the transfer of information between joint processes. The most appealing features of TE are that it has a solid foundation in information theory, and it naturally incorporates directional and dynamical information as it is inherently asymmetric (*i.e.*, different when computed over the two causal directions) and based on transition probabilities (*i.e.*, on the conditional probabilities associated with the transition of the observed system from its past states to its present state). Moreover, the formulation of TE does not assume any particular model as underlying the interaction between the considered processes, thus making it sensitive to all types of dynamical interaction. The popularity of this tool has grown even more with the recent elucidation of its close connection with the ubiquitous concept of Granger causality [2], which has led to formally bridge information-theoretic and predictive approaches to the evaluation of directional interactions between processes. Given all these advantages, the TE has been increasingly used to assess the transfer of information in physiological systems with typical applications in neurophysiology [3–6] and in cardiovascular physiology [7–9]. Nevertheless, in front of this widespread utilization of TE and other Granger causality measures, it should be remarked that these measures quantify "causality" from a statistical perspective which is quite distinct from the interventionist perspective that has to be followed to infer effectively the existence of real causal effects [10–12]. Accordingly in this study, when speaking of the transfer of information measured by TE we refer to the "predictive information transfer" intended as the amount of information added by the past states of a source process to the next state of a destination process, rather than to the causal information flow measured via interventional conditional probabilities [12].

The estimation of TE from the time series data taken as realizations of the investigated physiological processes is complicated by a number of practical issues. One major challenge is the estimation of the probability density functions involved in TE computation from datasets the length of which is limited by experimental constraints and/or by the need for stationarity [13,14]. Another critical point is that, to exploit the dynamical information contained in the transition probabilities, one should cover reasonably well the past history of the observed processes; since this corresponds to work with long conditioning vectors represented into high-dimensional spaces, TE estimation from short time series is further hampered, especially in the presence of multiple processes and long memory effects [15]. Moreover, an open issue in practical time series analysis is how to deal with instantaneous effects, which are effects occurring between two time series within the same time lag [16]. These effects may reflect fast (within sample) physiologically meaningful interactions, or be void of physiological meaning (e.g., may be due to unobserved confounders). In either case, instantaneous effects have an impact on the computation of any causality measure [17,18]. In particular, the presence of unmeasured exogenous inputs or latent variables which cannot be included in the observed data set (e.g., because they are not accessible) is a critical issue when investigating Granger causality in experimental data, as it may easily lead to the detection of spurious causalities [19–21]. Since an instantaneous correlation arises between two observed variables which are affected by latent variables with the same time delay, in the context of model-based analysis attempts have been made to counteract this problem by accounting for residual correlations which reflect zero-lag effects. Indeed, recent studies have proposed to incorporate terms from the covariance matrix of the model residuals into the so-called partial Granger causality measures [21,22], or to express the residual correlation in terms of model coefficients and exploit the resulting new model structure for defining extended Granger causality measures [17,18]. However, as similar approaches cannot be followed in the model-free context of TE analysis, instantaneous effects are usually not considered in the computation of TE on experimental data.

In the present study we describe an approach for the estimation of TE from short realizations of multivariate processes which is able to deal with the issues presented above. We develop an estimation framework that combines conditional entropy (CE) estimation, non-uniform embedding, and consideration of instantaneous causality. The framework is based on recognizing that TE can be interpreted as CE difference, and builds on an efficient CE estimator that compensates for the bias occurring for high dimensional conditioning vectors and follows a sequential embedding procedure whereby the conditioning vectors are formed progressively according to a criterion for CE minimization. This procedure realizes an approach for partial conditioning that follows the ideas first proposed in [15]. The novel contribution of the paper consists in the integration of the framework with a procedure for the inclusion of instantaneous effects. This is a crucial point in TE analysis because, even though it is now well recognized that instantaneous causality plays a key role in Granger causality analysis, instantaneous effects are commonly disregarded in the computation of TE on experimental data. While recent studies have started to unravel the issue of instantaneous causality in the linear parametric framework of multivariate autoregressive models [17,18,23,24], there is a paucity of works addressing the consequences of excluding instantaneous effects from the computation of model-free causality measures. In this paper, the issue of instantaneous causality is faced allowing for the possibility of zero-lag effects in TE computation, according to two alternative empirical procedures: if instantaneous effects are deemed as causally meaningful, the zero-lag term is assimilated to the lagged terms to make it causally relevant; if not, the zero-lag term is incorporated in both CE computations to obtain a compensation of its confounding effects. The resulting TE estimator, denoted as compensated TE (cTE), is first validated on simulations of linear stochastic and nonlinear deterministic systems. Then, the estimator is evaluated on representative examples of physiological time series which entail utilization of different strategies for compensating instantaneous causality and different procedures for significance assessment, *i.e.* cardiovascular variability series and multi-trial magnetoencephalography signals. The direct comparison between the proposed cTE and the traditional TE allows to make explicit the problem of disregarding instantaneous causality in the computation of the predictive information transfer in multivariate time series.

2. Methods

2.1. Transfer Entropy

Let us consider a composite physical system described by a set of M interacting dynamical (sub) systems and suppose that, within the composite system, we are interested in evaluating the information flow from the source system X to the destination system Y, collecting the remaining systems in the vector $\mathbf{Z} = \{Z^{(k)}\}_{k=1,...,M-2}$. We develop our framework under the assumption of stationarity, which allows to perform estimations replacing ensemble averages with time averages (for non-stationary formulations see, e.g., [10] and references therein). Accordingly, we denote x, y and $\mathbf{z}$ as the stationary stochastic processes describing the state visited by the systems X, Y and Z over time, and x_n, y_n and $\mathbf{z}_n$ as the stochastic variables obtained sampling the processes at the time n. Moreover, let $x_{t:n}$ represent the vector variable describing all the states visited by X from time t up to time n (assuming n as the present time and setting the origin of time at $t = 1$, $x_{1:n-1}$ represents the whole past history of the process x). Then, the transfer entropy (TE) from X to Y conditioned to $\mathbf{Z}$ is defined as:

$$TE_{X \to Y|\mathbf{Z}} = \sum p(y_{1:n}, x_{1:n-1}, \mathbf{z}_{1:n-1}) log \frac{p(y_n|x_{1:n-1}, y_{1:n-1}, \mathbf{z}_{1:n-1})}{p(y_n|y_{1:n-1}, \mathbf{z}_{1:n-1})} \tag{1}$$

where the sum extends over all states visited by the composite system, $p(\mathbf{a})$ is the probability associated with the vector variable $\mathbf{a}$, and $p(b \mid \mathbf{a}) = p(\mathbf{a},b)/p(\mathbf{a})$ is the probability of the scalar variable b conditioned to $\mathbf{a}$. The conditional probabilities used in (1) can be interpreted as transition probabilities, in the sense that they describe the dynamics of the transition of the destination system from its past states to its present state, accounting for the past of the other processes. Utilization of the transition probabilities as defined in (1) makes the resulting measure able to quantify the extent to which the transition of the destination system Y into its present state is affected by the past states visited by the source system X. Specifically, the TE quantifies the information provided by the past states of X about the present state of Y that is not already provided by the past of Y or any other system included in $\mathbf{Z}$. The formulation presented in (1) is an extension of the original TE measure proposed for bivariate systems [1] to the case of multiple interacting processes. The multivariate (conditional) TE formulation, also denoted as partial TE [25], rules out the information shared between X and Y that could be possibly triggered by their common interaction with $\mathbf{Z}$. As such, this formulation fulfills for multivariate systems the correspondence between TE and the concept of Granger causality [19], that refers to the exclusive consideration of direct effects between two processes after resolving the conditional effects of the other observed processes. Note that the conditional formulation has been shown essential for taking under control the effects of common confounders in experimental contexts such as cardiovascular variability analysis [24] or neural signal analysis [26]. In the following, we will indicate Granger causal effects from the system X to the system Y with the notation $X \to Y$ (or $x_{1:n-1} \to y_n$ if we refer to the corresponding processes).

Equivalently, the TE defined in (1) can be expressed in terms of mutual information (MI), as the conditional MI between the present state of the destination and the past states of the source given the past states of all systems except the source:

$$TE_{X \to Y|\mathbf{Z}} = I(y_n, x_{1:n-1} | y_{1:n-1}, \mathbf{z}_{1:n-1}) \tag{2}$$

or in terms of conditional entropy (CE), as the difference between the CE of the present state of the destination given the past states of all systems except the source and the CE of the present state of the destination given the past states of all systems including the source:

$$TE_{X \to Y|\mathbf{Z}} = H(y_n | y_{1:n-1}, \mathbf{z}_{1:n-1}) - H(y_n | x_{1:n-1}, y_{1:n-1}, \mathbf{z}_{1:n-1}) \tag{3}$$

These alternative compact formulations also favor the estimation of TE, as efficient estimators exist for both MI [27] and CE [28]. In Section 2.3 we propose an approach for the estimation of CE based on sequential non-uniform conditioning combined with bias compensation, which is exploited for estimating TE in short and noisy physiological time series. The CE, which constitutes the backbone of the presented approach for TE estimation, can be functionally defined as the difference between two Shannon entropies, e.g., according to (3), $H(y_n | y_{1:n-1}, \mathbf{z}_{1:n-1}) = H(y_{1:n}, \mathbf{z}_{1:n-1}) - H(y_{1:n-1}, \mathbf{z}_{1:n-1})$ and $H(y_n | x_{1:n-1}, y_{1:n-1}, \mathbf{z}_{1:n-1}) = H(x_{1:n-1}, y_{1:n}, \mathbf{z}_{1:n-1}) - H(x_{1:n-1}, y_{1:n-1}, \mathbf{z}_{1:n-1})$, where the entropy of any vector variable **a** is defined as $H(\mathbf{a}) = -\sum p(\mathbf{a}) \cdot \log p(\mathbf{a})$ and is usually measured in bits when the base of the logarithm is 2 or in *nats* when the base is e (as in the present study).

2.2. Compensated Transfer Entropy

An open issue in TE analysis is how to deal with instantaneous effects, which are effects occurring between two processes within the same time lag (e.g., with the notation above, $x_n \to y_n$). Instantaneous effects are the practical evidence of the concept of instantaneous causality, which is a known issue in causal analysis [16,19]. In practice, instantaneous causality between two time series may either have a proper causal meaning, when the time resolution of the measurements is lower than the time scale of the lagged causal influences between the underlying processes, or be void of such causal meaning, in the case of common driving effects occurring when an unmeasured process simultaneously affects the two processes under analysis [17]. In either case, instantaneous causality has an impact on the estimation of the TE: if it is causally meaningful, the analysis misses the zero-lag effect $x_n \to y_n$, if not, the analysis includes potential spurious effects taking the form $x_{1:n-1} \to x_n \to y_n$; these misleading detections may impair respectively the sensitivity and the specificity of TE estimation.

To counteract this problem from a practical perspective, we introduce a so-called compensated TE (cTE), which realizes a compensation for instantaneous causality in the computation of TE. This compensation exploits the representation of TE as CE difference and allows for the possibility of zero-lag interactions according to two alternative strategies. If instantaneous effects are deemed as causally meaningful, the zero-lag term of the source process, x_n, is incorporated in the second CE term used for TE computation:

$$cTE'_{X \to Y|\mathbf{Z}} = H(y_n | y_{1:n-1}, \mathbf{z}_{1:n}) - H(y_n | x_{1:n}, y_{1:n-1}, \mathbf{z}_{1:n}) \tag{4}$$

in this case, the zero-lag term is assimilated with the past states (x_n plays a similar role as $x_{1:n-1}$), so that the present state of the source system is taken as causally relevant to account for instantaneous causality in TE computation. If, on the contrary, instantaneous effects are deemed as non causally meaningful, the zero-lag term is incorporated both in the first and in the second CE terms used for TE computation:

$$cTE''_{X \to Y|\mathbf{Z}} = H(y_n | x_n, y_{1:n-1}, \mathbf{z}_{1:n}) - H(y_n | x_{1:n}, y_{1:n-1}, \mathbf{z}_{1:n}) \tag{5}$$

in this second case, the zero-lag term is considered as a conditioning factor (x_n plays a similar role as $y_{1:n-1}$ and $\mathbf{z}_{1:n}$), so that the present state of the source system is compensated to remove instantaneous causality from TE computation. The compensation performed in (5) is alternative to the test of time-shifted data recently proposed to detect instantaneous mixing between coupled processes [4]. Note that in both compensations in (4) and (5) instantaneous effects possibly occurring from any scalar element of $\mathbf{Z}$ towards Y are conditioned out considering the present term $\mathbf{z}_n$, in addition to the past terms $\mathbf{z}_{1:n-1}$, in the two CE computations; this is done to avoid that indirect effects $x_{1:n-1} \rightarrow \mathbf{z}_n \rightarrow y_n$ were misinterpreted as the presence of predictive information transfer from the system X to the system Y. Note that, in the absence of instantaneous causality among the observed processes, the two cTE measures defined in (4) and (5) reduce to the traditional TE.

2.3. Estimation Approach

The practical estimation of TE and cTE from finite length realizations of multiple processes faces the issue of reconstructing the state space of the observed multivariate dynamical system and then estimating probabilities within this multidimensional state space. In the context of TE/cTE estimation, state space reconstruction corresponds to identifying the multidimensional vector which more suitably represents the trajectory of the states visited by the composite system $\{X,Y,\mathbf{Z}\}$. The most commonly followed approach is to perform uniform time delay embedding, whereby each scalar process is mapped into trajectories described by delayed coordinates uniformly spaced in time [29]. In this way the past history of the source process, $x_{1:n-1}$, is approximated with the d-dimensional delay vector $[x_{n-u-(d-1)\tau}, ..., x_{n-u-\tau}, x_{n-u}]$, with τ and u representing the so-called embedding time and prediction time. This procedure suffers from many disadvantages: first, univariate embedding whereby coordinate selection is performed separately for each process does not guarantee optimality of the reconstruction for the multivariate state space [30]; second, selection of the embedding parameters d, τ and u is not straightforward, as many competing criteria exist which are all heuristic and somewhat mutually exclusive [31]; third, the inclusion of irrelevant coordinates consequent to the use of an uniform embedding exposes the reconstruction procedure to the so called "curse of dimensionality", a concept related to the sparsity of the available data within state spaces of increasing volume [32]. All these problems become more cumbersome when the available realizations are of short length, as commonly happens in physiological time series analysis due to lack of data or stationarity requirements. To counteract these problems, we describe in the following a TE/cTE estimation strategy based on the utilization of a non-uniform embedding procedure combined with a corrected CE estimator [15].

The basic idea underlying our estimation approach is to optimize the time-delay embedding to the estimation of CE, according to a sequential procedure which updates the embedding vector progressively, taking all relevant processes into consideration at each step and selecting the components that better describe the destination process. Specifically, a set of candidate terms is first defined including the past states (and, when relevant, also the present state) of all systems relevant to the estimation of the considered CE term; for instance, considering the terms in (4), the candidate set for the estimation of $H(y_n | y_{1:n-1}, \mathbf{z}_{1:n})$ will be the set $\Omega_1 = \{y_{n-1},...,y_{n-L}, \mathbf{z}_n, \mathbf{z}_{n-1},...,\mathbf{z}_{n-L}\}$, and the candidate set for the estimation of $H(y_n | x_{1:n}, y_{1:n-1}, \mathbf{z}_{1:n})$ in (4) will be the set $\Omega_2 = \{\Omega_1, x_n, x_{n-1},...,x_{n-L}\}$ (L is the number of time lagged terms to be tested for each scalar process). Given the generic candidate set Ω, the procedure for estimating the CE $H(y_n | \Omega)$ starts with an empty embedding vector $V_0 = [\cdot]$, and proceeds as follows: (i) at each step $k \geq 1$, form the candidate vector $[s, V_{k-1}]$, where s is an element of Ω not already included in V_{k-1}, and compute the CE of the destination process Y given the considered candidate vector, $H(y_n | [s, V_{k-1}])$; (ii) repeat step (i) for all possible candidates, and then retain the candidate for which the estimated CE is minimum, *i.e.*, set $V_k = [s', V_{k-1}]$ where $s' = \arg \min_s$ $H(y_n | [s, V_{k-1}])$; (iii) terminate the procedure when irrelevant terms begin to be selected, *i.e.* when the decrease of CE is no longer significant; according to the estimation procedure detailed below, this corresponds to stop the iterations at the step k' such that $H(y_n | V_{k'}) \geq H(y_n | V_{k'-1})$, and set $V_K = V_{k'-1}$ as embedding vector. With this procedure, only the components that effectively contribute to resolving the uncertainty of the target process (in terms of CE reduction) are included into the embedding vector, while the irrelevant components are left out. This feature, together with the termination criterion which prevents the selection of new terms when they do not bring further resolution of uncertainty for the destination process, help escaping the curse of dimensionality for multivariate CE estimation. Moreover the procedure avoids the nontrivial task of setting the embedding parameters (the only parameter is the number L of candidates to be tested for each process, which can be as high as allowed by the affordable computational times). It is worth noting that the proposed sequential procedure for candidate selection takes into account one term at a time, somehow disregarding joint effects that more candidates may have on CE reduction. As a consequence, the sequential instead of exhaustive strategy does not guarantee convergence to the absolute minimum of CE, and thus does not assure a semipositive value for the TE/cTE measures estimated according to (3), (4) and (5). However a sequential approach is often necessary in practical analysis, since exhaustive exploration of all possible combinations of candidate terms would become computationally intractable still at low embedding dimensions.

The application of the procedure described above relies on an efficient estimation of the CE. The problem amounts to estimating, at the k-th step of the procedure, the entropy of the scalar variable y_n conditioned to the vector variable V_k, seen as the difference of two Shannon entropies: $H(y_n \mid V_k)$ $= H(y_n, V_k) - H(V_k)$. A major problem in estimating CE is the bias towards zero which affects the estimates as the dimension of the reconstructed state space grows higher [33,34]. Since the bias increases progressively with the embedding dimension, its occurrence also prevents from being able to reveal the inclusion of irrelevant terms into the embedding vector by looking at the estimated CE; in other words, since the estimated CE decreases progressively as a result of the bias rather than of the inclusion of relevant terms, the iterations of the sequential procedure for nonuniform embedding cannot be properly stopped. To deal with this important problem, we propose to compensate the CE bias adding a corrective term as proposed by Porta *et al.* [28,34], in order to achieve a minimum in the estimated CE which serves as stopping criterion for the embedding procedure. The idea is based on the consideration that, for time series of limited length, the CE estimation bias is due to the isolation of the points in the k-dimensional state space identified by the vectors V_k; such an isolation becomes more and more severe as the dimension k increases. Since isolated points tend to give the same contribution to the two entropy terms forming CE (*i.e.*, $p(V_k) \approx p(y_n, V_k)$ if V_k is an isolated point), their contribution to the CE estimate will be null; therefore, the CE estimate decreases progressively towards zero at increasing the embedding dimension [*i.e.*, when k is high compared to the series length, $H(V_k) \approx H(y_n, V_k)$ and thus $H(y_n \mid V_k) \approx 0$], even for completely unpredictable processes for which conditioning should not decrease the information carried. This misleading indication of predictability in the analysis of short time series is counteracted introducing a corrective term for the CE. The correction is meant at quantifying the fraction of isolated points V_k in the k-dimensional state space, denoted as $n(V_k)$, and on substituting their null contribution with the maximal information amount carried by a white noise with the same marginal distribution of the observed process y_n [*i.e.*, with $H(y_n)$]. The resulting final estimate is obtained adding the corrective term $n(V_k)H(y_n)$ to the estimated CE $H(y_n \mid V_k)$. In the present study, practical implementation of the correction is performed in the context of entropy estimation through uniform quantization [15,28,34]. Briefly, each time series is coarse grained spreading its dynamics over Q quantization levels, so that the state space containing the vectors V_k is partitioned in Q^k disjoint hypercubes. As all points falling within the same hypercube are considered indistinguishable to each other, the Shannon entropy is estimated approximating the probabilities with the frequency of visitation of the hypercubes. Partitioning in disjoint hypercubes helps also in quantifying the fraction of isolated points $n(V_k)$, which is taken simply as the fraction of points found only once inside the hypercubes.

3. Validation

In this section we test the compensation for instantaneous causality in TE computation proposed in Section 2.2, as well as the approach for CE estimation described in Section 2.3, on numerical simulations reproducing different conditions of interaction between multivariate processes. The proposed simulations were devised, in terms of imposed dynamics, interaction conditions and series length, to mimic the conditions typical of the two applicative contexts which are then considered in Section 4, *i.e.*, short-term cardiovascular variability and magnetoencephalography. The reader is referred to [15,28,34] for more extensive validations which investigate the dependence of CE measures on a variety of dynamics, series length, noise conditions, and parameter settings. Here, we consider short realizations of linear stochastic and nonlinear deterministic coupled systems with and without instantaneous effects, and compare TE and cTE as regards their ability to detect the absence or presence of information transfer between pairs of systems. All TE and cTE computations were performed following the described procedure for nonuniform embedding, including in the initial set of candidates $L = 10$ past terms for each process (plus the zero-lag term when relevant); this choice was based on the necessity to cover the whole range of expected time lagged interactions, while at the same time keeping reasonably low the computational times. The number of quantization levels used for coarse-graining the dynamics of each process was set at $Q = 6$, in accordance with previous validation studies [15,28,34]; whereas in theory high values of Q would lead to finer state space partitions and more accurate TE estimates, in practice Q should remain as low as $Q^K \approx N$ for series of length N (with K the embedding dimension) [15,28,34].

3.1. Physiologically Meaningful Instantaneous Causality

In the first simulation we considered the case in which instantaneous effects are causally meaningful, *i.e.*, correspond to real causal effects between pairs of processes. While zero-lag causal effects are unattainable in physical systems because interactions take time to occur, in practical analysis instantaneous causality becomes meaningfully relevant when the time resolution of the measurements is lower than the time scale of the lagged effects occurring between the processes, or when the time series are built in a way that entails the existence of zero-lag effects. Situations like these are commonly modeled in the framework of Bayesian networks or structural vector autoregression models [18,23,35]. Within this context, we consider a simulation scheme with $M = 3$ linear stochastic processes X, Y, and Z which interact according to the equations:

$$
\begin{aligned}
x_n &= a_1 x_{n-1} + a_2 x_{n-2} + u_n \\
y_n &= b_1 y_{n-1} + b_2 y_{n-2} + c x_n - c x_{n-1} + v_n \\
z_n &= c y_n + c x_{n-1} + w_n
\end{aligned}
\tag{6}
$$

where u_n, v_n and w_n are independent white noises with zero mean and variance $\sigma^2_u = 5$, $\sigma^2_v = 1$, and $\sigma^2_w = 1$. According to (6), the processes X and Y are represented as second order autoregressive processes described by two complex-conjugate poles with modulus $\rho_{x,y}$ and phases $\varphi_{x,y} = \pm 2\pi f_{x,y}$; setting modulus and central frequency of the poles as $\rho_x = 0.95$, $\rho_y = 0.92$, $f_x = 0.3$, $f_y = 0.1$, the parameters quantifying the dependence of x_n and y_n on their own past in (6) are $a_1 = 2\rho_x \cos\varphi_x = 0.5871$, $a_2 = -\rho^2_x = -0.9025$, $b_1 = 2\rho_y \cos\varphi_y = 1.4886$, $a_2 = -\rho^2_y = -0.8464$. The other parameters, all set with a magnitude $c = 0.5$, identify causal effects between pairs of processes; the imposed effects are mixed instantaneous and lagged from X to Y, exclusively instantaneous from Y to Z, and exclusively lagged from X to Z. With this setting, self-dependencies and causal effects are consistent with rhythms and interactions commonly observed in cardiovascular and cardiorespiratory variability, showing an autonomous oscillation at the frequency of the Maier waves (f_y ~0.1 Hz) for Y, which is transmitted to Z mimicking feedback effects from arterial pressure to heart period, and an oscillation at a typical respiratory frequency (f_x ~0.3 Hz) for X, which is transmitted to both Y and Z mimicking respiratory-related effects on arterial pressure and heart period (a realization of the three processes is shown in Figure 1a).

The analysis was performed on 100 realizations of (6), each lasting $N = 300$ points. For each realization, we computed the TE according to (3) and the cTE according to (4). The statistical significance of each estimated information transfer was assessed by using surrogate time series.

Specifically, the TE or cTE of the original time series was compared with the distribution of its values obtained for a set of $S = 40$ realizations of time-shifted surrogates, obtained by shifting the source time series of a randomly selected lag (>20 points); then, the null hypothesis of absence of information transfer was rejected if the original TE or cTE took the first or second position in the descending ordered sequence of original and surrogate values (this corresponds to a type-I error probability of 0.0405 [36]).

An example of the analysis is depicted in Figure 1. Each panel reports the corrected CE estimated for the destination process after conditioning to all processes except the source process (black) and after conditioning to all processes including the source process (red), together with the term selected at each step of the conditioning procedure. Note that the two estimated CE profiles overlap whenever no terms from the source process are selected even if considered as possible candidates, so that the two CE minima are the same and the estimated TE or cTE is zero. For instance, considering the estimation of TE or cTE from Y to X conditioned to Z (lower left panel in Figure 1b,c) we see that the first repetition of the embedding procedure –which starts from the initial set of candidate terms $\Omega_1 = \{x_{n-1},...,x_{n-10}, z_{n-1},...,z_{n-10}\}$ – selects progressively the past terms of X with lags 5, 2, and 3, terminating at the third step with the embedding vector $V_3 = [x_{n-5}, x_{n-2}, x_{n-3}]$. The second repetition of the procedure, although starting with the enlarged set of candidates $\Omega_2 = \{\Omega_1, y_{n-1},...,y_{n-10},\}$ which includes also past terms from the source system Y, selects exactly the same candidates leading again to the embedding vector $V_3 = [x_{n-5}, x_{n-2}, x_{n-3}]$ and yielding no reduction in the estimated CE minimum, so that we have $TE_{Y \to X|Z} = cTE'_{Y \to X|Z} = 0$.

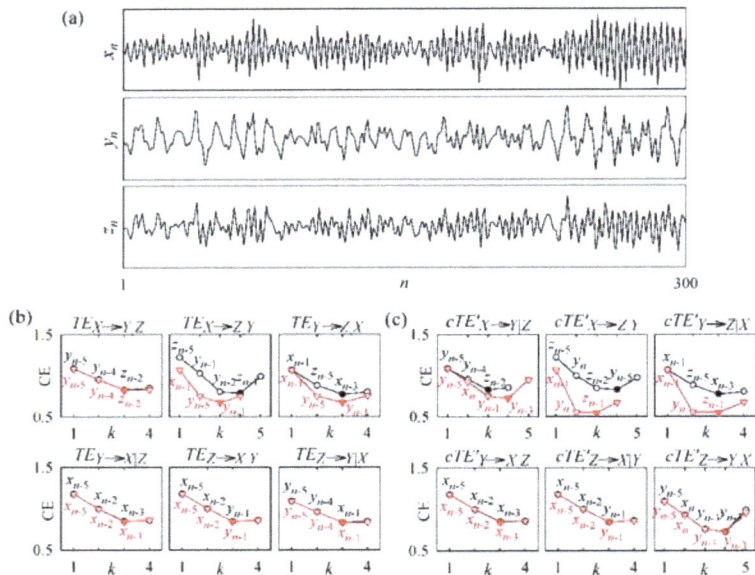

Figure 1. Example of transfer entropy analysis performed for the first simulation. **(a)** realization of the three processes generated according to (6). **(b)** TE estimation between pairs of processes based on nonuniform embedding; each panel depicts the CE estimated for the destination process through application of the non-uniform embedding procedure without considering the source process (black circles), or considering the source process (red triangles), in the definition of the set of candidates; the terms selected at each step k of the sequential embedding are indicated within the plots, while filled symbols denote each detected CE minimum. **(c)** Same of (b) for estimation of the compensated TE (cTE').

On the contrary, the selection of one or more terms from the input process during the second repetition of the procedure leads to a decrease in the CE minimum, and thus to the detection of a positive information transfer. For instance, considering the estimation of TE from Y to Z conditioned to X (upper right panel in Figure 1b) the vector resulting from the first embedding is $[x_{n-1}, z_{n-5}, x_{n-3}]$, while the second embedding selects at the second and third steps some past terms from the source process Y (*i.e.*, the terms y_{n-5} and y_{n-1}), so that the selected embedding vector changes to $[x_{n-1}, y_{n-5}, y_{n-1}]$ and this results in a reduction of the CE minimum with respect to the first embedding and in the detection of a nonzero information transfer ($TE_{Y \to Z | X} > 0$).

The difference between TE and cTE is in the fact that in cTE computation the zero-lag term of the source process is a possible candidate in the second repetition of the embedding procedure, so that when selected to enter the embedding vector, it may reduce the information carried by the target process and thus lead to detecting information transfer. In the example of Figure 1, this is the case of the analysis performed from X to Y: the traditional TE misses detection of the existing information transfer because the procedure selects at both repetitions the embedding vector $[y_{n-5}, y_{n-4}, z_{n-2}]$, failing to include any term from the source system X and thus returning $TE_{X \to Y | Z} = 0$ (Figure 1b, upper left panel); on the contrary the compensated TE captures the information transfer thanks to the fact that the zero-lag term x_n is in the set of candidates for the second embedding, and is selected determining a reduction in the estimated CE that ultimately leads to $cTE'_{X \to X | Z} > 0$ [Figure 1c, upper left panel].

Figure 2 reports the results of the analysis extended to all realizations. As seen in Figure 2a, the distributions of both *TE* and *cTE'* are close to zero when computed over the directions for which no information transfer was imposed (*i.e.*, $Y \to X$, $Z \to X$ and $Z \to Y$), and cover a range of larger positive values over the directions with imposed coupling ($X \to Y$, $X \to Z$ and $Y \to Z$). *cTE'* shows higher values than *TE* when computed over the coupled directions, while the two distributions substantially overlap when evaluated over the uncoupled directions. Note that markedly higher values are obtained for *cTE'* compared to *TE* even for the direction $X \to Z$ even though X does not contribute to Z in an instantaneously causal way; this is likely due to the fact that Y causes Z instantaneously, an effect that cannot be detected in the traditional analysis and ultimately leads to underestimation of the TE.

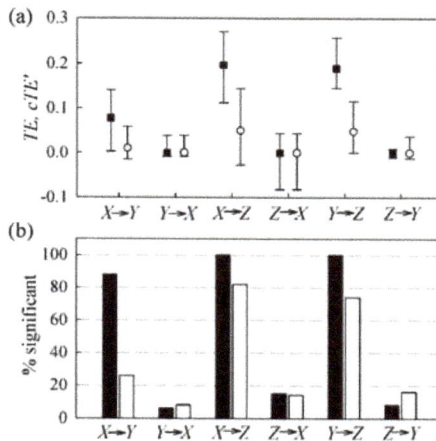

Figure 2. Results of transfer entropy analysis for the first simulation. **(a)** Distribution over 100 realizations of (6) (expressed as 5th percentile, median and 95th percentile) of the information transfer estimated between each pair of processes using the traditional TE (white) and the compensated TE (black). **(b)** Percentage of realizations for which the information transfer estimated using TE (white) and compensated TE (black) was detected as statistically significant according to the test based on time-shifted surrogates.

The results of Figure 2a are further supported by the percentage of significant information transfer of Figure 2b. Indeed, while over the uncoupled directions the number of detected significant causal couplings is low and comparable for TE and cTE' (the overall specificity is 87% for the TE and 90% for the cTE), over the coupled directions the number of detected significant couplings is substantially higher for cTE' than for TE (the overall sensitivity is 61% for the TE and 96% for the cTE). Thus, in this situation with causally meaningful instantaneous interactions, utilization of the cTE in place of the traditional TE yields a better sensitivity in the detection of information transfer between coupled processes.

3.2. Non-Physiological Instantaneous Causality

In the second simulation we considered the case in which instantaneous effects are not physiologically meaningful, reproducing a situation of cross-talk between two nonlinear processes which is typical in the analysis of neurophysiological settings where data acquired at the scalp level are the result of the instantaneous mixing of unmeasured cortical sources. Specifically, we considered the simulated systems X' and Y' described by two unidirectionally coupled logistic processes x' and y':

$$\begin{aligned} x'_n &= R_1 x'_{n-1} \left(1 - x'_{n-1}\right) \\ y'_n &= C x'_{n-1} + (1 - C)\left[R_2 y'_{n-1}\left(1 - y'_{n-1}\right)\right] \end{aligned} \tag{7}$$

which were then instantaneously mixed to obtain the processes x and y as:

$$\begin{aligned} x_n &= (1 - \varepsilon)x'_n + \varepsilon y'_n + u_n \\ y_n &= \varepsilon x'_n + (1 - \varepsilon)y'_n + w_n \end{aligned} \tag{8}$$

where u and w are independent additive noise processes with zero mean and variance set to get a signal-to-noise ratio of 20 dB. In (7), we set $R_1 = 3.86$ and $R_2 = 4$ to obtain a chaotic behavior for the two logistic maps describing the autonomous dynamics of X and Y; the parameters C and ε in (7) and (8) set respectively the strength of coupling from X to Y and the amount of instantaneous mixing between the two processes.

The analysis was performed at varying the coupling strength from $C = 0$ (absence of coupling) to $C = 1$ (full coupling, intended as absence of self-dependencies in Y with maximal dependence on X) in the absence of signal mixing ($\varepsilon = 0$), and at varying the mixing parameter from $\varepsilon = 0$ to $\varepsilon = 0.4$ either in the absence of coupling ($C = 0$) or with fixed coupling ($C = 0.2$). For each combination of the parameters, 50 realizations of (7-8) were generated, each lasting 100 points, and the TE and cTE were computed according to (3), (4) and (5), respectively. Since in this simulation the data were interpreted as having a trial structure, as typically happens in neurophysiological studies, the statistical significance of each estimated information transfer was assessed by means of a permutation test. The test consisted in performing repeatedly ($S = 100$ times in this study) a random shuffling of the relative ordering of the trials for the two processes to get S datasets with uncoupled trials; then, the null hypothesis of absence of information transfer was rejected if the median TE (or cTE'') computed for the original trials was outside the 95-th percentile of the distribution of the median TE (or cTE'') computed over the S datasets with shuffled trials (this corresponds to set a type-I error probability of 0.05).

Examples of the analysis performed with significant coupling but absence of signal cross-talk ($C = 0.2$, $\varepsilon = 0$) and significant cross-talk but absence of coupling ($C = 0$, $\varepsilon = 0.2$) are depicted in Figure 3a,b, respectively. In the first case, both TE and cTE seem able to detect correctly the imposed unidirectional coupling. Indeed, in the computation of $TE_{X \to Y}$ and $cTE''_{X \to Y}$ the second repetition of the conditioning procedure (red) selects a term from the input process (*i.e.*, x_{n-1}) determining a decrease in the estimated CE minimum and thus the detection of a positive information transfer; on the contrary, the analysis performed from Y to X does not select any term from the source process in the second repetition of the conditioning procedure, thus leading to unvaried CE and hence to null values of the information transfer ($TE_{Y \to X} = cTE''_{Y \to X} = 0$). The identical behavior of TE and cTE is explained by noting that, in this case with absence of instantaneous signal mixing, zero-lag effects are not present, and indeed the zero-lag term is not selected (although tested) during the embedding procedures for cTE. On the contrary, in the case of Figure 3b where the instantaneous mixing is not trivial, the two repetitions of the embedding procedure for cTE both select the zero lag-term (x_n in the analysis from X to Y and y_n in the analysis from Y to X); as a consequence, the cTE correctly reveals the absence of information transfer from X to Y and from Y to X, while the TE seems to indicate a false positive detection of information transfer over both directions because of the CE reduction determined by inclusion of a term from the input process during the second conditioning.

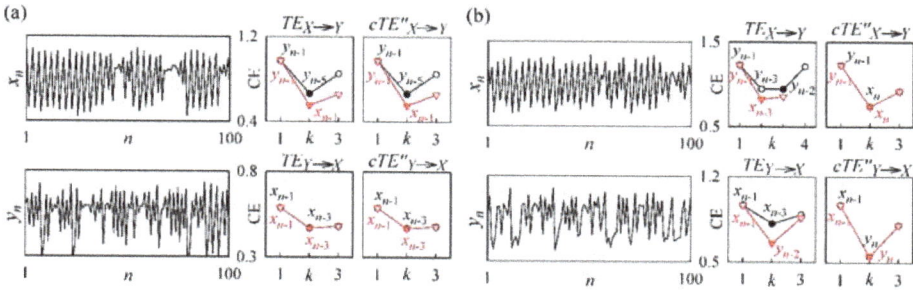

Figure 3. Example of transfer entropy analysis performed for the second simulation. **(a)** Presence of coupling and absence of instantaneous mixing ($C = 0.2$, $\varepsilon = 0$) **(b)** Absence of coupling and presence of instantaneous mixing ($C = 0$, $\varepsilon = 0.2$). Panels depict a realization of the two processes X and Y generated according to (7) and (8), together with the estimation of *TE* and *cTE''* over the two directions of interaction based on nonuniform embedding and conditional entropy (CE, see caption of Figure 1 for details).

Figure 4 reports the results of the overall analysis. As shown in Figure 4a, the traditional and compensated TE perform similarly in the absence of signal cross-talk, as the median values of *TE* and *cTE''* are statistically significant, according to the permutation test, for all values of $C > 0$ when computed from X to Y, and are never statistically significant when computed from Y to X. On the contrary, the presence of instantaneous mixing may induce the traditional TE to yield a misleading indication of information transfer for uncoupled processes. This erroneous indication occurs in Figure 4b where both $TE_{X \rightarrow Y}$ and $TE_{Y \rightarrow X}$ are statistically significant with $\varepsilon > 0$ even though X and Y are uncoupled over both the directions of interaction, and in Figure 4c where $TE_{Y \rightarrow X}$ is statistically significant with $\varepsilon = 0.2$ even though no coupling was imposed from Y to X (in total, false positive detections using the TE were five out of six negative cases with presence of instantaneous mixing). Unlike the traditional TE, the cTE does not take false positive values in the presence of signal cross-talk, as the detected information transfer is not statistically significant over both directions in the case of uncoupled systems of Figure 4b, and is statistically significant from X to Y but not from Y to X in the case of unidirectionally coupled systems of Figure 4c. Thus, in this simulation where instantaneous causality is due to common driving effects, utilization of *cTE''* in place of the traditional *TE* measure yields a better specificity in the detection of predictive information transfer.

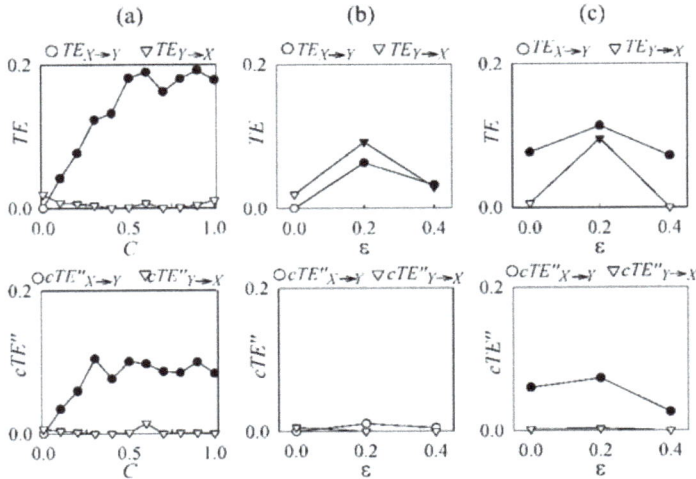

Figure 4. Results of transfer entropy analysis for the second simulation, showing the median values over 50 realizations of (7) and (8) of the TE (first panel row) and the compensated TE (second panel row) computed along the two directions of interactions ($X \to Y$, circles; $Y \to X$, triangles) **(a)** at varying the parameter C with parameter $\varepsilon = 0$; **(b)** at varying ε with $C = 0$ (b); and **(c)** varying ε with $C = 0.2$. Filled symbols denote statistically significant values of TE or cTE'' assessed by means of the permutation test.

4. Application Examples

This section describes the evaluation of the proposed TE/cTE estimation approach in physiological systems where commonly only short realizations of the studied processes (few hundred points) are available due to stationarity constraints. The considered applications are taken as examples of commonly performed time series analyses of physiological systems, *i.e.*, the study of short-term cardiovascular and cardiorespiratory interactions during a paced breathing protocol [7], and the study of neural interactions from magnetoencephalographic data during an experiment of visuo-motor integration [37].

4.1. Cardiovascular and Cardiorespiratory Variability

In the first application we studied cardiovascular and cardiorespiratory time series measured during an experiment of paced breathing [7]. The considered dynamical systems are the respiratory system, the vascular system, and the cardiac system, from which we take the respiratory flow, the systolic arterial pressure and the heart period as representative processes, respectively denoted as processes x, y and z. Realizations of these processes were obtained measuring in a healthy subject the beat-to beat time series of heart period, z_n, systolic pressure, y_n, and respiratory flow, x_n, respectively as the sequences of the temporal distances between consecutive heartbeats detected from the electrocardiogram, the local maxima of the arterial pressure signal (acquired through the Finapres device) measured inside each detected heart period, and the values of the airflow signal (acquired from the nose through a differential pressure transducer) sampled at the onset of each detected heart period. The measurement convention is illustrated in Figure 5. The experimental protocol consisted in signal acquisition, after subject stabilization in the resting supine position, for 15 min with spontaneous breathing, followed by further 15 min with the subject inhaling and exhaling in time with a metronome acting at 15 cycles/min (paced breathing at 0.25 Hz). Two artifact-free windows of $N = 300$ samples, measured synchronously for the $M = 3$ series during spontaneous breathing and during paced breathing, were considered for the analysis. Weak stationarity of each series was checked by means of a test checking the stability of the mean and variance over the analysis window [38]. The analyzed series are shown in Figure 6.

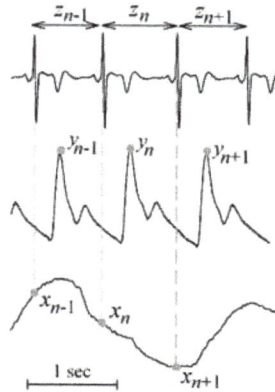

Figure 5. Measurement of heart period (series z), systolic arterial pressure (series y) and respiratory flow (series x) variability series from the electrocardiogram, arterial blood pressure and nasal flow signals.

In this application, instantaneous effects between the measured time series were considered as physiologically meaningful, since from the above described measurement convention we can infer that the occurrence of the present respiration value, x_n, precedes in time the occurrence of the present systolic pressure value, y_n, which in turn precedes in time the end of the present heart period, z_n (see Figure 5). Therefore, cTE analysis was performed for this application using the compensation proposed in (4). The statistical significance of each estimated TE and cTE' was assessed using time shifted surrogates. The results of the analysis for the spontaneous breathing and paced breathing conditions are depicted in Figure 6a,b, respectively. Utilization of the traditional TE led to detect as statistically significant the information transfer measured from respiration to heart period during spontaneous breathing ($TE_{X \to Z}$ in Figure 6a, and from respiration to systolic pressure during paced breathing ($TE_{X \to Y}$ in Figure 6b. The same analysis performed accounting for instantaneous causality effects led to detect a higher number of statistically significant interactions, specifically from respiration to heart period and from systolic pressure to heart period during both conditions ($cTE'_{X \to Z}$ and $cTE'_{Y \to Z}$ in Figure 6a,b), and also from respiration to systolic pressure during paced breathing ($cTE'_{X \to Y}$ in Figure 6b).

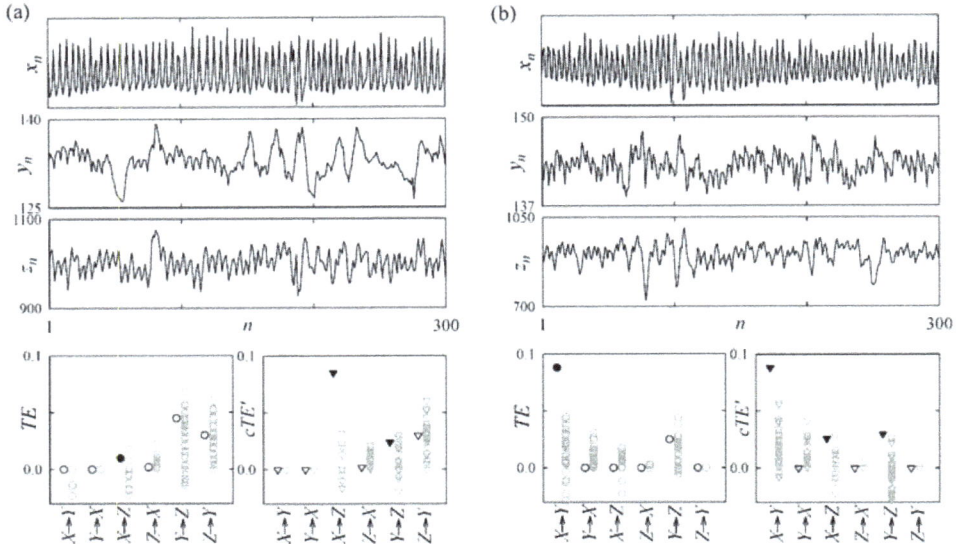

Figure 6. Transfer entropy analysis in cardiovascular and cardiorespiratory variability performed. (a) during spontaneous breathing and (b) during paced breathing. Plots depict the analyzed time series of respiratory flow (x_n, system X), systolic arterial pressure (y_n, system Y) and heart period (z_n, system Z) together with the corresponding TE (circles) and compensated TE (triangles) estimated between each pair of series. The gray symbols indicate the values of TE/cTE obtained over 40 pairs of time-shifted surrogates; filled symbols denote statistically significant *TE* or *cTE′*.

Though not conclusive as they are drawn on a single subject, these results suggest a higher sensitivity of the cTE, compared with the traditional TE, in the detection of information transfers that can be associated to known cardiovascular and cardiorespiratory mechanisms. These mechanisms, also recently investigated using tools based on transfer entropy [7], are the baroreflex modulation of heart rate, manifested through coupling from systolic pressure to heart period variability [39], and the effects of respiration on heart period (describing the so-called respiratory sinus arrhythmia [40]) and on arterial pressure (describing the mechanical perturbations of arterial pressure originating from respiration-related movements [41]). In particular, the higher sensitivity of cTE to the information transferred from systolic pressure to heart period, denoted in this example by the significant values observed for $cTE'_{Y \rightarrow Z}$ but not for $TE_{Y \rightarrow Z}$ in both conditions, could suggest a major role played by fast vagal effects −whereby the systolic pressure affects heart period within the same heartbeat—in the functioning of the baroreflex mechanism.

4.2. Magnetoencephalography

The second application is about quantification of the information transfer between different cerebral areas from the analysis of magnetoencephalographic (MEG) data. The analyzed MEG signals were taken from a database of neurobiological recordings acquired during a visuo-tactile cognitive experiment [42]. Briefly, a healthy volunteer underwent a recording session in which simultaneous visual and tactile stimuli were repeatedly presented (60 trials). At each trial, geometric patterns resembling letters of the Braille code were both shown on a monitor and embossed on a tablet, and the subject had to perceive whether the pattern seen on the screen was the same of that touched on the tablet. The MEG signals (VSM whole head system) were recorded with 293 Hz sampling frequency during two consecutive time frames of 1 s, before (rest window) and after (task window) the presentation of the combined stimuli.

The two dynamical systems considered for this application were the somatosensory cortex (system X) and the visual cortex (system Y). At each experimental trial, we considered two MEG sensors as representative of the two areas, and considered the signals measured from these sensors as realizations of the processes x and y. Sensor selection was performed trial by trial through a suitable event-related field analysis looking for the scalp locations, situated within the visual cortex and the somatosensory cortex, at which the signal magnitude was maximized in response to pure-visual or pure-tactile stimulation [42]. The considered signals were preprocessed applying a band-pass filter (FFT filter, 2–45 Hz); moreover, the event-related field was removed from each task window by subtraction of the average response over the 60 trials. An example of the analyzed signals is shown in Figure 7a,7b.

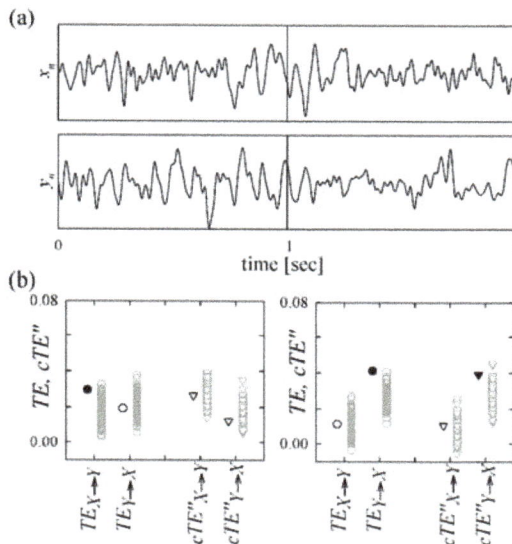

Figure 7. Transfer entropy analysis in magnetoencephalography performed before (left) and during (right) presentation of the combined visuo-tactile stimuli. (**a**) Representative MEG signals acquired from the somatosensory cortex (x_n, system X) and the visual cortex (y_n, system Y) for one of the experiment trials (n ranges from 1 to 293 samples before and during simulation). (**b**) Median over the 60 trials of TE (circles) and compensated TE (triangles) estimated for the two directions of interaction between X and Y before and during stimulation; gray symbols indicate the values of TE/cTE'' obtained over 100 trial permutations; filled symbols denote statistically significant TE or cTE''.

In this application, instantaneous effects were considered as non-physiological, because in large part they are the result by artifacts of volume conduction, *i.e.*, of the instantaneous mixing of unmeasured cortical sources which are simultaneously mapped onto the different MEG sensors [43]. Therefore, cTE analysis was performed using the compensation for signal cross-talk proposed in (5). The statistical significance of each estimated TE and cTE'' value was assessed using a permutation test applied to the 60 trials.

The results shown in Figure 7b indicate that the TE is statistically significant from the somatosensory area towards the visual area before stimulus presentation, and from the visual area to the somatosensory area during stimulation. On the other hand, the cTE was not statistically significant along any direction before task, and was significant from the visual area to the somatosensory area during task. Therefore, utilization of cTE'' seems to indicate in this exemplary application the emergence of causality $X{\rightarrow}Y$ with stimulus presentation, with a significant information transfer detected only during execution of the task. This result is compatible with the activation of mechanisms of sensory-motor integration moving from rest to task, with the posterior visual cortex driving the coherent activation of the somatosensory cortex during the combined visuo-tactile stimulation [44]. Moreover, the significant information transfer detected by the traditional TE over the opposite direction in the absence of stimulation, which is more difficult to interpret according to the paradigm proposed by this experiment, could be interpreted as a false positive detection of information transfer, thus confirming the lower specificity of non-compensated TE analysis evidenced by the simulation results. While the results reported here are certainly not conclusive, we believe that utilization of a nonlinear, model-free tool like TE, in conjunction with the compensation for instantaneous mixing realized by cTE, may deepen the interpretation of the mechanisms of multisensory integration involved in visuo-tactile experiments given by more standard tools, e.g., based on spectral analysis [37,42].

5. Discussion

Our results suggest that the framework proposed in this study for the practical estimation of multivariate TE can successfully deal with the issues arising in the conditions typical of physiological time series analysis. First, to counteract the problems related to high dimensionality and small sample size, we exploited a data-efficient estimation approach which combines a strategy for optimizing the embedding of multiple time series with a method for correcting the bias that affect conditional entropy estimates progressively at increasing the embedding dimension [15]. The reported simulation results indicate that using this approach together with appropriate statistical tests (*i.e.*, time-shifted surrogates or, when the dataset has a trial structure, permutation tests), detection of significant information transfer is possible even when the analyzed realizations are very short (a few hundred data points). Moreover, we devised a compensation strategy aimed at properly taking into account the concept of instantaneous causality in the computation of TE. In the presented simulated datasets utilization of this strategy led to an improvement in sensitivity of about 35% when instantaneous effects were physiologically meaningful, and to an improvement in specificity of about 85% when instantaneous effects were non physiological (*i.e.*, due to common driving from unobserved sources). These two kinds of improvement were suggested also by the reported representative applications to physiological time series. In cardiovascular and cardiorespiratory variability, where the construction of the time series suggests the existence of physiological causal effects occurring at lag zero, the compensated TE evidenced better than the traditional TE the presence of expected interaction mechanisms (e.g., the baroreflex). In magnetoencephalography, where instantaneous effects are likely the result of the simultaneous mapping of single sources of brain activity onto several recording sensors, utilization of the proposed compensation suggested the activation of multisensory integration mechanisms in response to a specific stimulation paradigm. Nevertheless, we emphasize that practical analysis was limited in the present study to preliminary investigations aimed at supporting the feasibility of the proposed approach in different fields of application, and that systematic tests performed on extensive databases need to be carried out to corroborate the validity of our experimental results.

While with the present study we have proposed feasible approaches to deal with the detrimental effects of instantaneous causality in the practical estimation of TE, it is important to remark that the proposed compensations constitute an empirical rather than a principle solution to the problem. In fact, from a theoretical perspective the compensation achieved in (4) through the index cTE' could not yield a better sensitivity than the traditional TE measure (3), because an instantaneous causal effect from X to Y can be detected by cTE' reflecting a direct effect $x_n{\rightarrow}y_n$, but by TE as well reflecting an indirect effect $x_{1:n-1}{\rightarrow}x_n{\rightarrow}y_n$, (provided that X has an internal memory structure). Therefore, the higher sensitivity observed for the cTE in this case should be explained in practical terms (*i.e.*, as an easier estimation of a direct than an indirect effect). Moreover, when instantaneous effects are causally meaningful, including them in TE computation as done in (4) might yield to a detection of information transfer not only over the direction of the actual causal effects, but also over the opposite direction. On the other hand, when instantaneous effects are not causally meaningful the full removal of zero-lag effects performed by (5) may be conservative when real causal effects taking place within the same sample are present besides the spurious effects to be removed. Another point regarding theoretical values of the index cTE'' is that conditioning to the zero-lag term as done in (5) may cause, in particular circumstances involving unobserved variables (e.g., due to latent confounders or resulting from inappropriate sampling), spurious detections of predictive information transfer reflecting an effect known as "selection bias" or "conditioning on a collider" [45]. Nevertheless it is likely that, in most practical situations in which real short data sequences are considered and significance tests are applied, the null hypothesis of absence of information transfer cannot be rejected solely as a consequence of spurious effects deriving from selection bias. Further studies should be aimed at assessing the real capability of these spurious effects to produce detectable predictive information transfer in practical estimation contexts.

As to the practical utilization of the cTE estimation framework developed in this study, we stress that the proposed compensation for instantaneous causality relies on prior knowledge about the nature of the zero-lag interactions among the observed physiological processes. Indeed, we have shown that the proposed compensation strategies work properly only when one can reasonably assume that instantaneous effects are the result of an improper sampling of actual physiological causal interactions, or of a simultaneous mapping of unobserved processes. In fact, using the index cTE' when instantaneous effects are not causally meaningful may exacerbate the false positive detection of information transfer, while using cTE'' in the presence of meaningful instantaneous effects does not improve the detection rate. Therefore, future studies should aim at integrating within our framework recently proposed approaches for the inference of the direction of instantaneous causality based on data structure rather than on prior assumptions [17,23]. Another interesting development would be to combine together the approach for partial conditioning recently proposed in [46], which selects the most informative subset of processes for describing the source process, with our nonuniform embedding procedure, which selects the most informative subset of lagged variables for describing the target process. Such an integrated approach for dimensionality reduction would further favor the development of a fully multivariate efficient TE estimator. Finally we remark that, whereas in this study we have followed a uniform quantization approach for estimating entropies, other approaches such as those using kernel density and nearest neighbor estimators have been proven more accurate [4,13,14]. Accordingly, future investigations will be directed towards the implementation of correction strategies realizing for these alternative estimators the compensation of the CE bias obtained here in the context of uniform quantization.

References

1. Schreiber, T. Measuring information transfer. *Phys. Rev. Lett.* **2000**, *2000*, 461–464.
2. Barnett, L.; Barrett, A.B.; Seth, A.K. Granger causality and transfer entropy are equivalent for Gaussian variables. *Phys. Rev. Lett.* **2009**, *103*, 238701.
3. Wibral, M.; Rahm, B.; Rieder, M.; Lindner, M.; Vicente, R.; Kaiser, J. Transfer entropy in magnetoencephalographic data: Quantifying information flow in cortical and cerebellar networks. *Progr. Biophys. Mol. Biol.* **2011**, *105*, 80–97.
4. Vicente, R.; Wibral, M.; Lindner, M.; Pipa, G. Transfer entropy-a model-free measure of effective connectivity for the neurosciences. *J. Comp. Neurosci.* **2011**, *30*, 45–67.
5. Vakorin, V.A.; Kovacevic, N.; McIntosh, A.R. Exploring transient transfer entropy based on a group-wise ICA decomposition of EEG data. *Neuroimage* **2010**, *49*, 1593–1600.
6. Gourevitch, B.; Eggermont, J.J. Evaluating information transfer between auditory cortical neurons. *J. Neurophysiol.* **2007**, *97*, 2533–2543.
7. Faes, L.; Nollo, G.; Porta, A. Information domain approach to the investigation of cardio-vascular, cardio-pulmonary, and vasculo-pulmonary causal couplings. *Front. Physiol.* **2011**, *2*, 1–13.
8. Faes, L.; Nollo, G.; Porta, A. Non-uniform multivariate embedding to assess the information transfer in cardiovascular and cardiorespiratory variability series. *Comput. Biol. Med.* **2012**, *42*, 290–297.

9. Vejmelka, M.; Palus, M. Inferring the directionality of coupling with conditional mutual information. *Phys. Rev. E* **2008**, *77*, 026214.

10. Chicharro, D.; Ledberg, A. Framework to study dynamic dependencies in networks of interacting processes. *Phys. Rev. E* **2012**, *86*, 041901.

11. Chicharro, D.; Ledberg, A. When two become one: The limits of causality analysis of brain dynamics. *PLoS One* **2012**.

12. Lizier, J.T.; Prokopenko, M. Differentiating information transfer and causal effect. *Eur. Phys. J. B* **2010**, *73*, 605–615.

13. Hlavackova-Schindler, K.; Palus, M.; Vejmelka, M.; Bhattacharya, J. Causality detection based on information-theoretic approaches in time series analysis. *Phys. Rep.* **2007**, *441*, 1–46.

14. Lee, J.; Nemati, S.; Silva, I.; Edwards, B.A.; Butler, J.P.; Malhotra, A. Transfer Entropy Estimation and Directional Coupling Change Detection in Biomedical Time Series. *Biomed. Eng.* **2012**.

15. Faes, L.; Nollo, G.; Porta, A. Information-based detection of nonlinear Granger causality in multivariate processes via a nonuniform embedding technique. *Phys. Rev. E* **2011**, *83*, 051112.

16. Lutkepohl, H. *New Introduction to Multiple Time Series Analysis*; Springer-Verlag: Heidelberg, Germany, 2005.

17. Faes, L.; Erla, S.; Porta, A.; Nollo, G. A framework for assessing frequency domain causality in physiological time series with instantaneous effects. *Philos. Transact. A* **2013**, in press.

18. Faes, L.; Nollo, G. Extended causal modelling to assess Partial Directed Coherence in multiple time series with significant instantaneous interactions. *Biol. Cybern.* **2010**, *103*, 387–400.

19. Granger, C.W.J. Investigating causal relations by econometric models and cross-spectral methods. *Econometrica* **1969**, *37*, 424–438.

20. Geweke, J. Measurement of linear dependence and feedback between multiple time series. *J. Am. Stat. Assoc.* **1982**, *77*, 304–313.

21. Guo, S.X.; Seth, A.K.; Kendrick, K.M.; Zhou, C.; Feng, J.F. Partial Granger causality—Eliminating exogenous inputs and latent variables. *J. Neurosci. Methods* **2008**, *172*, 79–93.

22. Barrett, A.B.; Barnett, L.; Seth, A.K. Multivariate Granger causality and generalized variance. *Phys. Rev. E* **2010**, *81*, 041907.

23. Hyvarinen, A.; Zhang, K.; Shimizu, S.; Hoyer, P.O. Estimation of a Structural Vector Autoregression Model Using Non-Gaussianity. *J. Machine Learn. Res.* **2010**, *11*, 1709–1731.

24. Porta, A.; Bassani, T.; Bari, V.; Pinna, G.D.; Maestri, R.; Guzzetti, S. Accounting for Respiration is Necessary to Reliably Infer Granger Causality From Cardiovascular Variability Series. *IEEE Trans. Biomed. Eng.* **2012**, *59*, 832–841.

25. Vakorin, V.A.; Krakovska, O.A.; McIntosh, A.R. Confounding effects of indirect connections on causality estimation. *J. Neurosci. Methods* **2009**, *184*, 152–160.

26. Chen, Y.; Bressler, S.L.; Ding, M. Frequency decomposition of conditional Granger causality and application to multivariate neural field potential data. *J. Neurosci. Methods* **2006**, *150*, 228–237.

27. Kraskov, A.; Stogbauer, H.; Grassberger, P. Estimating mutual information. *Phys. Rev. E* **2004**, *69*, 066138.

28. Porta, A.; Baselli, G.; Lombardi, F.; Montano, N.; Malliani, A.; Cerutti, S. Conditional entropy approach for the evaluation of the coupling strength. *Biol. Cybern.* **1999**, *81*, 119–129.

29. Takens, F. Detecting strange attractors in fluid turbulence. In *Dynamical Systems and Turbulence*; Rand, D., Young, S.L., Eds.; Springer-Verlag: Berlin, Germany, 1981; pp. 336–381.

30. Vlachos, I.; Kugiumtzis, D. Nonuniform state-space reconstruction and coupling detection. *Phys. Rev. E* **2010**, *82*, 016207.

31. Small, M. *Applied nonlinear time series analysis: Applications in physics, physiology and finance*; World Scientific Publishing: Singapore, 2005.

32. Runge, J.; Heitzig, J.; Petoukhov, V.; Kurths, J. Escaping the Curse of Dimensionality in Estimating Multivariate Transfer Entropy. *Phys. Rev. Lett.* **2012**, *108*, 258701.

33. Pincus, S.M. Approximate Entropy As A Measure of System-Complexity. *Proc. Nat. Acad. Sci. USA* **1991**, *88*, 2297–2301.

34. Porta, A.; Baselli, G.; Liberati, D.; Montano, N.; Cogliati, C.; Gnecchi-Ruscone, T.; Malliani, A.; Cerutti, S. Measuring regularity by means of a corrected conditional entropy in sympathetic outflow. *Biol. Cybern.* **1998**, *78*, 71–78.

35. Bollen, K.A. *Structural equations with latent variables*; John Wiley & Sons: NY, USA, 1989.

36. Yu, G.H.; Huang, C.C. A distribution free plotting position. *Stoch. Env. Res. Risk Ass.* **2001**, *15*, 462–476.

37. Erla, S.; Faes, L.; Nollo, G.; Arfeller, C.; Braun, C.; Papadelis, C. Multivariate EEG spectral analysis elicits the functional link between motor and visual cortex during integrative sensorimotor tasks. *Biomed. Signal Process. Contr.* **2011**, *7*, 221–227.

38. Magagnin, V.; Bassani, T.; Bari, V.; Turiel, M.; Maestri, R.; Pinna, G.D.; Porta, A. Non-stationarities significantly distort short-term spectral, symbolic and entropy heart rate variability indices. *Physiol Meas.* **2011**, *32*, 1775–1786.

39. Cohen, M.A.; Taylor, J.A. Short-term cardiovascular oscillations in man: measuring and modelling the physiologies. *J. Physiol* **2002**, *542*, 669–683.

40. Hirsch, J.A.; Bishop, B. Respiratory sinus arrhythmia in humans: how breathing pattern modulates heart rate. *Am. J. Physiol.* **1981**, *241*, H620–H629.

41. Toska, K.; Eriksen, M. Respiration-synchronous fluctuations in stroke volume, heart rate and arterial pressure in humans. *J. Physiol* **1993**, *472*, 501–512.

42. Erla, S.; Papadelis, C.; Faes, L.; Braun, C.; Nollo, G. Studying brain visuo-tactile integration through cross-spectral analysis of human MEG recordings. In *Medicon 2010, IFMBE Proceedings*; Bamidis, P.D., Pallikarakis, N., Eds.; Springer: Berlin, Germany, 2010; pp. 73–76.

43. Marzetti, L.; del Gratta, C.; Nolte, G. Understanding brain connectivity from EEG data by identifying systems composed of interacting sources. *Neuroimage* **2008**, *42*, 87–98.

44. Bauer, M. Multisensory integration: A functional role for inter-area synchronization? *Curr. Biol.* **2008**, *18*, R709–R710.

45. Cole, S.R.; Platt, R.W.; Schisterman, E.F.; Chu, H.; Westreich, D.; Richardson, D.; Poole, C. Illustrating bias due to conditioning on a collider. *Int. J. Epidemiol.* **2010**, *36*, 417–420.

46. Marinazzo, D.; Pellicoro, M.; Stramaglia, S. Causal information approach to partial conditioning in multivariate data sets. *Comput. Math. Methods Med.* **2012**, *2012*, 303601.

entropy

MDPI

Article

Information Theory Analysis of Cascading Process in a Synthetic Model of Fluid Turbulence

Massimo Materassi [1,*], Giuseppe Consolini [2], Nathan Smith [3] and Rossana De Marco [2]

[1]   Istituto dei Sistemi Complessi ISC-CNR, via Madonna del Piano 10, 50019 Sesto Fiorentino, Italy
[2]   INAF-Istituto di Astrofisica e Planetologia Spaziali Area di Ricerca Roma Tor Vergata,
      Via del Fosso del Cavaliere, 100, 00133 Roma, Italy; E-Mails: giuseppe.consolini@iaps.inaf.it (G.C.);
      rossana.demarco@iaps.inaf.it (R.D.M.)
[3]   Department of Electronic and Electrical Engineering, University of Bath, Claverton Down,
      Bath BA2 7AY, UK; E-Mail: N.Smith@bath.ac.uk

*   Author to whom correspondence should be addressed; E-Mail: massimo.materassi@isc.cnr.it;
     Tel.: +39-347-6113002.

Received: 29 September 2013; in revised form: 12 February 2014 / Accepted: 19 February 2014 /
Published: 27 February 2014

Abstract: The use of transfer entropy has proven to be helpful in detecting which is the verse of dynamical driving in the interaction of two processes, X and Y. In this paper, we present a different normalization for the transfer entropy, which is capable of better detecting the information transfer direction. This new normalized transfer entropy is applied to the detection of the verse of energy flux transfer in a synthetic model of fluid turbulence, namely the Gledzer–Ohkitana–Yamada shell model. Indeed, this is a fully well-known model able to model the fully developed turbulence in the Fourier space, which is characterized by an energy cascade towards the small scales (large wavenumbers k), so that the application of the information-theory analysis to its outcome tests the reliability of the analysis tool rather than exploring the model physics. As a result, the presence of a direct cascade along the scales in the shell model and the locality of the interactions in the space of wavenumbers come out as expected, indicating the validity of this data analysis tool. In this context, the use of a normalized version of transfer entropy, able to account for the difference of the intrinsic randomness of the interacting processes, appears to perform better, being able to discriminate the wrong conclusions to which the "traditional" transfer entropy would drive.

Keywords: transfer entropy; dynamical systems; turbulence; cascades; shell models

1. Introduction

This paper is about the use of quantities, referred to as information dynamical quantities (IDQ), derived from the Shannon information [1] to determine cross-predictability relationships in the study of a dynamical system. We will refer to as "cross-predictability" the possibility of predicting the (near) future behavior of a process, X, by observing the present behavior of a process, Y, likely to be interacting with X. Observing X given Y is of course better than observing only X, as far as predicting X is concerned: it will be rather interesting to compare how the predictability of X is increased given Y with the increase of predictability of Y given X. To our understanding, this cross-predictability analysis (CPA) gives an idea of the verse of dynamical driving between Y and X. In particular, the data analysis technique presented here is tested on a synthetic system completely known by construction. The system at hand is the Gledzer–Ohkitana–Yamada (GOY) shell model, describing the evolution of turbulence in a viscous incompressible fluid. In this model, the Fourier component interaction takes place locally in

Entropy **2014**, *16*, 1272–1286

the space of wavenumbers, and due to dissipation growing with k, a net flux of energy flows from the larger to the smaller scales (direct cascade). A dynamical "driving" of the large on the small scales is then expected, which is verified here through simulations: mutual information and transfer entropy analysis are applied to the synthetic time series of the Fourier amplitudes that are interacting.

The purpose of applying a certain data analysis technique to a completely known model is to investigate the potentiality of the analysis tool in retrieving the expected information, preparing it for future applications to real systems. The choice of the GOY model as a test-bed for the IDQ-based CPA is due both to its high complexity, rendering the test rather solid with respect to the possible intricacies expected in natural systems, and to its popularity in the scientific community, due to how faithfully it simulates real features of turbulence.

In order to focus on how IDQ-based CPA tools are applied in the study of coupled dynamical processes, let us consider two processes, X and Y, whose proxies are two physical variables, x and y, evolving with time, and let us assume that the only thing one measures are the values, $x(t)$ and $y(t)$, as time series. In general, one may suppose the existence of a stochastic dynamical system (SDS) governing the interaction between X and Y, expressed mathematically as:

$$\begin{cases} \dot{x} = f(x,y,t) \\ \dot{y} = g(x,y,t) \end{cases} \tag{1}$$

where the terms, f and g, contain stochastic forces rendering the dynamics of x and y probabilistic [2]. Actually, the GOY model studied here is defined as deterministic, but the procedures discussed are perfectly applicable, in principle, to any closed system in the form of (1). The sense of applying probabilistic techniques to deterministic processes is that such processes may be so complicated and rich, that a probabilistic picture is often preferable, not to mention the school of thought according to which physical chaos is stochastic, even if formally deterministic [3,4]. Indeed, since real-world measurements always have finite precision and most of the real-world systems are highly unstable (according to the definition of Prigogine and his co-workers), deterministic trajectories turn out to be unrealistic, hence an unuseful tool to describe reality.

Through the study of the IDQs obtained from $x(t)$ and $y(t)$, one may, for example, hope to deduce whether the dependence of $\dot{x}$ on y is "stronger" than the dependence of $\dot{y}$ on x, hence how Y is driving X.

The IDQs discussed here have been introduced and developed over some decades. After Shannon's work [1], where the information content of a random process was defined, Kullback used it to make comparisons between different probability distributions [5], which soon led to the definition of mutual information (MI) as a way to quantify how much the two processes deviate from statistical independence. The application of MI to time series analysis appears natural: Kantz and Schreiber defined and implemented a set of tools to deduce dynamic properties from observed time series (see [6] and the references therein), including time-delayed mutual information.

The tools of Kantz and Schreiber were augmented in [7] with the introduction of the so-called transfer entropy (TE): information was no longer describing the underdetermination of the system observed, but rather, the interaction between two processes studied in terms of how much information is gained on the one process observing the other, *i.e.*, in terms of information transfer. An early review about the aforementioned IDQs can be found in [8].

TE was soon adopted as a time series analysis tool in many complex system fields, such as space physics [9,10] and industrial chemistry [11], even if Kaiser and Schreiber developed a criticism and presented many caveats to the extension of the TE to continuous variables [12]. Since then, however, many authors have been using TE to detect causality, for example, in stock markets [13], symbolic dynamics [14], biology and genetics [15] and meteorology [16,17]. In the field of neuroscience, TE has been applied broadly, due to the intrinsic intricacy of the matter [18], and recently extended to multi-variate processes [19].

Entropy **2014**, *16*, 1272–1286

A very important issue is, namely, the physical meaning of the IDQs described before. Indeed, while the concept of Shannon entropy is rather clear and has been related to the thermodynamical entropy in classical works, such as [20], mutual information and transfer entropy have not been clearly given yet a significance relevant for statistical mechanics. The relationship between TE and the mathematical structure of the system (1) has been investigated in [16], while a more exhaustive study on the application of these information theoretical tools to systems with local dynamics is presented in [21] and the references therein. This "physical sense" of the IDQs will be the subject of our future studies.

The paper is organized as follows.

A short review of the IDQs is done in Section 2. Then, the use of MI and TE to discriminate the "driver" and the "driven" process is criticized, and new normalized quantities are introduced, more suitable for analyzing the cross-predictability in dynamical interactions of different "intrinsic" randomness (normalizing the information theoretical quantities, modifying them with respect to their initial definitions, is not a new thing: in [22], the transfer entropy is modified, so as to include some basic null hypothesis in its own definition; in [23], the role of information compression in the definitions of IDQs is stressed, which will turn out to emerge in the present paper, as well).

The innovative feature of the IDQs described here is the introduction of a variable delay, τ: the IDQs peak in the correspondence of some $\tilde{\tau}$ estimating the characteristic time scale(s) of the interaction, which may be a very important point in predictability matters [24]. The problem of inferring interaction delays via transfer entropy has also been given a rigorous treatment in [25], where ideas explored in [24] are discussed with mathematical rigor.

In Section 3, the transfer entropy analysis (TEA) is applied to synthetic time series obtained from the GOY model of turbulence, both in the form already described, e.g., in [10,24], and in the new normalized version, defined in Section 2. The advantages of using the new normalized IDQs are discussed, and conclusions are drawn in Section 4.

2. Normalized Mutual Information and Transfer Entropy

In all our reasoning, we will use four time series: those representing two processes, $x(t)$ and $y(t)$, and those series themselves time-shifted forward by a certain amount of time, τ. The convenient notation adopted reads:

$$x := x(t), \quad y := y(t), \quad x_\tau := x(t+\tau), \quad y_\tau := y(t+\tau) \tag{2}$$

(in Equation (2) and everywhere, ":=" means "equal by definition"). The quantity, $p_t(x,y)$, is the joint probability of having a certain value of x and y at time t; regarding notation, the convention:

$$p_t(x,y) := p(x(t), y(t)) \tag{3}$$

is understood.

Shannon entropy is defined for a stochastic process, A, represented by the variable, a,

$$I_t(A) := -\sum_a p_t(a) \log_2 p_t(a) \tag{4}$$

quantifying the uncertainty on A before measuring a at time t. Since, in practice, discretized continuous variables are often dealt with, all the distributions, $p_t(a)$, as in Equation (4), are then probability mass functions (pmfs) rather than probability density functions (pdfs).

For two interacting processes, X and Y, it is worth defining the *conditional Shannon information entropy of X given Y*:

$$I_t(X|Y) := -\sum_{x,y} p_t(x,y) \log_2 p_t(x|y) \tag{5}$$

The instantaneous MI shared by X and Y is defined as:

$$M_t(X,Y) := I_t(X) - I_t(X|Y) =$$

$$= \sum_{x,y} p_t(x,y) \log_2 \left[\frac{p_t(x,y)}{p_t(x)p_t(y)} \right] : \tag{6}$$

positive $M_t(X,Y)$ indicates that X and Y interact. The factorization of probabilities and non-interaction has an important dynamical explanation in stochastic system theory: when the dynamics in Equation (1) is reinterpreted in terms of probabilistic path integrals [26], then the factorization of probabilities expresses the absence of interaction terms in stochastic Lagrangians. Indeed, (stochastic) Lagrangians appear in the exponent of transition probabilities, and their non-separable addenda, representing interactions among sub-systems, are those terms preventing probabilities from being factorizable.

About $M_t(X,Y)$, one should finally mention that it is symmetric: $M_t(X,Y) = M_t(Y,X)$.

There may be reasons to choose to use the MI instead of, say, cross-correlation between X and Y: the commonly used cross-correlation encodes only information about the second order momentum, while MI uses all information defined in the probability distributions. Hence, it is more suitable for studying non-linear dependencies [27,28], expected to show up at higher order momenta.

In the context of information theory (IT), we state that a process, Y, drives a process, X, between t and $t + \tau$ (with $\tau > 0$) if observing y at the time, t; we are less ignorant of what x at the time, $t + \tau$, is going to be like, than how much we are on y at the time $t + \tau$ observing x at the time t. The delayed mutual information (DMI):

$$M_{Y \to X}(\tau;t) := I_t(X_\tau) - I_t(X_\tau|Y) =$$

$$= \sum_{x_\tau,y} p_t(x_\tau,y) \log_2 \left[\frac{p_t(x_\tau,y)}{p_t(x_\tau)p_t(y)} \right] \tag{7}$$

turns out to be very useful for this purpose. DMI is clearly a quantity with which cross-predictability is investigated.

In [9], a generalization of DMI is presented and referred to as transfer entropy (TE), by adapting the quantity originally introduced by Schreiber in [7] to dynamical systems, such as Equation (1), and to time delays τ that may be varied, in order to test the interaction at different time scales:

$$T_{Y \to X}(\tau;t) := I_t(X_\tau|X) - I_t(X_\tau|X,Y) =$$

$$= \sum_{x_\tau,x,y} p_t(x_\tau,x,y) \log_2 \left[\frac{p_t(x_\tau,x,y)p_t(x)}{p_t(x_\tau,x)p_t(x,y)} \right] \tag{8}$$

In practice, the TE provides the amount of knowledge added to X at time $t + \tau$, knowing $x(t)$, by the observation of $y(t)$.

The easiest way to compare the two verses of cross-predictability is of course that of taking the difference between the two:

$$\Delta M_{Y \to X}(\tau;t) = M_{Y \to X}(\tau;t) - M_{X \to Y}(\tau;t)$$

$$\Delta T_{Y \to X}(\tau;t) = T_{Y \to X}(\tau;t) - T_{X \to Y}(\tau;t) \tag{9}$$

as done in [9,10,24]. The verse of prevailing cross-predictability is stated to be that of information transfer. Some comments on quantities in Equation (9) are necessary.

Consider taking the difference between $T_{Y \to X}(\tau;t)$ and $T_{X \to Y}(\tau;t)$ in order to understand which is the prevalent verse of information transfer: if one of the two processes were inherently more noisy than the other, then the comparison between X and Y via such differences would be uneven, somehow.

Since cross-predictability via information transfer is about the efficiency of driving, then the quantities, $M_{Y \to X}(\tau; t)$ and $T_{Y \to X}(\tau; t)$, must be compared to the uncertainty induced on the "investigated system" by all the other things working on it and rendering its motion unpredictable, in particular the noises of its own dynamics (here, we are not referring only to the noisy forces on it, but also/mainly to the internal instabilities resulting in randomness). When working with $M_{Y \to X}(\tau; t)$, this "uncertainty" is quantified by $I(X_\tau)$, while when working with $T_{Y \to X}(\tau; t)$, the "uncertainty-of-reference" must be $I_t(X_\tau | X)$. One can then define a normalized delayed mutual information (NDMI):

$$R_{Y \to X}(\tau; t) := \frac{M_{Y \to X}(\tau; t)}{I_t(X_\tau)} \tag{10}$$

and a normalized transfer entropy (NTE):

$$K_{Y \to X}(\tau; t) := \frac{T_{Y \to X}(\tau; t)}{I_t(X_\tau | X)} \tag{11}$$

or equally:

$$R_{Y \to X}(\tau; t) = 1 - \frac{I_t(X_\tau | Y)}{I_t(X_\tau)}$$

$$K_{Y \to X}(\tau; t) = 1 - \frac{I_t(X_\tau | X, Y)}{I_t(X_\tau | X)} \tag{12}$$

These new quantities, $R_{Y \to X}(\tau; t)$ and $K_{Y \to X}(\tau; t)$, will give a measure of how much the presence of an interaction augments the predictability of the evolution, *i.e.*, will quantify cross-predictability.

The positivity of $\Delta R_{Y \to X}(\tau; t)$ or $\Delta K_{Y \to X}(\tau; t)$ is a better criterion than the positivity of $\Delta M_{Y \to X}(\tau; t)$ or $\Delta T_{Y \to X}(\tau; t)$ for discerning the driving direction, since the quantities involved in $R_{Y \to X}(\tau; t)$ and $K_{Y \to X}(\tau; t)$ factorize the intrinsic randomness of a process and try to remove it with the normalization, $I_t(X_\tau)$ and $I_t(X_\tau | X)$, respectively. Despite this, $\Delta M_{Y \to X}(\tau; t)$ and $\Delta T_{Y \to X}(\tau; t)$ can still be used for that analysis in the case that the degree of stochasticity of X and Y are comparable. Consider for instance that at $t + \tau$, the Shannon entropy of X and Y are equal both to a quantity, I_0, and the conditioned ones equal both to J_0; clearly, one has:

$$\Delta R_{Y \to X}(\tau; t) = \frac{\Delta M_{Y \to X}(\tau; t)}{I_0}, \quad \Delta K_{Y \to X}(\tau; t) = \frac{\Delta T_{Y \to X}(\tau; t)}{J_0}$$

and the quantity, $\Delta R_{Y \to X}(\tau; t)$, is proportional to $\Delta M_{Y \to X}(\tau; t)$ through a number I_0^{-1}, so they encode the same knowledge. The same should be stated for $\Delta K_{Y \to X}(\tau; t)$ and $\Delta T_{Y \to X}(\tau; t)$. This is why we claim that the diagnoses in [9,10,24] were essentially correct, even if we will try to show here that the new normalized quantities work better in general.

Before applying the calculation of the quantities, $\Delta T_{Y \to X}(\tau; t)$ and $\Delta K_{Y \to X}(\tau; t)$, to the turbulence model considered in Section 3, it is worth underlining again the dependence of all these IDQs on the delay, τ: the peaks of the IDQs on the τ axis indicate those delays after which the process, X, shares more information with the process, Y, *i.e.*, the characteristic time scales of their cross-predictability, due to their interaction.

3. Turbulent Cascades and Information Theory

This section considers an example in which we know what must be expected, and apply our analysis tools to it to check and refine them. In this case, the application of the normalized quantities instead of the traditional ones revised in Section 2 is investigated in some detail. In the chosen example, the IDQs are used to recognize the existence of cascades in a synthetic model of fluid turbulence [29,30].

Some theoretical considerations are worth being done in advance. The quantities described above are defined using instantaneous pmfs, *i.e.*, pmfs that exist at time t. As a result, the quantities may vary with time. Unfortunately, it is difficult to recover the statistics associated with such PMFs, except in artificial cases, when running ensemble simulations on a computer. When examining real-world systems, for example in geophysics, then this luxury is not available. Hence, in many cases, one can only calculate time statistics rather than ensemble statistics; since any analysis in terms of time statistics is only valid when the underlying system is sufficiently ergodic, what one can do is to restrict the analysis to locally ergodic cases, picking up data segments in which ergodicity apparently holds. In the following experiment, pmfs are calculated by collecting histograms from the time series, with appropriate choices of bin-width, e.g., see [31].

The system at hand is described in [30,32] and the references therein and is referred to as the Gledzer–Ohkitana–Yamada shell model (the GOY model, for short): this is a dynamic system model, which can be essentially considered as a discretization of the fluid motion problem, governed by the Navier–Stokes equation, in the Fourier space. The GOY model was one of the first available shell models for turbulence. Indeed, other, more advanced models exist (see e.g., [30]). However, we will limit our discussion to the GOY model, because all the other refined ones mainly do not substantially differ in the energy cascading mechanism in the inertial domain.

The physical variable that evolves is the velocity of the fluid, which is assigned as a value, V_h, at the h-th site of a 1D lattice; each of these V_h evolves with time as $V_h = V_h(t)$. The dependence upon the index, h, in V_h is the space-dependence of the velocity field. With respect to this space dependence, a Fourier transform can be performed: out of the real functions, $V_h(t)$, a set of complex functions $u_n = u_n(t)$ will be constructed, where u_n is the n-th Fourier amplitude of the velocity fluctuation field at the n-th shell characterized by a wavenumber, k_n. The n-th wavenumber k_n is given by:

$$k_n = k_0 q^n \tag{13}$$

k_0 being the fundamental, lowest wavenumber and q a magnifying coefficient relating the n-th to the $(n+1)$-th wavenumber as $k_{n+1} = q k_n$. In the case examined, the coefficient, q, is two, approximately meaning that the cascade takes place, halving the size of eddies from u_n to u_{n+1}.

Each Fourier mode, $u_n(t)$, is a physical process in its own right, and all these physical processes interact. The velocity field, V_h, is supposed to be governed by the usual Navier–Stokes equation, whose non-linearities yield a coupling between different modes [29]. The system is not isolated, but an external force stirs the medium. The force is assigned by giving its Fourier modes, and here, it is supposed to have only the $n = 4$ mode different from zero. The complex Fourier amplitude, f_n, of the stirring external force is hence given by $f_n = \delta_{4,n}(1+i)f$, f being a constant. The system of ordinary differential equations governing the u_ns according to the GOY model turns out to be written as:

$$\begin{cases} \dot{u}_n = i \left(k_n u_{n+1} u_{n+2} - \frac{1}{2} k_{n-1} u_{n+1} u_{n-1} - \frac{1}{2} k_{n-2} u_{n-1} u_{n-2} \right)^* - \nu k_n^2 u_n + f_n \\ \\ f_n = \delta_{4,n}(1+i)f \end{cases} \tag{14}$$

where $n = 1, 2, \dots$ and z^* is the complex conjugate of z. Each mode, u_n, is coupled to u_{n+1}, u_{n+2}, u_{n-1} and u_{n-2} in a non-linear way, and in addition, it possesses a linear coupling to itself via the dissipative term, $-\nu k_n^2 u_n$. There is also a coupling to the environment through f_n, which actually takes place only for the fourth mode. In the present simulations, the values of $f = 5 * 10^{-3}(1+i)$ and $\nu = 10^{-7}$ were used. The integration procedure is the one due to Adam and Bashfort, described in [32], with an integration step of 10^{-4}.

Even if the lattice is 1D, the equations in (14) show coefficients suitably adapted, so that the spectral and statistical properties of turbulence here are those of a real 3D fluid dynamics, so that we are really studying 3D turbulence.

The stirring force pumps energy and momentum into the fluid, injecting them at the fourth scale, and the energy and momentum are transferred from u_4 to all the other modes via the non-linear couplings. There is a scale for each k_n and eddy turnover time τ_n. At each k_n, a $\tau_n = \frac{2\pi}{k_n|u_n|}$ corresponds. The characteristic times of the system will be naturally assigned in terms of the zeroth mode eddy

Entropy **2014**, *16*, 1272–1286

turnover time, τ_0, or of other τ_ns. After a certain transitory regime, the system reaches an "equilibrium" from the time-average point of view, in which the Fourier spectrum appears for the classical energy cascade of turbulence, as predicted by Kolmogorov [29] (see Figure 1).

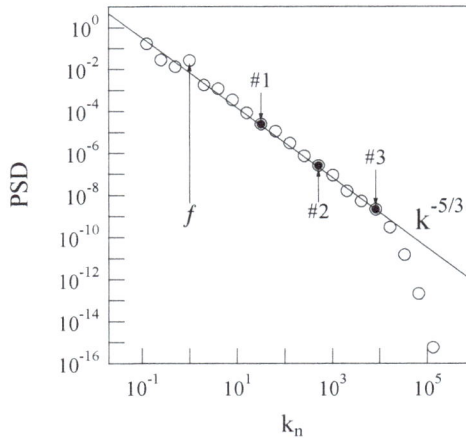

Figure 1. The time-average instantaneous power spectral density (PSD) of the velocity field of the Gledzer–Ohkitana–Yamada (GOY) model after a certain transitory regime. The modes chosen for the transfer entropy analysis (TEA) are indicated explicitly, together with the $n = 4$ scale at which the external force, f, is acting. The quantity on the ordinate is $PSD(k_n)$, as defined in Equation (15) below.

The quantity represented along the ordinate axis of Figure 1 is the average power spectral density (PSD), defined as follows:

$$PSD(k_n) := \frac{1}{t_2 - t_1} \int_{t_1}^{t_2} \frac{|u_n|^2}{k_n} dt \tag{15}$$

where $[t_1, t_2]$ is a time-interval taken after a sufficiently long time, such that the system (14) has already reached its stationary regime (*i.e.*, after some initial transient regime with heterogeneous fluctuations in which the turbulence is not fully developed yet). The evaluation of the duration of the transient regime is made "glancing at" the development of the plot of $\frac{|u_n|^2}{k_n}$ *versus* k_n as the simulation time runs and picking the moment after which this plot does not change any more. In terms of the quantities involved in Equation (15), this means t_1 is "many times" the largest eddy turnover time.

The energetic, and informatic, behavior of the GOY system is critically influenced by the form of the dissipative term, $-\nu k_n^2 u_n$, in Equation (14): the presence of the factor, k_n^2, implies that the energy loss is more and more important for higher and higher $|k_n|$, *i.e.*, smaller and smaller scales. The energy flows from any mode, u_n, both towards the smaller and the higher scales, since u_n is coupled both with the smaller scales and larger scales. Energy is dissipated at all the scales, but the dissipation efficiency grows with k_n^2, so that almost no net energy can really reflow back from the small scales to the large ones. In terms of cross-predictability, a pass of information in both verses is expected, but the direct cascade (*i.e.*, from small to large $|k_n|$s) should be prevalent. Not only this: since the ordinary differential equations Equation (14) indicate a k-local interaction (up to the second-adjacent n, *i.e.*, $n \pm 1$ and $n \pm 2$ coupling with n), one also expects the coupling between u_m and $u_{n \gg m}$ or $u_{n \ll m}$ to be almost vanishing and the characteristic interaction times to be shorter for closer values of m and n. Our program is to check all these expectations by calculating the transfer entropy and its normalized version for the synthetic data obtained running the GOY model (14). In particular, we would like to detect the verse of information transfer along the inertial domain between shells not directly coupled in the evolution Equation (14).

To get the above target, and to investigate the application of TEA to the GOY model and illustrate the advantages of using the new normalized quantities discussed in Section 2, we selected three non-consecutive shells. In particular, the choice:

$$\#1 \leftrightarrow n = 9, \quad \#2 \leftrightarrow n = 13, \quad \#3 \leftrightarrow n = 17$$

is made. The real parts of u_9, u_{13} and u_{17} are reported in Figure 2 as functions of time for a short time interval of 15 τ_4. For each of the selected shells, we considered very long time series of the corresponding energy $e_n = |u_n|^2$. The typical length of the considered time series is of many ($\simeq 1,000$) eddy turnover times of the injection scale.

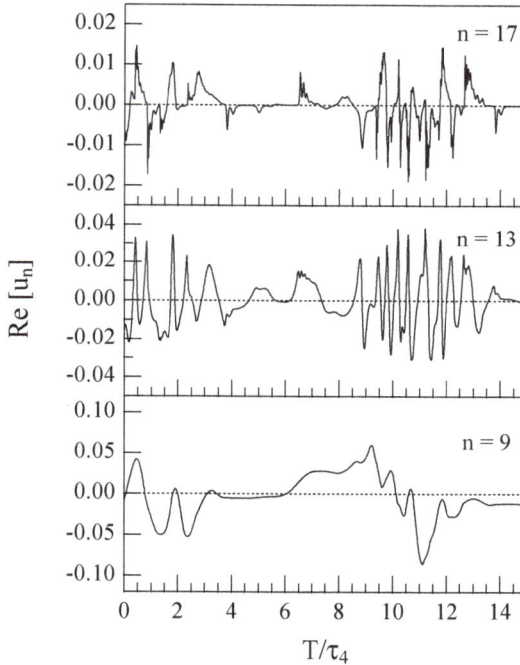

Figure 2. Time series ploTts showing the real part of the processes, $u_9(t)$, $u_{13}(t)$ and $u_{17}(t)$. The time is given in units of the eddy turnover time, τ_4, of the scale forced.

The quantities, $\Delta T_{1 \rightarrow 2}$ and $\Delta T_{1 \rightarrow 3}$, and $\Delta K_{1 \rightarrow 2}$ and $\Delta K_{1 \rightarrow 3}$, can be calculated as functions of the delay, τ. The difference $\Delta T_{i \rightarrow j}$ or $\Delta K_{i \rightarrow j}$ are calculated as prescribed in Section 2 using the time series, $e_i(t)$ and $e_j(t)$, in the place of $y(t)$ and $x(t)$, respectively. The calculations of the quantities, $T_{1 \rightarrow 2}$, $T_{2 \rightarrow 1}$, $T_{1 \rightarrow 3}$, $T_{3 \rightarrow 1}$ and the corresponding quantities normalized, *i.e.*, $K_{1 \rightarrow 2}$, $K_{2 \rightarrow 1}$, $K_{1 \rightarrow 3}$ and $K_{3 \rightarrow 1}$, give the results portrayed in Figure 3, where all these quantities are reported synoptically as a function of τ in units of the eddy turnover time, $\tau_{\#1}$, that pertains to the #1 mode (with $n = 9$).

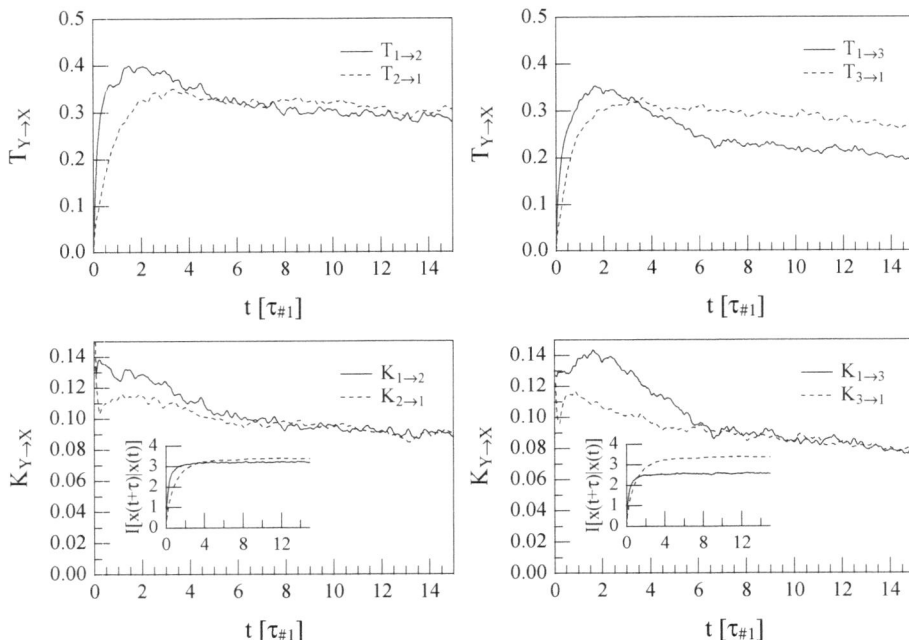

Figure 3. The quantities, $T_{1\to2}$, $T_{2\to1}$, $T_{1\to3}$, $T_{3\to1}$, $K_{1\to2}$, $K_{2\to1}$, $K_{1\to3}$ and $K_{3\to1}$ (see Section 2) calculated for the three modes of the GOY model chosen. In the case of the quantities $K_{a\to b}$, the inset shows the normalization factor. All the quantities are expressed as functions of the delay in units of the eddy turnover time $\tau_{\#1} = \tau_9$. Note that transfer entropy is always positive, indicating one always learns something from observing another mode; transfer entropy decreases as τ increases, since the distant past is not very helpful for predicting the immediate future (about the positiveness of these quantities, it should be mentioned that this has been tested against surrogate data series, as described in Figure 4 below).

The use of non-adjacent shells to calculate the transfer of information is a choice: the interaction between nearby shells is obvious from Equation (14), while checking the existence of an information transfer cascade down from the large to the small scales requires checking it non-locally in the k-space.

All the plots show clearly that there is a direct cascade for short delays. The first noticeable difference between the transfer entropies and the normalized transfer entropies is that in the #1 ↔ #3 coupling, a non-understandable inverse regime appears after about $4\tau_{\#1}$, when the "traditional" transfer entropy is used. Instead, the use of the normalized quantities suggests decoupling after long times (after about $6\tau_{\#1}$). A comparison between the #1 ↔ #2 and #1 ↔ #3 interactions is also interesting: the maximum of the "direct cascade" coupling is reached at less than $0.5\tau_{\#1}$ for both the interactions if the TEs are used. However in the plot of $K_{1\to2}$, $K_{2\to1}$, $K_{1\to3}$ and $K_{3\to1}$, some time differences appear; this is clarified when difference quantities are plotted, as in Figure 4.

Entropy **2014**, *16*, 1272–1286

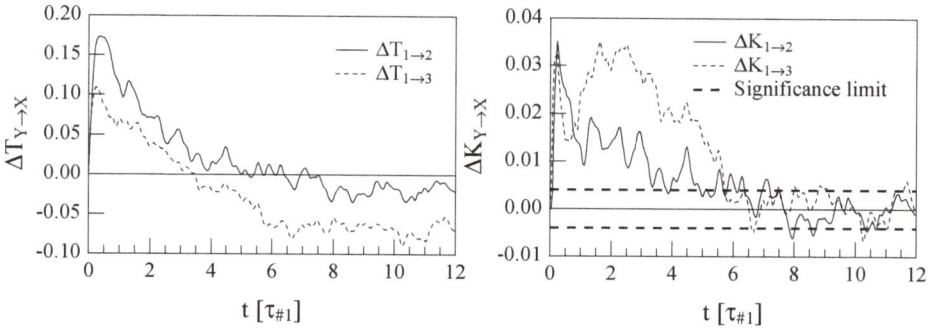

Figure 4. Comparison between TEA via the "traditional" transfer entropies (TEs) **(left)** and the new renormalized quantities **(right)**, in the case of the modes selected in the GOY model. The significance limit appearing in the plot on the right was obtained by a surrogate data test, through $N = 10^4$ surrogate data realizations, as described in the text.

Both the analyses diagnose a prevalence of the smaller-onto-larger wavenumber drive for sufficiently small delays: the $\Delta T_{1\to 2}$ indicates a driving of u_9 onto u_{13} (Mode #1 onto Mode #2) for $\tau \lesssim 5\tau_{\#1}$, while $\Delta T_{1\to 3}$ indicates a driving of u_9 onto u_{17} (Mode #1 onto Mode #3) for $\tau \lesssim 3.5\tau_{\#1}$. This is expected, due to how the system (14) is constructed. What is less understandable is the fact that for large values of τ, the quantities $\Delta T_{1\to 2}(\tau)$ and, even more, $\Delta T_{1\to 3}(\tau)$ become negative. This would suggest that after a long time a (weaker) "inverse cascade" prevails; however, this is not contained in the system (14) in any way and, hence, is either evidence of chaos-driven unpredictability or is an erroneous interpretation. For this reason, it is instructive to examine the plots of $\Delta_{1\to 2}K(\tau)$ and $\Delta_{1\to 3}K(\tau)$: after roughly $6.5\tau_{\#1}$, the modes appear to become decoupled, since $\Delta_{1\to 2}K \simeq 0$ and $\Delta_{1\to 3}K \simeq 0$, albeit with significant noise. The lack of evidence of an inverse cascade in these plots suggests that the interpretation of an inverse cascade as descending from the old TEA was wrong.

The misleading response of the ΔT analysis may be well explained looking at the insets in the lower plots of Figure 3, where the quantities reported as functions of τ are the normalization coefficients, $I_t(X_\tau|X)$, indicating the levels of inherent unpredictability of the two $e_n(t)$ compared. In the case of the $1 \to 2$ comparison, the levels of inherent unpredictability of the two time series, $e_9(t)$ and $e_{13}(t)$, become rather similar for large τ, while the asymptotic levels of inherent unpredictability are very different for $e_9(t)$ and $e_{17}(t)$ (indeed, one should expect that the more different is m from n, the more different level of inherent unpredictability will be shown by e_m and e_n). This means that $\Delta T_{1\to 2}(\tau)$ and $\Delta_{1\to 2}K(\tau)$ are expected to give a rather similar diagnosis, while the calculation of $\Delta_{1\to 3}K(\tau)$ will probably fix any misleading indication of $\Delta T_{1\to 3}(\tau)$.

Another observation that deserves to be made is about the maxima of $\Delta T_{1\to 2}(\tau)$, $\Delta T_{1\to 3}(\tau)$, $\Delta_{1\to 2}K(\tau)$ and $\Delta_{1\to 3}K(\tau)$ with respect to τ: this should detect the characteristic interaction time for the interaction, (e_9, e_{13}), and for the interaction, (e_9, e_{17}). In the plots of $\Delta T_{1\to 2}(\tau)$ and $\Delta T_{1\to 3}(\tau)$, one observes a maximum for $\Delta T_{1\to 2}$ at $\tau \simeq 0.6\tau_{\#1}$ and a maximum for $\Delta T_{1\to 3}$ just slightly before this. It appears that the characteristic time of interaction of e_9 with e_{13} is slightly larger than of e_9 with e_{17}: this is a little bit contradictory, because of the k-local hypothesis after which the energy is transferred from e_9 to e_{13} and then to e_{17}.

What happens in the plots of $\Delta_{1\to 2}K(\tau)$ and $\Delta_{1\to 3}K(\tau)$ is different and more consistent with what we know about the GOY model. First of all, the maximum for $\Delta_{1\to 3}K$ is not unique: there is a sharp maximum at about $0.5\tau_{\#1}$, which comes a little bit before the maximum of $\Delta_{1\to 2}K$, exactly as in the case of the ΔTs. However, now, the big maxima of $\Delta_{1\to 3}K$ are occurring between 1.5 and 3 times $\tau_{\#1}$, so that maybe different processes in the interaction, (e_9, e_{17}), are emerging. Actually, distant wavenumbers may interact through several channels, and more indirect channels enter the play as the wavenumbers become more distant. This might explain the existence *de facto* of two different characteristic times for the interaction, (e_9, e_{17}), as indicated by the plot of $\Delta K_{1\to3}(\tau)$: a very early one at about $0.5\tau_{\#1}$; and a later one within the interval $(1.5\tau_{\#1}, 3.5\tau_{\#1})$.

The differences between the TEAs performed with ΔT and ΔK appear to be better understandable considering that through the normalization in ΔK, one takes into account the self information entropy of each process (*i.e.*, the degree of unpredictability of each process). The motivation is essentially identical to the preference of relative error over absolute error in many applications.

As far as the surrogate data test used to produce the level of confidence in Figure 4, we use the method described in [33]. In this case, a set of more than 10^4 surrogate data copies have been realized, by randomizing the Fourier phases. In each of these surrogate datasets, the delayed transfer entropy was calculated. Than, a statistical analysis, with a confidence threshold of five percent, was performed. A similar level of confidence was obtained also for the results in Figure 3, but it is not reported explicitly on the plot for clarity.

4. Conclusions

Mutual information and transfer entropy are increasingly used to discern whether relationships exist between variables describing interacting processes, and if so, what is the dominant direction of dynamical influence in those relationships? In this paper, these IDQs are normalized in order to account for potential differences in the intrinsic stochasticity of the coupled processes.

A process, Y, is considered to influence a process, X, if their interaction reduces the unpredictability of X: this is why one chooses to normalize the IDQs with respect to the Shannon entropy of X, taken as a measure of its unpredictability.

The normalized transfer entropy is particularly promising, as has been illustrated for a synthetic model of fluid turbulence, namely the GOY model. The results obtained here about the transfer entropy and its normalized version for the interactions between Fourier modes of this model point towards the following conclusions.

The fundamental characteristics of the GOY model non-linear interactions, expected by construction, are essentially re-discovered via the TEA of its Fourier components, both using the unnormalized IDQs and the normalized ones: the prevalence of the large-to-small scale cascade; the locality of the interactions in the k-space; the asymptotic decoupling after a suitably long delay. The determination of the correct verse of dynamical enslaving is better visible using $K_{Y\to X}$ and $\Delta K_{Y\to X}$, in which the intrinsic randomness of the two processes is taken into account (in particular, the inspection of $\Delta T_{Y\to X}$ indicated the appearance of an unreasonable inverse cascade for large τ, which was ruled out by looking at $\Delta K_{Y\to X}$).

An indication is then obtained that for the irregular non-linear dynamics at hand, the use of the TEA via $\Delta K_{Y\to X}(\tau; t)$ is promising in order to single out relationships of cross-predictability (transfer of information) between processes.

The systematic application of the TEA via $\Delta K_{Y\to X}$ to models and natural systems is going to be done in the authors' forthcoming works.

Conflicts of Interest: The authors declare no conflict of interest.

References

1. Shannon, C.E. A mathematical theory of communication. *Bell Syst. Tech. J.* **1948**, *27*, 379–423, 623–656.
2. A Good Introduction on SDSs May Be Found in the Webiste. Available online: http://www.scholarpedia. org/article/Stochastic_dynamical_systems (accessed on 16 July 2013).
3. Klimontovich, Y.L. *Statistical Theory of Open Systems*; Kluwer Academic Publishers: Dordrecht, The Netherlands, 1995; Volume 1.
4. Elskens, Y.; Prigogine, I. From instability to irreversibility. *Proc. Natl. Acad. Sci. USA* **1986**, *83*, 5756–5760.
5. Kullback, S. *Information Theory and Statistics*; Wiley: New York, NY, USA, 1959.

Entropy **2014**, *16*, 1272–1286

6. Kantz, H.; Schreiber, T. *Nonlinear Time Series Analysis*; Cambridge University Press: Cambridge, MA, USA, 1997.
7. Schreiber, T. Measuring information transfer. *Phys. Rev. Lett.* **2000**, *85*, 461–464.
8. Hlaváčková-Schindler, K.; Paluš, M.; Vejmelkab, M.; Bhattacharya, J. Causality detection based on information-theoretic approaches in time series analysis. *Phys. Rep.* **2007**, *441*, 1–46.
9. Materassi, M.; Wernik, A.; Yordanova, E. Determining the verse of magnetic turbulent cascades in the Earth's magnetospheric cusp via transfer entropy analysis: Preliminary results. *Nonlinear Processes Geophys.* **2007**, *14*, 153–161.
10. De Michelis, P.; Consolini, G.; Materassi, M.; Tozzi, R. An information theory approach to storm-substorm relationship. *J. Geophys. Res.* **2011**, doi:10.1029/2011JA016535.
11. Bauer, M.; Thornhill, N.F.; Meaburn, A. Specifying the directionality of fault propagation paths using transfer entropy. In Proceedings of the 7th International Symposium on Dynamics and Control of Process Systems, Cambridge, MA, USA, 5–7 July 2004.
12. Kaiser, A.; Schreiber, T. Information transfer in continuous processes. *Phys. D* **2002**, *166*, 43–62.
13. Kwon, O.; Yang, J.-S. Information flow between stock indices. *Eur. Phys. Lett.* **2008**, doi:10.1209/0295-5075/82/68003.
14. Kugiumtzis, D. Improvement of Symbolic Transfer Entropy. In Proceedings of the 3rd International Conference on Complex Systems and Applications, Normandy, France, 29 June – 2 July 2009; pp. 338–342.
15. Tung, T.Q.; Ryu, T.; Lee, K.H.; Lee, D. Inferring Gene Regulatory Networks from Microarray Time Series Data Using Transfer Entropy. In Proceedings of the Twentieth IEEE International Symposium on Computer-Based Medical Systems (CBMS'07), Maribor, Slovenia, 20–22 June 2007.
16. Kleeman, R. Measuring dynamical prediction utility using relative entropy. *J. Atmos. Sci.* **2002**, *59*, 2057–2072.
17. Kleeman, R. Information flow in ensemble weather oredictions. *J. Atmos. Sci.* **2007**, *64*, 1005–1016.
18. Vicente, R.; Wibral, M.; Lindner, M.; Pipa, G. Transfer entropy—A model-free measure of effective connectivity for the neurosciences. *J. Comput. Neurosci.* **2011**, *30*, 45–67.
19. Lizier, J.T.; Heinzle, J.; Horstmann, A.; Haynes, J.-D.; Prokopenko, M. Multivariate information-theoretic measures reveal directed information structure and task relevant changes in fMRI connectivity. *J. Comput. Neurosci.* **2011**, *30*, 85–107.
20. Jaynes, E.T. Information theory and statistical mechanics. *Phys. Rev.* **1957**, *106*, 620–630.
21. Lizier, J.T. The Local Information Dynamics of Distributed Computation in Complex Systems. Ph.D. Thesis, The University of Sydney, Darlington, NSW, Australia, 11 October 2010.
22. Papana, A.; Kugiumtzis, D. Reducing the bias of causality measures. *Phys. Rev. E* **2011**, *83*, 036207.
23. Faes, L.; Nollo, G.; Porta, A. Information-based detection of nonlinear Granger causality in multivariate processes via a nonuniform embedding technique. *Phys. Rev. E* **2011**, *83*, 051112.
24. Materassi, M.; Ciraolo, L.; Consolini, G.; Smith, N. Predictive Space Weather: An information theory approach. *Adv. Space Res.* **2011**, *47*, 877–885.
25. Wibral, M.; Pampu, N.; Priesmann, V.; Siebenhühner, F.; Seiwart, H.; Lindner, M.; Lizier, J.T.; Vincente, R. Measuring information-transfer delays. *PLoS One* **2013**, *8*, e55809.
26. Phythian, R. The functional formalism of classical statistical dynamics. *J. Phys. A* **1977**, *10*, 777–789.
27. Bar-Yam, Y. A mathematical theory of strong emergence using mutiscale variety. *Complexity* **2009**, *9*, 15–24.
28. Ryan, A.J. Emergence is coupled to scope, not level. *Complexity* **2007**, *13*, 67–77.
29. Frisch, U. *Turbulence, the Legacy of A. N. Kolmogorov*; Cambridge University Press: Cambridge, UK, 1995.
30. Ditlevsen, P.D. *Turbulence and Shell Models*; Cambridge University Press: Cambridge, UK, 2011.
31. Knuth, K.H. Optimal data-based binning for histograms. 2006, arXiv:physics/0605197 v1.
32. Pisarenko, D.; Biferale, L.; Courvoisier, D.; Frisch, U.; Vergassola, M. Further results on multifractality in shell models. *Phys. Fluids A* **1993**, *5*, 2533–2538.
33. Theiler, J.; Eubank, S.; Longtin, A.; Galdrikian, B.; Doyne Farmer, J. Testing for nonlinearity in time series: The method of surrogate data. *Phys. D* **1992**, *58*, 77–94.

entropy

MDPI

Article

Inferring a Drive-Response Network from Time Series of Topological Measures in Complex Networks with Transfer Entropy

Xinbo Ai

Automation School, Beijing University of Posts and Telecommunications, Beijing 100876, China; axb@bupt.edu.cn; Tel.: +86-10-62283022; Fax: +86-10-62283022.

External Editor: Deniz Gencaga

Received: 19 August 2014; in revised form: 5 October 2014; Accepted: 28 October 2014; Published: 3 November 2014

Abstract: Topological measures are crucial to describe, classify and understand complex networks. Lots of measures are proposed to characterize specific features of specific networks, but the relationships among these measures remain unclear. Taking into account that pulling networks from different domains together for statistical analysis might provide incorrect conclusions, we conduct our investigation with data observed from the same network in the form of simultaneously measured time series. We synthesize a transfer entropy-based framework to quantify the relationships among topological measures, and then to provide a holistic scenario of these measures by inferring a drive-response network. Techniques from Symbolic Transfer Entropy, Effective Transfer Entropy, and Partial Transfer Entropy are synthesized to deal with challenges such as time series being non-stationary, finite sample effects and indirect effects. We resort to kernel density estimation to assess significance of the results based on surrogate data. The framework is applied to study 20 measures across 2779 records in the Technology Exchange Network, and the results are consistent with some existing knowledge. With the drive-response network, we evaluate the influence of each measure by calculating its strength, and cluster them into three classes, *i.e.*, driving measures, responding measures and standalone measures, according to the network communities.

Keywords: network inference; topological measures; transfer entropy

PACS: 05.45.Tp; 05.90.+m

---

1. Introduction

1.1. Problem Statement

The last decade has witnessed a flourishing progress of network science in many interdisciplinary fields [1,3]. It is proved both theoretical and practically that topological measures are essential to complex network investigations, including representation, characterization, classification and modeling [4,8]. Over the years, scientists have constantly introduced new measures in order to characterize specific features of specific networks [9,13]. Each measure alone is of practical importance and can capture some meaningful properties of the network under study, but when so many measures are put together we will find that they are obviously not "Mutually Exclusive and Collectively Exhaustive", namely, some measures fully or partly capture the same information provided by others while there are still properties that cannot be captured by any of the existing measures. Having an overwhelming number of measures complicates attempts to determine a definite measure-set that would form the basis for analyzing any network topology [14,15]. With the increasing popularity

Entropy **2014**, *16*, 5753–5772

of network analyses, the question which topological measures offer complementary or redundant information has become more important [13]. Although it might be impossible to develop a "Mutually Exclusive and Collectively Exhaustive" version of measure-set at present, there is no doubt that efforts to reveal the relationships among these measures could give valuable guidance for a more effective selection and utilization of the measures for complex network investigations.

1.2. Related Works

The relationship of topological measures has been a research topic for several years [4,14,28], and there mainly exist two paradigms, *i.e.*, analytical modeling and data-driven modeling. For a few topological measures of model networks, *i.e.*, networks generated with certain algorithm, some analytical interrelationships are found. For example, the clustering coefficient C of generalized random graphs are functions of the first two moments of the degree distribution [16], and for the small world model, the clustering coefficient C could be related to the mean degree and rewiring probability [17]. The relationship between the average path length and its size in a star-shaped network can be derived as: $L_{star}=2-2/N$ [18], while for a Barabási-Albert scale-free network, the relationship between them is: $L\sim\ln N/\ln\ln N$ [19]. The advantage of the analytical modeling is that the resulting relationships are of rigorous mathematical proofs, but this paradigm imposes limitations in that only a small part of the relationships can be derived analytically, and it is not sure whether these conclusions still hold true for real-life networks. If enough is known about the measures and the way in which they interact, a fruitful approach is to construct mechanism models and compare such models to experimental data. If less is known, a data-driven approach is often needed where their interactions are estimated from data [20]. In other words, when the intrinsic mechanism of real-life network is not clear, the situation we will be concerned with here, the data-driven paradigm might be more suitable. Some of the relevant papers following the data-driven paradigm are reviewed as follows.

Jamakovic *et al.* [14] collected data from 20 real-life networks from technological, social, biological and linguistic systems, and calculated the correlation coefficients between 14 topological measures. It was observed that subsets of measures were highly correlated, and Principal Component Analysis (PCA) showed that only three dimensions were enough to retain most of the original variability in the data, capturing more than 99% of the total data set variance. Li *et al.* [21] investigated the linear correlation coefficients between nine widely studied topological measures in three classical complex network models, namely, Erdős-Rényi random graphs, Barabási-Albert graphs, and Watts-Strogatz small-world graphs. They drew a similar conclusion, namely that the measure correlation pattern illustrated the strong correlations and interdependences between measures, and argued that the both these three types of networks could be characterized by a small set of three or four measures instead of by the nine measures studied. Costa *et al.* [4] summarized dozens of topological measures in their review paper and conducted correlations analysis between some of the most traditional measures for Barabási-Albert (BA) network, Erdős-Rényi (ER), and Geographical Networks (GN). They found that particularly high absolute values of correlations had been obtained for the BA model, with low absolute values observed for the ER and GN cases. Further, they found that the correlations obtained for specific network models not necessarily agreed with that obtained when the three models were considered together. Roy *et al.* [22] studied 11 measures across 32 data sets in biological networks, and created a heat map based on paired measures correlations. They concluded that the correlations were not very strong overall. Filkov *et al.* [23] also used a heat map and multiple measure correlations to compare networks of various topologies. They correlated 15 measures across 113 real data sets which represented systems from social, technical, and biological domains. They also found that the 15 measures were not coupled strongly. Garcia-Robledo *et al.* presented an experimental study on the correlation between several topological measures of the Internet. By drawing bar plot of the average correlation for each measure, they recognized the average neighbor connectivity as the most correlated measure; with the correlation heat map, they concluded that distance measures were highly correlated [24]. Bounova and de Weck proposed an overview of network topology measures and a computational approach to analyze graph topology via multiple-metric analysis on graph ensembles, and they found that density-related measures and graph distance-based measures were orthogonal to each other [25].

More recently, Li *et al.* explored the linear correlation between the centrality measures using numerical simulations in both Erdős-Rényi networks and scale-free networks as well as in real-world networks. Their results indicated that strong linear correlations did exist between centrality measures in both ER and SF networks, and that network size had little influence on the correlations [26]. Sun and Wandelt performed regression analysis in order to detect the functional dependencies among the network measures, and used the coefficient of determination to explain how well the measures depended on each other. They built a graph for the network measures: each measure was a node and a link existed if there was a functional dependency between two measures. By setting a threshold, they got a functional network of the measures with six connected components [27]. Lin and Ban focused on the evolution of the US airline system from a complex network perspective. By plotting scatter diagrams and calculating linear correlations, they found that there was a high correlation between "strength" and "degree", while "betweenness" did not always keep consistent with "degree" [28].

The abovementioned researches all follow the data-driven paradigm, and provide convincing arguments in favor of using the statistical approach to correlate the measures. The correlations between topological measures strongly depend on the graph under study [14], and results from these studies differ greatly. Some of them argue that most of the measures are strongly correlated and thus can be redundant, while others argue that these correlations are not strong overall. Even the resulting correlation patterns of the same measures in different networks are not consistent. Just as Bounova and de Weck pointed out that pulling networks from different domains together for statistical analysis might provide incorrect conclusions, because there often exists considerable diversity among graphs that share any given topology measures, patterns vary depending on the underlying graph construction model, and many real data sets are not actual statistical ensembles [25].

1.3. Primary Contribution of This Work

To address the issue mentioned above, we resort to two research strategies:

1. On the one hand, our investigation will be based on data observed from the same network, instead of data pieced together from different networks in several fields. More specially, we will record the trajectories of the measures of the same system, and try to infer their relationships from simultaneously measured time series of these measures.
2. On the other hand, our investigation will adopt another data-driven method, *i.e.*, transfer entropy. Since our data is in the form of time series, transfer entropy, instead of correlation coefficients or other model-based methods, might be a better choice for our purpose. There are at least two reasons. For one thing, the correlation measure is designed for static data analysis and when applying to time series data, all dynamical properties of the series are discarded [29]. For another, correlations, linear or nonlinear, only indicate the extent to which two variables behave similarly. They cannot establish relationships of influence, since the interactions between the measures are not necessarily symmetric. Neither can they indicate if two measures are similar not because they interact with each other, but because they are both driven by a third [30]. We need a novel tool not only to detect synchronized states, but also to identify drive-response relationships. These issues can be addressed by measures of information transfer. One such measure is Schreiber's Transfer Entropy [31], which is with minimum of the assumption of the dynamic of the system and the nature of their coupling.

This paper will follow the data-driven paradigm and employ transfer entropy as a quantitative description of interactions among topological measures. Transfer entropy has been proposed to distinguish effectively driving and responding elements and to detect asymmetry in the interaction of subsystems. It is it widely applicable because it is model-free and sensitive to nonlinear signal properties [32]. Thus the transfer entropy is able to measure the influences that one measure can exert over another. On the basis of these pair-wise relationships, we will construct a so-called drive-response network with the measures as its nodes and the pair-wise relationships as its edges. The resulting network will enable us to gain a deeper insight into patterns and implications of relationships among network topological measures. In this paper, we mainly consider the following fundamental questions:

1.   Whether or not there exist drive-response relationships between topological measures? For example, will the network diameter influence the average path length? If that is the case, how to measure the strength of this relationship?
2.   What does the overall picture look like when measures are put together? What is the structure of the measure-set? Can the measures be grouped into different communities?
3.   Are all the measures equally important? If not so, how to identify the pivotal ones?

In order to conduct our investigation, high-quality data is necessary. It is usually difficult to obtain the evolutional record of complex network [33]. Thanks to the advanced information systems in Beijing Technology Market Management Office (or BTMMO, for short), we are able to collect a complete data set which describe the evolution of the Technology Exchange Network day by day. Our proposed method will take the Technology Exchange Network as an empirical application, which allowing us to study several measures across as many as 2779 datasets.

The remainder of this paper is organized as follows: the next section will synthesize a transfer entropy-based framework to infer and analyze the drive-response network. The emphasis is on how to quantify the relationships among time series which are continuous, non-stationary, and of finite sample effect and indirect effect. Section 3 will apply the proposed framework to an empirical investigation on the Technology Exchange Network. Some concluding remarks are made in Section 4.

2. Methodology

2.1. Main Principle

The proposed method is to mine the overall pattern of the relationships among network topological measures from their time series, which is shown in Figure 1.

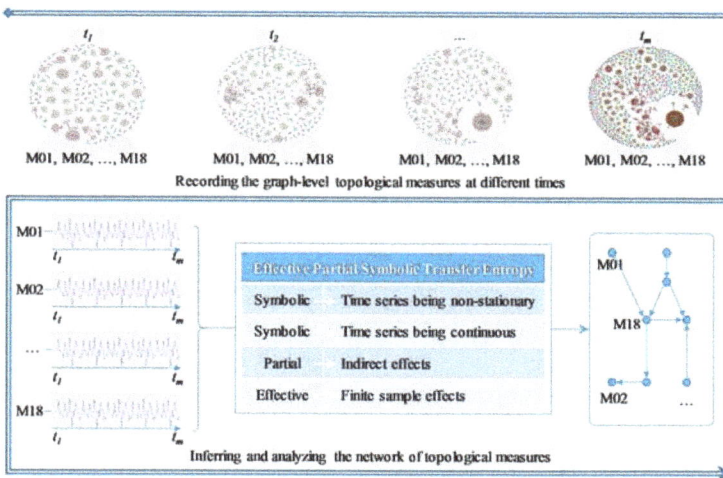

Figure 1. Entropy-based framework to infer drive-response network of topological measures from their time series.

The relationships here refer in particular to drive-response relationships. A convenient way to represent drive-response relations between two variables is to connect these two with a directed edge, and correspondingly the overall relation pattern can be illustrated in the form of a network. The tool for network inference is the transfer entropy, which is proposed by several researchers for revealing networks from dynamics [34,35].

It is worth noting that what is to be constructed here is the network of measures, which should not be confused with the original network of units. As shown in the upper half of Figure 1, we will trace the original network at successive time points, acquiring time series for each topological measure. And we will quantify the relationships between measures based on these time series with transfer entropy, and then construct a drive-response network which stands for the relation pattern among these topological measures, as shown in the lower half of Figure 1.

The process of inferring the drive-response network is as follows: The network can be presented by a set V of nodes and a set E of edges, connected together as a directed graph denoted $G = (V, E)$. The nodes here are the measures, and the edges are the drive-response couplings between any two measures. In our study, the couplings are detected by transfer entropy. Namely, connectivity is based on the estimation of the influence one measure v exerts on another measure u. If there exists significant coupling, there will be a directed edge from v to u. The resulting network is also a weighted one, with the transfer entropy value as the weight of each edge.

Once the drive-response network is constructed, we may gain in-depth understanding of the relationships among the measures by analyzing the network. For example, we can calculate the prestige of each node to reveal which measures are more influential, and we can cluster the measures into different groups by detecting the communities in the network.

The main steps of the proposed method are as follows:

Step 1: Time Series Observation on Topological Measures. Record the graph-level topological trajectories in form of simultaneously measured time series

Step 2: Drive-response Network Inference. Calculate the transfer entropy between each pair of measures based on their time series, assess the statistical significance and construct the drive-response network.

Step 3: Drive-response Network Analysis. Calculate the prestige of each node and detect community to gain a deep understanding.

The process will be explained step by step in more details in the following section.

2.2. Main Steps

2.2.1. Time Series Observation on Topological Measures

In our study, the data is collected from observation of the same system. We will track the topological measures of an evolving network at successive points in time spaced at uniform time intervals, resulting in sequences of observations on topological measures which are ordered in time. The topological measures to be investigated in our study are discussed as follows.

Topological measures can be divided in two groups, *i.e.*, measures at global network level and measures at local node level [27,36], corresponding to the measurable element. Local topological measures characterize individual network components while global measures describe the whole network [37]. Since the observed object in our study is the network as a whole, only those graph-level measures will be selected. In other words, node-level such as the degree of a certain node will not be taken into account.

Due to the fact that the number of proposed measures is overwhelming and new measures are introduced every day, there is no consensus on a definitive set of measures that provides a "complete" characterization of real-world complex networks [24] and no classifications of these measures are universally accepted. On the basis of several important and influential works such as [4,9,25], we will classify these measures into four categories, *i.e.*, Category I, Distance Relevant Measures; Category II, Centralization Measures; Category III, Connection Measures; Category IV, Entropy and other Complexity Measures. We will select a few measures from each of the four categories and conduct our investigation on these selected measures.

Measures selected from Category I: M01, Average Path Length; M02, Diameter; M03, Eccentricity; M04, Integration; M05, Variation. All of them are based on distance. For example, the average path length is defined as the average number of steps along the shortest paths for all possible pairs of network nodes. Diameter is the greatest distance between any pair of vertices. The eccentricity in the local node level is defined as the greatest distance between v and any other vertex, and eccentricity in the global network level is the sum of all the vertices eccentricities. Graph integration is based on vertex centrality while variation is based on vertex distance deviation.

Measures selected from Category II: M06, Centralization; M07, Degree Centralization; M08, Closeness Centralization; M09, Betweenness Centralization; M10, Eigenvector Centralization. In local node level, centrality is to quantify the importance of a node. Historically first and conceptually simplest is degree centrality, which is defined as the number of links incident upon a node. The closeness of a node is defined as the inverse of the farness, which is the sum of its distances to all other nodes. Betweenness centrality quantifies the number of times a node acts as a bridge along the shortest path between two other nodes. The corresponding concept of centrality at the global network level is centralization. In our study, we will employ Freeman's formula [38] to calculate graph-level centralization scores based on node-level centrality.

Measures selected from Category III: M11, Vertex Connectivity; M12, Edge Connectivity; M13, Connectedness; M14, Global Clustering Coefficient; M15, Assortativity Coefficient. These measures refer to connection. For example, the vertex connectivity is defined as the minimum number of nodes whose deletion from a network disconnects it. Similarly, the edge connectivity is defined as the minimum number of edges whose deletion from a network disconnects it. Connectedness is defined as the ratio of the number of edges and the number of possible edges. It measures how close the network is to complete. Global Clustering coefficient is to quantify the overall probability for the network to have adjacent nodes interconnected. It is also called second order extended connectivity, which can be calculated by counting the edges between the second neighbors of vertex, and again comparing that count to the number of edges in the complete graph that could be formed by all second neighbors. Assortativity is defined as the Pearson correlation coefficient of the degrees at both ends of the edges. This measure reveals whether highly connected nodes tend to be connected with other high degree nodes.

Measures selected from Category IV: M16, Radial Centric Information Index; M17, Compactness Measure Based on Distance Degrees; M18, Complexity Index B. All these measures refer to entropy, information and other complexity in the network. Radial Centric Information Index and Compactness Measure Based on Distance Degrees are both information-theoretic measures to determine the structural information content of a network, and both of them are based on Shannon's entropy. Complexity Index B is a more recently developed measure due to Bonchev [39]. The complexity index b_v is the ratio of the vertex degree and its distance degree. The sum over all b_v indices is the convenient measure of network complexity, *i.e.*, the complexity index B.

A systematic discussion about the topological measures in complex networks is out of the scope of this paper. Detailed description of these measures can be found in [4,18,40,41].

Entropy **2014**, *16*, 5753–5772

Table 1. Commonly used topological measures for undirected and un-weighted graph $G=(N(G),E(G))$. $N(G)$ and $E(G)$ are called vertex and edge set respectively.

ID	Name	Definition	Ref.
I. Distance Relevant Measures			
M01	Average Path Length	$L := \frac{2}{N(N-1)} \sum\limits_{u,v \in N(G)} d(u,v)$, here N denotes the number of the nodes, and $d(u,v)$ denotes the steps along the shortest path between nodes u and v	[18]
M02	Diameter	$D := \max\limits_{u,v} d(u,v)$	[18]
M03	Eccentricity	$e(G) := \sum\limits_{v \in N(G)} \max_{u \in N(G)} d(u,v)$, here, $d(u,v)$ stands for the distances between $u,v \in N(G)$	[40]
M04	Integration	$D(G) := \frac{1}{2} \sum\limits_{v \in N(G)} D(v)$, here, $D(v)$ is the vertex centrality which is defined as $D(v) := \sum\limits_{v \in N(G)} d(v,u)$	[40]
M05	Variation	$\mathrm{var}(G) := \max_{u \in N(G)} \Delta D^*(v)$, here, $\Delta D^*(v)$ is the distance vertex deviation $\Delta D^*(v) := D(v) - D^*(G)$	[40]
II. Centralization Measures			
M06	Centralization	$\Delta G^* := \sum\limits_{v \in N(G)} \Delta D^*(v)$, here $\Delta D^*(v)$ is the distance vertex deviation which is defined as $\Delta D^*(v) := D(v) - D^*(G)$	[40]
M07	Degree Centralization *	$C_D(G) := \sum\limits_{v \in N(G)} \left(\max_{u \in N(G)} C_D(u) - C_D(v) \right)$, here $C_D(v)$ is the degree centrality of vertex v, $C_D(v) := \frac{k_v}{N-1}$, and k_v is the vertex degree	[41]
M08	Closeness Centralization *	$C_C(G) := \sum\limits_{v \in N(G)} \left(\max_{u \in N(G)} C_C(u) - C_C(v) \right)$, here $C_C(v)$ is the closeness centrality of vertex v, $C_C(v) := \left[\frac{1}{N-1} \sum\limits_{w \in N(G)} d(w,v) \right]^{-1}$	[41]
M09	Betweenness Centralization *	$C_B(G) := \sum\limits_{v \in N(G)} \left(\max_{u \in N(G)} C_B(u) - C_B(v) \right)$, here $C_B(v)$ is the closeness centrality of vertex v, which is defined as $C_B(v) := \frac{2}{(n-1)(n-2)} \sum\limits_{u,w \in N(G)} \frac{g_{u,w}(v)}{g_{u,w}}$, and $g_{u,w}$ is the number of paths connecting u and w, $g_{u,w}(v)$ is the number of paths that v is on	[41]
M10	Eigenvector Centralization*	$C_E(G) := \sum\limits_{v \in N(G)} \left(\max_{u \in N(G)} C_E(u) - C_E(v) \right)$, here $C_E(v)$ is the centrality of vertex v, which can be calculated by the formula $C_E(v) = \frac{1}{\lambda} \sum\limits_{u \in N(G)} a_{v,u} C_E(u)$, and the vector of the centralities of vertices is the eigenvector of adjacency matrix $A=(a_{ij})$	[41]
III. Connection Measures			
M11	Vertex Connectivity	$\kappa(G) := \min \{\kappa(u,v)\mid \text{unordered pair } u, v \in N(G)\}$, here, $\kappa(u,v)$ is defined as the least number of vertices, chosen from $N(G) - \{u,v\}$, whose deletion from G would destroy every path between u and v	[41]
M12	Edge Connectivity	$\lambda(G) := \min\{\lambda(u,v)\mid \text{unordered pair } u,v \in N(G)\}$, here, $\lambda(u,v)$ is the least number of edges whose deletion from G would destroy every path between u and v	[41]
M13	Connectedness	$E_N(G) := \frac{A(G)}{N^2}$, here $A(G)$ is the index of total adjacency $A(G) := \frac{1}{2} \sum\limits_{i=1}^{N} \sum\limits_{j=1}^{N} a_{ij}$ and a_{ij} is the entry lies in row i and column j in the adjacent matrix $A=(a_{ij})$	[40]
M14	Global Clustering Coefficient	$C(G) := \frac{1}{N} \sum\limits_{v \in N(G)} \frac{2E_{N(v)}}{V_{N(v)} * (V_{N(v)} - 1)}$, where, $(V_{N(v)}, E_{N(v)})$ is the sub-graph of G that contains all neighborhood vertices and their edges	[18]
M15	Assortativity Coefficient	$r = \frac{\frac{1}{M} \sum_{j>i} k_i k_j a_{ij} - \left[\frac{1}{M} \sum_{j>i} \frac{1}{2}(k_i + k_j) a_{ij} \right]^2}{\frac{1}{M} \sum_{j>i} \frac{1}{2}\left(k_i^2 + k_j^2\right) a_{ij} - \left[\frac{1}{M} \sum_{j>i} \frac{1}{2}(k_i + k_j) a_{ij} \right]^2}$, here, M is the total number of edges, and a_{ij} is the entry lies in row i and column j in the adjacent matrix $A=(a_{ij})$	[4]

Table 1. *Cont.*

ID	Name	Definition	Ref.
IV. Entropy and Other Complexity Measures			
M16	Radial Centric Information Index	$I_{C,R}(G) := \sum_{i=1}^{k} \frac{|N_i^e|}{N} \log\left(\frac{|N_i^e|}{N}\right)$, here, $|N_i^e|$ is the number of vertices having the same eccentricity	[40]
M17	Compactness Measure Based on Distance Degrees	$I_{C,\delta_D}(G) := 2W \log(2W) - \sum_k q_k \log q_k$ here, W is the Wiener Index $W = \frac{1}{2} \sum_{u,v \in N(G)} d(u,v)$ and q_k is the sum of the distance degrees of all vertices located at a topological distance of k from the center of the graph	[40]
M18	Complexity Index B	$B(G) := \sum_{v \in N(G)} b_v = \sum_{v \in N(G)} \frac{k_v}{\mu(v)}$. Here, b_i is the ratio of the vertex degree k_v and its distance degree $\mu(v) := \sum_{u \in N(G)} d(v,u)$	[40]

* In our study some of the centralization measures are normalized by dividing by the maximum theoretical score for a graph with the same number of vertices. For degree, closeness and betweenness the most centralized structure is an undirected star. For eigenvector centrality the most centralized structure is the graph with a single edge.

Besides the 18 topological measures mentioned above, we also track two performance measures, *i.e.* "P01: Technological Volume" and "P02: Contract Turnover", which will be described in Section 2.3.2.

2.2.2. Drive-response Network Inference

The drive-response network to be constructed can be denoted as $G=(V,E)$, here, $V=\{V_1,V_2,\ldots,V_n\}$ is the set of vertices/nodes, *i.e.*, the measures, and E is the set of edges, *i.e.*, pair-wise relations between any two measures. The adjacency matrix A of the drive-response network is defined as follows:

$$a_{ij} = \begin{cases} EPSTE^*_{v_i \rightarrow v_j}, & (v_i, v_j) \in E \\ 0, & (v_i, v_j) \notin E \end{cases} \tag{1}$$

Here $EPSTE^*_{v_i \rightarrow v_j}$ is the effective partial symbolic transfer entropy from measure v_i to measure v_j that is of statistical significance. The calculation of $EPSTE^*_{v_i \rightarrow v_j}$ is the most complicated step in the proposed method, and we will depict it in details as follows.

The Transfer Entropy from a time series Y to a times series X as the average information contained in the source Y about the next state of the destination X that was not already contained in the destination's past [31,35]:

$$TE_{Y \rightarrow X} = \sum_{x_{t+1}, x_t, y_t} p(x_{t+1}, x_t, y_t) \log \frac{p(x_{t+1}|x_t, y_t)}{p(x_{t+1}|x_t)} \tag{2}$$

Here, *t+1* indicates a given point in time, *t* indicates the previous point, x_t is element *t* of the time series of variable X and y_t is element *t* of the time series of variable Y. $p(A,B)$ and $p(A|B)$ are the joint and conditional distribution respectively, and $p(A|B)=p(A,B)/p(B)$. In order to calculate $p(x_{t+1},x_t)$ we have to count how many times a particular combination of symbols, (a,b) appears in the joint columns X_{t+1} and X_t, then divide by the total number of occurrences of all possible combinations. For example, if there are two possible combinations, (a,b) and (a',b') appear 21 and seven times, respectively, then $p(a,b)=21/(21+7)=0.75$ and $p(a',b')=7/(21+7)=0.25$. $p(x_t+1,x_t,y_t)$ can be calculated in the same way.

Though the analytic form of transfer entropy is relatively simple, but its application to investigation on time series of topological measures is not so easy. There are five major practical challenges:

(1) Time series being non-stationary: The probabilities are estimated from observations of a single instance over a long time series. It is very important therefore that the time series is statistically

stationary over the period of interest, which can be a practical problem with transfer entropy calculations [42]. In most cases the time series of topological measures are non-stationary.

(2)   Time series being continuous: It is problematic to calculate the transfer entropy on continuous-valued time series such as we have here. Kaiser and Schreiber developed a criticism and presented many caveats to the extension of the transfer entropy to continuous variables [43]. Here we will resort to another solution.

(3)   Finite sample effects: A finite time series will result in fewer examples of each combination of states from which to calculate the conditional probabilities. When used to analyze finite experimental time series data, there is a strong risk of overestimating the influence, a problem that is known from the literatures [44,45].

(4)   Indirect effects: When evaluating the influence between two time series from a multivariate data set, the case in our study, it is necessary to take the effects of the remaining variables into account, and distinguish between direct and indirect effects [46].

(5)   Statistical significance: A small value of transfer entropy suggests no relation while a large value does. Two irrelevant series can have non-zero transfer entropy due to finite sample size of the time series [47], thus it is not a good choice to simply select a threshold value to judge whether there exists drive-response relationship between two measures.

In the last few years, several improved transfer entropy algorithms have been proposed to deal with some of these challenges. For example, Symbolic Transfer Entropy [48] is a solution for (1) and (2), Effective Transfer Entropy [44] is for (3), while Partial Transfer Entropy [49] is for (4). In order to deal with these practical challenges all at once, techniques from Symbolic Transfer Entropy, Effective Transfer Entropy, and Partial Transfer Entropy should be synthesized, resulting an effective, partial, symbolic version of Transfer Entropy as follows:

Let us consider $\{v_{1,t}\}$, $\{v_{2,t}\}$, $t=1,2,\cdots k$ as the denotations for the time series of measures v_1 and v_2 respectively. The embedding parameters in order to form the reconstructed vector of the time series of v_1 are the embedding dimension m_1 and the time delay τ_1. The reconstructed vector of v_1 is defined as:

$$v_{1,t} = \left(v_{1,t}, v_{1,t-\tau_1}, \cdots, v_{1,t-(m_1-1)\tau_1} \right)'$$ (3)

where $t=1,\cdots,k'$ and $k' = k - \max\{(m_1-1)\tau_1, (m_2-1)\tau_2\}$.

The reconstructed vector for v_2 is defined accordingly, with parameters m_2 and τ_2. For each vector $v_{1,t}$, the ranks of its components assign a rank-point $\hat{v}_{1,t} = [r_{1,t}, r_{2,t}, \cdots, r_{m_1,t}]$ where $r_{j,t} \in \{1, 2, \cdots, m_1\}$ for $j=1,\cdots,m_1$, and $\hat{v}_{2,t}$ is defined accordingly.

The symbolic transfer entropy is defined as [48]:

$$STE_{v_2 \to v_1} = \sum p(\hat{v}_{1,t+1}, \hat{v}_{1,t}, \hat{v}_{2,t}) \log \frac{p(\hat{v}_{1,t+1}|\hat{v}_{1,t}, \hat{v}_{2,t})}{p(\hat{v}_{1,t+1}|\hat{v}_{1,t})}$$ (4)

Here, symbolic transfer entropy uses a convenient rank transform to find an estimate of the transfer entropy on continuous data without the need for kernel density estimation. Since slow drifts do not have a direct effect on the ranks, it still works well for non-stationary time series.

The partial symbolic transfer entropy is defined conditioning on the set of the remaining time series $z=\{v_3,v_4,\cdots,v_N\}$.

$$PSTE_{v_2 \to v_1} = \sum p(\hat{v}_{1,t+1}, \hat{v}_{1,t}, \hat{v}_{2,t}, \hat{z}_t) \log \frac{p(\hat{v}_{1,t+1}|\hat{v}_{1,t}, \hat{v}_{2,t}, \hat{z}_t)}{p(\hat{v}_{1,t+1}|\hat{v}_{1,t}, \hat{z}_t)}$$ (5)

where the rank vector $\hat{z}_t$ is defined as the concatenation of the rank vectors for each of the embedding vectors of the time series in z. The partial symbolic transfer entropy is the pure or direct information flow between them, information transmitted indirectly by the environment (the other measures) is eliminated.

Finally, we will define effective partial symbolic transfer entropy as follows:

$$EPSTE_{v_2 \to v_1} = PSTE_{v_2 \to v_1} - \frac{1}{M} \sum PSTE_{v_2 \to v_1}^{shuffled} \tag{6}$$

where M is the times to shuffle the series (we set $M=200$ in our study) of $\hat{v}_2$ and

$$PSTE_{v_2 \to v_1}^{shuffled} = \sum p\left(\hat{v}_{1,t+1}, \hat{v}_{1,t}, \hat{v}_{2,t}^{shuffled}, \hat{z}_t\right) \log \frac{p\left(\hat{v}_{1,t+1} \middle| \hat{v}_{1,t}, \hat{v}_{2,t}^{shuffled}, \hat{z}_t\right)}{p(\hat{v}_{1,t+1} | \hat{v}_{1,t}, \hat{z}_t)} \tag{7}$$

Here, the elements of $\hat{v}_2$ is randomly shuffled, which implies that all statistical dependencies between the two series have been destroyed. $PSTE_{v_2 \to v_1}^{shuffled}$ consequently converges to zero with increasing sample size and any nonzero value of $PSTE_{v_2 \to v_1}^{shuffled}$ is due to small sample effects representing the bias in the standard entropy measure.

By now, we have coped with the practical challenges (1), (2), (3) and (4) with effective partial symbolic transfer entropy. For challenge (5), *i.e.*, statistical significance, it may be evaluated by using bootstrapping strategies, surrogate data or random permutations [50,51]. Under the surrogate-based testing scheme, we will assess the significance with kernel density estimation.

By shuffling the time series $\hat{v}_2$ for M times, we now get M different $PSTE_{v_2 \to v_1}^{shuffled}$ values and we will denote them as $p_1, p_2, \ldots ,p_M$, and we denote $PSTE_{v_2 \to v_1}$ as p_0. We build with $M+1$ values a probability distribution function using a kernel approach, known as Parzen-Rosenblat method [52,53], which can be expressed as:

$$\hat{f}_h(x) = \frac{1}{(M+1)h} \sum_{i=0}^{M} K\left(\frac{p_i - x}{h}\right) \tag{8}$$

Here, $K(\bullet)$ is the kernel function and h is the bandwidth. We will employ the most widely used Gaussian kernel $K(x) = \frac{1}{\sqrt{2\pi}} e^{-\frac{1}{2}x^2}$ here, and the bandwidth will be selected using pilot estimation of derivatives [54].

The existence of a drive-response link between two measures is then determined using this probability and a pre-defined significant level. The final $EPSTE_{v_2 \to v_1}^*$, a_{21}, is defined as:

$$a_{21} = EPSTE_{v_2 \to v_1}^* = \begin{cases} EPSTE_{v_2 \to v_1}, & p \le p_{threshold} \\ 0, & p > p_{threshold} \end{cases} \tag{9}$$

Here, $p = \int_{p_0}^{\infty} \hat{f}_h(x)dx$ (one-side test is adopted here and obviously p_0 is expected to be bigger than other p_i and lies in the right side) and we set $p_{threshold}=0.01$ in our study.

Other entries a_{ij} in the adjacency matrix A can be calculated in the same way.

2.2.3. Drive-response Network Analysis

The resulting network can be analyzed from a two-fold perspective:

At the local node level, we are going to calculate the in-strength and out-strength [55] of each node to assess how influential and comprehensive it is:

$$S_{in}(i) = \sum_{j=1}^{N} a_{ji} \tag{10}$$

$$S_{out}(i) = \sum_{j=1}^{N} a_{ij} \tag{11}$$

Here, a_{ij} is the entry of the adjacency matrix of drive-response network as described in the previous section.

In the context of drive-response relationships, the specific implications of the in-strength and out-strength are: the out-strength is the sum of information flowing from the measure to others, which stands for its prestige and reveals its influence to others. The in-strength is the sum of information flowing from others to the measure. The greater the in-strength a measure has, the more comprehensive the measure is, since there are more information flows to it from other measures.

On the global network level, we are going to cluster the measures into different groups by detecting communities in the resulting network. Measures in the same community are clustered into one group. If the resulting network is complicated, tools for detecting communities in the network science can be employed.

2.3. Further Remarks

2.3.1. Choice of Coupling Measure

In fact, the choice of coupling measure between pairs of time series permits many alternatives, ranging from correlation and partial correlation to mutual information and causality measures [56], with the cross correlation [57] and Granger causality [58] being the famous ones. Some of these popular tools are non-directional, e.g. correlation or partial correlation, and mutual information measures, thus these measures cannot provide satisfactory results for our study since the interactions between the measures are not necessarily symmetric. Granger causality has acquired preeminent status in the study of interactions and is able to detect asymmetry in the interaction. However, its limitation is that the model should be appropriately matched to the underlying dynamics of the examined system, otherwise model misspecification may lead to spurious causalities [46]. Given a complex system with a priori unknown dynamics, the first choice might be Transfer Entropy [59]. Its advantages are obvious: (1) it makes minimal assumptions about the dynamics of the time series and does not suffer from model misspecification bias; (2) it can captures both linear and nonlinear effects; (3) it is numerically stable even for reasonably small sample sizes [60].

2.3.2. Validation of the Proposed Method

The proposed method is an integration of techniques from Symbolic Transfer Entropy, Effective Transfer Entropy, and Partial Transfer Entropy. All these relevant techniques have been proved theoretically and practically by numerical simulation and empirical investigations, respectively in [44,48,49], and these techniques are compatible, which means they can be synthesized to provide more comprehensive solutions. In fact, there already exist some synthesized methods such as Partial Symbolic Transfer Entropy [46], Corrected Symbolic Transfer Entropy with surrogate series to make the results more effective [61], and Effective Transfer Entropy based on symbolic encoding techniques [62]. To our best knowledge, research taking into account of all these five practical issues mentioned above and synthesizing all these techniques all at once still lacks.

Since our investigation is applied here to purely observational data, we have no way to validate the proposed framework with simulated signals or outside intervention. To valid the feasibility of the proposed method, we will resort to another strategy which is based on experiential evidence:

Suppose that the relationship between the topological measures here and some other measures are experientially approved. We then embed these extra measures into our data, deduce the drive-response relationship between these extra measures and the topological measures, and test if the results of our proposed method are consistent with the existing knowledge. Obviously the consistence will give us more confidence on the feasibility of our method.

Here, the extra measures are two performance indices of technology exchange, *i.e.*, P01: Technological Volume and P02: Contract Turnover, which have been adopted by BTMMO for several years [63]. In the context of System Theory, it is generally believed that system structure determines its function/the performance. Thus, what is expected is that these two embedded measures will be responding ones while some of the topological measures will drive them.

3. Empirical Application and Results

3.1. Time Series of Topological Measures in Technology Exchange Network

Technology exchange holds a great potential for promoting innovation and competitiveness at regional and national levels [64]. In order to expedite scientific and technological progress and innovation, it is stipulated in the Regulation of Beijing Municipality on Technology Market that the first party of the contract can get tax relief if the contract is identified as a technique-exchanging one and registered by BTMMO. Because of the preferential taxes, most of the technology exchange activities are recorded by BTMMO, in the form of technology contracts. Thus it offers the chance for us to obtain high-quality technology exchange records from BTMMO. We are able to capture the total evolutional scenario of the Technology Exchange Network.

Networks serve as a powerful tool for modeling the structure of complex systems, and there is no exception for technology exchange. Intuitively we can model technology exchange as a network with the contracting parties as the nodes and contracts as the edges which linking the two contracting parties together. However, in the complex network literature, it is often assumed that no self-connections or multiple connections exist [4]. In other words, we will model the technology exchange as a simple graph as follows: we take the contracting parties as the nodes, and if contractual relationship between any two parties exists, regardless of how many contracts they signed, there will be (only) one edge linking them together. Here, the resulting technology exchange network is treated as an undirected and un-weighted one.

We observed the 18 topological and two performance measures of the Technology Exchange Network in Beijing day by day from 24 May 2006 to 31 December 2013, obtaining 2779 records in total. This is the most fine-grained data that we can get, because technology exchange activities can only be accurate at the day level, rather than to hours or seconds as in stock exchanges. The complete dataset is provided as a supplementary.

Since the technology exchange network is not a connected one, all the measures are calculated on the giant component of these networks. The measures M01, M02, M07, M08, M09, M10, M11, M12, M13, M14, M15 are calculated with the *igraph* packages [41], while measures M03, M04, M05, M06, M16, M17, M18 are calculated by the QuACN package [40]. All these algorithms are implemented in the R language [65], and we visualize our results mainly with the *ggplot2* package [66]. The time series of all these measures are shown in Figure 2.

3.2. Drive-Response Network Inference and Analysis

Calculating the time series depicted in Section 3.1 with the method proposed in Section 2.2.2, we get the adjacent matrix, which is shown in Figure 3 (Since the EPSTE values are rather small, all the entries are multiplied by 10,000 for ease of plotting).

In Figure 3, the diagonal entries are marked with "X" because we will not study the self-correlations and these entries are omitted. Red-filled entries are not only greater than zero but also statistically significant, thus each red-filled entry stands for a drive-response relationship. The darker the color, the more significant the relationship is. It can be seen that the darkest entries are all located in row M07, M10, M15, and M16, which means the strong drive-response relationships share these four common drives. We will further analyze these rows after constructing the drive-response network.

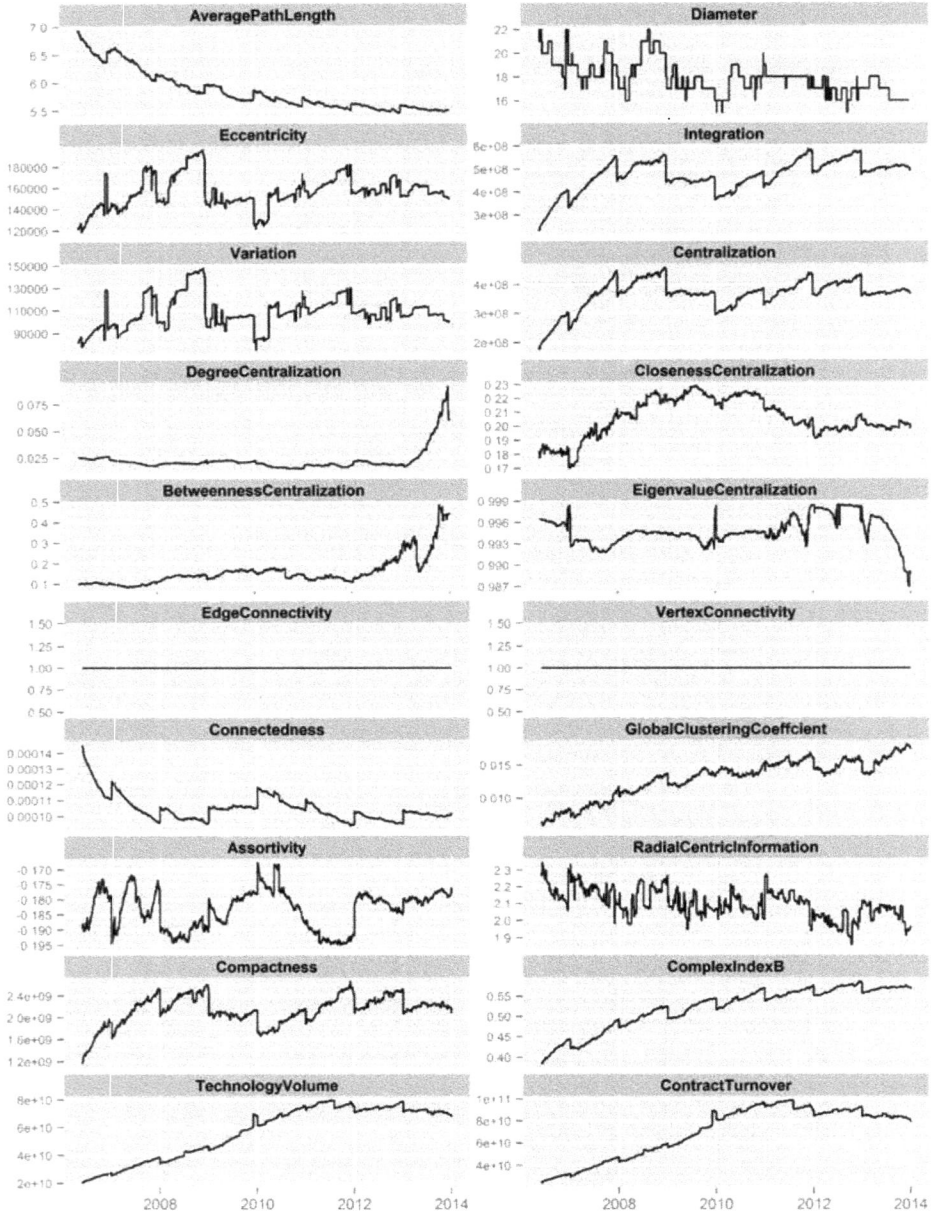

Figure 2. Time series of topological and performance measures of Technology Exchange Network in Beijing ranging from 24 May 2006 to 31 December 2013.

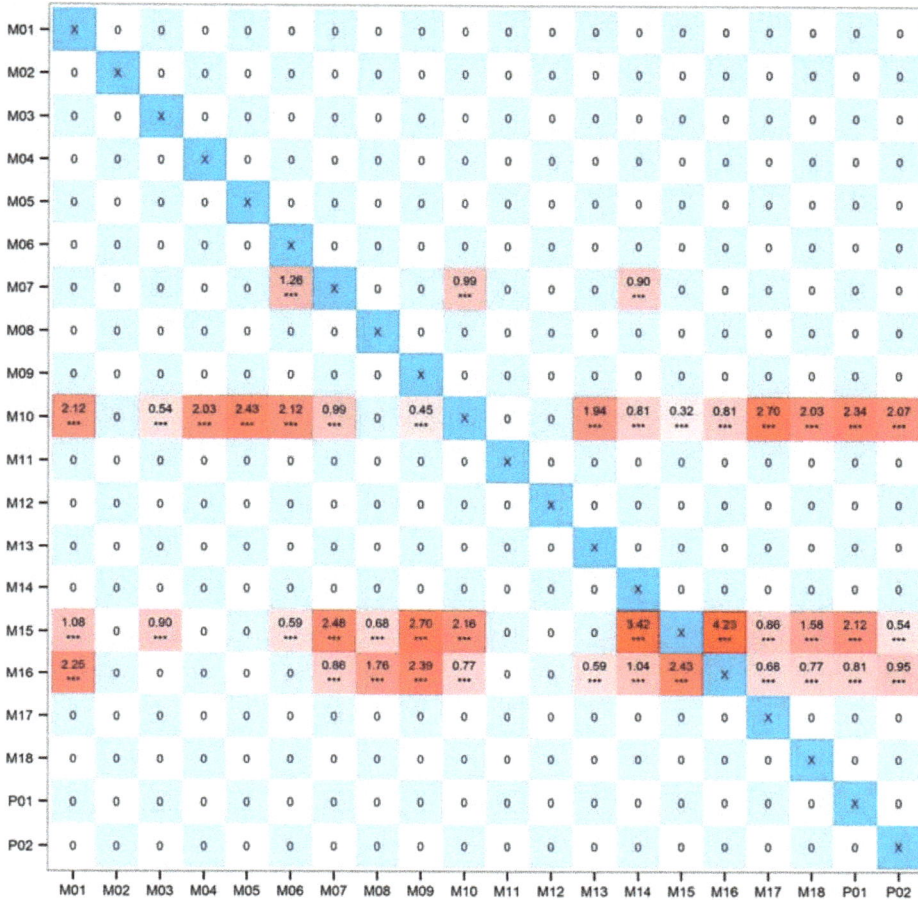

	M01	M02	M03	M04	M05	M06	M07	M08	M09	M10	M11	M12	M13	M14	M15	M16	M17	M18	P01	P02
M01	X	0	0	0	0	0	0	0	0	0	0	0	0	0	0	0	0	0	0	0
M02	0	X	0	0	0	0	0	0	0	0	0	0	0	0	0	0	0	0	0	0
M03	0	0	X	0	0	0	0	0	0	0	0	0	0	0	0	0	0	0	0	0
M04	0	0	0	X	0	0	0	0	0	0	0	0	0	0	0	0	0	0	0	0
M05	0	0	0	0	X	0	0	0	0	0	0	0	0	0	0	0	0	0	0	0
M06	0	0	0	0	0	X	0	0	0	0	0	0	0	0	0	0	0	0	0	0
M07	0	0	0	0	0	1.26	X	0	0	0.99	0	0	0.90	0	0	0	0	0	0	0
M08	0	0	0	0	0	0	0	X	0	0	0	0	0	0	0	0	0	0	0	0
M09	0	0	0	0	0	0	0	0	X	0	0	0	0	0	0	0	0	0	0	0
M10	2.12	0	0.54	2.03	2.43	2.12	0.99	0	0.45	X	0	0	1.94	0.81	0.32	0.81	2.70	2.03	2.34	2.07
M11	0	0	0	0	0	0	0	0	0	0	X	0	0	0	0	0	0	0	0	0
M12	0	0	0	0	0	0	0	0	0	0	0	X	0	0	0	0	0	0	0	0
M13	0	0	0	0	0	0	0	0	0	0	0	0	X	0	0	0	0	0	0	0
M14	0	0	0	0	0	0	0	0	0	0	0	0	0	X	0	0	0	0	0	0
M15	1.08	0	0.90	0	0	0.59	2.48	0.68	2.70	2.16	0	0	3.42	0	X	4.23	0.86	1.58	2.12	0.54
M16	2.25	0	0	0	0	0	0.86	1.76	2.39	0.77	0	0	0.59	1.04	2.43	X	0.68	0.77	0.81	0.95
M17	0	0	0	0	0	0	0	0	0	0	0	0	0	0	0	0	X	0	0	0
M18	0	0	0	0	0	0	0	0	0	0	0	0	0	0	0	0	0	X	0	0
P01	0	0	0	0	0	0	0	0	0	0	0	0	0	0	0	0	0	0	X	0
P02	0	0	0	0	0	0	0	0	0	0	0	0	0	0	0	0	0	0	0	X

Figure 3. The adjacent matrix of drive-response network of topological and performance measures in Technology Exchange Network.

Finish the steps proposed in Section 2.2.3, and we can draw the drive-response network, which is shown in Figure 4. In Figure 4, each measure is mapped as a node, and each arrow stands for a drive-response relationship, and we associate each edge with a weight value, *i.e.*, the effective partial symbolic transfer value, which is mapped as the width of the lines.

Some basic features of the resulting network can be mentioned: there are 20 nodes and 43 edges in the network. The connectedness is 0.1131579, with the average vertex degree 4.3. It can been seen that the resulting network is not a connected one, with three isolated measures which are relatively independent.

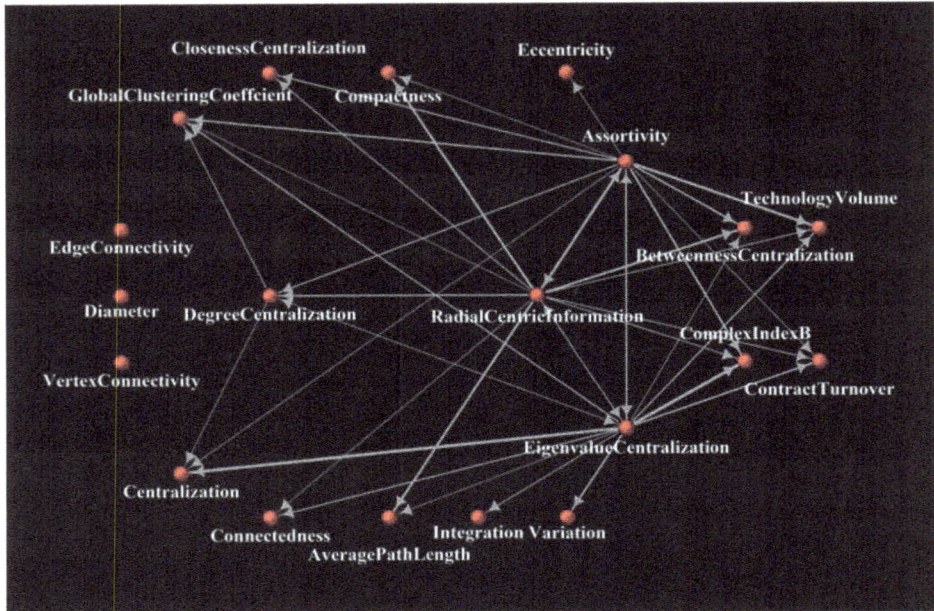

Figure 4. The drive-response network of topological and performance measures in Technology Exchange Network.

Although most of the vertices are connected, it is certainly not a dense graph. The diameter of giant component of the resulting is 3, and the average path length is 1.34375. Thus, the information flowing on this network is relatively simple. Further investigation on the drive-response network is two-fold. On the one hand, we will calculate the out-strength and in-strength of each measure to uncover how influential and comprehensive it is. On the other hand, we will cluster the measures into different groups. The out-strength and in-strength values of each measure is shown in Table 2.

It can be seen from Table 2 that the most influential measures are Eigenvalue Centralization, Assortativity Coefficient, Radial Centric Information and Degree Centralization. Among these measures, two of them are centralization measures, one is connection measure and the remaining other is entropy measure. There is no distance relevant measure to be influential ones. In other words, distance relevant measures are usually driven by others. Graph integration, variation, eccentricity and average path length are influenced by the graph assortativity and eigenvalue centralization.

It can also be seen from Table 2 that measure M14: Global Clustering Coefficient is the most comprehensive one since it takes up the most information from others. Another popular measure M01: Average Path Length, also has a relatively great in-strength value. These two measures are often employed to characterize and class networks [1,4]. Except the three isolated measures, the in-strength values of all the other measures are greater than zero, which implies that most measures are influenced by others.

Table 2. The out-strength and in-strength values of each measure.

ID	Measure	Out-strength	In-strength
M01	Average Path Length	0.00	0.0005448487(#03)
M02	Diameter	0.00	0.00
M03	Eccentricity	0.00	0.0001440922(#15)
M04	Integration	0.00	0.0002026297(#14)
M05	Variation	0.00	0.0002431556(#12)
M06	Centralization	0.00	0.0003962536(#08)
M07	Degree Centralization	0.0003152017(#04)	0.0004322766(#06)
M08	Closeness Centralization	0.00	0.0002431556(#13)
M09	Betweenness Centralization	0.00	0.0005538545(#02)
M10	Eigenvector Centralization	0.0023685158(#01)	0.0003917507(#09)
M11	Vertex Connectivity	0.00	0.00
M12	Edge Connectivity	0.00	0.00
M13	Connectedness	0.00	0.0002521614(#11)
M14	Global Clustering Coefficient	0.00	0.0006168948(#01)
M15	Assortativity Coefficient	0.0023324927(#02)	0.0002746758(#10)
M16	Radial Centric Information Index	0.0015264769(#03)	0.0005043228(#04)
M17	Compactness Measure Based on Distance Degrees	0.00	0.0004232709(#07)
M18	Complexity Index B	0.00	0.0004367795(#05)

According to the network structure, we can cluster the 18 measures into three groups:

Driving measures:

- M07, Degree Centralization;
- M10, Eigenvalue Centralization;
- M15, Assortativity Coefficient;
- M16, Radial Centric Information.

Responding measures:

- M01, Average Path Length;
- M03, Eccentricity;
- M04, Integration;
- M05, Variation;
- M06, Centralization;
- M08, Closeness Centralization;
- M09, Betweenness Centralization;
- M13, Connectedness;
- M14, Global Clustering Coefficient;
- M17, Compactness;
- M18: Complexity Index B.

Standalone measures:

- M02, Diameter;
- M11, Vertex Connectivity;
- M12, Edge Connectivity.

The isolation implies that these measures have no information flow with other measures. It doesn't mean that these measures are trivial ones; rather, they should be treated as non-redundant ones because they contain special information that is not include by other measures, indicating that some of them may reveal different topological aspects of real-world networks.

Entropy **2014**, *16*, 5753–5772

Now, we will tend to the two embedding measures: "P01, Technological Volume" and "P02, Contract Turnover". In our resulting network, both the two are identified as responding measures, which is consistent with the principle of System Theory that "The structure of the system determines its function". Further, the in-strength value of "P01, Technological Volume" (0.0005268372) is greater than that of "P02, Contract Turnover" (0.0003557277). This result is also consistent with our experience that what we are investigating on is the technology exchange network, it stands to reason that the technological volume gains more information from its own structure. Both these two results can serve as evidence for the feasibility of our proposed method.

4. Conclusions and Discussions

Taking into account that pulling networks from different domains and topologies together for statistical analysis might provide incorrect conclusions [25], we conduct our investigation with the data observed from the same network in the form of simultaneously measured time series. In order to reveal the relationships among topological measures from their time series, we synthesize a practical framework comprising techniques from Symbolic Transfer Entropy, Effective Transfer Entropy, and Partial Transfer Entropy, which is able to deal with the challenges such as time series being non-stationary, time series being continuous, finite sample effects and indirect effects. Using a surrogate-based testing scheme, we assess the statistical significance of the resulting drive-response relationships with kernel density estimation. Thus, the synthesized framework can serve as a complete solution for the application of transfer entropy in complicated issues. Furthermore, the framework doesn't stop at the pair-wise relationships, but makes further efforts to provide a holistic scenario in the form of a drive-response network. The transfer entropy-based framework not only quantifies the pair-wise influence one measures exerts on another, but also reveals the overall structure of the measures.

We select 18 topological measures and apply the proposed method to the empirical investigation on Technology Exchange Network. After calculating the drive-response relationships and inferring the network of these measures, we identify the most influential and most comprehensive measures according to their in-strength and out-strength values. We also cluster these measures into three groups, *i.e.*, driving measures, responding measures, and standalone measures. By embedding two performance measures, *i.e.*, technological volume and contract turnover and calculating the relationships between topological and performance measures, we find that our results are consistent with the principle of System Theory and some existing knowledge, which validates the feasibility of our proposed method.

Our conclusion is based on the purely observational data from Technology Exchange Network in Beijing, thus the resulting drive-response network should not be simply generalized. In other words, the drive-response network may not hold true for other networks. However, the proposed method is applicable to other types of network, in case that the time series of topological measures in that network can be observed.

It is to be mentioned that although we can divide the measures into driving and responding ones, it is not to say that the driving measures can determine the responding measures. The drive-response relationships are not equal to deterministic relationships. In general, approaches based on observational quantities alone are not able to disclose a deterministic picture of the system, and interventional techniques will ultimately be needed. Nevertheless, the proposed method can serve as a heuristic tool in detecting directed information transfer, and the detected drive-response relationships can be viewed as a justifiable inferential statistic for true relationships, which give us more evidence and confidence to reveal intrinsic relationship among the topological measures.

Acknowledgments: We are grateful to Beijing Technology Market Management Office for its permission to collect data from Beijing Technical Contract Registration System. The authors would like to thank the two anonymous reviewers for their valuable comments and suggestions to improve the quality of the paper.

Conflicts of Interest: The author declares no conflict of interest.

References

1. Watts, D.J.; Strogatz, S.H. Collective dynamics of small-world networks. *Nature* **1998**, *393*, 440–442.
2. Barabási, A.L.; Albert, R. Emergence of scaling in random networks. *Science* **1999**, *286*, 509–512.
3. Costa, L.D.F.; Oliveira, O.N.; Travieso, G.; Rodrigues, F.A.; Boas, P.R.V.; Antiqueira, L.; Viana, M.P.; Rocha, L.E.C. Analyzing and modeling real-world phenomena with complex networks: A survey of applications. *Adv. Phys.* **2011**, *60*, 329–412.
4. Costa, L.D.F.; Rodrigues, F.A.; Travieso, G.; Boas, P.R.V. Characterization of complex networks: A survey of measurements. *Adv. Phys.* **2007**, *56*, 167–242.
5. Koç, Y.; Warnier, M.; van Mieghem, P.; Kooij, R.E.; Braziera, F.M.T. A topological investigation of phase transitions of cascading failures in power grids. *Physica A* **2014**, *415*, 273–284.
6. Peres, R. The impact of network characteristics on the diffusion of innovations. *Physica A* **2014**, *402*, 330–343.
7. Lee, D.; Kim, H. The effects of network neutrality on the diffusion of new Internet application services. *Telemat. Inform.* **2014**, *31*, 386–396.
8. Albert, R.; DasGupta, B.; Mobasheri, N. Topological implications of negative curvature for biological and social networks. *Phys. Rev. E* **2014**, *89*.
9. Hernández, J.M.; van Mieghem, M. *Classification of graph metrics.* Available online: http://naseducation.et. tudelft.nl/images/stories/javier/report20110617_Metric_List.pdf accessed on 2 October 2014.
10. Estrada, E.; de la Pena, J.A.; Hatano, N. Walk entropies in graphs. *Linear Algebr. Appl.* **2014**, *443*, 235–244.
11. Topirceanu, A.; Udrescu, M.; Vladutiu, M. Network fidelity: A metric to quantify the similarity and realism of complex networks, Proceedings of the Third International Conference on Cloud and Green Computing, Karlsruhe, Germany, 30 September–2 October 2013.
12. Vianaa, M.P.; Fourcassié, V.; Pernad, A.; Costa, L.D.F.; Jost, C. Accessibility in networks: A useful measure for understanding social insect nest architecture. *Chaos Solitons Fractals* **2013**, *46*, 38–45.
13. Bearden, K. *Inferring complex networks from time series of dynamical systems: Pitfalls, misinterpretations, and possible solutions* **2012**, *arXiv*, 1208.0800.
14. Jamakovic, A.; Uhlig, S. On the relationships between topological metrics in real-world networks. *Netw. Heterog. Media (NHM)* **2008**, *3*, 345–359.
15. Krioukov, D.; Chung, F.; Claffy, K.C.; Fomenkov, M.; Vespignani, A.; Willinger, W. The workshop on internet topology (WIT) report. *Comput. Commun. Rev.* **2007**, *37*, 69–73.
16. Barrat, A.; Barthelemy, M.; Vespignani, A. *Dynamical Processes on Complex Networks*; Cambridge University Press: New York, NY, USA, 2008.
17. Barrat, A.; Weigt, M. On the properties of small-world network models. *Eur. Phys. J. B* **2000**, *13*, 547–560.
18. Chen, G.; Wang, X.; Li, X. *Introduction to Complex Networks: Models, Structures and Dynamics*; Higher Education Press: Beijing, China, 2012.
19. Cohen, R.; Havlin, S. Scale-free networks are ultrasmall. *Phys. Rev. Lett.* **2003**, *86*, 3682–3685.
20. Chicharro, D.; Ledberg, A. Framework to study dynamic dependencies in networks of interacting processes. *Phys. Rev. E* **2012**, *86*.
21. Li, C.; Wang, H.; de Haan, W.; Stam, C.J.; van Mieghem, P. The correlation of metrics in complex networks with applications in functional brain networks. *J. Stat. Mech. Theory Exp.* **2011**, *11*.
22. Roy, S.; Filkov, V. Strong associations between microbe phenotypes and their network architecture. *Phys. Rev. E* **2009**, *80*.
23. Filkov, V.; Saul, Z.; Roy, S.; D'Souza, R.M.; Devanbu, P.T. Modeling and verifying a broad array of network properties. *Europhys. Lett. (EPL)* **2009**, *86*.
24. Garcia-Robledo, A.; Diaz-Perez, A.; Morales-Luna, G. Correlation analysis of complex network metrics on the topology of the Internet, Proceedings of the International Conference and Expo. on Emerging Technologies for a Smarter World, Melville, NY, USA, 21–22 October 2013.
25. Bounova, G.; de Weck, O. Overview of metrics and their correlation patterns for multiple-metric topology analysis on heterogeneous graph ensembles. *Phys. Rev. E* **2012**, *85*.
26. Li, C.; Li, Q.; van Mieghem, P.; Stanley, H.E.; Wang, H. *Correlation between centrality metrics and their application to the opinion model* **2014**, *arXiv*, 1409.6033.
27. Sun, X.; Wandelt, S. Network similarity analysis of air navigation route systems. *Transp. Res. E* **2014**, *70*, 416–434.

28. Lin, J.; Ban, Y. The evolving network structure of US airline system during 1990–2010. *Physica A* **2014**, *410*, 302–312.

29. Tung, T.Q.; Ryu, T.; Lee, K.H.; Lee, D. Inferring gene regulatory networks from microarray time series data using transfer entropy, Proceedings of the IEEE International Symposium on Computer-Based Medical Systems, Maribor, Slovenia, 20–22 June 2007.

30. Albano1, A.M.; Brodfuehrer, P.D.; Cellucci, C.J.; Tigno, X.T.; Rapp, P.E. Time series analysis, or the quest for quantitative measures of time dependent behavior. *Philipp. Sci. Lett.* **2008**, *1*, 18–31.

31. Schreiber, T. Measuring information transfer. *Phys. Rev. Lett.* **2000**, *85*, 461–464.

32. Vejmelka, M.; Paluš, M. Inferring the directionality of coupling with conditional mutual information. *Phys. Rev. E* **2008**, *77*.

33. Wu, J.; Di, Z. Complex network in statistical physics. *Prog. Phys.* **2004**, *24*, 18–46.

34. Timme, M.; Casadiego, J. Revealing networks from dynamics: An introduction. *J. Phys. A* **2014**, *47*.

35. Sandoval, L. Structure of a global network of financial companies based on transfer entropy. *Entropy* **2014**, *16*, 4443–4482.

36. Durek, P.; Walther, D. The integrated analysis of metabolic and protein interaction networks reveals novel molecular organizing principles. *BMC Syst. Biol.* **2008**, *2*.

37. Rubina, T. Tools for analysis of biochemical network topology. *Biosyst. Inf. Technol.* **2012**, *1*, 25–31.

38. Freeman, L.C. Centrality in social networks conceptual clarification. *Soc. Netw.* **1979**, *1*, 215–239.

39. Bonchev, D.; Rouvray, D.H. *Complexity in Chemistry, Biology, and Ecology*; Springer: New York, NY, USA, 2005.

40. Mueller, L.A.J.; Kugler, K.G.; Dander, A.; Graber, A.; Dehmer, M. QuACN: An R package for analyzing complex biological networks quantitatively. *Bioinformatics* **2011**, *27*, 140–141.

41. Csardi, G.; Nepusz, T. The igraph software package for complex network research. *InterJ. Complex Syst.* 2006, 1695. Available online: http://www.necsi.edu/events/iccs6/papers/c1602a3c126ba822d0bc4293371c.pdf accessed on 28 October 2014.

42. Thorniley, J. An improved transfer entropy method for establishing causal effects in synchronizing oscillators, Proceedings of the Eleventh European Conference on the Synthesis and Simulation of Living Systems, Winchester, UK, 5–8 August 2008.

43. Kaiser, A.; Schreiber, T. Information transfer in continuous processes. *Physica D* **2002**, *166*, 43–62.

44. Marschinski, R.; Kantz, H. Analysing the information flow between financial time series. *Eur. Phys. J. B* **2002**, *30*, 275–281.

45. Lizier, J.T.; Prokopenko, M. Differentiating information transfer and causal effect. *Eur. Phys. J. B* **2010**, *73*, 605–615.

46. Diks, C.G.H.; Kugiumtzis, D.; Kyrtsou, K.; Papana, A. *Partial Symbolic Transfer Entropy*; CeNDEF Working Paper 13–16; University of Amsterdam: Amsterdam, The Netherlands, 2013.

47. Jo, S. Computational Studies of Glycan Conformations in Glycoproteins. Ph.D. Thesis, University of Kansas, Lawrence, KS, USA, 31 May 2013.

48. Staniek, M.; Lehnertz, K. Symbolic transfer entropy. *Phys. Rev. Lett.* **2008**, *100*.

49. Vakorin, V.A.; Krakovska, O.A.; McIntosh, A.R. Confounding effects of indirect connections on causality estimation. *J. Neurosci. Methods* **2009**, *184*, 152–160.

50. Amblard, P.O.; Michel, O.J.J. On directed information theory and Granger causality graphs. *J. Comput. Neurosci.* **2010**, *30*, 7–16.

51. Good, P.I. *Permutation, Parametric and Bootstrap Tests of Hypotheses*, 3rd ed.; Springer: New York, NY, USA, 2005.

52. Rosenblatt, M. Remarks on some nonparametrice stimates of a density function. *Annal. Inst. Stat. Math.* **1956**, *27*, 832–837.

53. Parzen, E. On estimation of a probability density function and mode. *Annal. Inst. Stat. Math.* **1962**, *33*, 1065–1076.

54. Sheather, S.J.; Jones, M.C. A reliable data-based bandwidth selection method for kernel density estimation. *J. R. Stat. Soc.* **1991**, *53*, 683–690.

55. Guo, S.; Lu, Z.; Chen, Z.; Luo, H. Strength-strength and strength-degree correlation measures for directed weighted complex network analysis. *IEICE Trans. Inf. Syst.* **2011**, *E94-D*, 2284–2287.

56. Kramer, M.A.; Eden, U.T.; Cash, S.S.; Kolaczyk, E.D. Network inference with confidence from multivariate time series. *Phys. Rev. E* **2009**, *79*.

57. Gozolchiani, A.; Yamasaki, K.; Gazit, O.; Havlin, S. Pattern of climate network blinking links follows El Niño events. *Europhys. Lett. (EPL)* **2008**, *83*.

58. Granger, C.W.J. Some aspects of causal relationships. *J. Econom.* **2003**, *112*, 69–71.

59. Lungarella, M.; Ishiguro, K.; Kuniyoshi, Y.; Otsu, N. Methods for quantifying the causal structure of bivariate time series. *Int. J. Bifurcat. Chaos* **2007**, *17*, 903–921.

60. Lungarella, M.; Sporns, O. Mapping information flow in sensorimotor networks. *Comput. Biol.* **2006**, *2*.

61. Chernihovskyi, A. Information-theoretic approach for the characterization of interactions in nonlinear dynamical systems. *2011, urn:nbn:de:hbz:5N-25132*; Universitäts-und Landesbibliothek Bonn. Available online: http://hss.ulb.uni-bonn.de/2011/2513/2513.htm accessed on 19 August 2014.

62. Peter, F.J.; Dimpfl, T.; Huergo, L. Using transfer entropy to measure information flows between financial markets. Proceedings of Midwest Finance Association 2012 Annual Meetings, New Orleans, LA, USA, 22–25 February 2012.

63. *2012 Report of Beijing Technology Market*; Beijing Technology Market Management Office: Beijing, China, 2012.

64. Bennett, D.J.; Vaidya, K.G. Meeting technology needs of enterprises for national competitiveness. *Int. J. Technol. Manag.* **2005**, *32*, 112–153.

65. R Core Team, *R: A Language and Environment for Statistical Computing*; R Foundation for Statistical Computing: Vienna, Austria, 2014.

66. Wickham, H. *Ggplot2: Elegant Graphics for Data Analysis*; Springer: New York, NY, USA, 2009.

entropy

MDPI

Article

Risk Contagion in Chinese Banking Industry: A Transfer Entropy-Based Analysis

Jianping Li [1,*], Changzhi Liang [1,2], Xiaoqian Zhu [1,2], Xiaolei Sun [1] and Dengsheng Wu [1]

[1] Institute of Policy & Management, Chinese Academy of Sciences, Beijing 100190, China; liangchangzhi12345@gmail.com (C.L.); xiaoqian@mail.ustc.edu.cn (X.Z.); xlsun@casipm.ac.cn (X.S.); wds@casipm.ac.cn (D.W.)

[2] University of Chinese Academy of Sciences, Beijing 100190, China

* Author to whom correspondence should be addressed; ljp@casipm.ac.cn; Tel.: +86-(10)-59358805.

Received: 7 September 2013; in revised form: 19 October 2013; Accepted: 9 December 2013; Published: 16 December 2013

Abstract: What is the impact of a bank failure on the whole banking industry? To resolve this issue, the paper develops a transfer entropy-based method to determine the interbank exposure matrix between banks. This method constructs the interbank market structure by calculating the transfer entropy matrix using bank stock price sequences. This paper also evaluates the stability of Chinese banking system by simulating the risk contagion process. This paper contributes to the literature on interbank contagion mainly in two ways: it establishes a convincing connection between interbank market and transfer entropy, and exploits the market information (stock price) rather than presumptions to determine the interbank exposure matrix. Second, the empirical analysis provides an in depth understanding of the stability of the current Chinese banking system.

Keywords: interbank exposure matrix; risk contagion; transfer entropy

1. Introduction

The Basel III Accord published in 2009 proposed for the first time an additional capital requirement for inter-financial sector exposures, indicating that regulators have been aware of the necessity to prevent the occurrence of risk contagion among banks. As a matter of fact, the 2008 subprime mortgage crisis has triggered a global financial crisis through the contagion among banks. Hoggarth *et al.* [1] studied 47 banking crisis over the 1977–1998 period in both developing and developed countries and find that the resulting cumulative output loss reached as much as 15%–20% of annual GDP. These cases show that such banking crises can have substantial impacts on the economy.

Traditional micro-prudential regulation focuses on the risk management of a specific bank, which has been proved insufficient from a systemic perspective. In extreme circumstances, a single bank failure can lead to massive bank failures because the initial shock can spread to other banks through the interbank market. Considering the possibility that bank interdependencies magnify the risk, regulators are trying to push bank supervision more towards a system-wide framework. Banks are also required to not only look at the risk of individual exposures, but also account for correlations of the exposures when assessing their investment portfolios [2].

2. Literature Review

Quite a few research papers on risk contagion among banks regard the banking system as a network, and the contagion process is simulated using network dynamics. For example, Nier *et al.* [3] constructed a banking system whose structure is described by parameters including the level of capitalization, the degree to which banks are connected, the size of interbank exposures and concentration degree of the system, and then they analyzed the resilience of the system to an initial bank failure by varying the structural parameters, and identified a negative and non-linear relationship between contagion and capital.

Insightful studies by Allen and Gale [4] and Freixas *et al.* [5] illustrate that the possibility of contagion largely depends on the structure of the interbank market. Allen and Gale consider a simplified banking system consisted of 4 banks, and initialize the system with different market structures, their results indicate that for the same shock, a complete structure is more robust than incomplete structure. Freixas *et al.* discuss three scenarios of interbank exposures through credit lines: a credit chain, diversified lending and a money center case, and they conclude that contagious failures occur more easily in the credit chain case than in the diversified lending case; as for the money center case, the probability of contagion is determined by the values of model parameters. Both researches uncover critical issues concerning how interbank market structure affects risk contagion among banks, yet the models still have room for improvement given the complexity of the real interbank market.

Recently, a series of papers have revealed the latest progress in banking network studies. Berman *et al.* [6] formalized a model for propagation of an idiosyncratic shock on the banking network, and constructed the stability index, which can be used to indicate the stability of the banking network. Haldane and May [7] draw analogies with the dynamics of ecological food webs and with networks within which infectious diseases spread. Minoiu and Reyes [8] investigated the properties of global banking networks with bank lending data for 184 countries, and find that the 2008–2009 crisis perturbed the network significantly. DasGupta and Kaligounder [9] investigated the global stability of financial networks that arise in the OTC derivatives market.

Subsequently, a considerable amount of simulations and empirical researches on interbank contagion were performed, as surveyed by Upper [10]. Examples include Sheldon and Maurer [11] for the Swiss banking system, Blåvarg and Nimander [12] for Sweden, Furfine [13] for the US Federal Funds market, Upper and Worms [14] for Germany, Elsinger *et al.* [15] for Austria, Van Lelyveld and Liedorp [16] and Mistrulli [17] for Italy. These papers follow a similar routine: first estimate the actual interbank exposure matrix (a $N \times N$ square matrix reflecting the credit exposure of N banks to each other in the system), then simulate the impact of a single bank failure or multiple bank failures on the system. The key step of this routine is the estimation of the interbank exposure matrix, because the matrix depicts the structure of the interbank market, and will eventually determine the possibility of contagion in the banking system.

Owing to the limitations of data sources, interbank exposure matrices can only be estimated indirectly. Currently, the maximum entropy estimation with balance sheet data is the most widely used method in determining the interbank exposure matrix. In this method, the aggregated interbank assets and liabilities disclosed in balance sheets are the only input information, and the matrix can be derived by maximizing its entropy. Some authors claim that this method is the least biased given that only limited information of the interbank market structure, namely the aggregated interbank assets and liabilities are available. However, considering the fact that there may be other available data concerning the interbank market structure and the maximization of the matrix entropy probably deviates from reality, the assumption of the method can be problematic. Mistrulli [18] shows that for the Italian banking system the use of maximum entropy techniques underestimates contagion risk relative to an approach that uses information on actual bilateral exposures.

Transfer entropy is a relatively new concept introduced by Schreiber in 2000 [19], and it measures the information transfer between two time series. Compared with other cross-correlation statistics, transfer entropy is an asymmetric measure and takes into account only statistical dependencies truly originating in the "source" series, but not those deriving from a shared history, like in the case of a common external driver. These characteristics of transfer entropy make it a superior tool to analyze the casual interactions among variables of a complex system. In the last decade, transfer entropy has been applied to studies within the context of financial markets. Marschinski and Matassini [20] designed a procedure to apply transfer entropy to the detection of casual effect between two financial time series. Kwon and Yang [21] calculated the transfer entropy between 135 NYSE stocks and identified the leading companies by the directionality of the information transfer. In a separate paper [22], they analyzed the information flow between 25 stock markets worldwide, their results show that America is the biggest source of information flow.

This paper aims to establish a new method to determine the interbank exposure matrix, within a transfer entropy context. Furthermore, the stability of Chinese banking industry is investigated. The remainder of this paper is organized as follows: in Section 2, a detailed description of the method is given. Section 3 presents the empirical study and results. Section 4 concludes the presentation.

3. Method

3.1. Definition of Transfer Entropy

When considering the interactions between two systems evolving in time, the linear correlation coefficient, Kendall rank correlation coefficient, and mutual information [23] are the most commonly used statistics. However, they are incapable of distinguishing information that is actually exchanged from shared information due to common history and input signals. Schreiber proposed transfer entropy to exclude these influences by appropriate conditioning of the transition probabilities.

Consider two processes I and J, the transfer entropy is defined as:

transfer entropy from J to I =

information about future observation $I(t+1)$ *gained from past observations of* I *and* J

$-$*information about future observation* $I(t+1)$ *gained from past observations of* I *only* $\qquad$ (1)

Equation (1) measures how much additional information does J provide for the prediction of $I(t+1)$ apart from the historical information provided by I itself. A mathematical expression of Equation (1) is:

$$T_{J \to I} = \sum p\left(i_{t+1}, i_t^{(k)}, j_t^{(l)}\right) \log p\left(i_{t+1} \mid i_t^{(k)}, j_t^{(l)}\right) - \sum p\left(i_{t+1}, i_t^{(k)}\right) \log p\left(i_{t+1} \mid i_t^{(k)}\right)$$

$$= \sum p\left(i_{t+1}, i_t^{(k)}, j_t^{(l)}\right) \log \frac{p\left(i_{t+1} \mid i_t^{(k)}, j_t^{(l)}\right)}{p\left(i_{t+1} \mid i_t^{(k)}\right)}$$

(2)

Here, i_t represents the state of I at time t, and $i_t^{(k)}$ is a k dimensional vector representing the most recent k states of I before i_{t+1}, $j_t^{(l)}$ is a l dimensional vector representing the most recent l states of J before j_{t+1}. Additionally, $i_t, j_t \in D = \{ d_1, d_2, d_3, \ldots \ldots d_{N-1}, d_N \}$. The transfer entropy from I to J can be derived by exchanging i and j in Equation (2).

3.2. Numerical Solution for Transfer Entropy

Though the analytic form of transfer entropy is relatively simple, there is still a distance between numerical and practical application. In most cases, we need to obtain I and J by coarse graining a continuous system at resolution ε. Usually, when $\varepsilon \to 0$, we will get a more accurate transfer entropy, but the computational cost will grow rapidly as well. For this consideration, an appropriate resolution should be determined to balance the accuracy and computational cost. In this paper, we set the resolution according to the length of dataset, for a dataset of N samples, the continuous interval of the sample is discretized into $(N/4)^{1/3}$ parts, which balances the accuracy and efficiency.

Another difficulty lies in that the conditional probabilities in Equation (2) can't be estimated directly given I and J. To solve this problem, we propose a transformation on Equation (2). According to the definition of conditional probability, $p\left(i_{t+1}|i_t^{(k)}, j_t^{(l)}\right)$ and $p\left(i_{t+1}|i_t^{(k)}\right)$ can be rewritten as:

$$p\left(i_{t+1} \mid i_t^{(k)}, j_t^{(l)}\right) = \frac{p\left(i_{t+1}, i_t^{(k)}, j_t^{(l)}\right)}{p\left(i_t^{(k)}, j_t^{(l)}\right)}$$

$$p\left(i_{t+1} \mid i_t^{(k)}\right) = \frac{p\left(i_{t+1}, i_t^{(k)}\right)}{p\left(i_t^{(k)}\right)}$$

(3)

Substituting Equation (3) into Equation (2), we have:

$$T_{J \to I} = \sum p\left(i_{t+1}, i_t^{(k)}, j_t^{(l)}\right) \log \frac{p\left(i_{t+1}, i_t^{(k)}, j_t^{(l)}\right) \cdot p\left(i_t^{(k)}\right)}{p\left(i_t^{(k)}, j_t^{(l)}\right) \cdot p\left(i_t^{(k)}, j_t^{(l)}\right)}$$

(4)

This new expression contains only joint probability, and thus simplifies the calculation of transfer entropy. Generally speaking, the parameter k and l should be as large as possible so that the information introduced by the history of process I itself can be excluded to the most extent. However, as the amount of data required grows like $N^{(k+l)}$ [24], the finite sample effects would be quite significant if k and l is excessively large, so reasonable values of both k and l is of crucial importance in practice. In this paper, since we have limited sample, both k and l are set to be 1.

3.3. Determine the Interbank Exposure Matrix with Transfer Entropy

As mentioned in the Introduction section, the widely used maximum entropy estimation of interbank exposure matrix suffers from biased assumptions and can significantly deviate from practice. In this paper, we determine the interbank market structure by calculating the transfer entropy matrix of the banking industry with daily stock closing price, and then an adjustment on the transfer entropy matrix is made by using the RSA algorithm [17] as well as the aggregated interbank assets and liabilities, after which we derive the interbank exposure matrix. The interbank market may be represented by the following $N \times N$ matrix:

$$X = \begin{bmatrix} x_{11} & \cdots & x_{1j} & \cdots & x_{1N} \\ \vdots & \ddots & \vdots & \ddots & \vdots \\ x_{i1} & \cdots & x_{ij} & \cdots & x_{iN} \\ \vdots & \ddots & \vdots & \ddots & \vdots \\ x_{N1} & \cdots & x_{Nj} & \cdots & x_{NN} \end{bmatrix}$$

Here, x_{ij} represents the amount of money bank i lends to bank j. Since a bank can't lend to itself, we have N diagonal elements equal to 0. But to identify the matrix, other $N^2 - N$ elements have to be estimated.

Previous studies on the movement of stock prices such as Levine and Zeros [25], Chiarella and Gao [26] and Hooker [27] have proved that stock markets' return are affected by macroeconomic indicators such as GDP, productivity, employment and interest rates. In terms of the correlation between two stocks, especially when they belong to the same sector, we can see two types of mechanisms to generate significant correlation between them [28]:

- External effect (e.g. economic, political news, *etc.*) that influences both stock prices simultaneously. In this case the change for both prices appears at the same time.
- One of the companies has an influence on the other (e.g. one of the company's operations depends on the other). In this case the price change of the influenced stock appears later in time because it

needs some time to react on the price change of the first stock, in other words one of the stocks pulls the other.

According to the definition of transfer entropy, it measures the information flow between two time series. The transfer entropy from stock price of bank I to stock price of bank J measures the information flow from I to J, which depicts how much influence does the stock price of bank I has on the stock price of bank J.For the two types of correlation mechanisms for stocks, the external effect that influences two stock prices of banks at the same time generates no information flow between bank I and bank J, thus such effect does not contribute to $T_{I \to J}$, where $T_{I \to J}$ measures the second type of correlation between two stock prices.

Since stock price reflects investors' expectation of a company's future earnings [29,30], we infer that $T_{I \to J}$ measures the influence of earning condition of bank I to bank J, that is to say, such influence is realized mainly through interbank lending and borrowing between I and J, so $T_{I \to J}$ depicts the lending and borrowing activity between I and J, and can be used to estimate the interbank exposure.

We use a transfer entropy matrix to depict the structure of interbank exposure matrix. Define $\vec{s_i} = \{s_{i1}, s_{i2}, s_{i3}, \ldots \ldots s_{iT}\}$ as time series of the stock price for bank. The transfer entropy from $\vec{s_i}$ to $\vec{s_j}$ is:

$$T_{I \to J} = \sum p\left(j_{t+1}, j_t^{(k)}, i_t^{(l)}\right) \log \frac{p\left(j_{t+1}, j_t^{(k)}, i_t^{(l)}\right) \cdot p\left(j_t^{(k)}\right)}{p\left(j_t^{(k)}, i_t^{(l)}\right) \cdot p\left(j_t^{(k)}, i_t^{(l)}\right)}$$

(5)

The structure of resulting transfer entropy matrix $T = \{T_{I \to J}\}$ serves as an approximation of interbank exposure matrix structure. To determine the interbank exposure matrix, we need to adjust the transfer entropy matrix so that the resulted matrix meets the following constraints:

$$\sum_{j=1}^{N} x_{ij} = a_i, \quad \sum_{i=1}^{N} x_{ij} = l_j$$

(6)

where, a_i represents the amount of money bank i lends to other banks and l_j represents the amount of money bank j raised from other banks.

The adjustment can be described by the following optimization problem:

$$\min \sum_{i=1}^{N} \sum_{j=1}^{N} x_{ij} \ln\left(\frac{x_{ij}}{T_{ij}}\right)$$

$$s.t. \quad \sum_{j=1}^{N} x_{ij} = a_i$$

$$\sum_{i=1}^{N} x_{ij} = l_j$$

$$x_{ij} \geq 0$$

(7)

where x_{ij} represent the interbank exposure matrix.

This problem can be solved numerically using RAS algorithm, the process is summarized as the following iterations:

Step 1: (row adjustment): $T_{ij}^u \to T_{ij}^u \rho_i^u$, where $\rho_i^u = \frac{a_i}{\sum\limits_{\forall j | T_{ij}^u > 0} T_{ij}^u}$

Step 2: (column adjustment): $T_{ji}^u \to T_{ji}^u \sigma_j^u$, where $\sigma_j^u = \frac{l_j}{\sum\limits_{\forall i | T_{ij}^u > 0} T_{ij}^u}$

Step 3: return to step 1

3.4. The Contagion Process Modeling

According to the literature review of risk contagion among banking systems by Upper, the most widely used mechanism of the contagion process is the fictitious default algorithm developed by Eisenberg and Noe [31] and the sequential default algorithm developed by Furfine [13]. We will describe both models in this section, which will be used in the contagion process simulation in Section 4.4.

3.4.1. Eisenberg-Noe's Mechanism

Eisenberg and Noe [31] developed a clearing mechanism that solves the interbank payment vectors of all banks in the system simultaneously. The interbank market structure is represented by (L,e), where L is a $n \times n$ nominal interbank liabilities matrix, and e is the exogenous operating cash flow vector. Let $\overline{p_i}$ represent total nominal liability of bank i to all other banks, that is $\overline{p_i} = \sum_{j=1}^{n} L_{ij}$. Let:

$$\Pi_{ij} \equiv \begin{cases} \dfrac{L_{ij}}{\overline{p_i}} & \text{if } \overline{p_i} > 0 \\ 0 & \text{otherwise} \end{cases} \tag{8}$$

be the relative liabilities matrix.

The mechanism sets three criteria in the clearing process, namely: (1) Limited liability, a bank could pay no more than its available cash flow; (2) The priority of debt, stock holders of a bank receive no value until it pays off its outstanding liabilities; (3) Proportionality, if default occurs, creditors are paid in proportion to the size of their nominal claim on the defaulted bank's assets. Eisenberg and Noe demonstrate that there exists a unique clearing payment vector under the three criteria and the regular financial system assumption. For a payment vector $p^* \in [0, \overline{p}]$, it is a clearing payment vector if and only if the following condition holds: $p_i^* = \min[e_i + \sum_{j=1}^{n} \Pi_{ij}^T p_j^*, \overline{p_i}]$.

The number of banks defaulted can be obtained by comparing the clearing payment vector with nominal liability vector. A fictitious default algorithm is implemented to calculate the clearing payment vector, which can be summarized by the following steps:

- Initialize $p_i = \overline{p_i}$, and calculate the net value of bank i, $V_i = \sum_{j=1}^{n} \Pi_{ij}^T p_j + e_i - p_i$. If $\forall i$, $V_i \geq 0$, it means no bank defaults and the clearing payment vector is $p_i = \overline{p_i}$, the algorithm terminates; otherwise go to step 2.
- Find banks with net value $V_i < 0$, these banks can only pay part of the liabilities to other banks, and the ratio is $\theta_i = (\sum_{j=1}^{n} \Pi_{ij}^T p_j + e_i)/p_i$, we denote these banks by U. Under the assumption that only banks in U default, we replace L_{ij} by $\theta * L_{ij}$ so that the limited liability criterion is met, and thus get new L_{ij}, Π_{ij}, P_i and V_i. Repeat step 2 while U is not empty.

The procedure gives us the clearing payment vector for the banking system which satisfies $p_i^* = \min[e_i + \sum_{j=1}^{n} \Pi_{ij}^T p_j^*, \overline{p_i}]$. By tracing the fictitious default process, we obtain the sequence of defaults (keep in mind that the process is fictitious, and in reality both the clearing process and defaults are simultaneous) and can distinguish between defaults caused by bad economic situation (defaults in the first round)-and defaults caused by the defaults of other banks (defaults after the first round).

3.4.2. Furfine's Sequential Default Algorithm

Another contagion process model is developed by Furfine [13], we refer to it as the sequential default algorithm. He regards the contagion as a sequential process. For example, the initial failure of bank i will result in capital loss of its creditors, which is calculated by multiplying the loss rate of a given default by the exposure of its creditors. If the loss is large enough, its creditors will go bankrupt, and this may trigger another round of bank failures. We denote the loss rate given default as α, the equity capital of bank i as C_i, external loss resulting from non-interbank market is assume to be proportional to C_i and the ratio is set to be a constant β ,the contagion mechanism can be summarized as below [13]:

Round 1: bank i fails because of an external shock;
Round 2: bank j suffers a total capital loss of αX_{ji} (X_{ji} is the exposure bank j to bank i), and $\alpha X_{ji} > (1-\beta)C_j$, which leads to the failure of bank j;
Round 3: the failures of bank i and j results in a total capital loss of $\alpha(X_{ki}+X_{kj}) > (1-\beta)C_k$ for bank k and eventually leads to the failure of bank k.
Round 4: similar to round 3, and the contagion process will continue until all surviving banks can absorb the capital loss with their equity capital, which means no banks go bankruptcy.

We illustrate the procedure of the method in Figure 1.

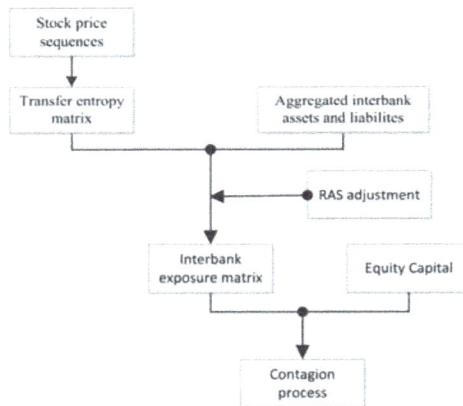

Figure 1. Procedure of the method.

4. Empirical Research and Results

In this section, we investigate the possibility of contagion in Chinese banking industry based on the transfer entropy method. As illustrated in Section 3, stock price, aggregated interbank assets and liabilities are essential in the determination of interbank exposure matrix. Generally speaking, the information is only available for listed banks. According to a survey conducted by China Banking Regulatory Commission in 2012, the Chinese banking system consisted of more than 300 banks, among which 16 are listed banks. The 16 listed banks are the largest banks in China, with total assets amounting to 65% of the total banking industry assets in 2012. The remaining unlisted banks are much smaller scale, thus can be merged into a single bank, which we call it edian bank. That is to say, we have 17 banks in Chinese banking system.

4.1. Data Description

The data used in this study stems from two sources. Daily stock closing prices of the 16 listed banks are drawn from the Wind database, a leading integrated service provider of financial data in China. The time interval for each stock sequence is from the first trading day to 31 December 2012. Aggregated interbank assets and liabilities are drawn from the annual reports of the 16 listed banks. Instead of publishing aggregated interbank assets and liabilities directly, the annual reports give us their sub-items, namely deposits in other banks, due from banks, financial assets purchased under resale, deposits from other banks, interbank borrowing, and repurchase agreements. The annual reports also provide us with the equity capital of banks. The 16 listed banks are listed in Table 1:

Table 1. The 16 Chinese listed banks

Names of the banks	
ChinaMinsheng Bank (CMBC)	Bank of Beijing (BCCB)
Spd Bank (SPDB)	Bank of Ningbo (BONB)
Industrial Bank (IB)	China Construction Bank (CCB)
China Merchants Bank (CMB)	ChinaEverbright Bank (CEB)
Bank of Communications (BOCOM)	Bank of Nanjing (BONJ)
Agricultural Bank of China (ABC)	Bank of China (BOC)
Huaxia Bank (HXB)	ChinaCitic Bank (CITIC)
Industrial And Commercial Bank of China (ICBC)	Pingan Bank (PAB)

4.2. Data Preprocessing

The transfer entropy matrix calculation requires the stock price of each bank to have the same time interval, which is not the case in the original samples. To unify the time interval, starting points of all listed banks are changed to be the same as Pingan Bank, who was the latest to go public among the 16 listed banks. The new stock price sequences cover the period 2011/1/4–2012/12/31, with 487 records.

Exploiting the information extracted from annual reports, we obtain the aggregated interbank assets by summing up deposits in other banks, due from banks, and financial assets purchased under resale. The interbank liabilities are derived by summing up deposit from other banks, interbank borrowing, and repurchase agreements.

As for the median bank, its stock price is supposed to be the weighted average of the other 16 listed bank, total assets of each bank is chosen as the weighting coefficient. Its aggregated assets and liabilities are obtained by subtracting the total aggregated assets and liabilities of the banking system from the sum of the 16 listed banks.

4.3. Transfer Entropy Matrix & Interbank Exposure Matrix Calculation

The stock price sequence is divided into two separate parts, the first part is from 2011/1/4 to 2011/12/30, and the second part is from 2012/1/4 to 2012/12/31. By applying the procedure we have described above, we obtain the interbank exposure matrix of 2011 and 2012. The matrices are presented as Figures 2 and 3, in the form of heat maps.

Figure 2. Heat map of interbank exposure matrix in 2011.

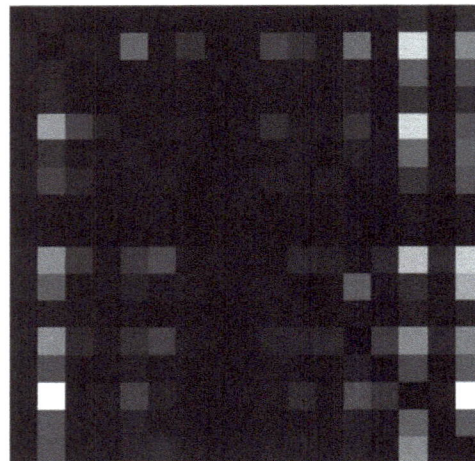

Figure 3. Heat map of interbank exposure matrix in 2012.

The heat map is consisted of 17×17 grids, with each corresponding to an element in the interbank exposure matrix. The grey scale of each grid is proportional to the value of the element in the matrix, the brighter the grid is, the larger value the element has. Here we introduce the concept of contrast rate; it is defined as the luminosity ratio of the brightest part to the darkest part of a map. Obviously, the heat map in Figure 3 shows a higher contrast rate than that in Figure 2, this means that the interbank market structure in 2011 is more diversified than that in 2012. Considering the whole interbank exposure matrix is too big to present here, only part of the matrix is shown in Tables 2 and 3.

Table 2. Interbank exposure matrix of five major Chinese banks in 2011.

billion	ICBC	CCB	BOCOM	ABC	BOC
ICBC	0.0	130.9	36.4	51.8	103.1
CCB	128.2	0.0	29.9	34.6	62.9
BOCOM	44.0	24.9	0.0	19.8	77.6
ABC	151.0	72.0	30.6	0.0	72.1
BOC	183.6	49.8	51.7	64.3	0.0

Table 3. Interbank exposure matrix of five major Chinese banks in 2012.

billion	ICBC	CCB	BOCOM	ABC	BOC
ICBC	0.0	127.8	21.6	88.8	253.5
CCB	191.5	0.0	28.1	65.6	251.6
BOCOM	65.7	37.4	0.0	19.5	131.1
ABC	171.5	74.2	96.7	0.0	227.0
BOC	304.9	89.0	43.3	27.5	0.0

4.4. The Contagion Process Simulation

In the contagion process simulation, we present the results given by both Furfine's sequential default algorithm and Eisenberg-Noe's mechanism.

4.4.1. Furfine's Sequential Default Algorithm

Note that both the loss rate given default and the non-interbank capital loss caused by a shock outside the interbank market have an influence on the contagion process, so we construct scenarios with different loss rates given default α and different non-interbank capital loss rate β; the loss rate given default ranges from 0.1 to 1 with a step of 0.1, the initial equity capital loss rate are set to be 0, 0.3 and 0.5. Tables 4 and 5 give the simulation results.

In Tables 4 and 5, the first column represents initially failed banks, and only banks that could trigger a contagion process are listed. The elements in each row are the amount of banks failed in the contagion process under different α and β values. If an element equals 1, it means the corresponding bank will not cause other bank failures given its own failure. The results show that given specific α and β, only a few of the 17 banks can trigger a contagion process in the system, which reflects that Chinese banking industry is resistant to contagion to a great extent. With greater β, the number of banks capable of triggering a contagion process is increasing, along with the amount of banks failed in the process. Such trend is also identified when we raise the value of α. This is in accordance with the findings of Nier *et al.* [3].

Another interesting finding is the existence of a threshold value for α. Given the initial failure of a specific bank at certain β, contagion won't occur if α is lower than the threshold value. But when α is greater than the threshold value, the amount of bank failures will increase sharply. A reasonable inference is that interbank market can effectively diversify the risk caused by an initial bank failure in normal condition, thus prevent contagion from happening. However, the linkages between banks also serve as channels through which risk may spread. Under severe conditions, risk will be transferred to all banks in the system and result in knock-on defaults.

Table 4. Number of failed banks in the 2011 contagion process simulation.

α	0.1	0.2	0.3	0.4	0.5	0.6	0.7	0.8	0.9	1.0
					$\beta = 0.5$					
ICBC	1	2	2	9	10	10	10	17	17	17
BOC	1	1	1	1	1	1	1	12	17	17
IB	1	1	1	1	1	1	2	9	10	10
SPDB	1	1	1	1	1	1	1	1	1	10
CITIC	1	1	1	1	1	1	1	1	1	10
					$\beta = 0.3$					
ICBC	1	1	2	2	2	10	10	10	10	10
IB	1	1	1	1	1	1	1	1	2	3
					$\beta = 0$					
ICBC	1	1	1	2	2	2	2	9	10	10

Table 5. Number of failed banks in the 2012 contagion process simulation.

α	0.1	0.2	0.3	0.4	0.5	0.6	0.7	0.8	0.9	1.0
					$\beta = 0.5$					
ICBC	1	1	1	1	11	11	17	17	17	17
BOC	1	1	1	1	1	11	17	17	17	17
IB	1	1	1	1	1	1	1	10	10	17
BOCOM	1	1	1	1	1	1	1	1	1	17
					$\beta = 0.3$					
ICBC	1	1	1	1	1	1	11	11	17	17
BOC	1	1	1	1	1	1	1	11	14	17
					$\beta = 0$					
ICBC	1	1	1	1	1	1	1	1	1	11

To give a concrete example of the contagion process among Chinese banking system, we describe the contagion process triggered by ICBC when α=0.5 and $\beta = 0.5$ in Figure 4. This is a 3-stage process which begins with the failure of ICBC. In the second stage, IB suffers a capital loss including half the money lent to ICBC and half the equity capital, which exceeds the total equity capital IB holds and leads to its failure. The failures of ICBC and IB lead to the subsequent failures of eight other banks. In the final stage, the contagion ends and the remaining seven banks survive.

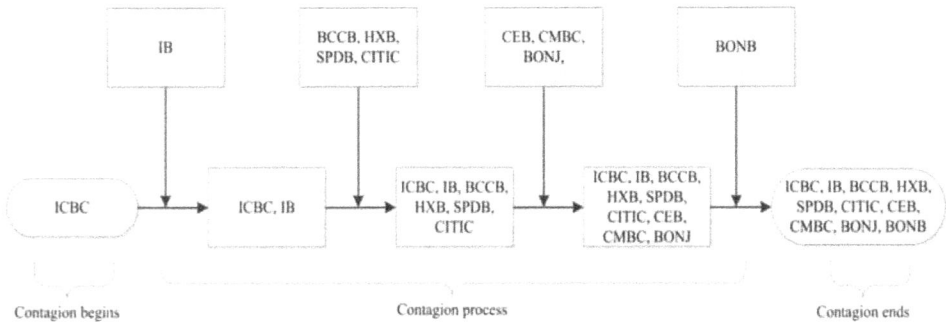

Figure 4. 2011 contagion process given the initial bankruptcy of ICBC when *α=0.5* and *β = 0.5*.

4.4.2. Eisenberg-Noe's Mechanism

In Eisenberg-Noe's mechanism, the interbank exposure matrix and the exogenous operating cash flow vector e are necessary to clear the system. For bank i, e_i is represented by its equity capital under the risk free assumption. However, considering that its non-interbank assets face market risk, credit risk *etc.*, e_i should be adjusted by subtracting potential exogenous capital loss caused by these risks. Similar to the simulation above, we suppose the adjusted exogenous capital loss is proportional to equity capital, and the ratio is a constant β. We clear the system in scenarios of $\beta = 0.5, 0.3, 0$ respectively, and obtain the clearing payment vectors as well as the defaulted banks. Table 6 lists the number of defaulted banks under different β, Table 7 gives the percentage of debt repaid by defaulted banks.

Table 6. Defaults under different exogenous capital loss rate

		2011	2012
	0.5	6	7
β	0.3	4	3
	0	0	1

With the decreasing exogenous capital loss, the economic situation of banks in the system is ameliorated, thus the number of banks defaulted in the process drops as well. Unlike Furfine's sequential default algorithm, loss rate given default here is dynamically determined in the clearing process, which seems more consistent with the true contagion process. In Table 7, it is natural to find that the percentage of debt repaid by defaulted banks is increasing with decreasing β, since the exogenous cash flow increases.

Generally speaking, Eisenberg-Noe's model makes fewer assumptions about the system than Furfine's sequential default algorithm. In Eisenberg-Noe's model clearing process of all banks and defaults are simultaneous; the loss rate given default for a specific bank is determined by its solvency. while in Furfine's sequential default algorithm, banks are cleared sequentially. The results in Table 7 indicate that in reality the loss rate given default is usually quite low, even when the whole system suffers an shock of exogenous capital loss of 50%, so the scenarios of quite large α in Furfine's model simulation is unlikely to happen in reality. Despite these differences, both simulations reveal something in common, that the Chinese banking system is resistant to exogenous shock, and massive defaults won't happen unless under extreme situations.

Table 7. Recovery rate of interbank liabilities

		BCCB	ICBC	BONJ	BONB	PAB	BOC	BOI
2011								
	0.5	0.926	0.852	0.912	0.972	0.764	-	0.592
β	0.3	-	0.999	0.996	-	0.895	-	0.820
	0	-	-	-	-	-	-	-
2012								
	0.5	0.956	0.943	0.959	0.906	0.864	0.762	0.283
β	0.3	-	-	-	-	0.952	0.878	0.393
	0	-	-	-	-	-	-	0.557

Notes: columns 3–9 are the percentage of debt repaid by defaulted banks, for example 0.926 in column 3 means BCCB paid 92.6% of its debt to other banks in the scenario of $\beta = 1$. Banks denoted by "-" and those not listed in this table means they are solvent in corresponding scenario.

5. Conclusions and Further Direction

In this paper, we investigate the risk contagion due to interbank exposure among the Chinese banking industry with a transfer entropy method. By reviewing previous work, we find that maximum entropy method is widely used in determining the interbank exposure matrix, although it is theoretically problematic. We propose a transfer entropy based method to estimate the interbank exposure matrix and present a qualitative analysis to validate it. Then two widely used mechanisms of the contagion process are presented and are adopted to simulate the contagion process.

The empirical analysis is based on 2011 and 2012 stock price sequences of 16 listed Chinese banks from the Wind database, and the corresponding annual reports. We calculate the corresponding transfer entropy matrix at first, and then the interbank exposure matrix is obtained after a RAS adjustment. We run Furfine's sequential default algorithm by varying loss rate given default and non-interbank capital loss rate, and Eisenber-Noe's fictitious default algorithm by varying non-interbank capital loss rate. Both results indicate that the chance of Chinese banking system suffering a systemic crisis is quite low, or in other words, the Chinese banking system is rather stable. Systemically important banks are identified in the simulations, ICBC and BOC tend to trigger massive defaults under various scenarios of non-interbank capital loss rates, and ICBC, BONJ, PAB is more likely to default than other banks, revealing the unbalanced state of their interbank liabilities and interbank assets. This gives regulators implications on which banks require additional regulation.

In the end, it's worth noting that our research is primary and there is still much room for improvement. In this paper, we use transfer entropy between stocks prices to approximate the interbank market structure, which is based on the foundation of our qualitative analysis that transfer entropy only contains the correlation of two stock prices due to interbank links. However, this qualitative analysis is not quantitatively exact, and further mathematical proof is required to address this issue.

Acknowledgments: This research is supported by National Natural Science Foundation of China (No.71071148, No.71003091), Key Research Program of Institute of Policy and Management, Chinese Academy of Sciences (Y201171Z05), the Youth Innovation Promotion Association of Chinese Academy of Sciences.

Conflicts of Interest: The authors declare no conflict of interest.

References

1. Hoggarth, G.; Reis, R.; Saporta, V. Costs of banking system instability: Some empirical evidence. *J. Bank. Finance* **2002**, *26*, 825–855.
2. Lehar, A. Measuring systemic risk: A risk management approach. *J. Bank. Finance* **2005**, *29*, 2577–2603.
3. Nier, E.; Yang, J.; Yorulmazer, T.; Alentorn, A. Network models and financial stability. *J. Econ. Dyn. Control* **2007**, *31*, 2033–2060.
4. Allen, F.; Gale, D. Financial contagion. *J. Polit. Econ.* **2000**, *108*, 1–33.
5. Freixas, X.; Parigi, B.M.; Rochet, J.C. Systemic risk, interbank relations, and liquidity provision by the Central bank. *J. Money Credit Bank.* **2000**, *32*, 611–638.
6. Berman, P.; DasGupta, B.; Kaligounder, L.; Karpinski, M. On the computational complexity of measuring global stability of banking networks. *Algorithmica* **2013**, 1–53.
7. Haldane, A.G.; May, R.M. Systemic risk in banking ecosystems. *Nature* **2011**, *469*, 351–355.
8. Minoiu, C.; Reyes, J.A. A network analysis of global banking: 1978–2010. *J. Financ. Stabil.* **2013**, *9*, 168–184.
9. DasGupta, B.; Kaligounder, L. Contagion in financial networks: Measure, evaluation and implications. 2012; arXiv:1208.3789 [q-fin.GN].
10. Upper, C. Simulation methods to assess the danger of contagion in interbank markets. *J. Financ. Stabil.* **2011**, *7*, 111–125.
11. Sheldon, G.; Maurer, M. Interbank lending and systemic risk: An empirical analysis for Switzerland. *Revue Suisse D'economie Politique Et De Statistique* **1998**, *134*, 685–704.
12. Blåvarg, M.; Nimander, P. Interbank exposures and systemic risk. *Sveriges Riksbank Econ. Rev.* **2002**, *2*, 19–45.
13. Furfine, C.H. Interbank exposures: Quantifying the risk of contagion. *J. Money Credit Bank.* **2003**, 111–128.
14. Upper, C.; Worms, A. Estimating bilateral exposures in the German interbank market: Is there a danger of contagion? *Eur. Econ. Rev.* **2004**, *48*, 827–849.
15. Elsinger, H.; Lehar, A.; Summer, M. Risk assessment for banking systems. *Manage. Sci.* **2006**, *52*, 1301–1314.

16. Van Lelyveld, I.; Liedorp, F. Interbank contagion in the Dutch banking sector: A sensitivity analysis. *Int. J. Cent. Bank.* **2006**, *2*, 99–133.

17. Mistrulli, P.E. Assessing financial contagion in the interbank market: Maximum entropy *versus* observed interbank lending patterns. *J. Bank. Finance* **2011**, *35*, 1114–1127.

18. Mistrulli, P. Interbank Lending Patterns and Financial Contagion. Available online: http://www.econ.upf. edu/docs/seminars/mistrulli.pdf (accessed on 27 May 2005).

19. Schreiber, T. Measuring information transfer. *Phys. Rev. Lett.* **2000**, *85*, 461–464.

20. Marschinski, R.; Matassini, L. *Financial Markets as a Complex System: A Short Time Scale Perspective, Research Notes in Economics & Statistics 01-4*; Eliasson, A., Ed.; Deutsche Bank Research: Frankfurt, Germany, 2001.

21. Baek, S.K.; Jung, W.-S.; Kwon, O.; Moon, H.-T. Transfer entropy analysis of the stock market. 2005; arXiv:physics/0509014 [physics.soc-ph].

22. Kwon, O.; Yang, J.S. Information flow between stock indices. *Europhys. Lett.* **2008**, *82*, 68003.

23. Hu, Q.; Zhang, L.; Zhang, D.; Pan, W.; An, S.; Pedrycz, W. Measuring relevance between discrete and continuous features based on neighborhood mutual information. *Expert Syst. Appl.* **2011**, *38*, 10737–10750.

24. Marschinski, R.; Kantz, H. Analysing the information flow between financial time series. *Eur. Phys. J. B* **2002**, *30*, 275–281.

25. Levine, R.; Zervos, S. Stock markets, banks, and economic growth. *Am. Econ. Rev.* **1998**, *88*, 537–558.

26. Chiarella, C.; Gao, S. The value of the S&P 500—A macro view of the stock market adjustment process. *Global Financ. J.* **2004**, *15*, 171–196.

27. Hooker, M.A. Macroeconomic factors and emerging market equity returns: A Bayesian model selection approach. *Emerg. Mark. Rev.* **2004**, *5*, 379–387.

28. Kullmann, L.; Kertész, J.; Kaski, K. Time-dependent cross-correlations between different stock returns: A directed network of influence. *Phys. Rev. E* **2002**, *66*, 026125.

29. Abarbanell, J.S.; Bushee, B.J. Fundamental analysis, future earnings, and stock prices. *J. Accounting Res.* **1997**, *35*, 1–24.

30. Allen, E. J.; Larson, C. R.; Sloan, R. G. Accrual reversals, earnings and stock returns. *J. Accounting Econ.* **2013**, *56*, 113–129.

31. Eisenberg, L.; Noe, T.H. Systemic risk in financial systems. *Manage. Sci.* **2001**, *47*, 236–249.

entropy

MDPI

Article

Structure of a Global Network of Financial Companies Based on Transfer Entropy

Leonidas Sandoval Jr.

Insper Instituto de Ensino e Pesquisa, Rua Quatá, 300, São Paulo 04546-042, Brazil;
E-Mail: leonidassj@insper.edu.br; Tel.: +55-11-4504-2300

Received: 16 May 2014; in revised form: 30 June 2014 / Accepted: 1 August 2014 /
Published: 7 August 2014

Abstract: This work uses the stocks of the 197 largest companies in the world, in terms of market capitalization, in the financial area, from 2003 to 2012. We study the causal relationships between them using Transfer Entropy, which is calculated using the stocks of those companies and their counterparts lagged by one day. With this, we can assess which companies influence others according to sub-areas of the financial sector, which are banks, diversified financial services, savings and loans, insurance, private equity funds, real estate investment companies, and real estate trust funds. We also analyze the exchange of information between those stocks as seen by Transfer Entropy and the network formed by them based on this measure, verifying that they cluster mainly according to countries of origin, and then by industry and sub-industry. Then we use data on the stocks of companies in the financial sector of some countries that are suffering the most with the current credit crisis, namely Greece, Cyprus, Ireland, Spain, Portugal, and Italy, and assess, also using Transfer Entropy, which companies from the largest 197 are most affected by the stocks of these countries in crisis. The aim is to map a network of influences that may be used in the study of possible contagions originating in those countries in financial crisis.

Keywords: financial markets; propagation of crises; transfer entropy

JEL Classification: G1; G15

1. Introduction

In his speech delivered at the Financial Student Association in Amsterdam [1], in 2009, Andrew G. Haldane, Executive Director of Financial Stability of the Bank of England, called for a rethinking of the financial network, that is the network formed by the connections between banks and other financial institutions. He warned that, in the last decades, this network had become more complex and less diverse, and that these facts may have led to the crisis of 2008.

According to him, it was the belief of theoreticians and practitioners of the financial market that connectivity between financial companies meant risk diversification and dispersion, but further studies showed that networks of certain complexity exhibit a robust but fragile structure, where crises may be dampened by sharing a shock among many institutions, but where they may also spread faster and further due to the connections between companies. Other issue to be considered was the fact that some nodes in the financial network were very connected to others, while some were less connected. The failure of a highly connected node could, thus, spread a small crisis to many other nodes in the network. Another factor was the small-world property of the financial network, where one company was not very far removed from another, through relations between common partners, or common partners of partners.

Such a connected network was also more prone to panic, tightening of credit lines, and distress sales of assets, some of them caused by uncertainties about who was a counterpart to failing companies.

Due to some financial innovations, risk was now shared among many parties, some of them not totally aware of all the details of a debt that was sectorized, with risk being decomposed and then reconstituted in packages that were then resold to other parties. This made it difficult to analyze the risk of individual institutions, whose liabilities were not completely known even to them, since they involved the risks of an increasingly large number of partners.

The other important aspect, the loss of diversity, increased when a large number of institutions adopted the same strategies in the pursuit of return and in the management of risk. Financial companies were using the same models and using the same financial instruments, with the same aims.

In the same speech, Haldane pointed at some directions that could improve the stability of the financial network. The first one was to map the network, what implied the collection, sharing and analysis of data. This analysis needed to include techniques that didn't focus only on the individual firms, like most econometric techniques do, but also on the network itself, using network techniques developed for other fields, like ecology or epidemiology. The second was to use this knowledge to properly regulate this network. The third was to restructure the financial network, eliminating or reinforcing weak points. All these need a better understanding of the connections between financial institutions and how these connections influence the very topology of the financial network.

This article contributes to the first direction pointed by Haldane, that of understanding the international financial network. We do it by calculating a network based on the daily returns of the stocks of the 197 largest financial companies across the world in terms of market capitalization that survive a liquidity filter. These include not just banks, but also diversified financial services, insurance companies, one investment company, a private equity, real estate companies, REITS (Real Estate Investment Trusts), and savings and loans institutions. We use the daily returns in order to build the network because we believe that the price of a stock encodes a large amount of information about the company to which it is associated that goes beyond the information about the assets and liabilities of the company. Also, we believe that it is more interesting to study the effects of stock prices on other stock prices, as in the propagation of a financial crisis, rather than the spreading of defaults, since defaults are events that are usually avoided by injecting external capital into banks.

The network is built using Transfer Entropy, a measure first developed in information science. The network is a directed one, which reveals the transfer of information between the time series of each stock. This network is used in order to determine which are the most central nodes, according to diverse centrality criteria. The identification of these central stocks is important, since in most models of the propagation of shocks, highly central nodes are often the major propagators. We also enlarge the original network obtained by Transfer Entropy to include the most liquid stocks belonging to financial companies in some European countries that have been receiving much attention recently due to the fact that they are facing different degrees of economic crises, and determine who are the major financial companies in the world that are most affected by price movements of those stocks, and which of those stocks belonging to countries in crisis are the most influent ones.

1.1. Propagation of Socks in Financial Networks

The work that is considered the first that deals with the subject is the one of Allen and Gale [2], where the authors modeled financial contagion as an equilibrium phenomenon, and concluded that equilibrium is fragile, that liquidity shocks may spread through the network, and that cascade events depend on the completeness of the structure of interregional claims between banks. In their model, they used four different regions, which may be seen as groups of banks with some particular specializations. They focused in one channel of contagion, which are the overlapping claims that different regions or sectors of the banking system have on one another. According to them, another possible channel of contagion that was not considered is incomplete information among agents. As an example, the information of a shock in one region may create a self-fulfilling shock in another region if that information is used as a prediction of shocks in other regions. Another possible channel of contagion is the effect of currency markets in the propagation of shocks from one country to another.

In their results, the spreading of a financial crisis depends crucially on the topology of the network. A completely connected network is able to absorb shocks more efficiently, and a network with strong connections limited to particular regions which are not themselves well connected is more prone to the dissemination of shocks. In a work previous to theirs, Kirman [3] built a network of interacting agents and made the network evolve with the probability of each of the links dependent on the experience of the agents involved, obtaining results that were very different from those which might have been predicted by looking at the individuals in isolation.

Later and Allen *et al.* [4] made a review of the progress of the network approach to the propagation of crises in the financial market. They concluded that there is an urgent need for empirical work that maps the financial network, so that the modern financial systems may be better understood, and that a network perspective would not only account for the various connections within the financial sector or between the financial sector and other sectors, but also would consider the quality of such links. Upper [5] made a survey of a diversity of simulation methods that have been used with a variety of financial data in order to study contagion in financial networks, and made a comparison between the various methods used.

There is an extensive literature on the propagation of shocks in networks of financial institutions, and describing all the published works in this subject is beyond the scope of this article. Most of the works in this field can be divided into theoretical and empirical ones, most of them considering networks of banks where the connections are built on the borrowing and lending between them. In most theoretical works [6–23], networks are built according to different topologies (random, small world, or scale-free), and the propagation of defaults is studied on them. The conclusions are that small world or scale-free networks are, in general, more robust to cascades (the propagation of shocks) than random networks, but they are also more prone to propagations of crises if the most central nodes (usually, the ones with more connections) are not themselves backed by sufficient funds. Most empirical works [24–38] are also based on the structure derived from the borrowing and lending between banks, and they show that those networks exhibit a core-periphery structure, with few banks occupying central, more connected positions, and others populating a less connected neighborhood. Those articles showed that this structure may also lead to cascades if the core banks are not sufficiently resistant, and that the network structures changed considerably after the crisis of 2008, with a reduction on the number of connected banks and a more robust topology against the propagation of shocks.

1.2. Transfer Entropy

The networks based on the borrowing and lending between banks are useful for determining the probabilities of defaults, but they are not useful in the study of how the stock price of one company relates with the stock price of another company. Such a relation may be obtained using the correlation between each stock price (or better, on its log-return) but, although useful for determining which stocks behave similarly to others, the correlations between them cannot establish a relation of causality or of influence, since the action of a stock on another is not necessarily symmetric. A measure that has been used in a variety of fields, and which is both dynamic and non-symmetric, is *Transfer Entropy*, developed by Schreiber [39] and based on the concept of *Shannon Entropy*, first developed in the theory of information by Shannon [40]. Transfer entropy has been used in the study of cellular automata in Computer Science [41–43], in the study of the neural cortex of the brain [44–49], in the study of social networks [50], in Statistics [51–54], and in dynamical systems [55–57], and received a thermodynamic interpretation in [58].

In terms of the applications of Transfer Entropy to finance, Marschinski and Kantz [59] analyzed the information flow between the S&P500 index of the New York Stock Exchange (USA) and the DAX index of the Frankfurt Stock Exchange (Germany) and detected a nonlinear information transfer between both indices at the one minute scale. They also introduced a measure called *Effective Transfer Entropy*, which subtracts from Transfer Entropy some of the effects of noise or of a highly volatile time series. This concept is now amply used, particularly in the study of the cerebral cortex, and is also used in the present article.

Baek and Jung *et al.* [60] applied Transfer Entropy to the daily returns of 135 stocks listed on the New York Stock Exchange (NYSE) from 1983 to 2003, and concluded that companies of the energy industries influence the whole market. Kwon and Yang [61] applied Transfer Entropy to the S&P500 and Dow Jones indices of the New York Stock Exchange and to the stocks of 125 companies negotiated at this stock exchange in order to analyze the flow of information between them, concluding that there is more information flow from the indices to the stocks than from the stocks to the indices. Kwon and Yang [62] used the stock market indices of 25 countries and discovered that the Transfer Entropy from the American markets is high, followed by that of the European markets, and that the information flows mainly to the Asia Pacific stock markets.

Jizba and Kleinert *et al.* [63] used both Transfer Entropy (based on Shannon's entropy) and a variant version of Transfer Entropy based on Rényi's entropy, which is able to examine different regions of the probability density functions of time series by the variation of a parameter, in the study of the Transfer Entropy and of the Rényi Transfer Entropy between 11 stock market indices sampled in a daily basis in the period 1990–2009 and also between the DAX and the S&P 500 indices based on minute tick data gathered in the period from April, 2008 to November, 2009. Their results show that the information flow between world markets is strongly asymmetric with a distinct information surplus flowing from the Asia-Pacific region to both the European and the US markets, with a smaller excess of information also flowing from Europe to the US, what is clearly seen from a careful analysis of the Rényi information flow between the DAX and S& P500 indices. The results obtained by them are very similar for different choices of the parameter that specifies the sector of the probability distribution functions that is highlighted in the calculations.

Peter and Dimpfl *et al.* [64,65] used Transfer Entropy in order to analyze the information flows between the CDS (Credit Default Swap) market and the corporate bond market using data on 27 iTraxx companies, showing that, although there is information flowing in both directions, the CDS market sends more information to the bond market than vice-versa. Their work also shows that the information flow between both markets has been growing in time, and that the importance of the CDS market as source of information is higher during the crisis of 2008. They also analyzed the dynamic relation between the market risk (proxied by the VIX) and the credit risk (proxied by the iTraxx Europe), showing that information flows mainly from the VIX to the iTraxx Europe, and that, although the transfer of information was mostly bidirectional, the excess information flowing from the VIX to the iTraxx Europe was highest during the crisis of 2008.

Kim and An *et al.* [66] used Transfer Entropy on five monthly macro-economic variables (industrial production index, stock market index, consumer price index, exchange rate, and trade balance) for 18 countries, during the 1990s and the 2000s. They first applied Transfer Entropy in order to study the inter-relations of each of the five variables inside each country, and then the Transfer Entropy between the same variable across countries, for each of the five variables. Besides the relationship between variables inside countries, with some variations of results, they discovered that more influence transfers among the countries in Europe than in Asia or the Americas, most likely reflecting the formation of the European Union, that the stock market indices of Germany and Italy are strong information receivers from other European countries, and that one can expect that signs of the financial crisis originating from some European countries in crisis will be transmitted, with either positive or negative annotation, to the rest of Europe. They also discovered that the Americas, most notably the USA, are sources of information for the stock market indices of Brazil and Mexico, and for the exchange rate in Canada, and receivers of information of trade balance from Mexico and of industrial production index from Argentina. As for Asia, there is a cluster of information transfer formed by China, India and Japan in terms of exchange rate, and another one, in terms of industrial production index, between South Korea, Indonesia and Japan. China and South Korea are large receivers of information of the industrial production index from Indonesia and India, respectively, and Japan influences South Korea in terms of the consumer price index and acts as a receiver of information of the exchange rate from India and Indonesia.

Li and Liang *et al.* [67] used data of the stocks of 16 Chinese banks between 2011 and 2012 and applied Transfer Entropy in order to determine an interbank exposure matrix, using it to evaluate the stability of the Chinese banking system by simulating the risk contagion process using the resulting network. The results show that the Chinese banking system is quite stable with respect to systemic risk, and the study also identifies systemically important banks, what gives regulators information for the development of policies.

Dimpfl and Peter [68] applied Rényi's Transfer Entropy to high frequency data (at one minute intervals) from July, 2003 to April, 2010, of the S&P 500 (USA), the DAX (Germany), the CAC 40 (France), and the FTSE (UK) indices at the intervals in time when all stock exchanges were operating, in order to analyze the information flow across the Atlantic Ocean. Their results show that the information transfer between Europe and America increased during the 2008 financial crisis, and has remained higher than before the crisis occurred. The dominant role of the USA as a source of information to the European markets diminished after the crisis, except in the case of France. They also found that the collapse of the Lehman Brothers led to a significant increase in information flow among the countries that were part of the study. The comparison of results using different parameters for the Rényi Transfer Entropy did not show important differences between them.

1.3. How This Article Is Organized

Section 2 explains the data used in the article and some of the methodology. Section 3 explains Transfer Entropy and uses it in order to study the information flows between the stocks of financial institutions. Section 4 highlights which are the most central stocks according to different centralities criteria. Section 5 studies the dynamics of Transfer Entropy for the stock markets in moving windows in time. Section 6 studies the relationships between countries in crisis in Europe with the largest financial institutions, analyzing which stocks are more affected by movements in the stocks belonging to those countries in crisis. Finally, Section 7 shows some conclusions and possible future work.

2. Data and Methodology

In order to choose appropriate time series of the top stocks in terms of market capitalization belonging to the financial sector, we used the S&P 1200 Global Index as in 2012, which is a free-float weighted stock market index of stocks belonging to 31 countries. The stocks belonging to the index are responsible for approximately 70 percent of the total world stock market capitalization and 200 of them belong to the financial sector, as classified by Bloomberg. From those, we extracted 197 stocks that had enough liquidity with respect to the working days of the New York Stock Exchange (NYSE). From the 197 stocks, 79 belong to the USA, 10 to Canada, 1 to Chile, 21 to the UK, 4 to France, 5 to Germany, 7 to Switzerland, 1 to Austria, 2 to the Netherlands, 2 to Belgium, 5 to Sweden, 1 to Denmark, 1 to Finland, 1 to Norway, 6 to Italy, 4 to Spain, 1 to Portugal, 1 to Greece, 12 to Japan, 9 to Hong Kong, 1 to South Korea, 1 to Taiwan, 3 to Singapore, and 18 to Australia. The time series were collected from January, 2003, to December, 2012, thus covering a period of ten years. The stocks and their classification according to industry and sub industry are listed in Appendix A.

Some of the limitations of our choice of variables are that, first, some companies like Lehman Brothers or Bear-Stearns, which were key players prior to and during the crisis of 2008, are not present, since their stocks do not exist anymore. Second, there are companies that are major players in the financial industry, and particularly some funds, which are not listed in any stock exchange, and so are not in our data set. Such limitations are consequences of our choice of data set, and their effects might be lessened by the number of stocks being considered, but only up to a certain extent.

We took the daily closing prices of each stock, and the resulting time series of all 197 stocks were compared with the time series of the NYSE, which was taken as a benchmark, since it is by far the major stock exchange in the world. If an element of the time series of a stock occurred for a day in which the NYSE wasn't opened, then this element was deleted from the time series, and if an element of the time series of a stock did not occur in a day in which the NYSE functioned, then we repeated the closing price of the previous day. The idea was not to eliminate too many days of the time series by, as an example, deleting all closing prices in a day one of the stock exchanges did not operate. The methodology which we chose would be particularly bad for stocks belonging to countries where weekends occur on different days than for Western countries, like Muslim countries or Israel, but since no stocks from our set belong to those countries, differences on weekends are not relevant here.

The data are organized so as to place stocks of the same country together, and then to discriminate stocks by industry and sub industry, according to the classification used by Bloomberg. From the 197 stocks, 80 belong to Banks, 27 to Diversified Financial Services, 50 to Insurance Companies, 1 to an Investment Company, 1 to a Private Equity, 8 to Real Estate Companies, 28 are REITS (Real Estate Investment Trusts), and 2 belong to Savings and Loans.

In order to reduce non-stationarity of the time series of the daily closing prices, we use the log-returns of the closing prices, defined as

$$R_t = \ln(P_t) - \ln(P_{t-1}) , \tag{1}$$

where P_t is the closing price of the stock at day t and P_{t-1} is the closing price of the same stock at day $t-1$.

Since the stocks being considered belong to stock markets that do not operate at the same times, we run into the issue of lagging or not some stocks. Sandoval [69], when dealing with stock market indices belonging to stock markets across the globe, showed that it is not very clear that an index has to be lagged with respect to another, except in cases like Japan and the USA. A solution is to use both original and lagged indices in the same framework, and to do all calculations as if the lagged indices were different ones. The same procedure is going to be followed here with the log-returns of the closing prices of the stocks that have been selected, so we shall deal with $2 \times 197 = 394$ time series.

3. Transfer Entropy

In this section, we shall describe the concept of Transfer Entropy (TE), using it to analyze the data concerning the 197 stocks of companies of the financial sector and their lagged counterparts. We will start by describing briefly the concept of Shannon entropy.

3.1. Shannon Entropy

The American mathematician, electronic engineer and cryptographer, Claude Elwood Shannon (1916–2001), founded the theory of information in his work "A Mathematical Theory of Communication" [40], in which he derived what is now known as the *Shannon entropy*. According to Shannon, the main problem of information theory is how to reproduce at one point a message sent from another point. If one considers a set of possible events whose probabilities of occurrence are p_i, $i = 1, \cdots, n$, then a measure $H(p_1, p_2, \cdots, p_n)$ of the uncertainty of the outcome of an event given such distribution of probabilities should have the following three properties:

- $H(p_i)$ should be continuous in p_i;
- if all probabilities are equal, what means that $p_i = 1/n$, then H should be a monotonically increasing function of n (if there are more choices of events, then the uncertainty about one outcome should increase);
- if a choice is broken down into other choices, with probabilities c_j, $j = 1, \cdots, k$, then $H = \sum_{j=1}^{k} c_j H_k$, where H_k is the value of the function H for each choice.

Shannon proved that the only function that satisfies all three properties is given by

$$H = -\sum_{i=1}^{N} p_i \log_2 p_i , \tag{2}$$

where the sum is over all states for which $p_i \neq 0$ (Shannon's definition had a constant k multiplied by it, which has been removed here). The base 2 for the logarithm is chosen so that the measure is given in terms of bits of information. As an example, a device with two positions (like a flip-flop circuit) can store one bit of information. The number of possible states for N such devices would then be 2^N, and $\log_2 2^N = N$, meaning that N such devices can store N bits of information, as should be expected. This definition bears a lot of resemblance to Gibbs' entropy, but is more general, as it can be applied to any system that carries information.

The Shannon entropy represents the average uncertainty about measures i of a variable X (in bits), and quantifies the average number of bits needed to encode the variable X. In the present work, given the time series of the log-returns of a stock, ranging over a certain interval of values, one may divide such possible values into N different bins and then calculate the probabilities of each state i, what is the number of values of X that fall into bin i divided by the total number of values of X in the time series. The Shannon entropy thus calculated will depend on the number of bins that are selected. After selecting the number of bins, one associates a symbol (generally a number) to each bin.

Using the stocks of the J.P. Morgan (code JPM), classified as a Diversified Banking Institution, we shall give an example of the calculation of the Shannon Entropy for two different choices of bins. In Figure 1, we show the frequency distributions of the log-returns for the stocks of the J.P. Morgan from 2007 to 2012, which varied from -0.2323 to 0.2239 during that period, with two different binning choices. The first choice results in 24 bins of size 0.02, and the second choice results in 6 bins of size 0.1.

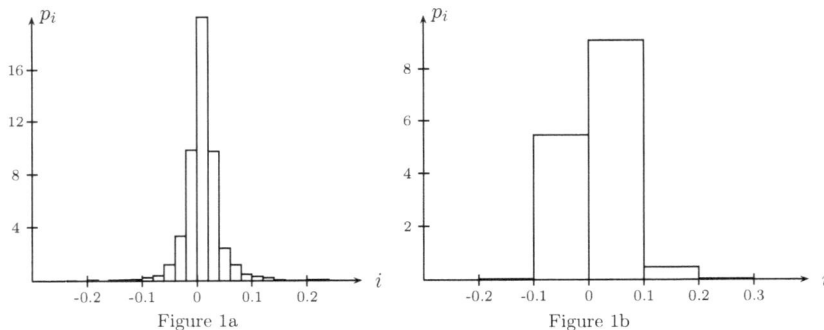

Figure 1a   Figure 1b

Figure 1. Histograms of the log-returns of the stocks of the J.P. Morgan for two different binnings. In Figure 1a, we have 24 bins in intervals of size 0.02, and in Figure 1b, 6 bins in intervals of size 0.1.

To each bin is assigned a symbol, which, in our case, is a number, from 1 to 24 in the first case and from 1 to 6 in the second case. Figure 2 shows the assigning of symbols for the two choices of binning for the first log-returns of the stocks of the J.P. Morgan. Then, we calculate the probability that a symbol appears in the time series and then use (2) in order to calculate the Shannon entropy, which, in our case, is $H = 2.55$ for bins of size 0.02 and $H = 0.59$ for bins of size 0.1. The second result is smaller than the first one because there is less information for the second choice of binning due to the smaller number of possible states of the system. The difference in values, though, is not important, since we shall use the Shannon entropy as a means of comparing the amount of information in different time series.

Date	Log-return	Symbol
01/03/2007	−0.0048	12
01/04/2007	0.0025	13
01/05/2007	−0.0083	12
01/08/2007	0.0033	13
01/09/2007	−0.0042	12
01/10/2007	0.0073	13
⋮	⋮	⋮

Date	Log-return	Symbol
01/03/2007	−0.0048	3
01/04/2007	0.0025	4
01/05/2007	−0.0083	3
01/08/2007	0.0033	4
01/09/2007	−0.0042	3
01/10/2007	0.0073	4
⋮	⋮	⋮

Figure 2. The assigning of symbols to the first values of the log-returns of the J.P. Morgan according to binning. On the left, for 24 bins and, on the right, for 6 bins.

Figure 3 shows the Shannon Entropy calculated for each stock in this study (the lagged stocks are not represented, since their entropies are nearly the same as the entropies of the original stocks). The results for both choices of binning are in fact very similar, and their correlation is 0.97. Stocks with higher Shannon Entropy are the most volatile ones. As one can see, the second choice, with larger bin sizes, shows the differences more sharply, which is one of the reasons why larger binnings are usually favored in the literature.

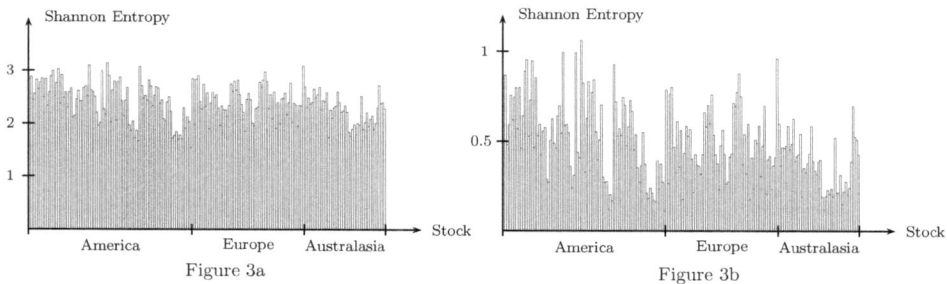

Figure 3a

Figure 3b

Figure 3. Shannon entropies of the 197 stocks, in the same order as they appear in Appendix A. Figure 3a is the Shannon Entropy for bins of size 0.02, and Figure 3b is the Shannon Entropy for bins of size 0.1.

3.2. Transfer Entropy

When one deals with variables that interact with one another, then the time series of one variable Y may influence the time series of another variable X in a future time. We may assume that the time series of X is a Markov process of degree k, what means that a state i_{n+1} of X depends on the k previous states of the same variable. This may be made more mathematically rigorous by defining that the time series of X is a Markov state of degree k if

$$p\left(i_{n+1}|i_n, i_{n-1}, \cdots, i_0\right) = p\left(i_{n+1}|i_n, i_{n-1}, \cdots, i_{n-k+1}\right), \tag{3}$$

where $p(A|B)$ is the conditional probability of A given B, defined as

$$p(A|B) = \frac{p(A, B)}{p(B)}. \tag{4}$$

What expression (3) means is that the conditional probability of state i_{n+1} of variable X on all its previous states is the same as the conditional probability of i_{n+1} on its k previous states, meaning that it does not depend on states previous to the kth previous states of the same variable.

One may also assume that state i_{n+1} of variable X depends on the ℓ previous states of variable Y. The concept is represented in Figure 4, where the time series of a variable X, with states i_n, and the time series of a variable Y, with states j_n, are identified.

The Transfer Entropy from a variable Y to a variable X is the average information contained in the source Y about the next state of the destination X that was not already contained in the destination's past. We assume that element i_{n+1} of the time series of variable X is influenced by the k previous states of the same variable and by the ℓ previous states of variable Y. The values of k and ℓ may vary, according to the data that is being used, and to the way one wishes to analyze the transfer of entropy of one variable to the other.

The Transfer Entropy from a variable Y to a variable X is defined as

$$
\begin{aligned}
TE_{Y \to X}(k,\ell) &= \sum_{i_{n+1},i_n^{(k)},j_n^{(\ell)}} p\left(i_{n+1},i_n^{(k)},j_n^{(\ell)}\right) \log_2 p\left(i_{n+1}|i_n^{(k)},j_n^{(\ell)}\right) \\
&\quad - \sum_{i_{n+1},i_n^{(k)},j_n^{(\ell)}} p\left(i_{n+1},i_n^{(k)},j_n^{(\ell)}\right) \log_2 p\left(i_{n+1}|i_n^{(k)}\right) \\
&= \sum_{i_{n+1},i_n^{(k)},j_n^{(\ell)}} p\left(i_{n+1},i_n^{(k)},j_n^{(\ell)}\right) \log_2 \frac{p\left(i_{n+1}|i_n^{(k)},j_n^{(\ell)}\right)}{p\left(i_{n+1}|i_n^{(k)}\right)} ,
\end{aligned}
\tag{5}
$$

where i_n is element n of the time series of variable X and j_n is element n of the time series of variable Y, $p(A,B)$ is the joint probability of A and B, and

$$
p\left(i_{n+1},i_n^{(k)},j_n^{(\ell)}\right) = p\left(i_{n+1},i_n,\cdots,i_{n-k+1},j_n,\cdots,j_{n-\ell+1}\right)
\tag{6}
$$

is the joint probability distribution of state i_{n+1} with its $k+1$ predecessors, and with the ℓ predecessors of state j_n, as in Figure 4.

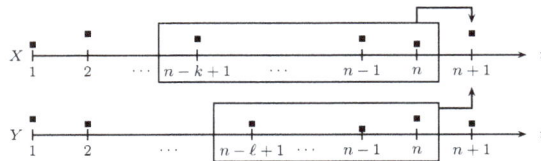

Figure 4. Schematic representation of the transfer entropy $T_{Y \to X}$.

This definition of Transfer Entropy assumes that events on a certain day may be influenced by events of k and ℓ previous days. We shall assume, with some backing from empirical data for financial markets [69], that only the day before is important, since log-returns of the prices of stocks were shown to have low memory (what is not the case for the volatility of the log-returns of prices). By doing so, formula (5) for the Transfer Entropy of Y to X becomes simpler:

$$
\begin{aligned}
TE_{Y \to X} &= \sum_{i_{n+1},i_n,j_n} p\left(i_{n+1},i_n,j_n\right) \log_2 \frac{p\left(i_{n+1}|i_n,j_n\right)}{p\left(i_{n+1}|i_n\right)} \\
&= \sum_{i_{n+1},i_n,j_n} p\left(i_{n+1},i_n,j_n\right) \log_2 \frac{p\left(i_{n+1},i_n,j_n\right) p\left(i_n\right)}{p\left(i_{n+1},i_n\right) p\left(i_n,j_n\right)} ,
\end{aligned}
\tag{7}
$$

where we took $k = \ell = 1$, meaning we are using lagged time series of one day, only.

In order to exemplify the calculation of Transfer Entropy, we will now show some steps for the calculation of the Transfer Entropy from the Deutsche Bank to the J.P. Morgan. In Figure 5, first table, we show the initial part of the time series for the log-returns of the J.P. Morgan, which we call vector X_{n+1} (first column), for its values lagged by one day, vector X_n (second column), and the log-returns of the Deutsche Bank lagged by one day, vector Y_n (third column). Calculating the minimum and maximum returns of the entire set of time series, we obtain a minimum value $m = -1.4949$ and a maximum value $M = 0.7049$. Considering then an interval $[-1.5, 0.8]$ with increments 0.1, we obtain 24 bins to which we assign numeric symbols going from 1 to 24. Then, we associate one symbol to each log-return, depending on the bin it belongs to. As seen in Figure 5, second table, most of the symbols orbit around the intervals closest to zero (corresponding to symbols 15 and 16), since most of the variations of the time series are relatively small.

In order to calculate the simplest probabilities, $p(i_n)$, appearing in (7), we just need to count how many times each symbol appears in vector X_n and then divide by the total number of occurrences. As an example, from the first 10 lines of data shown in Figure 5, symbol 15 appears four times. In order to calculate $p(i_{n+1}, i_n)$, we must count how many times a particular combination of symbols, (a, b), appears in the joint columns X_{n+1} and X_n. As an example, in the first 10 lines of such columns, the combination $(15, 15)$ appears zero times, the combination $(15, 16)$ appears four times, the combination $(16, 15)$ appears four times, and the combination $(16, 16)$ appears two times.

Date	X_{n+1}	X_n	Y_n		X_{n+1}	X_n	Y_n
04/01/2007	0.0025	−0.0048	0.0044		16	15	16
05/01/2007	−0.0083	0.0025	0.0001		15	16	16
08/01/2007	0.0033	−0.0083	−0.0127		16	15	15
09/01/2007	−0.0042	0.0033	−0.0053		15	16	15
10/01/2007	0.0073	−0.0042	0.0056		16	15	16
11/01/2007	0.0044	0.0073	−0.0106	$\longrightarrow$	16	16	15
12/01/2007	−0.0066	0.0044	0.0177		15	16	16
16/01/2007	0.0083	−0.0066	0.0137		16	15	16
17/01/2007	0.0008	0.0083	−0.0012		16	16	15
18/01/2007	−0.0058	0.0008	−0.0048		15	16	15
⋮	⋮	⋮	⋮		⋮	⋮	⋮

Figure 5. Table on the left: first log-returns of the time series of the J.P. Morgan (X_{n+1}), of its lagged values by one day (X_n), and of the log-returns of the Deutsche Bank (Y_n) lagged by one day. Table on the right: symbols are associated to each value of the log-return, inside an interval $[-1.5, 0.8]$ with increments 0.1.

For the whole data, we have the following probabilities and joint probabilities shown in Figure 6. Here, it becomes clearer why, sometimes, it is best to use a binning of larger size in order to calculate Transfer Entropy, since when one has too many binnings, the chance of having particular combinations drop very quickly, making the calculation of probabilities less informing.

We now sum over all combinations of the components of X_{n+1}, X_n, and Y_n using definition (7), obtaining as a result $TE_{177 \to 4} = 0.0155$. This result indicates the average amount of information transferred from the Deustche Bank to the J.P. Morgan which was not already contained in the information of the past state of the J.P. Morgan one day before. Doing the same for all possible combinations of stocks, one obtains a Transfer Entropy matrix, which is represented in terms of false colors in Figure 7a.

X_n	$Freq$	$p(i_n)$
13	1	0.0007
14	13	0.0086
15	757	0.5020
16	720	0.4775
17	14	0.0093
18	3	0.0020

X_{n+1}	X_n	$Freq$	$p(i_{n+1}, i_n)$
13	15	1	0.0007
14	15	7	0.0046
14	16	3	0.0020
14	17	2	0.0013
15	14	5	0.0033
15	15	338	0.2241
15	16	408	0.2706
15	17	5	0.0033
15	18	1	0.0007
16	14	5	0.0033
16	15	404	0.2679
16	16	304	0.2016
16	17	5	0.0033
16	18	2	0.0013
17	14	2	0.0013
17	15	5	0.0033
17	16	5	0.0033
17	17	2	0.0013
18	13	1	0.0007
18	15	2	0.0013

X_n	Y_n	$Freq$	$p(i_n, j_n)$
13	14	1	0.0007
14	14	2	0.0013
14	15	11	0.0073
15	14	10	0.0066
15	15	473	0.3137
15	16	271	0.1797
15	17	3	0.0020
16	15	289	0.1916
16	16	421	0.2792
16	17	10	0.0066
17	14	2	0.0013
17	15	4	0.0027
17	16	6	0.0040
17	17	1	0.0007
17	18	1	0.0007
18	16	2	0.0013
18	17	1	0.0007

X_{n+1}	X_n	Y_n	$Freq$	$p(i_{n+1}, i_n, j_n)$
13	15	15	1	0.0007
14	14	15	1	0.0007
14	15	14	1	0.0007
14	15	15	3	0.0020
14	15	16	3	0.0020
14	16	15	1	0.0007
14	16	16	1	0.0007
14	17	15	1	0.0007
14	17	17	1	0.0007
15	14	14	1	0.0007
15	14	15	4	0.0027
15	15	14	5	0.0033
15	15	15	216	0.1432
15	15	16	115	0.0763
15	15	17	2	0.0013
15	16	15	154	0.1021
15	16	16	247	0.1638
15	16	17	7	0.0046
15	17	14	1	0.0007
15	17	15	1	0.0007
15	17	16	3	0.0020
15	18	16	1	0.0007
16	14	14	1	0.0007

X_{n+1}	X_n	Y_n	$Freq$	$p(i_{n+1}, i_n, j_n)$
16	14	15	3	0.0020
16	15	14	1	0.0020
16	15	15	249	0.1651
16	15	16	151	0.1001
16	15	17	1	0.0007
16	16	15	132	0.0875
16	16	16	170	0.1127
16	16	17	2	0.0013
16	17	14	1	0.0007
16	17	15	1	0.0007
16	17	16	2	0.0013
16	17	18	1	0.0007
16	18	16	1	0.0007
16	18	17	1	0.0007
17	14	15	2	0.0013
17	15	14	1	0.0007
17	15	15	3	0.0020
17	15	16	1	0.0007
17	16	15	2	0.0013
17	16	16	3	0.0020
17	17	15	1	0.0007
17	17	16	1	0.0007
18	13	14	1	0.0007
18	15	15	1	0.0007
18	15	16	1	0.0007

Figure 6. Probabilities and joint probabilities of the times series X_{n+1}, X_n, and Y_n.

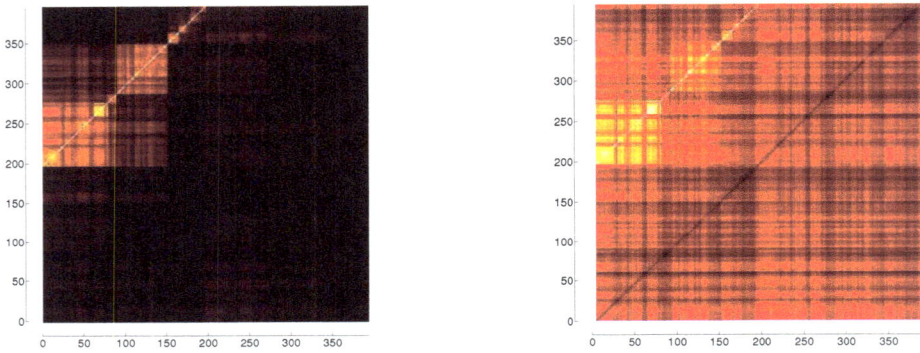

Figure 7a

Figure 7b

Figure 7. False color representations of the Transfer Entropy (TE) matrix. In Figure 7a, we have the representation of the TE for a binning of size 0.1; in Figure 7b, we have the representation of the TE for a binning of size 0.02. The brightness of Figure 7a has been enhanced in comparison with the brightness of Figure 7b, to facilitate visualization.

Here, like in the calculation of the Shannon Entropy, the size of the bins used in the calculations of the probabilities changes the resulting Transfer Entropy (TE). The calculations we have shown in Figures 5 and 6 are relative to a choice of binning of size 0.1. In order to compare the resulting TE matrix with that of another choice for binning, we calculated the TE for binning size 0.02, what leads to a much larger number of bins and to a much longer calculation time. The resulting TE matrix for binning 0.02 is plotted in Figure 7b. The two TE matrices are not very different, with the main dissimilarities being due to scale, and the visualization for binning size 0.1 is sharper than the one obtained using binning size 0.02. In what follows, we shall consider binning size 0.1 throughout the calculations, since it demands less computation time and delivers clearer results in comparison with the ones obtained for some smaller sized binnings.

3.3. *Effective Transfer Entropy*

Transfer Entropy matrices usually contain much noise, due to the finite size of data used in their calculation, non-stationarity of data, and other possible effects, and we must also consider that stocks that have more entropy, what is associated with higher volatility, naturally transfer more entropy to the others. We may eliminate some of these effects [59] if we calculate the Transfer Entropy of randomized time series, where the elements of each time series are individually randomly shuffled so as to break any causality relation between variables but maintain the individual probability distributions of each time series. The original Transfer Entropy matrix is represented in Figure 8a. The result of the average of 25 simulations with randomized data appears in Figure 8b. We only calculated 25 simulations because the calculations are very computationally demanding, and because the results for each simulation are very similar. Then, an Effective Transfer Entropy matrix (ETE) may be calculated by subtracting the Randomized Transfer Entropy matrix (RTE) from the Transfer Entropy matrix (TE):

$$ETE_{Y \to X} = TE_{Y \to X} - RTE_{Y \to X} \,.$$ (8)

The result is shown in Figure 8c.

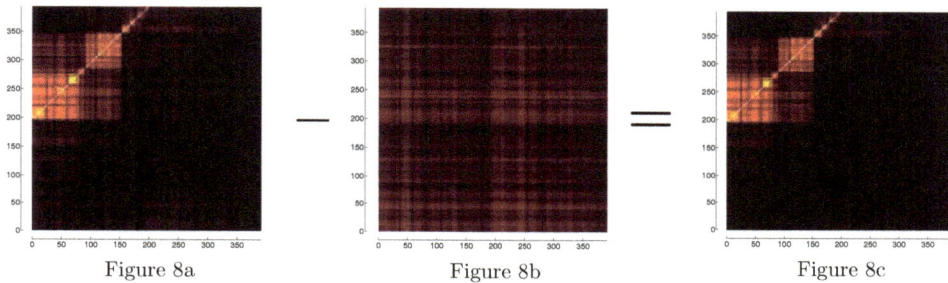

Figure 8a   Figure 8b   Figure 8c

Figure 8. False color representations of the Transfer Entropy matrix (Figure 8a), of the Randomized Transfer Entropy matrix (Figure 8b, the average of 25 simulations with randomized data), and of the Effective Transfer Entropy matrix (Figure 8c). The brightness of the Randomized Transfer Entropy Matrix was enhanced with respect to the other two matrices in order to facilitate visualization.

One effect of calculating Randomized Transfer Entropy matrices is that we may then define a limit where noise is expected to take over. The values calculated for the average of 25 simulations with randomized time series are in between 0 and 0.0523, while the values of the Transfer Entropy matrix calculated with the original time series range from 0 to 1.3407. So, values of TE smaller than around 0.05 are more likely to be the result of noise. The Effective Transfer Entropy matrix has values that range from −0.0202 to 1.3042.

The main feature of the representation of the Effective Transfer Entropy matrix (or of the Transfer Entropy matrix) is that it is clearly not symmetric. The second one is that the highest results are all in the quadrant on the left topmost corner (Quadrant 12). That is the quadrant related with the Effective Transfer Entropy (ETE) from the lagged stocks to the original ones. The main diagonal expresses the ETE from one stock to itself on the next day, which, by the very construction of the measure being used, is expected to be high. But Quadrant 12 also shows that there are larger transfers of entropy from lagged stocks to the other ones than between stocks on the same day. We must remind ourselves that we are dealing here with the daily closing prices of stocks, and that the interactions of prices of stocks, and their reactions to news, usually occur at high frequency. Here, we watch the effects that a whole day of negotiations of a stock has on the others. Figure 9a shows a closer look at the ETE of the stocks on stocks on the same day, what corresponds to the quadrant on the bottom left (Quadrant 11), and from lagged to original stocks, in Figure 9b (Quadrant 12).

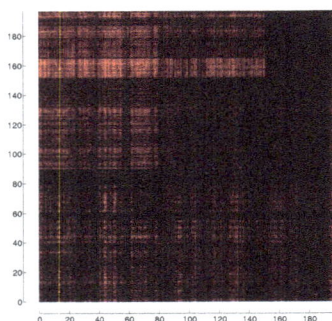

Figure 9a

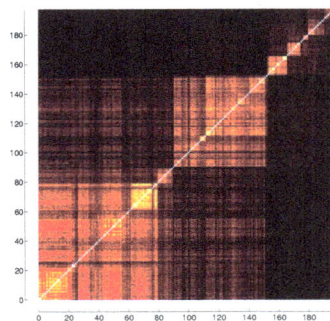

Figure 9b

Figure 9. False color representations of two quadrants of the Transfer Entropy matrix. Figure 9a shows the quadrant of the Effective Transfer Entropies (ETEs) from stocks to the stocks at the same day (Quadrant 11), and Figure 9b shows the quadrant of ETEs from lagged stocks to original ones (Quadrant 12). The brightness of Figure 9a has been enhanced with respect to the brightness of Figure 9b, for better visualization.

Analyzing Quadrant 12 (Figure 9b), we may see again the structures due to geographical positions, with clusters related with stocks from the USA (1 to 79), Canada (80 to 89), Europe (91 to 152), Japan (153 to 165), Hong Kong (166 to 174), Singapore (177 to 179), and Australia (180 to 197). We also detect some ETE from lagged stocks from the USA to stocks from Canada and Europe, from lagged stocks from Europe to stocks from the USA and Canada and, with a smaller strength, from lagged stocks from Europe to stocks from Australasia, and transfer of entropy within the Australasian stocks.

Quadrant 11 (Figure 9a) shows much smaller values, but one can see a clear influence of Japan (153–165) on North America (1–89) and Europe (91–152), and also some influence from Europe to the USA. A very light influence may be seen from the USA to itself on the next day, Canada, and Europe, but it is already hard to distinguish this influence from noise. There are negative values of ETE, what means that the Transfer Entropy calculated is smaller than what would be expected from noise. These are the same results found in [63], who used only same day time series in their calculations.

The Effective Transfer Entropy relations may be used in order to define a network where each stock is represented by a *node* and each ETE between two stocks is an *edge*. The network defined by the ETE matrix is one in which each edge has a label attached to it, which is the ETE it represents [70]. Another type of network may be obtained if one defines a threshold value for the ETE and then represents only the ETEs above this threshold as edges, and only the nodes connected by edges thus defined are represented in the network. The representation of such network is called an *asset graph*, and by using the concept of asset graph, we may choose values for a threshold and represent only the edges that are above that threshold and the nodes connected by them. By choosing appropriate threshold values for the ETE, above which edges and nodes are removed, we may obtain some filtered representations of the ETE structure between the stocks. This is more clearly visible if one plots only the elements of the ETE matrix that are above a certain threshold. In Figure 10, we take a closer look at the relationships between the stocks at threshold 0.4.

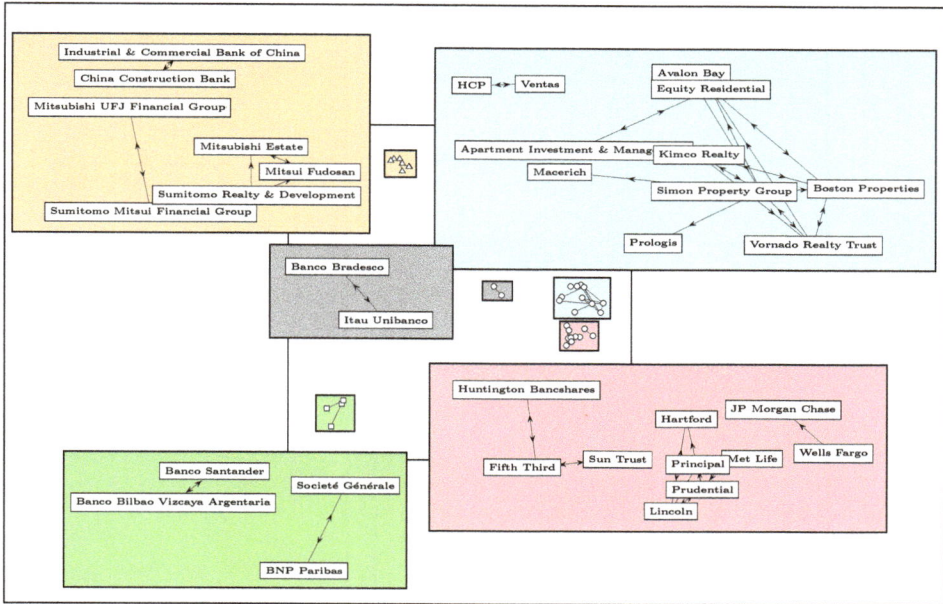

Figure 10. Detailed look at the ETEs between stocks at threshold 0.4. Each group of stocks is located in a magnified window, with the names of each stock close to the position it occupies in the complete network.

At the lower right corner, there are three small clusters of stocks from the USA in the same rectangle: the first one is the transfer entropy between stocks of two banks, the second one is a cluster of insurance companies, and the third one is a small cluster of super-regional banks. At the top right rectangle, there are two clusters of stocks from the USA, the first one a large cluster of REITS (Real Estate Investment Trusts), and the second one a pair of two REITS of Health Care. At the center of the graph, we have a rectangle with stocks of two major commercial banks based in Brazil negotiated in the New York Stock Exchange. At the lower left of the graph, there are two pairs: one of diversified banking institutions from France and one of major commercial banks from Spain. At the top left, we have the last clusters; the first one, a pair of stocks from Japan, and the second one is a cluster of Real Estate operations, management and services firms; the third one is a pair of two commercial banks from Hong Kong. It is to be noticed that most relations are reciprocate, although the ETE between stocks is rarely very similar.

We shall not make a deeper analysis of the remaining asset graphs, but one can see from the ETE matrix in figures 8 and 9 that integration begins inside countries, with the exception of certain countries from Europe, and then goes continental. Only at threshold 0.1 and below do we start having intercontinental integration. This may be due to differences in operation hours of the stock exchanges, to geographical, economic and cultural relations, or to other factors we failed to contemplate (see, for instance, [71] for a discussion and for further references).

3.4. Aggregate Data

After [66], we now aggregate data so as to compare first the ETEs among countries and then among continents. We do this by first calculating the correlation matrix of the stocks belonging to each country (we use Pearson's correlation) and then by calculating its eigenvalues and eigenvectors. As an example, we take the time series of log-returns of the 79 stocks belonging to the USA and calculate their correlations, and the eigenvalues and eigenvectors of the resulting correlation matrix. The largest eigenvalue is usually much larger than the others (it amounts to around 61% of the sum of the eigenvalues for the US data), and its first eigenvector (when normalized so that the sum of its components equals to 1) provides the weights of a vector that, when multiplied by the log-returns, results in an index which is associate with a "market mode" for that particular market. For countries with just one stock represented in the data, the index was the same time series of the single stock. So, by using this procedure, we created one index for each country, based on the stocks of the financial sector, only. Each index has a time series that is then used to calculate an ETE matrix where each line and column corresponds to one country, according to Table 1, in the same order as in this table.

Table 1. Countries used for the calculation of the ETE with aggregate data by country.

Countries					
1 - USA	5 - France	9 - Netherlands	13 - Finland	17 - Portugal	21 - South Korea
2 - Canada	6 - Germany	10 - Belgium	14 - Norway	18 - Greece	22 - Taiwan
3 - Chile	7 - Switzerland	11 - Sweden	15 - Italy	19 - Japan	23 - Singapore
4 - UK	8 - Austria	12 - Denmark	16 - Spain	20 - Hong Kong	24 - Australia

The ETE matrix for this data is depicted in Figure 11, together with the matrix corresponding to the ETEs from lagged to original variables. Again, we can notice a flow of information from Pacific Asia and Oceania to Europe and America on the same day (lower left quadrant), and higher values of ETE among European countries (top left quadrant and Figure 9b). There are particularly high values of ETE from lagged France and lagged Switzerland to the Netherlands and, almost symmetrically, from the lagged Netherlands to France and Switzerland.

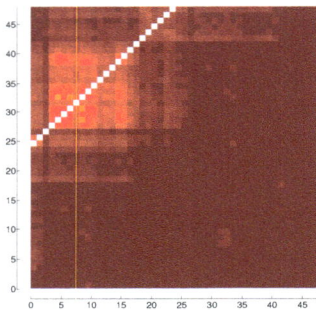

Figure 11a

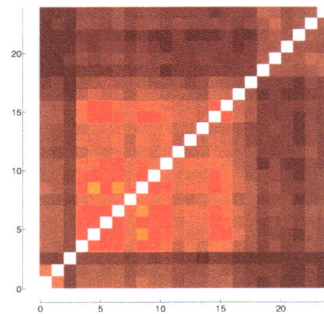

Figure 11b

Figure 11. (**Left**) Effective Transfer Entropy (ETE) matrix of the aggregate data by country; (**Right**) sector of the ETE matrix corresponding to ETEs from lagged to original variables. Brighter colors represent higher values of ETE and darker colors correspond to lower values of ETE.

We may do the same for continents, aggregating data now by continent, with three continents, America, Europe, and Asia if we consider Oceania together with Asia. The resulting ETE matrix is plotted in Figure 12. There is some transfer of entropy from Europe to America on the same day, and also from Asia and Oceania to America and Europe on the same day. We also have higher values of ETE from lagged America to Europe and Asia on the next day, from lagged Europe to America and Asia on the next day, and from lagged Asia to Europe on the next day. These results confirm the ones obtained in [63] and [66].

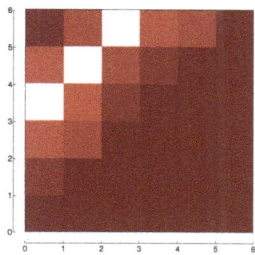

Figure 12a

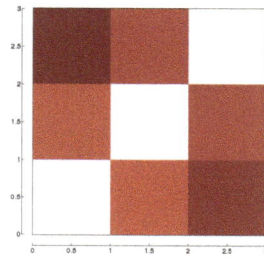

Figure 12b

Figure 12. (**Left**) Effective Transfer Entropy (ETE) matrix of the aggregate data by continent; (**Right**) sector of the ETE matrix corresponding to ETEs from lagged to original variables. Brighter colors represent higher values of ETE and darker colors correspond to lower values of ETE.

Also based on [66], we now aggregate data by industry (according to Bloomberg), in the same way that it was done for countries and continents, in order to study more directly the flow of information between the industries of the financial sector. The industries are the same as the ones we used in order to classify stocks within a country, and are displayed in Table 2.

Table 2. Industries used for the calculation of the ETE with aggregate data by industry.

Industries		
1 - Banks	4 - Investment Companies	7 - REITS
2 - Diversified Financial Services	5 - Private Equity Funds	8 - Savings and Loans
3 - Insurance	6 - Real State	

Figure 13 (left) shows the ETE matrix from industry to industry, and also (right) the sector of ETEs from lagged to original variables. Looking at the lower left quadrant, one may see that there is some transfer of entropy from Private Equity Funds and Real State Investments to Insurance and REITS, and also to Banks, Diversified Financial Services, and to Savings and Loans, all in the same day of negotiation.

Now, looking at the top left quadrant and in Figure 13b, we have the ETEs from lagged to original variables. There is clearly a cluster of Banks, Diversified Financial Services and Insurance Companies exchanging much information, also exchanging information in a lesser degree with Investment Companies, REITS, and Savings and Loans, which exchange some information with one another; finally, there is a very strong connection between Private Equity Funds and Real State Companies.

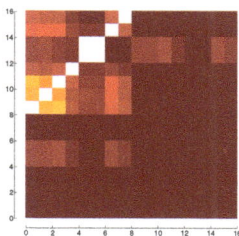

Figure 13a

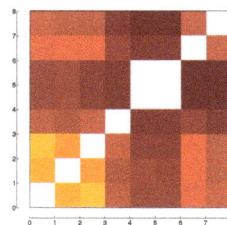

Figure 13b

Figure 13. (**Left**) Effective Transfer Entropy (ETE) matrix of the aggregate data by industry; (**Right**) sector of the ETE matrix corresponding to ETEs from lagged to original variables. Brighter colors represent higher values of ETE and darker colors correspond to lower values of ETE.

3.5. The Rényi Transfer Entropy

By adopting a different definition of entropy, we may define another measure of how entropy is transferred from a variable to another in such a way as to highlight the influence of extreme events in the time series of the variables concerned. Rényi entropy was first defined by the Hungarian mathematician Alfréd Rényi [72] as

$$H = \frac{1}{1-q} \log_2 \sum_{i=1}^{N} p_i^q ,$$ (9)

where the logarithm may be of any base, but we adopt base 2 so as to obtain measures in bits, and q is a positive parameter that may be varied in order to emphasize different regions of the probability distribution p_i. For $q < 1$, small values of p_i (the tails of the probability distribution) assume greater importance, and for $q > 1$, the larger values (more central) of the probability distribution assume greater importance. It may be shown that the Shannon entropy is the limiting case of the Rényi entropy when $q \to 1$.

Jizba and Kleinert *et al.* [63] proposed a variant of Transfer Entropy based on the Rényi entropy, and applied it to financial data. This variant may be written like

$$RTE_{Y \to X}(k, \ell) = \frac{1}{1-q} \log_2 \frac{\sum\limits_{i_{n+1}, i_n^{(k)}} \phi_q \left(i_n^{(k)} \right) p^q \left(i_{n+1} | i_n^{(k)} \right)}{\sum\limits_{i_{n+1}, i_n^{(k)}, j_n^{(\ell)}} \phi_q \left(i_n^{(k)}, j_n^{(\ell)} \right) p^q \left(i_{n+1} | i_n^{(k)}, j_n^{(\ell)} \right)} ,$$ (10)

where ϕ_q is the escort distribution [73] given by

$$\phi_q(i) = \frac{p^q(i)}{\sum\limits_i p_i^q} .$$ (11)

Rényi Transfer Entropy may be negative, and it may be zero without implying that processes X and Y are independent, and it also depends on the parameter q that enhances different regions of the probability distributions. Dimpfl and Peter [68] applied Rényi's Transfer Entropy to high frequency financial data and compared results using different values of the parameter q. In the particular case of $k = \ell = 1$, we obtain

$$RTE_{Y \to X} = \frac{1}{1-q} \log_2 \frac{\sum\limits_{i_{n+1}, i_n} \phi_q \left(i_n \right) p^q \left(i_{n+1} | i_n \right)}{\sum\limits_{i_{n+1}, i_n, j_n} \phi_q \left(i_n, j_n \right) p^q \left(i_{n+1} | i_n, j_n \right)} ,$$ (12)

which is the version we shall use here, since log-returns of financial data rarely depend on more than one day of data in the past, and because the calculations are also much simpler and faster using this simplification.

Figure 14 shows the Rényi Transfer Entropy calculated for $q = 0.1$, $q = 0.5$, $q = 0.9$, and $q = 1.3$. One may see that, although the local details are clearer for lower values of q, which favor the low probabilities of the probability distributions used in the calculations, the overall detail is best for higher values of q. One may also notice that, for $q = 0.9$, the results are very similar to the ones obtained with Transfer Entropy based on Shannon's entropy.

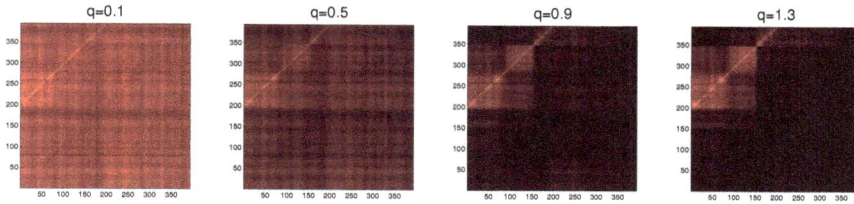

Figure 14. Rényi Transfer Entropy for different values of the parameter q. Brighter colors indicate higher values of transfer entropy, and darker colors indicate lower values of transfer entropy.

4. Centralities

In all studies on the propagation of shocks in financial networks discussed in the introduction [2–38], the centrality of a node (generally a bank in most studies) is one of the single most important factors in the capacity of that node in the propagation a crisis. In network theory, the centrality of a node is important in the study of which nodes are, by some standard, more influential than others. Such measures may be used, for instance, in the study of the propagation of epidemics, or the propagation of news, or, in the case of stocks, in the spreading of high volatility. There are various centrality measures [70], tending to different aspects of what we may think of "central". For undirected networks, for instance, we have Node Degree (ND), which is the total number of edges between a node and all others to which it is connected. This measure is better adapted to asset graphs, where not all nodes are connected between them, and varies according to the choice of threshold [71]. Another measure that can be used for asset graphs is Eigenvector Centrality (EC), which takes into account not just how many connections a node has, but also if it is localized in a region of highly connected nodes. There is also a measure called Closeness Centrality (CC) that measures the average distance (in terms of number of edges necessary to reach another node) of a certain node. This measure is larger for less central nodes, and if one wants a measure that, like the others, is larger for more central nodes, like the others we cited, then one may use Harmonic Closeness (HC), that is built on the same principles as Closeness Centrality, but is calculated using the inverse of the distances from one node to all others. The Betweenness Centrality (BC) of a node is another type of measure, that calculates how often a certain node is in the smaller paths between all other nodes. Still another measure of centrality, called Node Strength (NS), works for fully connected networks, and so is independent of thresholds in asset graphs, and takes into account the strength of the connections, which, in our case, are the correlations between the nodes. It measures the sum of the correlations of a node with all the others.

These measures of centrality are appropriate for an undirected network, like one that could be obtained by using correlation, but the networks built using Effective Transfer Entropy are directed nodes, that have either ingoing edges to a node, outgoing edges from the node, or both. So, centrality measures often break down into ingoing and outgoing ones. As an example, a node may be highly central with respect to pointing at other nodes, like the Google search page; these are called *hubs*. Other nodes may have many other nodes pointing at it, as in the case of a highly cited article in a network of citations; these are called *authorities*. Each one is central in a different way, and a node may be central according to both criteria. Node degree, for example, may be broken in two measures: In Node Degree (ND_{in}), which measures the sum of all ingoing edges to a certain node, and Out Node Degree (ND_{out}), which measures the sum of all outgoing edges from a node. In a similar way, one defines In Eigenvector Centrality (EC_{in}) and Out Eigenvector Centrality (EC_{out}), and In Harmonic Closeness (HC_{in}) and Out Harmonic Closeness (HC_{out}). Betweenness Centrality is now calculated along directed paths only, and it is called Directed Betweenness Centrality, (BC_{dir}).

As we said before, when applying centrality measures to asset graphs, those measures vary according to the chosen value for the threshold. As extreme examples, if the threshold is such that the network has very few nodes, Node Centrality, for example, will also be low. If the threshold value is such that every node is connected to every other node, then all Node Degrees will be the same: the number of all connections made between the nodes. It has been shown empirically [71] that one gets the most information about a set of nodes if one considers asset graphs whose thresholds are close to the minimum or the maximum of the values obtained through simulations with randomized data. We may rephrase it by saying that we obtain more information of a network when we consider its limit to results obtained from noise. From the simulations we have made in order to calculate the Effective Transfer Entropy, we could check that the largest values of Transfer Entropy for randomized data are close to 0.05 for the choice of bins with size 0.1 (Figure 1a). So, we shall consider here the centrality measures that were mentioned applied to the directed networks obtained from the Effective Transfer Entropy with threshold 0.05. The results are plotted in Figure 15. As the values of different centralities may vary a lot (from 3 to 153 for ND_{in} and from 0 to 1317 for BC_{dir}), we normalize all centrality measures by setting their maxima to one. For all but Directed Betweenness Centrality, stocks belonging to the Americas and to Europe appear more central.

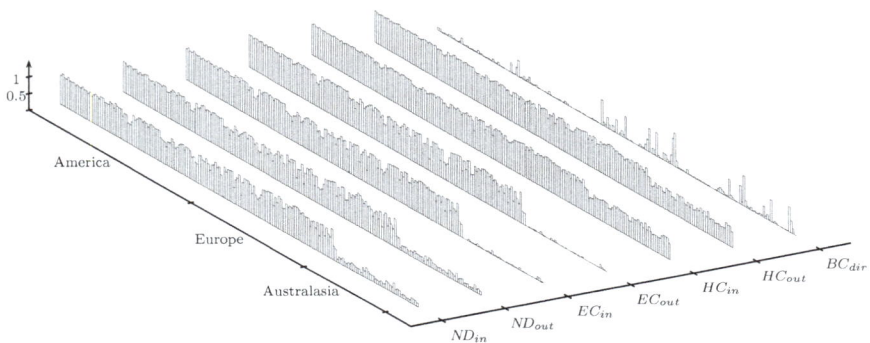

Figure 15. Centrality measures of stocks for the asset graph with threshold 0.05. All measures were normalized so as to have maximum one.

Table 3 presents the most central stocks according to each centrality measure. Only the first five stocks are shown (more, in case of draws). Lagged stocks appear with an $*$ besides the names of the companies. Since we are considering only the strong values of Effective Transfer Entropy, and since asset graphs do not involve the nodes that are not connected, this excludes all connections, except the ones between lagged and original log-returns. So, all in degrees are of original stocks and all out degrees (including Directed Betweenness) are of lagged stocks. For out degrees, insurance companies occupy the top positions, together with some banks, all of them belonging to European or to U.S. companies. For in degrees, we see a predominance of banks, but insurance companies also occupy top positions. This means there is a tendency of entropy being transferred from insurance companies to banks. For Directed Betweeenness, the top positions are occupied by major European banks and also by other types of companies.

By inspection, we may see that the companies with the largest centralities are also the ones with the larger values in terms of market capitalization. This same result has been found in the empirical results relating to the networks found by using the borrowing and lending between banks [24–38]. These networks, with a structure in which some few nodes have large centralities and most nodes have low centralities, are the ones that, in theoretical simulations [6–23], show more stability than most random networks, but also lead to more propagation of shocks when the most central nodes are not strong enough to act as appropriate buffers for the shocks.

Table 3. Classification of stocks with highest centrality measures, the countries they belong to, their industry and sub industry classifications, for asset graphs based on threshold 0.05. Only the five stocks with highest centrality values are shown (more, in case of draws).

Centrality	Company	Country	Industry	Sub Industry
		In Node Degree		
153	Credit Suisse Group AG	Switzerland	Banks	Diversified Banking Inst
150	Deutsche Bank AG	Germany	Banks	Diversified Banking Inst
149	Invesco	USA	Diversified Finan Serv	Invest Mgmnt/Advis Serv
149	ING Groep NV	Netherlands	Insurance	Life/Health Insurance
149	KBC Groep NV	Belgium	Banks	Commer Banks Non-US
		Out Node Degree		
160	ING Groep NV*	Netherlands	Insurance	Life/Health Insurance
158	Hartford Financial Services Group*	USA	Insurance	Multi-line Insurance
154	KBC Groep*	Belgium	Banks	Commer Banks Non-US
152	Genworth Financial*	USA	Insurance	Multi-line Insurance
151	Lincoln National Corp*	USA	Insurance	Life/Health Insurance
		In Eigenvector		
11.99	Invesco	USA	Diversified Finan Serv	Invest Mgmnt/Advis Serv
11.91	Credit Suisse Group AG	Switzerland	Banks	Diversified Banking Inst
11.86	Hartford Financial Services Group	USA	Insurance	Multi-line Insurance
11.85	Lincoln National Corp	USA	Insurance	Life/Health Insurance
11.83	MetLife	USA	Insurance	Multi-line Insurance
		Out Eigenvector		
0.094	Hartford Financial Services Group*	USA	Insurance	Multi-line Insurance
0.094	Lincoln National Corp*	USA	Insurance	Life/Health Insurance
0.093	Invesco*	USA	Diversified Finan Serv	Invest Mgmnt/Advis Serv
0.093	MetLife*	USA	Insurance	Multi-line Insurance
0.093	ING Groep*	Netherlands	Insurance	Life/Health Insurance
0.093	Genworth Financial*	USA	Insurance	Multi-line Insurance
0.093	Principal Financial Group*	USA	Insurance	Life/Health Insurance
0.093	UBS*	Switzerland	Banks	Diversified Banking Inst
0.093	Prudential Financial*	USA	Insurance	Life/Health Insurance
0.093	Ameriprise Financial*	USA	Diversified Finan Serv	Invest Mgmnt/Advis Serv
		In Harmonic Closeness		
174.00	Credit Suisse Group AG	Switzerland	Banks	Diversified Banking Inst
172.5	Deutsche Bank AG	Germany	Banks	Diversified Banking Inst
171.8	KBC Groep NV	Belgium	Banks	Commer Banks Non-US
171.2	ING Groep NV	Netherlands	Insurance	Life/Health Insurance
170.5	Commerzbank AG	Germany	Banks	Commer Banks Non-US
		Out Harmonic Closeness		
178	ING Groep*	Netherlands	Insurance	Life/Health Insurance
177	Hartford Financial Services Group*	USA	Insurance	Multi-line Insurance
175	KBC Groep*	Belgium	Banks	Commer Banks Non-US
174	Genworth Financial*	USA	Insurance	Multi-line Insurance
173	Barclays*	UK	Banks	Diversified Banking Inst
		Directed Betweenness		
1317	KBC Groep*	Belgium	Banks	Commer Banks Non-US
1202	China Construction Bank Corp*	Hong Kong	Banks	Commer Banks Non-US
1074	ING Groep*	Netherlands	Insurance	Life/Health Insurance
998	Goodman Group*	Australia	REITS	REITS-Diversified
984	Barclays*	UK	Banks	Diversified Banking Inst

For thresholds 0.1 and 0.2, with results not displayed here, there is a preponderance of insurance companies and banks from the USA, and for thresholds 0.3 and 0.4, also not displayed here, there are mostly banks and REITS occupying the first positions, also due to the fact that they are some of the only nodes that are part of the asset graphs at these threshold values.

The centrality measures we have considered thus far in this section do not take into account the strength of the connections between the nodes. There are centrality measures that take that into account, being the main one called *Node Strength* (NS), which, in undirected networks, is the sum of all connections made by a node. For directed networks, we have the *In Node Strength* (NS_{in}), which measures the sum of all ingoing connections to a node, and the *Out Node Strength* (NS_{out}), which measures the sum of all outgoing connections from a node. These are centrality measures that can be applied to the whole network, including all nodes. Figure 16 shows the results for both centrality measures, and Table 4 shows the top five stocks according to each node centrality. We used ETE in the calculations. Had we used TE instead, the results would be the same.

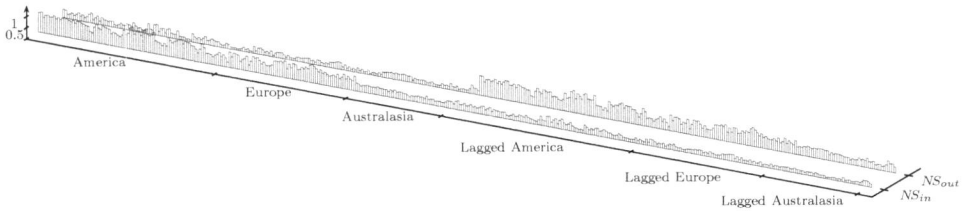

Figure 16. Node Strengths (in and out) for the whole network. Both measures were normalized so as to have maximum one.

Table 4. Top five stocks according to In Node Strength and to Out Node Strength, the countries they belong to, their industry and sub industry classifications. Nodes related with lagged stocks have an asterisk beside their names. Calculations were based on the ETEs between stocks.

Centrality	Company	Country	Industry	sub industry
		In Node Strength		
30.34	Hartford Financial Services Group	USA	Insurance	Multi-line Insurance
29.86	Lincoln National Corp	USA	Insurance	Life/Health Insurance
29.77	Prudential Financial	USA	Insurance	Life/Health Insurance
29.22	Principal Financial Group	USA	Insurance	Life/Health Insurance
27.87	Citigroup	USA	Banks	Diversified Banking Inst
		Out Node Strength		
30.16	Hartford Financial Services Group *	USA	Insurance	Multi-line Insurance
28.71	Prudential Financial *	USA	Insurance	Life/Health Insurance
27.83	Lincoln National *	USA	Insurance	Life/Health Insurance
27.31	Principal Financial Group *	USA	Insurance	Life/Health Insurance
26.57	ING Groep NV *	Netherlands	Insurance	Life/Health Insurance

The five top stocks for In Node Strength are those of Insurance Companies, qualified as authorities, which are nodes to which many other nodes point, and with high values of ETE, what means that there is a large amount of information flowing into the log-returns of those stocks. For Out Node Strength, again insurance companies dominate, what means that they send much information into the prices of the other stocks (they are also hubs).

5. Dynamics

We now look at the network of stocks of financial companies evolving in time. In order to do so, we use moving windows, each comprising data of one semester of a year, moving one semester at a time. The number of days of a semester ranges from 124 to 128, so that there is some small sample effect on the results of ETE, but this choice makes it possible to analyze the evolution of the transfer of information between the stocks at a more localized level in time. Figure 17 shows the ETEs calculated at each semester, and one can see that there are brighter colors, indicating higher levels of ETE, in times of crisis, like in the second semester of 2008 (Subprime Mortgage Crisis) and in the second semester of 2011 (European Sovereign Debt Crisis). Looking at the ETEs from original to original variables (bottom left sector of each ETE matrix), one can see that the largest ETEs on the same day are from Japanese stocks to American and European ones, as it was the case when we used the whole data, but now one may follow a growth in ETE from Japan to America and Europe on the second semester of 2007, on the first semester of 2010, on the second semester of 2011, and on the first semester of 2012.

Figure 18 shows the ETEs only from lagged to original variables, corresponding to the top, left sector of each ETE matrix, with the self-interactions removed for better visualization. The exchange of information between the time series of the stocks is low for the first semester of 2007, except for most US REITS. It increases, mostly among US banks and stocks from Europe, in the second semester of the same year. Prior to the crisis of 2008, we have high ETEs among US stocks, among some European stocks, and also between some stocks from Japan and from Hong Kong. During the height of the Subprime crisis (second semester of 2008 and first semester of 2009), the high exchange of information (represented by high ETE) seems to be restricted mostly to US and Canadian stocks. One interesting result is that the ETE is higher for the second semester of 2011, the height of the European Sovereign Debt Crisis, than during the crisis of 2008. The ETE lowers soon afterwards to normal levels. Something else to be noticed is that the exchange of information among REITS decreases in time. Since REITS (Real Estate Investment Trusts) represent the interest of investors in the real state market, a decrease in ETE is associated with lower volatility in this particular market. On a more local level, one can also detect an increase in ETE among mainly Japanese stocks and also among stocks from Hong Kong after the second semester of 2011.

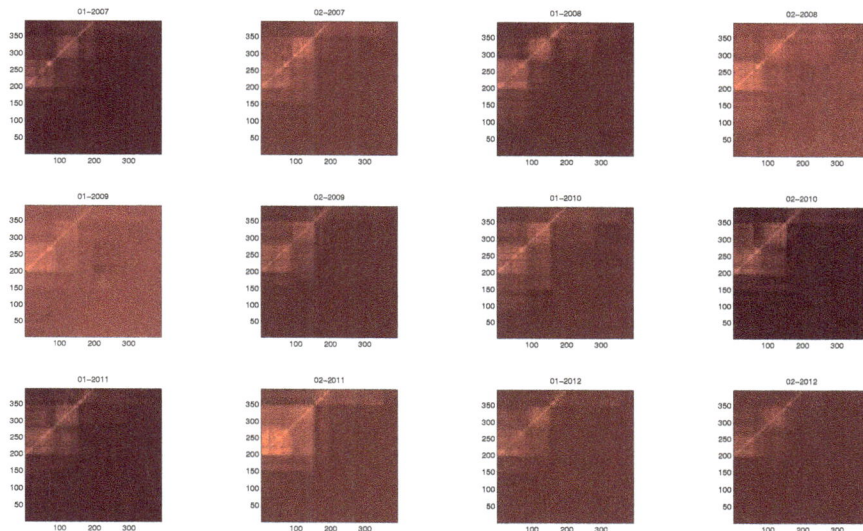

Figure 17. Effective Transfer Entropy (ETE) calculated for each semester, from 2007 to 2012. Brighter colors represent higher values of ETE and darker colors represent lower values of ETE.

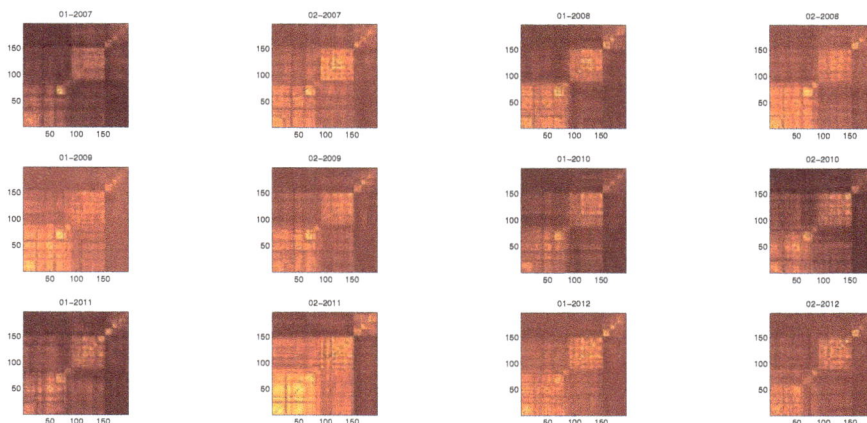

Figure 18. Effective Transfer Entropy (ETE) from lagged to original variables, calculated for each semester, from 2007 to 2012. Brighter colors represent higher values of ETE and darker colors represent lower values of ETE. The main diagonal, representing the ETE from a lagged variable to itself, has been removed for clarity of vision.

6. Relations with Economies in Crisis

Economic broadcasts of the past few years constantly warned of the dangers of a new global financial crisis that may be triggered by the failure of some European countries to pay their sovereign debts. It is not completely clear how far reaching a default by one of those countries could be, and which institutions are more vulnerable to that. Using networks based on financial loans and debts between banks, researchers can try to evaluate some of the consequences of defaults in banks, but, as said in the introduction, networks built on loans and debts do not account for a myriad of other economical facts that define the relationships between financial institutions. So, in order to attempt to study those relations, we shall build networks based on the ETEs between the 197 major financial institutions considered until now together with all financial institutions listed in Bloomberg of some of those countries in crisis, after a liquidity filter. The aim is to investigate which of the main financial institutions receive more entropy from the financial institutions of those countries, meaning that the prices of stocks from those target institutions are much influenced by the prices of institutions that might be in danger of collapse. Of course, we are not saying here that the institutions being considered that belong to one of the countries in crisis might default; we just analyze what could happen if the prices of their stocks would go substantially down.

The countries we shall consider here are Greece, Cyprus, Spain, Portugal, Italy, and Ireland, although Italy is not considered as a country in crisis, but is usually pointed at as being a fragile economy at the moment. We will do a separate analysis for each country, following the same procedures. First, we remove the stocks belonging to the country in crisis from the original network of financial institutions; then we add to this network all stocks that belong to the country in crisis and that are listed in Bloomberg. The number of stocks from each country is restrained by the data available and by the liquidity of those stocks. The second condition eliminates many of the time series available, particularly in less developed stock markets.

Greece is represented by 17 stocks, including the Bank of Greece, which is removed from the 197 original stocks of financial companies. For Cyprus, we obtain the time series of 20 stocks, after removing the less liquid ones. Spain is one of the main players in the international fears for the world economic market; we remove the stocks belonging to Spanish companies (four of them) from the bulk of main stocks and then add 26 stocks of financial companies from that country, including the ones that had been previously removed. Portugal is also an important country in the monitoring for an economic crisis since its institutions have deep connections with Spanish companies. In order to study the influence of its stocks on other stocks of main financial companies, we first remove the one stock belonging to Portugal in that group, that of the Banco Espírito Santo. Then we add to the data the log-returns of five major Portuguese banks, including the one that had been removed from the main block. The country in this group with the largest number of companies that take part of the original data set, 6 of them, is Italy, for which we start by removing those stocks from the main block, including the 6 original ones. Then we add 61 stocks belonging to the financial sector which are negotiated in Italy and which survive the liquidity filter. For Ireland, we have four stocks that survive the liquidity filter.

Table 5 shows the first five stocks that receive the most ETE from the stocks of each country in crisis. Almost all stocks that receive the most ETE are banks, with the exception of the ING Groep, which is a Dutch corporation that specializes in general banking services and in insurance, and so is not just an insurance company, but also a bank. The stocks that are most affected by Greek stocks are well spread among European banks, with the most affected one being the ING Groepe from the Netherlands. The stock most affected by Cypriot stocks is the one of the National Bank of Greece, what is expected due to the economic and financial relations between Cyprus and Greece. The remaining influence is evenly divided by some other European stocks. The ETE transmitted from Spain to the five most influenced stocks is larger than the ETE transmitted by Greece and Cyprus, and the influence is evenly divided among the European stocks. Portuguese stocks transmit more entropy to two of the largest Spanish banks, and also to some other European stocks. The influence of Italian stocks is much larger than the influence of other stocks belonging to the group of countries in crisis, and it spreads rather evenly among some European stocks. The influence from Irish stocks is low, and evenly distributed among European stocks, including two from the UK.

Table 5. Five stocks that receive more ETE from the stocks of each country in crisis. In the table, are shown the name of the company, the total ETE received from the stocks of countries in crisis, the country the stock belongs to, the industry and sub industry.

Stock	ETE	Country	Industry	Sub industry
Greece				
ING Groep	1.04	Netherlands	Insurance	Life/Health Insurance
KBC Groep	1.04	Belgium	Banks	Commercial Banks
Deutsche Bank	0.98	Germany	Banks	Diversified Banking Institution
Société Générale	0.98	France	Banks	Diversified Banking Institution
Crédit Agricole	0.94	France	Banks	Diversified Banking Institution
Cyprus				
National Bank of Greece	0.68	Greece	Banks	Commercial Banks
KBC Groep NV	0.34	Belgium	Banks	Commercial Banks
Deutsche Bank AG	0.33	Germany	Banks	Diversified Banking Institution
ING Groep NV	0.30	Netherlands	Insurance	Life/Health Insurance
DANSKE DC	0.28	Denmark	Banks	Commercial Banks
Spain				
Deutsche Bank	2.34	Germany	Banks	Diversified Banking Institution
BNP Paribas	2.33	France	Banks	Diversified Banking Institution
AXA	2.31	France	Insurance	Multi-line Insurance
ING Groep	2.21	Netherlands	Insurance	Life/Health Insurance
KBC Groep	2.17	Belgium	Banks	Commercial Bank
Portugal				
Banco Santander	0.91	Spain	Banks	Commercial Bank
Banco Bilbao Vizcaya Argentaria	0.72	Spain	Banks	Commercial Bank
BNP Paribas	0.62	France	Banks	Diversified Banking Institution
Deutsche Bank	0.60	Germany	Banks	Diversified Banking Institution
AXA	0.60	France	Insurance	Multi-line Insurance
Italy				
AXA	6.37	France	Insurance	Multi-line Insurance
Deutsche Bank AG	6.29	Germany	Banks	Diversified Banking Institution
BNP Paribas	6.18	France	Banks	Diversified Banking Institution
Banco Bilbao Vizcaya Argentaria	5.90	Spain	Banks	Commercial Bank
Société Générale	5.84	France	Banks	Diversified Banking Institution
Ireland				
ING Groep NV	0.39	Netherlands	Insurance	Life/Health Insurance
Barclays	0.37	UK	Banks	Diversified Banking Institution
Lloyds Banking Group	0.37	UK	Banks	Diversified Banking Institution
Aegon NV	0.36	Netherlands	Insurance	Multi-line Insurance
KBC Groep NV	0.36	Belgium	Banks	Commercial Bank

One must keep in mind that what we are measuring is the sum of ETEs to a particular company, and so the number of companies that send the ETEs is important, but since the number of relevant financial companies a country has is an important factor of its influence, we here consider the sum of ETEs as a determinant of the influence of one country on another.

It is interesting to see that there are some stocks that are consistently more influenced by the stocks of countries in crisis. The Deutsche Bank appears in five lists, and the ING Groep and the KBC Groep appear in four lists. Most of the stocks listed are also some of the more central ones according to different centrality criteria.

Table 6 shows the first five stocks that send the most ETE from the stocks of each country in crisis (four, in the case of Ireland). The most influential stocks are mainly those of banks, but we also have highly influent stocks belonging to insurance companies and to investment companies. The influence of Greece is distributed among some banks, and the influence of Cyprus is also mainly distributed among banks. The Spanish influence also comes from commercial banks, and is concentrated on the top three ones. The same applies to Portugal, with the main ETE being transmitted from a stock that belongs to a Spanish bank but that is also negotiated in Portugal. The most influential stocks from Italy are those of companies that are originally from other European countries, but whose stocks are also negotiated in Italy. The influence of Ireland is mainly distributed among two banks and one insurance company.

Table 6. Five stocks that send more ETE from each country in crisis. In the table, are shown the name of the company, the total ETE sent to the stocks of main financial companies, the industry and sub industry.

Stock	ETE	Industry	Sub industry
		Greece	
National Bank of Greece	5.95	Banks	Commercial Bank
Piraeus Bank	4.68	Banks	Commercial Bank
Cyprus Popular Bank	4.48	Banks	Commercial Bank
Eurobank Ergasias	4.38	Banks	Commercial Bank
Bank of Cyprus	4.28	Banks	Commercial Bank
		Cyprus	
Cyprus Popular Bank	5.18	Banks	Commercial Banks
Bank of Cyprus	4.01	Banks	Commercial Banks
Hellenic Bank	3.02	Banks	Commercial Banks
Interfund Investments	2.12	Investment Companies	Investment Companies
Demetra Investments	1.88	Investment Companies	Investment Companies
		Spain	
Banco Santander	15.90	Banks	Commercial Bank
Banco Bilbao Vizcaya Argentaria	14.74	Bank	Commercial Bank
Banco Popular Espanol	11.35	Banks	Commercial Bank
Banco de Sabadell	10.47	Bank	Commercial Bank
Banco Bradesco	9.99	Banks	Commercial Bank
		Portugal	
Banco Santander	12.67	Banks	Commercial Banks
Banco Espírito Santo	8.60	Banks	Commercial Banks
Banco BPI	8.32	Banks	Commercial Banks
Banco Comercial Portugues	3.08	Banks	Commercial Banks
Espirito Santo Financial Group	4.08	Banks	Commercial Banks
		Italy	
ING Groep NV	15.91	Insurance	Life - Health Insurance
Deutsche Bank AG	15.43	Banks	Diversified Banking Institution
AXA	15.23	Insurance	Multi-line Insurance
BNP Paribas	14.51	Banks	Diversified Banking Institution
UniCredit SpA	14.09	Banks	Diversified Banking Institution
		Ireland	
Bank of Ireland	12.67	Banks	Commercial Bank
Permanent TSB Group Holdings	8.60	Insurance	Property - Casualty Insurance
Allied Irish Banks	8.32	Banks	Commercial Bank
FBD Holdings	3.08	Insurance	Property - Casualty Insurance

So we may conclude that the most influenced stocks by stocks of the countries in crisis according to ETE are those of European companies, and mainly some stocks belonging to some particular banks. The stocks that influence the most, also according to the ETE criterium, are those of banks belonging to the countries in crisis, in particular if the banks are native to other countries, but their stocks are negotiated in the country in crisis.

In order to study the dynamics of the influences of the countries in crisis with the countries in the original sample, once more we aggregate data using the eigenvector corresponding to the largest eigenvalue of the correlation matrix of the stocks belonging to each country, as described in Section 3. By doing this, we calculate an ETE matrix such that the first 24 variables are the original countries in the sample, in the same order as in Section 3, and the remaining 6 variables are the aggregate time series for stocks belonging to Greece, Cyprus, Spain, Portugal, Italy, and Ireland, in this same order. This is done for each semester, from 20007 to 2012, and in Figure 19 we print only the ETEs from the lagged variables corresponding to the countries in crisis (vertical axis) to the original variables corresponding to the affected countries (horizontal axis).

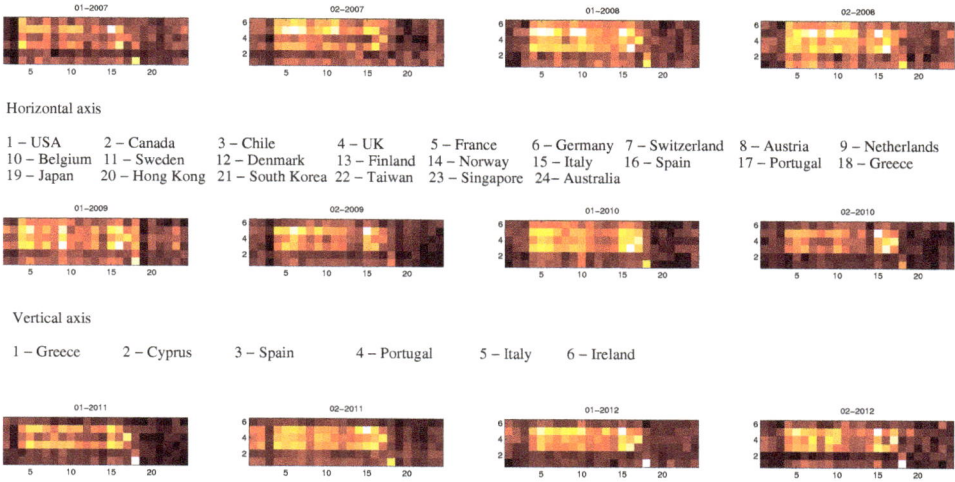

Horizontal axis

1 – USA   2 – Canada   3 – Chile   4 – UK   5 – France   6 – Germany 7 – Switzerland 8 – Austria   9 – Netherlands
10 – Belgium 11 – Sweden   12 – Denmark   13 – Finland 14 – Norway   15 – Italy   16 – Spain   17 – Portugal 18 – Greece
19 – Japan   20 – Hong Kong 21 – South Korea 22 – Taiwan   23 – Singapore 24– Australia

Vertical axis

1 – Greece   2 – Cyprus   3 – Spain   4 – Portugal   5 – Italy   6 – Ireland

Figure 19. Effective Transfer Entropy (ETE) from lagged variables of countries in crisis (vertical axis) to original variables of the 197 large financial companies (horizontal axis), calculated for each semester, from 2007 to 2012. Brighter colors represent higher values of ETE and darker colors represent lower values of ETE.

Analyzing the graphs, we may see that, as expected, there is a rise of ETE during the crisis of 2008 and the crisis of 2010. Looking at each country that is sending information, and ignoring the information going from one country to itself, we see that Greece and Cyprus do not send much information to other countries, and that the largest sources of information are Italy, Spain, and Portugal, in this same order. On the first semester of 2010, we see a lot of ETE between Italy, Spain and Portugal. By the first semester of 2011, the transfer of volatility was mainly due to two countries: Spain and Italy. The transfer of volatility rose again on the second semester of 2011, going down ever since, probably due to the efficacy of the austerity measures adopted by some of the countries in crisis and the policy of the European Union, which chose to sustain the strength of the Euro and the unity of the Eurozone.

We may also see that the transfer of entropy is mainly to European countries, as expected. So, according to ETE, the influence of these countries in crisis is mainly on Europe. Any crisis triggered by them would first hit other countries in Europe, most of them with more solid economies, and, only then, could affect other continents.

In Figure 20, we plot the average ETE that was sent, in each semester from 2007 to 2012, from the six countries in crisis to the 24 original countries in our data set, as a percentage of the average ETE sent from each of these original countries to themselves. So, whenever the percentage is above one, that means that the average information sent from the stocks of these countries was above the average information exchanged between the target countries. From the figure, we see that the average ETE sent from Italy and from Spain is always above the average, and that the average ETE sent from Portugal has also been above the average most of the time. Greece, Cyprus and Ireland have had ETEs sent below the average almost all of the time, and the average ETE sent from those countries has been going down in time, when compared with the average ETE between the target countries. This is evidence that, according to ETE, the stocks of Greece, Cyprus and Ireland have little effect on other stocks of the world, and this is not a result that depends on the number of stock considered for each country, since the result is from aggregate data. So, Italy, Spain and Portugal, in this order, seem to be the most influent countries in crisis.

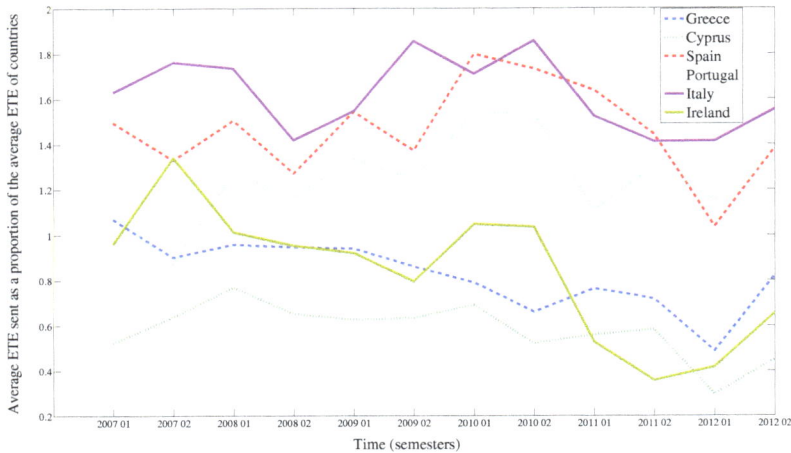

Figure 20. Average ETE sent from the countries in crisis to the original 24 countries as a percentage of the ETE sent from those countries to themselves, calculated each semester.

7. Conclusions

We have seen in this work how the stocks of the top 197 financial companies, in market volume, relate to one another, using the Effective Transfer Entropy between them. We saw that they are related first by country where the stocks are negotiated, and then by industry and sub industry. The network structure for Transfer Entropy is very different from one obtained by correlation, being the network obtained using Transfer Entropy a directed one, with causal influences between the stocks. The use of original and lagged log-returns also revealed some relationships between stocks, with the stocks of a previous day influencing the stocks of the following day. A study of the centralities of the stocks revealed that the most central ones are those of insurance companies of Europe and of the USA, or of banks of the USA and Europe. Since insurance and reinsurance companies are major CDS (Credit Default Securities) sellers, and banks are both major CDS buyers and sellers, some of this centrality of insurance companies, followed by banks, might be explained by the selling and buying of CDS.

A further study of the influence relations between stocks of companies belonging to countries in crisis, namely Greece, Cyprus, Spain, Portugal, Italy, and Ireland, reveal which are the most affected financial companies belonging to the group of largest financial stocks. This calls attention to liabilities of those companies to possible defaults or fall of stocks prices of companies belonging to those countries in crisis.

This work plants the seeds for the study of contagion among financial institutions, but now based on a real network, showing which companies are most central for the propagation of crises and which ones are more dependent on failing economies. This may be used to develop policies for avoiding the spread of financial crises.

Acknowledgments: This article was written using LATEX, all figures were made using PSTricks and Matlab, and the calculations were made using Matlab, Ucinet and Excel. All data and codes used are available with the author. Special thanks to Raul Ikeda, who determined and collected the data, and to Nanci Romero, who helped organizing and classifying the data according to industry and sub industry. Particular thanks go to the anonymous referees, who corrected mistakes, pointed at directions and bibliography, and made many useful suggestions. My thanks also go to the organizers and participants of the Econophysics Colloquium 2013 (in South Korea), and the organizers and participants of the first Paris Financial Management Conference (Paris), where this work has been presented in its earlier stages, and for colleagues of Insper, for useful suggestions.

Conflicts of Interest: The author declares no conflict of interest.

Appendix

A. List of Stocks Used

Here are displayed, in order of country and of industry and sub industry, the stocks that are used in the present work, not considering stocks from particular countries in crisis.

Country	Company	Industry	Sector
USA 1	Bank of America Corp	Banks	Diversified Banking Inst
USA 2	Citigroup Inc	Banks	Diversified Banking Inst
USA 3	Goldman Sachs Group Inc/The	Banks	Diversified Banking Inst
USA 4	JPMorgan Chase & Co	Banks	Diversified Banking Inst
USA 5	Morgan Stanley	Banks	Diversified Banking Inst
USA 6	Comerica Inc	Banks	Super-Regional Banks-US
USA 7	Capital One Financial Corp	Banks	Super-Regional Banks-US
USA 8	KeyCorp	Banks	Super-Regional Banks-US
USA 9	PNC Financial Services Group Inc/The	Banks	Super-Regional Banks-US
USA 10	SunTrust Banks Inc	Banks	Super-Regional Banks-US
USA 11	US Bancorp	Banks	Super-Regional Banks-US
USA 12	Wells Fargo & Co	Banks	Super-Regional Banks-US
USA 13	Fifth Third Bancorp	Banks	Super-Regional Banks-US
USA 14	Huntington Bancshares Inc/OH	Banks	Super-Regional Banks-US
USA 15	BB&T Corp	Banks	Commer Banks-Southern US
USA 16	First Horizon National Corp	Banks	Commer Banks-Southern US
USA 17	Regions Financial Corp	Banks	Commer Banks-Southern US
USA 18	M&T Bank Corp	Banks	Commer Banks-Eastern US
USA 19	Zions Bancorporation	Banks	Commer Banks-Western US
USA 20	Bank of New York Mellon Corp/The	Banks	Fiduciary Banks
USA 21	State Street Corp	Banks	Fiduciary Banks
USA 22	Northern Trust Corp	Banks	Fiduciary Banks
USA 23	Banco Bradesco SA	Banks	Commer Banks Non-US
USA 24	Itau Unibanco Holding SA	Banks	Commer Banks Non-US
USA 25	Banco Santander Chile	Banks	Commer Banks Non-US
USA 26	Credicorp Ltd	Banks	Commer Banks Non-US
USA 27	American Express Co	Diversified Finan Serv	Finance-Credit Card
USA 28	Ameriprise Financial Inc	Diversified Finan Serv	Invest Mgmnt/Advis Serv
USA 29	Franklin Resources Inc	Diversified Finan Serv	Invest Mgmnt/Advis Serv
USA 30	BlackRock Inc	Diversified Finan Serv	Invest Mgmnt/Advis Serv
USA 31	Invesco Ltd	Diversified Finan Serv	Invest Mgmnt/Advis Serv
USA 32	Legg Mason Inc	Diversified Finan Serv	Invest Mgmnt/Advis Serv
USA 33	T Rowe Price Group Inc	Diversified Finan Serv	Invest Mgmnt/Advis Serv
USA 34	E*TRADE Financial Corp	Diversified Finan Serv	Finance-Invest Bnkr/Brkr
USA 35	IntercontinentalExchange Inc	Diversified Finan Serv	Finance-Other Services
USA 36	NYSE Euronext	Diversified Finan Serv	Finance-Other Services
USA 37	NASDAQ OMX Group Inc/The	Diversified Finan Serv	Finance-Other Services
USA 38	Hudson City Bancorp Inc	Savings & Loans	S&L/Thrifts-Eastern US
USA 39	People's United Financial Inc	Savings & Loans	S&L/Thrifts-Eastern US
USA 40	ACE Ltd	Insurance	Multi-line Insurance
USA 41	American International Group Inc	Insurance	Multi-line Insurance
USA 42	Assurant Inc	Insurance	Multi-line Insurance
USA 43	Allstate Corp/The	Insurance	Multi-line Insurance
USA 44	Genworth Financial Inc	Insurance	Multi-line Insurance
USA 45	Hartford Financial Services Group Inc	Insurance	Multi-line Insurance
USA 46	Loews Corp	Insurance	Multi-line Insurance
USA 47	MetLife Inc	Insurance	Multi-line Insurance
USA 48	XL Group PLC	Insurance	Multi-line Insurance
USA 49	Cincinnati Financial Corp	Insurance	Multi-line Insurance
USA 50	Principal Financial Group Inc	Insurance	Life/Health Insurance
USA 51	Lincoln National Corp	Insurance	Life/Health Insurance
USA 52	Aflac Inc	Insurance	Life/Health Insurance
USA 53	Torchmark Corp	Insurance	Life/Health Insurance
USA 54	Unum Group	Insurance	Life/Health Insurance
USA 55	Prudential Financial Inc	Insurance	Life/Health Insurance
USA 56	Travelers Cos Inc/The	Insurance	Property/Casualty Ins
USA 57	Chubb Corp/The	Insurance	Property/Casualty Ins
USA 58	Progressive Corp/The	Insurance	Property/Casualty Ins
USA 59	Aon PLC	Insurance	Insurance Brokers
USA 60	Marsh & McLennan Cos Inc	Insurance	Insurance Brokers

Country	Company	Industry	Sector
USA 61	Berkshire Hathaway Inc	Insurance	Reinsurance
USA 62	CBRE Group Inc	Real Estate	Real Estate Mgmnt/Servic
USA 63	Apartment Investment & Management Co	REITS	REITS-Apartments
USA 64	AvalonBay Communities Inc	REITS	REITS-Apartments
USA 65	Equity Residential	REITS	REITS-Apartments
USA 66	Boston Properties Inc	REITS	REITS-Office Property
USA 67	Host Hotels & Resorts Inc	REITS	REITS-Hotels
USA 68	Prologis Inc	REITS	REITS-Warehouse/Industr
USA 69	Public Storage	REITS	REITS-Storage
USA 70	Simon Property Group Inc	REITS	REITS-Regional Malls
USA 71	Macerich Co/The	REITS	REITS-Regional Malls
USA 72	Kimco Realty Corp	REITS	REITS-Shopping Centers
USA 73	Ventas Inc	REITS	REITS-Health Care
USA 74	HCP Inc	REITS	REITS-Health Care
USA 75	Health Care REIT Inc	REITS	REITS-Health Care
USA 76	American Tower Corp	REITS	REITS-Diversified
USA 77	Weyerhaeuser Co	REITS	REITS-Diversified
USA 78	Vornado Realty Trust	REITS	REITS-Diversified
USA 79	Plum Creek Timber Co Inc	REITS	REITS-Diversified
Canada 1	Bank of Montreal	Banks	Commer Banks Non-US
Canada 2	Bank of Nova Scotia	Banks	Commer Banks Non-US
Canada 3	Canadian Imperial Bank of Commerce/Canada	Banks	Commer Banks Non-US
Canada 4	National Bank of Canada	Banks	Commer Banks Non-US
Canada 5	Royal Bank of Canada	Banks	Commer Banks Non-US
Canada 6	Toronto-Dominion Bank/The	Banks	Commer Banks Non-US
Canada 7	Manulife Financial Corp	Insurance	Life/Health Insurance
Canada 8	Power Corp of Canada	Insurance	Life/Health Insurance
Canada 9	Sun Life Financial Inc	Insurance	Life/Health Insurance
Canada 10	Brookfield Asset Management Inc	Real Estate	Real Estate Oper/Develop
Chile	Banco de Chil	Banks	Commer Banks Non-US
UK 1	Barclays PLC	Banks	Diversified Banking Inst
UK 2	HSBC Holdings PLC	Banks	Diversified Banking Inst
UK 3	Lloyds Banking Group PLC	Banks	Diversified Banking Inst
UK 4	Royal Bank of Scotland Group PLC	Banks	Diversified Banking Inst
UK 5	Standard Chartered PLC	Banks	Commer Banks Non-US
UK 6	Aberdeen Asset Management PLC	Diversified Finan Serv	Invest Mgmnt/Advis Serv
UK 7	Man Group PLC	Diversified Finan Serv	Invest Mgmnt/Advis Serv
UK 8	Schroders PLC	Diversified Finan Serv	Invest Mgmnt/Advis Serv
UK 9	Old Mutual PLC	Diversified Finan Serv	Invest Mgmnt/Advis Serv
UK 10	Provident Financial PLC	Diversified Finan Serv	Finance-Consumer Loans
UK 11	London Stock Exchange Group PLC	Diversified Finan Serv	Finance-Other Services
UK 12	Aviva PLC	Insurance	Life/Health Insurance
UK 13	Legal & General Group PLC	Insurance	Life/Health Insurance
UK 14	Prudential PLC	Insurance	Life/Health Insurance
UK 15	Standard Life PLC	Insurance	Life/Health Insurance
UK 16	RSA Insurance Group PLC	Insurance	Property/Casualty Ins
UK 17	3i Group PLC	Private	Private
UK 18	Hammerson PLC	REITS	REITS-Shopping Centers
UK 19	British Land Co PLC	REITS	REITS-Diversified
UK 20	Land Securities Group PLC	REITS	REITS-Diversified
UK 21	Segro PLC	REITS	REITS-Diversified
France 1	Credit Agricole SA	Banks	Diversified Banking Inst
France 2	BNP Paribas SA	Banks	Diversified Banking Inst
France 3	Societe Generale SA	Banks	Diversified Banking Inst
France 4	AXA SA	Insurance	Multi-line Insurance
Germany 1	Commerzbank AG	Banks	Commer Banks Non-US
Germany 2	Deutsche Bank AG	Banks	Diversified Banking Inst
Germany 3	Deutsche Boerse AG	Diversified Finan Serv	Finance-Other Services
Germany 4	Allianz SE	Insurance	Multi-line Insurance
Germany 5	Muenchener Rueckversicherungs AG	Insurance	Reinsurance

Country	Company	Industry	Sector
Switzerland 1	Credit Suisse Group AG	Banks	Diversified Banking Inst
Switzerland 2	UBS AG	Banks	Diversified Banking Inst
Switzerland 3	GAM Holding AG	Diversified Finan Serv	Invest Mgmnt/Advis Serv
Switzerland 4	Baloise Holding AG	Insurance	Multi-line Insurance
Switzerland 5	Zurich Insurance Group AG	Insurance	Multi-line Insurance
Switzerland 6	Swiss Life Holding AG	Insurance	Life/Health Insurance
Switzerland 7	Swiss Re AG	Insurance	Reinsurance
Austria	Erste Group Bank AG	Banks	Commer Banks Non-US
Netherlands 1	Aegon NV	Insurance	Multi-line Insurance
Netherlands 2	ING Groep NV	Insurance	Life/Health Insurance
Belgium 1	KBC Groep NV	Banks	Commer Banks Non-US
Belgium 2	Ageas	Insurance	Life/Health Insurance
Sweden 1	Nordea Bank AB	Banks	Commer Banks Non-US
Sweden 2	Skandinaviska Enskilda Banken AB	Banks	Commer Banks Non-US
Sweden 3	Svenska Handelsbanken AB	Banks	Commer Banks Non-US
Sweden 4	Swedbank AB	Banks	Commer Banks Non-US
Sweden 5	Investor AB	Investment Companies	Investment Companies
Denmark	Danske Bank A/S	Banks	Commer Banks Non-US
Finland	Sampo	Insurance	Multi-line Insurance
Norway	DNB ASA	Banks	Commer Banks Non-US
Italy 1	Banca Monte dei Paschi di Siena SpA	Banks	Commer Banks Non-US
Italy 2	Intesa Sanpaolo SpA	Banks	Commer Banks Non-US
Italy 3	Mediobanca SpA	Banks	Commer Banks Non-US
Italy 4	Unione di Banche Italiane SCPA	Banks	Commer Banks Non-US
Italy 5	UniCredit SpA	Banks	Diversified Banking Inst
Italy 6	Assicurazioni Generali SpA	Insurance	Multi-line Insurance
Spain 1	Banco Bilbao Vizcaya Argentaria SA	Banks	Commer Banks Non-US
Spain 2	Banco Popular Espanol SA	Banks	Commer Banks Non-US
Spain 3	Banco de Sabadell SA	Banks	Commer Banks Non-US
Spain 4	Banco Santander SA	Banks	Commer Banks Non-US
Portugal	Banco Espírito Santo SA	Banks	Commer Banks Non-US
Greece	National Bank of Greece SA	Banks	Commer Banks Non-US
Japan 1	Shinsei Bank Ltd	Banks	Commer Banks Non-US
Japan 2	Mitsubishi UFJ Financial Group Inc	Banks	Diversified Banking Inst
Japan 3	Sumitomo Mitsui Trust Holdings Inc	Banks	Commer Banks Non-US
Japan 4	Sumitomo Mitsui Financial Group Inc	Banks	Commer Banks Non-US
Japan 5	Mizuho Financial Group Inc	Banks	Commer Banks Non-US
Japan 6	Credit Saison Co Ltd	Diversified Finan Serv	Finance-Credit Card
Japan 7	Daiwa Securities Group Inc	Diversified Finan Serv	Finance-Invest Bnkr/Brkr
Japan 8	Nomura Holdings Inc	Diversified Finan Serv	Finance-Invest Bnkr/Brkr
Japan 9	ORIX Corp	Diversified Finan Serv	Finance-Leasing Compan
Japan 10	Tokio Marine Holdings In	Insurance	Property/Casualty Ins
Japan 11	Mitsui Fudosan Co Ltd	Real Estate	Real Estate Oper/Develop
Japan 12	Mitsubishi Estate Co Ltd	Real Estate	Real Estate Mgmnt/Servic
Japan 13	Sumitomo Realty & Development Co Ltd	Real Estate	Real Estate Oper/Develop
Hong Kong 1	Hang Seng Bank Ltd	Banks	Commer Banks Non-US
Hong Kong 2	Industrial & Commercial Bank of China Ltd	Banks	Commer Banks Non-US
Hong Kong 3	BOC Hong Kong Holdings Ltd	Banks	Commer Banks Non-US
Hong Kong 4	China Construction Bank Corp	Banks	Commer Banks Non-US
Hong Kong 5	Hong Kong Exchanges and Clearing Ltd	Diversified Finan Serv	Finance-Other Services
Hong Kong 6	Ping An Insurance Group Co of China Ltd	Insurance	Multi-line Insurance
Hong Kong 7	China Life Insurance Co Ltd	Insurance	Life/Health Insurance
Hong Kong 8	Cheung Kong Holdings Ltd	Real Estate	Real Estate Oper/Develop
Hong Kong 9	Sun Hung Kai Properties Ltd	Real Estate	Real Estate Oper/Develop
South Korea	Shinhan Financial Group Co Ltd	Diversified Finan Serv	Diversified Finan Serv
Taiwan	Cathay Financial Holding Co Ltd	Insurance	Life/Health Insurance
Singapore 1	DBS Group Holdings Ltd	Banks	Commer Banks Non-US
Singapore 2	Oversea-Chinese Banking Corp Ltd	Banks	Commer Banks Non-US
Singapore 3	United Overseas Bank Ltd	Banks	Commer Banks Non-US

Country	Company	Industry	Sector
Australia 1	Australia & New Zealand Banking Group Ltd	Banks	Commer Banks Non-US
Australia 2	Commonwealth Bank of Australia	Banks	Commer Banks Non-US
Australia 3	National Australia Bank Ltd	Banks	Commer Banks Non-US
Australia 4	Westpac Banking Corp	Banks	Commer Banks Non-US
Australia 5	Macquarie Group Ltd	Diversified Finan Serv	Finance-Invest Bnkr/Brkr
Australia 6	ASX Ltd	Diversified Finan Serv	Finance-Other Services
Australia 7	AMP Ltd	Insurance	Life/Health Insurance
Australia 8	Suncorp Group Ltd	Insurance	Life/Health Insurance
Australia 9	Insurance Australia Group Ltd	Insurance	Property/Casualty Ins
Australia 10	QBE Insurance Group Ltd	Insurance	Property/Casualty Ins
Australia 11	Lend Lease Group	Real Estate	Real Estate Mgmnt/Servic
Australia 12	CFS Retail Property Trust Group	REITS	REITS-Shopping Centers
Australia 13	Westfield Group	REITS	REITS-Shopping Centers
Australia 14	Dexus Property Group	REITS	REITS-Diversified
Australia 15	Goodman Group	REITS	REITS-Diversified
Australia 16	GPT Group	REITS	REITS-Diversified
Australia 17	Mirvac Group	REITS	REITS-Diversified
Australia 18	Stockland	REITS	REITS-Diversified

References

1. Haldane, A.G. *Rethinking the financial network*; Speech Delivered at the Financial Student Association on 28 April 2009; Financial Student Association: Amsterdam, The Netherlands, 2009.
2. Allen, F.; Gale, D. Financial contagion. *J. Polit. Econ.* **2000**, *108*, 1–33.
3. Kirman, A. The economy as an evolving network. *J. Evol. Econ.* **1997**, *7*, 339–353.
4. Allen, F.; Babus, A. Networks in finance. In *The Network Challenge: Strategy, Profit, and Risk in an Interlinked World*; Kleindorfer, P., Wind, Y., Gunther, R., Eds.; Wharton Scool Publishing: Bergen County, NJ, USA, 2009.
5. Upper, C. Simulation methods to assess the danger of contagion in interbank markets. *J. Financ. Stab.* **2011**, *7*, 111–125.
6. Watts, D.J. A simple model of global cascades on random networks. *Proc. Natl. Acad. Sci. USA* **2002**, *99*, 5766–5771.
7. Vivier-Lirimont, S. Interbanking networks: Towards a small financial world? In *Cahiers de la Maison des Sciences Economiques*; Université Panthon-Sorbonne (Paris 1): Paris, France, 2004; v04046.
8. Leitner, Y. Financial networks: Contagion, commitment and private sector bailouts. *J. Financ.* **2005**, *60*, 2925–2953.
9. Nier, E.; Yang, J.; Yorulmazer, T.; Alentorn, A. Network models and financial stability. *J. Econ. Dyn. Control* **2007**, *31*, 2033–2060.
10. Castiglionesi, F.; Navarro, N. *Optimal Fragile Financial Networks*; Tilburg University: Tilburg, The Netherlands, 2007.
11. Cossin, D.; Schellhorn, H. Credit risk in a network economy. *Manag. Sci.* **2007**, *53*, 1604–1617.
12. Lorenz, J.; Battiston, S.; Schweitzer, F. Systemic risk in a unifying framework for cascading processes on networks. *Eur. Phys. J. B* **2009**, *71*, 441–460.
13. Schweitzer, F.; Fagiolo, G.; Sornette, D.; Vega-Redondo, F.; White. D.R. Economic Networks: What do we know and what do we need to know? *Adv. Complex Syst.* **2009**, *12*, 407–422.
14. Gai, P.; Kapadia, S. Contagion in Financial Networks. *Proc. R. Soc. A* **2010**, *466*, rspa20090410.
15. Georg, C.-P. *The Effect of the Interbank Network Structure on Contagion and Financial Stability*, No. 12; Working Papers on Global Financial Markets, Universität Jena und Universität Halle; EconStor: Hamburg, Germany, 2010.
16. Canedo, J.M.D.; Martínez-Jaramillo, S. Financial contagion: A network model for estimating the distribution of loss for the financial system. *J. Econ. Dyn. Control* **2010**, *34*, 2358–2374.
17. Gai, P.; Haldane. A.; Kapadia, S. Complexity, concentration and contagion. *J. Monet. Econ.* **2011**, *58*, 453–470.
18. Tabak, B.M.; Takami, M.; Rocha, J.M.C.; Cajueiro, D.O. Directed Clustering Coefficient as a Measure of Systemic Risk in Complex Banking Networks. *Physica A* **2014**, *394*, 211–216.
19. Battiston, S.; Gatti, D.D.; Gallegati, M.; Greenwald, B.; and Stiglitz. J. Liaisons dangereuses: Increasing connectivity, risk sharing, and systemic risk. *J. Econ. Dyn. Control* **2012**, *36*, 1121–1141.
20. Battiston, S.; Gatti, D.D.; Gallegati, M.; Greenwald, B.; Stiglitz, J. Default cascades: When does risk diversification increase stability? *J. Financ. Stab.* **2012**, *8*, 138–149.

21. Amini, H.; Cont, R.; Minca, A. Resilience to contagion in financial networks. *Math. Financ.* **2013**, doi:10.1111/mafi.12051.

22. Elliott, M.; Golub, B.; Jackson, M.O. Financial Networks and Contagion. 2013, 2175056. Social Science Research Network (SSRN). Available online: http://ssrn.com/abstract=2175056 (accessed on 6 August 2014).

23. Acemoglu, D.; Osdaglar, A.; Tahbaz-Salehi, A. *Systemic risk and stability in financial networks*, No. w18727; The National Bureau of Economic Research: Cambridge, MA, USA, 2013.

24. Boss, M.; Elsinger, H.; Summer, M.; Thurner, S. Network topology of the interbank market. *Quant. Financ.* **2004**, *4*, 677–684.

25. Müller, J. Interbank credit lines as a channel of contagion. *J. Financ. Serv. Res.* **2006**, *29*, 37–60.

26. Soramäki, K.; Bech, M.L.; Arnold, J.; Glass, R.J.; Beyeler, W.E. The topology of interbank payment flows. *Physica A* **2007**, *379*, 317–333.

27. Hattori, M; Suda, Y. Developments in a cross-border bank exposure "network". In Proceedings of a CGFS Workshop Held at the Bank for International Settlements, Basel, Switzerland, December 2007.

28. Iori, G.; Masi, G.; Precup, O.V.; Gabbi, G.; Caldarelli, G. A network analysis of the Italian overnight money market. *J. Econ. Dyn. Control* **2008**, *32*, 259–278.

29. Markose, S.; Giansante, S.; Gatkowski, M.; Shaghaghi, A.R. Too Interconnected To Fail: Financial Contagion and Systemic Risk In Network Model of CDS and Other Credit Enhancement Obligations of US Banks. 2010, Working Papers Series WPS-033 21/04/2010, No. 683. Computational Optimization Methods in Statistics, Econometrics and Finance (COMISEF). Available online: http://comisef.eu/files/wps033.pdf (accessed on 6 August 2014).

30. Kubelec, C.; Sá, F. *The geographical composition of national external balance sheets: 1980-2005*; Bank of England Working Paper No.384; Bank of England: London, UK, 2010.

31. Minoiu, C.; Reyes, J.A. *A network analysis of global banking: 1978-2009*; IMF Working Paper WP/11/74; International Monetary Fund: Washington, DC, USA, 2011.

32. Lee, K.-M.; Yang, J.-S.; Kim G.; Lee J.; Goh K.-I.; Kim, I.-M. Impact of the Topology of Global Macroeconomic Network on the Spreading of Economic Crises. *PLoS ONE* **2011**, *6*, e18443.

33. Battiston, S.; Puliga, M.; Kaushik, R.; Tasca, P.; Caldarelli, G. DebtRank: too central to fail? Financial networks, the FED and systemic risk. *Sci. Rep.* **2012**, *2*, doi:10.1038/srep00541, 1–6.

34. Martínez-Jaramillo, S.; Alexandrova-Kabadjova, B.; Bravo-Benítez, B.; Solórzano-Margain, J.P. An empirical study of the Mexican banking system's network and its implications for systemic risk. *J. Econ. Dyn. Control* **2014**, *40*, 242–265.

35. Hale, G. Bank relationships, business cycles, and financial crises. *J. Int. Econ.* **2012**, *88*, 312–325.

36. Kaushik, R.; Battiston, S. Credit Default Swaps drawup networks: too interconnected to be stable? *PLoS ONE* **2013**, *8*, doi:10.1371/journal.pone.0061815, e61815.

37. Chinazzi, M.; Fagiolo, G.; Reyes, J.A.; Schiavo, S. Post-mortem examination of the international financial network. *J. Econ. Dyn. Control* **2013**, *37*, 1692–1713.

38. Memmel, C.; Sachs, A. Contagion in the interbank market and its determinants. *J. Financ. Stab.* **2013**, *9*, 46–54.

39. Schreiber, T. Measuring information transfer. *Phys. Rev. Lett.* **2000**, *85*, 461–464.

40. Shannon, C.E. A Mathematical Theory of Communication. *Bell Syst. Tech. J.* **1948**, *27*, 379–423, 623–656.

41. Lizier, J.T.; Prokopenko, M.; Zomaya, A.Y. Local information transfer as a spatiotemporal filter for complex systems. *Phys. Rev. E* **2008**, *77*, doi: 10.1103/PhysRevE.77.026110.

42. Lizier, J.T.; Prokopenko, M. Differentiating information transfer and causal effect. *Euro. Phys. J. B* **2010**, *73*, 605–615.

43. Lizier, J.T.; Mahoney, J.R. Moving frames of referrence, relativity and invariance in Transfer Entropy and information dynamics. *Entropy* **2013**, *15*, 177–197.

44. Sumioka, H.; Yoshikawa, Y.; Asada, M. Causality detected by transfer entropy leads acquisition of joint attention. In Proceedings of 6th IEEE International Conference on Development and Learning, London, UK, 11–13 July 2007; pp. 264–269.

45. Papana, A.; Kugiumtzis, D.; Larsson, P.G. Reducing the bias of causality measures. *Phys. Rev. E* **2011**, *83*, doi:10.1103/PhysRevE.83.036207.

46. Shew, W.L.; Yang, H.; Yu, S.; Roy, R.; Plenz, D. Information capacity and transmission are maximized in balanced cortical networks with neuronal avalanches. *J. Neurosci.* **2011**, *31*, 55–63.

47. Vicente, R.; Wibral, M.; Lindner, M.; Pipa, G. Transfer entropy—a model-free measure of effective connectivity for the neurosciences. *J. Comput. Neurosci.* **2011**, *30*, 45–67.

48. Stetter, O.; Battaglia, D.; Soriano, J.; Geisel, T. Model-free reconstruction of excitatory neuronal connectivity from calcium imaging signals. *PLoS Comput. Biol.* **2012**, *8*, doi:10.1371/journal.pcbi.1002653.

49. Faes, L.; Nollo, G.; Porta, A. Compensated Transfer Entropy as a tool for reliably estimating information transfer in physiological time series. *Entropy* **2013**, *15*, 198–219.

50. Ver Steeg, G,; Galstyan, A. Information transfer in social media. In Proceedings of the 21st International Conference on World Wide Web, Lyon, France, 16–20 April 2012; ACM: New York, NY, USA, 2012; pp. 509–518.

51. Barnett, L.; Bossomaier, T. Transfer entropy as a log-likelihood ratio. *Phys. Rev. Lett.* **2012**, *109*, doi;10.1103/PhysRevLett.109.138105.

52. Amblard, P.-O.; Michel, O.J.J. The relation between Granger causality and directed information theory: A review. *Entropy* **2013**, *15*, 113–143.

53. Hahs, D.W.; Pethel, S.D. Transfer Entropy for coupled autoregressive processes. *Entropy* **2013**, *15*, 767–788.

54. Liu, L.F.; Hu, H.P.; Deng, Y.S.; Ding, N.D. An entropy measure of non-stationary processes. *Entropy* **2014**, *16*, 1493–1500.

55. Liang, X.S. The Liang–Kleeman information flow: Theory and applications. *Entropy* **2013**, *15*, 327–360.

56. Nichols, J.M.; Bucholtz, F.; Michalowicz, J.V. Linearized Transfer Entropy for continuous second order systems. *Entropy* **2013**, *15*, 3186–3204.

57. Materassi, M.; Consolini, G.; Smith, N.; De Marco, R. Information theory analysis of cascading process in a synthetic model of fluid turbulence. *Entropy* **2014**, *16*, 1272–1286.

58. Prokopenko, M.; Lizier, J.T.; Price, D.C. On thermodynamic interpretation of Transfer Entropy. *Entropy* **2013**, *15*, 524–543.

59. Marschinski, R.; Kantz, H. Analysing the information flow between financial time series - an improved estimator for transfer entropy. *Euro. Phys. J. B* **2002**, *30*, 275–281.

60. Baek, S.K.; Jung, W.-S.; Kwon, O.; Moon, H.-T. Transfer Entropy Analysis of the Stock Market. **2005**, ArXiv.org: physics/0509014v2.

61. Kwon, O.; Yang, J.-S. Information flow between composite stock index and individual stocks. *Physica A* **2008**, *387*, 2851–2856.

62. Kwon, O.; Yang, J.-S. Information flow between stock indices. *Euro. Phys. Lett.* **2008**, *82*, doi:10.1209/0295-5075/82/68003.

63. Jizba, P.; Kleinert, H.; Shefaat, M. Rényi's information transfer between financial time series. *Physica A* **2012**, *391*, 2971–2989.

64. Peter, F.J.; Dimpfl, T.; Huergo, L. Using transfer entropy to measure information flows from and to the CDS market. In Proceedings of the European Economic Association and Econometric Society, Oslo, Norway, 25–29 August 2011; pp. 25–29.

65. Dimpfl, T.; Peter, F.J. Using transfer entropy to measure information flows between financial markets. In Proceedings of Midwest Finance Association 2012 Annual Meetings, New Orleans, LA, USA, 21–24 February 2012.

66. Kim, J.; Kim, G.; An, S.; Kwon, Y.-K.; Yoon, S. Entropy-based analysis and bioinformatics-inspired integration of global economic information transfer. *PLOS One* **2013**, *8*, doi:10.1371/journal.pone.0051986.

67. Li, J.; Liang, C.; Zhu, X.; Sun, X.; Wu, D. Risk contagion in Chinese banking industry: A Transfer Entropy-based analysis. *Entropy* **2013**, *15*, 5549–5564.

68. Dimpfl, T.; Peter, F.J. The impact of the financial crisis on transatlantic information flows: An intraday analysis. *J. Int. Financ. Mark. Inst. Money* **2014**, *31*, 1–13.

69. Sandoval, L., Jr. To lag or not to lag? How to compare indices of stock markets that operate at different times. *Physica A* **2014**, *403*, 227–243.

70. Newman, M.E.J. *Networks, and introduction*; Oxford University Press: New York, NY, USA, 2010.

71. Sandoval, L., Jr. Cluster formation and evolution in networks of financial market indices. *Algorithm. Financ.* **2013**, *2*, 3–43.

72. Rényi, A. On measures of information and entropy. In Proceedings of the fourth Berkeley Symposium on Mathematics, Statistics and Probability, Berkeley, CA, USA, 20–30 June 1960; University of California Press: Berkeley, CA, USA, 1961; pp. 547–561.

73. Beck, C.; Schögl, F. *Thermodynamics of Chaotic Systems: An Introduction*; Cambridge Nonlinear Science Series; Cambridge University Press: New York, NY, USA, 1993.

entropy

MDPI

Article

The Liang-Kleeman Information Flow: Theory and Applications

X. San Liang [1,2]

[1] School of Marine Sciences and School of Mathematics and Statistics, Nanjing University of Information
Science and Technology (Nanjing Institute of Meteorology), 219 Ningliu Blvd, Nanjing 210044, China;
E-Mail: sanliang@courant.nyu.edu; Tel.: +86-25-5869-5695; Fax: +86-25-5869-5698

[2] China Institute for Advanced Study, Central University of Finance and Economics, 39 South College Ave,
Beijing 100081, China

Received: 17 October 2012; in revised form: 22 November 2012 / Accepted: 28 December 2012 /
Published: 18 January 2013

Abstract: Information flow, or information transfer as it may be referred to, is a fundamental notion in
general physics which has wide applications in scientific disciplines. Recently, a rigorous formalism
has been established with respect to both deterministic and stochastic systems, with flow measures
explicitly obtained. These measures possess some important properties, among which is flow or
transfer asymmetry. The formalism has been validated and put to application with a variety of
benchmark systems, such as the baker transformation, Hénon map, truncated Burgers-Hopf system,
Langevin equation, *etc*. In the chaotic Burgers-Hopf system, all the transfers, save for one, are
essentially zero, indicating that the processes underlying a dynamical phenomenon, albeit complex,
could be simple. (Truth is simple.) In the Langevin equation case, it is found that there could be
no information flowing from one certain time series to another series, though the two are highly
correlated. Information flow/transfer provides a potential measure of the cause–effect relation
between dynamical events, a relation usually hidden behind the correlation in a traditional sense.

Keywords: Liang-Kleeman information flow; causation; emergence; Frobenius-Perron operator;
time series analysis; atmosphere-ocean science; El Niño; neuroscience; network dynamics; financial
economics

1. Introduction

Information flow, or information transfer as it sometimes appears in the literature, refers to the
transference of information between two entities in a dynamical system through some processes, with
one entity being the source, and another the receiver. Its importance lies beyond its literal meaning in
that it actually carries an implication of causation, uncertainty propagation, predictability transfer, *etc.*,
and, therefore, has applications in a wide variety of disciplines. In the following, we first give a brief
demonstration of how it may be applied in different disciplines; the reader may skip this part and go
directly to the last two paragraphs of this section.

According to how the source and receiver are chosen, information flow may appear in two types of
form. The first is what one would envision in the usual sense, *i.e.*, the transference between two parallel
parties (for example, two chaotic circuits [1]), which are linked through some mechanism within a
system. This is found in neuroscience (e.g., [2–4]), network dynamics (e.g., [5–7]), atmosphere–ocean
science (e.g., [8–11]), financial economics (e.g., [12,13]), to name but a few. For instance, neuroscientists
focus their studies on the brain and its impact on behavior and cognitive functions, which are associated
with flows of information within the nervous system (e.g., [3]). This includes how information flows
from one neuron to another neuron across the synapse, how dendrites bring information to the
cell body, how axons take information away from the cell body, and so forth. Similar issues arise

in computer and social networks, where the node–node interconnection, causal dependencies, and directedness of information flow, among others, are of concern [6,14,15]. In atmosphere–ocean science, the application is vast, albeit newly begun. An example is provided by the extensively studied El Niño phenomenon in the Pacific Ocean, which is well known through its linkage to global natural disasters, such as the floods in Ecuador and the droughts in Southeast Asia, southern Africa and northern Australia, to the death of birds and dolphins in Peru, to the increased number of storms over the Pacific, and to the famine and epidemic diseases in far-flung parts of the world [16–18]. A major focus in El Niño research is the predictability of the onset of the irregularly occurring event, in order to issue in-advance warning of potential hazardous impacts [19–21]. It has now become known that the variabilities in the Indian Ocean could affect the El Niño predictability (e.g., [22]). That is to say, at least a part of the uncertainty source for El Niño predictions is from the Indian Ocean. Therefore, to some extent, the El Niño predictability may also be posed as an information flow problem, *i.e.*, a problem on how information flows from the Indian Ocean to the Pacific Ocean to make the El Niño more predictable or more uncertain.

Financial economics provides another field of application of information flow of the first type; this field has received enormous public attention since the recent global financial crisis triggered by the subprime mortgage meltdown. A conspicuous example is the cause–effect relation between the equity and options markets, which reflects the preference of traders in deciding where to place their trades. Usually, information is believed to flow unidirectionally from equity to options markets because informed traders prefer to trade in the options markets (e.g., [23]), but recent studies show that the flow may also exist in the opposite way: informed traders actually trade both stocks and "out-of-the-money" options, and hence the causal relation from stocks to options may reverse [12]. More (and perhaps the most important) applications are seen through predictability studies. For instance, the predictability of asset return characteristics is a continuing problem in financial economics, which is largely due to the information flow in markets. Understanding the information flow helps to assess the relative impact from the markets and the diffusive innovation on financial management. Particularly, it helps the prediction of jump timing, a fundamental question in financial decision making, through determining information covariates that affect jump occurrence up to the intraday levels, hence providing empirical evidence in the equity markets, and pointing us to an efficient financial management [13].

The second type of information flow appears in a more abstract way. In this case, we have one dynamical event; the transference occurs between different levels, or sometimes scales, within the same event. Examples for this type are found in disciplines such as evolutionary biology [24–26], statistical physics [27,28], turbulence, *etc.*, and are also seen in network dynamics. Consider the transitions in biological complexity. A reductionist, for example, views that the emergence of new, higher level entities can be traced back to lower level entities, and hence there is a "bottom-up" causation, *i.e.*, an information flow from the lower levels to higher levels. Bottom-up causation lays the theoretical foundation for statistical mechanics, which explains macroscopic thermodynamic states from a point of view of molecular motions. On the other hand, "top-down" causation is also important [29,30]. In evolution (e.g., [31]), it has been shown that higher level processes may constrain and influence what happens at lower levels; particularly, in transiting complexity, there is a transition of information flow, from the bottom-up to top-down, leading to a radical change in the structure of causation (see, for example [32]). Similar to evolutionary biology, in network dynamics, some simple computer networks may experience a transition from a low traffic state to a high congestion state, beneath which is a flow of information from a bunch of almost independent entities to a collective pattern representing a higher level of organization (e.g., [33]). In the study of turbulence, the notoriously challenging problem in classical physics, it is of much interest to know how information flows over the spectrum to form patterns on different scales. This may help to better explain the cause of the observed higher moments of the statistics, such as excess kurtosis and skewness, of velocity components and velocity derivatives [34]. Generally, the flows/transfers are two-way, *i.e.*, both from small scales to large scales, and from large scales to small scales, but the flow or transfer rates may be quite different.

Apart from the diverse real-world applications, information flow/transfer is important in that it offers a methodology for scientific research. In particular, it offers a new way of time series analysis [35–37]. Traditionally, correlation analysis is widely used for identifying the relation between two events represented by time series of measurements; an alternative approach is through mutual information analysis, which may be viewed as a type of nonlinear correlation analysis. But both correlation analysis and mutual information analysis put the two events on an equal stance. As a result, there is no way to pick out the cause and the effect. In econometrics, Granger causality [38] is usually employed to characterize the causal relation between time series, but the characterization is just in a qualitative sense; when two events are mutually causal, it is difficult to differentiate their relative strengths. The concept of information flow/transfer is expected to remedy this deficiency, with the mutual causal relation quantitatively expressed.

Causality implies directionality. Perhaps the most conspicuous observation on information flow/transfer is its asymmetry between the involved parties. A typical example is seen in our daily life when a baker is kneading a dough. As the baker stretches, cuts, and folds, he guides a unilateral flow of information from the horizontal to the vertical. That is to say, information goes only from the stretching direction to the folding direction, not *vice versa*. The one-way information flow (in a conventional point of view) between the equity and options markets offers another good example. In other cases, such as in the aforementioned El Niño event, though the Indian and Pacific Oceans may interact with each other, *i.e.*, the flow route could be a two-way street, the flow rate generally differs from one direction to another direction. For all that account, transfer asymmetry makes a basic property of information flow; it is this property that distinguishes information flow from the traditional concepts such as mutual information.

As an aside, one should not confuse dynamics with causality, the important property reflected in the asymmetry of information flow. It is temptating to think that, for a system, when the dynamics are known, the causal relations are determined. While this might be the case for linear deterministic systems, in general, however, this need not be true. Nonlinearity may lead a deterministic system to chaos; the future may not be predictable after a certain period of time, even though the dynamics is explicitly given. The concept of emergence in complex systems offers another example. It has long been found that irregular motions according to some simple rules may result in the emergence of regular patterns (such as the inverse cascade in the planar turbulence in natural world [39,40]). Obviously, how this instantaneous flow of information from the low-level entities to high-level entities, *i.e.*, the patterns, cannot be simply explained by the rudimentary rules set *a priori*. In the language of complexity, emergence does not result from rules only (e.g., [41–43]); rather, as said by Corning (2002) [44], "Rules, or laws, have no causal efficacy; they do not in fact 'generate' anything... the underlying causal agencies must be separately specified."

Historically, quantification of information flow has been an enduring problem. The challenge lies in that this is a real physical notion, while the physical foundation is not as clear as those well-known physical laws. During the past decades, formalisms have been established empirically or half-empirically based on observations in the aforementioned diverse disciplines, among which are Vastano and Swinney's time-delayed mutual information [45], and Schreiber's transfer entropy [46,47]. Particularly, transfer entropy is established with an emphasis of the above transfer asymmetry between the source and receiver, so as to have the causal relation represented; it has been successfully applied in many real problem studies. These formalisms, when carefully analyzed, can be approximately understood as dealing with the change of marginal entropy in the Shannon sense, and how this change may be altered in the presence of information flow (see [48], section 4 for a detailed analysis). This motivates us to think about the possibility of a rigorous formalism when the dynamics of the system is known. As such, the underlying evolution of the joint probability density function (pdf) will also be given, for deterministic systems, by the Liouville equation or, for stochastic systems, by the Fokker-Planck equation (cf. §4 and §5 below). From the joint pdf, it is easy to obtain the marginal density, and hence the marginal entropy. One thus expects that the concept of information flow/transfer may be built on a rigorous footing when the dynamics are known, as is the case with many real world problems like those in atmosphere–ocean science. And, indeed, Liang and Kleeman (2005) [49] find that, for two-dimensional (2D) systems, there is a concise law on entropy evolution that makes the hypothesis come true. Since then, the formalism has been extended to systems in different forms and of arbitrary dimensionality, and has been applied with success in benchmark dynamical systems and more realistic problems. In the following sections, we will give a systematic introduction of the theories and a brief review of some of the important applications.

In the rest of this review, we first set up a theoretical framework, then illustrate through a simple case how a rigorous formalism can be achieved. Specifically, our goal is to compute within the framework, for a continuous-time system, the transference rate of information, and, for a discrete-time system or mapping, the amount of the transference upon each application of the mapping. To unify the terminology, we may simply use "information flow/transfer" to indicate either the "rate of information flow/transfer" or the "amount of information flow/transfer" wherever no ambiguity exists in the context. The next three sections are devoted to the derivations of the transference formulas for three different systems. Sections 3 and 4 are for deterministic systems, with randomness limited within initial conditions, where the former deals with discrete mappings and the latter with continuous flows. Section 5 discusses the case when stochasticity is taken in account. In the section that follows, four major applications are briefly reviewed. While these applications are important *per se*, some of them also provide validations for the formalism. Besides, they are also typical in terms of computation; different approaches (both analytical and computational) have been employed in computing the flow or transfer rates for these systems. We summarize in Section 7 the major results regarding the formulas and their corresponding properties, and give a brief discussion on the future research along this line. As a convention in the history of development, the terms "information flow" and "information transfer" will be used synonymously. Throughout this review, by entropy we always mean Shannon or absolute entropy, unless otherwise specified. Whenever a theorem is stated, generally only the result is given and interpreted; for detailed proofs, the reader is referred to the original papers.

2. Mathematical Formalism

2.1. Theoretical Framework

Consider a system with n state variables, $x_1, x_2, ..., x_n$, which we put together as a column vector $x = (x_1, ..., x_n)^T$. Throughout this paper, x may be either deterministic or random, depending on the context where it appears. This is a notational convention adopted in the physics literature, where random and deterministic states for the same variable are not distinguished. (In probability theory, they are usually distinguished with lower and upper cases like x and X.) Consider a sample space of x, $\Omega \subset \mathbb{R}^n$. Defined on Ω is a joint probability density function (pdf) $\rho = \rho(x)$. For convenience, assume that ρ and its derivatives (up to an order as high as enough) are compactly supported. This makes sense, as in the real physical world, the probability of extreme events vanishes. Thus, without loss of generality, we may extend Ω to $\mathbb{R}^n$ and consider the problem on $\mathbb{R}^n$, giving a joint density in $L^1(\mathbb{R}^n)$ and n marginal densities $\rho_i \in L^1(\mathbb{R})$:

$$\rho_i(x_i) = \int_{\mathbb{R}^{n-1}} \rho(x_1, x_2, ..., x_n)\, dx_1...dx_{i-1}dx_{i+1}...dx_n, \qquad i = 1,...n$$

Correspondingly, we have an entropy functional of ρ (joint entropy) in the Shannon sense

$$H = -\int_{\mathbb{R}^n} \rho(x) \log \rho(x)\, dx \tag{1}$$

and n marginal entropies

$$H_i = -\int_{\mathbb{R}} \rho(x_i) \log \rho(x_i)\, dx_i, \qquad i = 1,...,n \tag{2}$$

Consider an n-dimensional dynamical system, autonomous or nonautonomous,

$$\frac{dx}{dt} = F(x, t) \tag{3}$$

where $F = (F_1, F_2, ..., F_n)^T$ is the vector field. With random inputs at the initial stage, the system generates a continuous stochastic process $\{x(t), t \geq 0\}$, which is what we are concerned with. In many cases, the process may not be continuous in time (such as that generated by the baker transformation, as mentioned in the introduction). We thence also need to consider a system in the discrete mapping form:

$$x(\tau + 1) = \Phi(x(\tau)) \tag{4}$$

with τ being positiver integers. Here Φ is an n-dimensional transformation

$$\Phi : \mathbb{R}^n \to \mathbb{R}^n, \qquad (x_1, x_2, ..., x_n) \mapsto (\Phi_1(x), \Phi_2(x), ..., \Phi_n(x)) \tag{5}$$

the counterpart of the vector field F. Again, the system is assumed to be perfect, with randomness limited within the initial conditions. Cases with stochasticity due to model inaccuracies are deferred to Section 5. The stochastic process thus formed is in a discrete time form $\{x(\tau), \tau\}$, with $\tau > 0$ signifying the time steps. Our formalism will be established henceforth within these frameworks.

2.2. Toward a Rigorous Formalism—A Heuristic Argument

First, let us look at the two-dimensional (2D) case originally studied by Liang and Kleeman [49]

$$\frac{dx_1}{dt} = F_1(x_1, x_2, t) \tag{6}$$

$$\frac{dx_2}{dt} = F_2(x_1, x_2, t) \tag{7}$$

This is a system of minimal dimensionality that admits information flow. Without loss of generality, examine only the flow/transfer from x_2 to x_1.

Under the vector field $F = (F_1, F_2)^T$ x evolves with time; correspondingly its joint pdf $\rho(x)$ evolves, observing a Liouville equation [50]:

$$\frac{\partial \rho}{\partial t} + \frac{\partial}{\partial x_1}(F_1 \rho) + \frac{\partial}{\partial x_2}(F_2 \rho) = 0 \tag{8}$$

As argued in the introduction, what matters here is the evolution of H_1 namely the marginal entropy of x_1. For this purpose, integrate (8) with respect to x_2 over $\mathbb{R}$ to get:

$$\frac{\partial \rho_1}{\partial t} + \frac{\partial}{\partial x_1} \int_{\mathbb{R}} F_1 \rho \, dx_2 = 0 \tag{9}$$

Other terms vanish, thanks to the compact support assumption for ρ. Multiplication of (9) by $-(1 + \log \rho_1)$ followed by an integration over $\mathbb{R}$ gives the tendency of H_1:

$$\frac{dH_1}{dt} = \int_{\mathbb{R}^2} \left[\log \rho_1 \frac{\partial(\rho F_1)}{\partial x_1} \right] dx_1 dx_2 = -E \left(\frac{F_1}{\rho_1} \frac{\partial \rho_1}{\partial x_1} \right) \tag{10}$$

where E stands for mathematical expectation with respect to ρ. In the derivation, integration by parts has been used, as well as the compact support assumption.

Now what is the rate of information flow from x_2 to x_1? In [49], Liang and Kleeman argue that, as the system steers a state forward, the marginal entropy of x_1 is replenished from two different sources: one is from x_1 itself, another from x_2. The latter is through the very mechanism namely information flow/transfer. If we write the former as dH_1^*/dt, and denote by $T_{2\to1}$ the rate of information flow/transfer from x_2 to x_1 (T stands for "transfer"), this gives a decomposition of the marginal entropy increase according to the underlying mechanisms:

$$\frac{dH_1}{dt} = \frac{dH_1^*}{dt} + T_{2\to1} \tag{11}$$

Here dH_1/dt is known from Equation (10). To find $T_{2\to1}$, one may look for dH_1^*/dt instead. In [49], Liang and Kleeman find that this is indeed possible, based on a heuristic argument. To see this, multiply the Liouville Equation (8) by $-(1 + \log\rho)$, then integrate over $\mathbb{R}^2$. This yields an equation governing the evolution of the joint entropy H which, after a series of manipulation, is reduced to

$$\frac{dH}{dt} = \int_{\mathbb{R}^2} \nabla \cdot (\rho \log \rho \boldsymbol{F}) dx_1 dx_2 + \int_{\mathbb{R}^2} \rho \nabla \cdot \boldsymbol{F} dx_1 dx_2$$

where ∇ is the divergence operator. With the assumption of compact support, the first term on the right hand side goes to zero. Using E to indicate the operator of mathematical expectation, this becomes

$$\frac{dH}{dt} = E\left(\nabla \cdot \boldsymbol{F}\right) \tag{12}$$

That is to say, the time rate of change of H is precisely equal to the mathematical expectation of the divergence of the vector field. This remarkably concise result tells that, as a system moves on, the change of its joint entropy is totally controlled by the contraction or expansion of the phase space of the system. Later on, Liang and Kleeman show that this is actually a property holding for deterministic systems of arbitrary dimensionality, even without invoking the compact assumption [51]. Moreover, it has also been shown that, the local marginal entropy production observes a law in the similar form, if no remote effect is taken in account [52].

With Equation (12), Liang and Kleeman argue that, apart from the complicated relations, the rate of change of the marginal entropy H_1 due to x_1 only (*i.e.*, dH_1^*/dt as symbolized above), should be

$$\frac{dH_1^*}{dt} = E\left(\frac{\partial F_1}{\partial x_1}\right) = \int_{\mathbb{R}^2} \rho \frac{\partial F_1}{\partial x_1}\, dx_1 dx_2 \tag{13}$$

This heuristic reasoning makes the separation (11) possible. Hence the information flows from x_2 to x_1 at a rate of

$$\begin{aligned}
T_{2\to1} &= \frac{dH_1}{dt} - \frac{dH_1^*}{dt} = -E\left(\frac{F_1}{\rho_1}\frac{\partial\rho_1}{\partial x_1}\right) - E\left(\frac{\partial F_1}{\partial x_1}\right) \\
&= -E\left[\frac{1}{\rho_1}\frac{\partial(F_1\rho_1)}{\partial x_1}\right] \\
&= -\int_{\mathbb{R}^2} \rho_{2|1}(x_2|x_1)\frac{\partial(F_1\rho_1)}{\partial x_1} dx_1 dx_2
\end{aligned} \tag{14}$$

where $\rho_{2|1}$ is the conditional pdf of x_2, given x_1. The rate of information flow from x_1 to x_2, written $T_{1\to2}$, can be derived in the same way. This tight formalism (called "LK2005 formalism" henceforth), albeit based on heuristic reasoning, turns out to be very successful. The same strategy has been applied again in a similar study by Majda and Harlim [53]. We will have a chance to see these in Sections 4 and 6.

2.3. Mathematical Formalism

The success of the LK2005 formalism is remarkable. However, its utility is limited to systems of dimensionality 2. For an n-dimensional system with $n > 2$, the so-obtained Equation (14) is not the transfer from x_2 to x_1, but the cumulant transfer to x_1 from all other components $x_2, x_3,..., x_n$. Unless one can screen out from Equation (14) the part contributed from x_2, it seems that the formalism does not yield the desiderata for high-dimensional systems.

To overcome the difficulty, Liang and Kleeman [48,51] observe that, the key part in Equation (14) namely dH_1^*/dt actually can be alternatively interpreted, for a 2D system, as the evolution of H_1 with the effect of x_2 excluded. In other words, it is the tendency of H_1 with x_2 frozen instantaneously at time t. To avoid confusing with dH_1^*/dt, denote it as $dH_{1\cancel{2}}/dt$, with the subscript $\cancel{2}$ signifying that the effect of x_2 is removed. In this way dH_1/dt is decomposed into two disjoint parts: $T_{2\to1}$ namely the rate of information flow and $dH_{1\cancel{2}}/dt$. The flow is then the difference between dH_1/dt and $dH_{1\cancel{2}}/dt$:

$$T_{2\to1} = \frac{dH_1}{dt} - \frac{dH_{1\cancel{2}}}{dt} \tag{15}$$

For 2D systems, this is just a restatement of Equation (14) in another set of symbols; but for systems with dimensionality higher than 2, they are quite different. Since the above partitioning does not have any restraints on n, Equation (15) is applicable to systems of arbitrary dimensionality.

In the same spirit, we can formulate the information transfer for discrete systems in the form of Equation (4). As x is mapped forth under the transformation Φ from time step τ to $\tau+1$, correspondingly its density ρ is steered forward by an operator termed after Georg Frobenius and Oskar Perron, which we will introduce later. Accordingly the entropies H, H_1, and H_2 also change with time. On the interval $[\tau, \tau+1]$, let H_1 be incremented by ΔH_1 from τ to $\tau+1$. By the foregoing argument, the evolution of H_1 can be decomposed into two exclusive parts according to their driving mechanisms, *i.e.*, the information flow from x_2, $T_{2\to1}$, and the evolution with the effect of x_2 excluded, written as $\Delta H_{1\cancel{2}}$. We therefore obtain the discrete counterpart of Equation (15):

$$T_{2\to1} = \Delta H_1 - \Delta H_{1\cancel{2}} \tag{16}$$

Equations (15) and (16) give the rates of information flow/transfer from component x_2 to component x_1 for systems (3) and (4), respectively. One may switch the corresponding indices to obtain the flow between any component pair x_i and x_j, $i \neq j$. In the following two sections we will be exploring how these equations are evaluated.

3. Discrete Systems

3.1. Frobenius-Perron Operator

For discrete systems in the form of Equation (4), as x is carried forth under the transformation Φ, there is another transformation, called Frobenius–Perron operator $\mathcal{P}$ (F-P operator hereafter), steering $\rho(x)$, *i.e.*, the pdf of x, to $\mathcal{P}\rho$ (see a schematic in Figure 1). The F-P operator governs the evolution of the density of x.

A rigorous definition requires some ingredients of measure theory which is beyond the scope this review, and the reader may consult with the reference [50]. Loosely speaking, given a transformation $\Phi : \Omega \to \Omega$ (in this review, $\Omega = \mathbb{R}^n$), $x \mapsto \Phi x$, it is a mapping $\mathcal{P} : L^1(\Omega^n) \to L^1(\Omega^n)$, $\rho \mapsto \mathcal{P}\rho$, such that

$$\int_\omega \mathcal{P}\rho(x)dx = \int_{\Phi^{-1}(\omega)} \rho(x)dx \tag{17}$$

for any $\omega \subset \Omega$. If Φ is nonsingular and invertible, it actually can be explicitly evaluated. Making transformation $y = \Phi(x)$, the right hand side is, in this case,

$$\int_{\Phi^{-1}(\omega)} \rho(x) \, dx = \int_\omega \rho\left[\Phi^{-1}(y)\right] \cdot \left|J^{-1}\right| \, dy$$

where J is the Jacobian of Φ:

$$J = \det \left[\frac{\partial(y_1, y_2, ..., y_n)}{\partial(x_1, x_2, ..., x_n)} \right]$$

and J^{-1} its inverse. Since ω is arbitrarily chosen, we have

$$\mathcal{P}\rho(x) = \rho\left[\Phi^{-1}(x)\right] \cdot \left|J^{-1}\right| \tag{18}$$

If no nonsingularity is assumed for the transformation Φ, but the sample space Ω is in a Cartesian product form, as is for this review, the F-P operator can also be evaluated, though not in an explicit form. Consider a domain

$$\omega = [a_1, x_1] \times [a_2, x_2] \times ... \times [a_n, x_n]$$

where $a = (a_1, ..., a_n)$ is some constant point (usually can be set to be the origin). Let the counterimage of ω be $\Phi^{-1}(\omega)$, then it has been proved (c.f. [50]) that

$$\mathcal{P}\rho(x) = \frac{\partial^n}{\partial x_n...\partial x_2 \partial x_1} \int_{\Phi^{-1}(\omega)} \rho(\xi_1, \xi_2, ..., \xi_n)\, d\xi_1 d\xi_2 ... d\xi_n$$

In this review, we consider a sample space $\mathbb{R}^n$, so essentially all the F-P operators can be calculated this way.

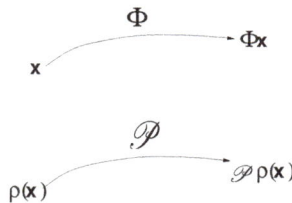

Figure 1. Illustration of the Frobenius-Perron operator $\mathcal{P}$, which takes $\rho(x)$ to $\mathcal{P}\rho(x)$ as Φ takes x to Φx.

3.2. Information Flow

The F-P operator $\mathcal{P}$ allows for an evaluation of the change of entropy as the system evolves forth. By the formalism (16), we need to examine how the marginal entropy changes on a time interval $[\tau, \tau+1]$. Without loss of generality, consider only the flow from x_2 to x_1. First look at increase of H_1. Let ρ be the joint density at step τ, then the joint density at step $\tau + 1$ is $\mathcal{P}\rho$, and hence

$$\begin{aligned} \Delta H_1 &= H_1(\tau+1) - H_1(\tau) \\ &= -\int_{\mathbb{R}} (\mathcal{P}\rho)_1(y_1) \cdot \log(\mathcal{P}\rho)_1(y_1)\, dy_1 + \int_{\mathbb{R}} \rho_1(x_1) \cdot \log \rho_1(x_1)\, dx_1 \end{aligned} \tag{19}$$

Here $(\mathcal{P}\rho)_1$ means the marginal density of x_1 at $\tau + 1$; it is equal to $\mathcal{P}\rho$ with all components of x but x_1 being integrated out. The independent variables with respect to which the integrations are taken are dummy; but for the sake of clarity, we use different notations, *i.e.*, x and y, for them at time step τ and $\tau + 1$, respectively.

The key to the formalism (16) is the finding of

$$\Delta H_{1\,2} = H_{1\,2}(\tau+1) - H_1(\tau) \tag{20}$$

namely the increment of the marginal entropy of x_1 on $[\tau, \tau+1]$ with the contribution from x_2 excluded. Here the system in question is no longer Equation (4), but a system with a mapping modified from Φ:

$$\Phi_{\setminus 2}: \begin{cases} y_1 = & \Phi_1(x_1, x_2, x_3, ..., x_n) \\ y_3 = & \Phi_3(x_1, x_2, x_3, ..., x_n) \\ \vdots & \vdots \\ y_n = & \Phi_n(x_1, x_2, x_3, ..., x_n) \end{cases} \tag{21}$$

with x_2 frozen instantaneously at τ as a parameter. Again, we use $x_i = x_i(\tau)$, $y_i = \Phi(x(\tau)) = x_i(\tau+1)$, $i = 1, ..., n$, to indicate the state variables at steps τ and $\tau+1$, respectively, to avoid any possible confusion. In the mean time, the dependence on τ and $\tau+1$ are suppressed for notational economy. Corresponding to the modified transformation $\Phi_{\setminus 2}$ is a modified F-P operator, written $\mathcal{P}_{\setminus 2}$. To find $H_{1 \setminus 2}(\tau+1)$, examine the quantity $h = -\log(\mathcal{P}_{\setminus 2}\rho)_1(y_1)$, where the subscript 1 indicates that this is a marginal density of the first component, and the dependence on y_1 tells that this is evaluated at step $\tau+1$. Recall how Shannon entropy is defined: $H_{1 \setminus 2}(\tau+1)$ is essentially the mathematical expectation, or "average" in loose language, of h. More specifically, it is h multiplied with some pdf followed by an integration over $\mathbb{R}^n$, i.e., the corresponding sample space. The pdf is composed of several different factors. The first is, of course, $(\mathcal{P}_{\setminus 2}\rho)_1(y_1)$. But h, as well as $(\mathcal{P}_{\setminus 2})_1$, also has dependence on x_2, which is embedded within the subscript $\setminus 2$. Recall how x_2 is treated during $[\tau, \tau+1]$: It is frozen at step τ and kept on as a parameter, given all other components at τ. Therefore, the second part of the density is $\rho(x_2 | x_1, x_3, ..., x_n)$, i.e., the conditional density of x_2 on $x_1, x_3, ..., x_n$. (Note again that x_i means variables at time step τ.) This factor introduces extra dependencies: $x_3, x_4, ..., x_n$ (that of x_1 is embedded in y_1), which must also be averaged out, so the third factor of the density is $\rho_{3...n}(x_3, ..., x_n)$ namely the joint density of $(x_3, x_4, ..., x_n)$. Put all these together,

$$H_{1 \setminus 2}(\tau+1) = -\int_{\mathbb{R}^n} (\mathcal{P}_{\setminus 2}\rho)_1(y_1) \cdot \log(\mathcal{P}_{\setminus 2}\rho)_1(y_1) \cdot \rho(x_2|x_1, x_3, ..., x_n)$$
$$\cdot \rho_{3...n}(x_3, ..., x_n) \, dy_1 dx_2 dx_3 ... dx_n \tag{22}$$

Subtraction of $H_{1 \setminus 2}(\tau+1) - H_1(\tau)$ from Equation (19) gives, eventually, the rate of information flow/transfer from x_2 to x_1:

$$T_{2 \to 1} = -\int_{\mathbb{R}} (\mathcal{P}\rho)_1(y_1) \cdot \log(\mathcal{P}\rho)_1(y_1) \, dy_1$$
$$+ \int_{\mathbb{R}^n} (\mathcal{P}_{\setminus 2}\rho)_1(y_1) \cdot \log(\mathcal{P}_{\setminus 2}\rho)_1(y_1) \cdot \rho(x_2|x_1, x_3, ..., x_n) \cdot$$
$$\rho_{3...n}(x_3, ..., x_n) \, dy_1 dx_2 dx_3 ... dx_n \tag{23}$$

Notice that the conditional density of x_2 is on x_1, not on y_1. (x_1 and y_1 are the same state variable evaluated at different time steps, and are connected via $y_1 = \Phi_1(x_1, x_2, ..., x_n)$.)

Likewise, it is easy to obtain the information flow between any pair of components. If, for example, we are concerned with the flow from x_j to x_i ($i, j = 1, 2, ..., n, i \neq j$), replacement of the indices 1 and 2 in Equation (23) respectively with i and j gives

$$T_{j \to i} = -\int_{\mathbb{R}} (\mathcal{P}\rho)_i(y_i) \cdot \log(\mathcal{P}\rho)_i(y_i) \, dy_i$$
$$+ \int_{\mathbb{R}^n} (\mathcal{P}_{\setminus j}\rho)_i(y_i) \cdot \log(\mathcal{P}_{\setminus j}\rho)_1(y_i) \cdot \rho(x_j \mid x_1, x_2, ..., x_{j-1}, x_{j+1}, ..., x_n) \cdot$$
$$\rho_{\setminus i}\, dx_1 dx_2 ... dx_{i-1} dy_i dx_{i+1} ... dx_n \tag{24}$$

Here the subscript $\backslash j$ of $\mathcal{P}$ means the F-P operator with the effect of the jth component excluded through freezing it instantaneously as a parameter. We have also abused the notation a little bit for the density function to indicate the marginalization of that component. That is to say,

$$\rho_{\backslash i} = \rho_{\backslash i}(x_1, ..., x_{i-1}, x_{i+1}, ..., x_n) = \int_{\mathbb{R}} \rho(x) \, dx_i \tag{25}$$

and $\rho_{\backslash ij}$ is the density after being marginalized twice, with respect to x_i and x_j. To avoid this potential notation complexity, alternatively, one may reorganize the order of the components of the vector $x = (x_1, ..., x_n)^T$ such that the pair appears in the first two slots, and modify the mapping Φ accordingly. In this case, the flow/transfer is precisely the same in form as Equation (23). Equations (23) and (24) can be evaluated explicitly for systems that are definitely specified. In the following sections we will see several concrete examples.

3.3. Properties

The information flow obtained in Equations (23) or (24) has some nice properties. The first is a concretization of the transfer asymmetry emphasized by Schreiber [47] (as mentioned in the introduction), and the second a special property for 2D systems.

Theorem 3.1. *For the system Equation (4), if Φ_i is independent of x_j, then $T_{j \to i} = 0$ (in the mean time, $T_{i \to j}$ need not be zero).*

The proof is rather technically involved; the reader is referred to [48] for details. This theorem states that, if the evolution of x_i has nothing to do with x_j, then there will be no information flowing from x_j to x_i. This is in agreement with observations, and with what one would argue on physical grounds. On the other hand, the vanishing $T_{j \to i}$ yields no clue on $T_{i \to j}$, i.e., the flow from x_i to x_j need not be zero in the mean time, unless Φ_j does not rely on x_i. This is indicative of a very important physical fact: information flow between a component pair is not symmetric, in contrast to the notion of mutual information ever existing in information theory. As emphasized by Schreiber [47], a faithful formalism must be able to recover this asymmetry. The theorem shows that our formalism yields precisely what is expected. Since transfer asymmetry is a reflection of causality, the above theorem is also referred to as *property of causality* by Liang and Kleeman [48].

Theorem 3.2. *For the system Equation (4), if $n = 2$ and Φ_1 is invertible, then $T_{2 \to 1} = \Delta H_1 - E \log |J_1|$, where $J_1 = \partial \Phi_1 / \partial x_1$.*

A brief proof will help to gain better understanding of the theorem. If $n = 2$, the modified system has a mapping $\Phi_{\backslash 2}$ which is simply Φ_1 with x_2 as a parameter. Equation (22) is thus reduced to

$$H_{1\backslash 2}(\tau + 1) = -\int_{\mathbb{R}^2} (\mathcal{P}_{\backslash 2}\rho)_1(y_1) \cdot \log(\mathcal{P}_{\backslash 2}\rho)_1(y_1) \cdot \rho(x_2|x_1) \, dy_1 dx_2$$

where $y_1 = \Phi_1(x_1, x_2)$, and $(\mathcal{P}_{\backslash 2}\rho)_1$ the marginal density of x_1 evolving from $\rho_{\backslash 2} = \rho_1$ upon one transformation of $\Phi_{\backslash 2} = \Phi_1$. By assumption Φ_1 is invertible, that is to say, $J_1 = \frac{\partial \Phi_1}{\partial x_1} \neq 0$. The F-P operator hence can be explicitly written out:

$$\begin{aligned} (\mathcal{P}_{\backslash 2}\rho)_1(y_1) &= \rho\left[\Phi_1^{-1}(y_1, x_2)\right] \cdot \left|J_1^{-1}\right| \\ &= \rho_1(x_1) \left|J_1^{-1}\right| \end{aligned} \tag{26}$$

So

$$\Delta H_{1\backslash 2} = H_{1\backslash 2}(\tau + 1) - H_1(\tau)$$

$$
\begin{aligned}
&= -\int_{\mathbb{R}^2} \rho_1(x_1) \left| J_1^{-1} \right| \log \left(\rho_1(x_1) \left| J_1^{-1} \right| \right) \rho(x_2|x_1) |J_1| \, dx_1 dx_2 + \int_{\mathbb{R}} \rho_1 \log \rho_1 \, dx_1 \\
&= -\int_{\mathbb{R}^2} \rho_1(x_1) \, \rho(x_2|x_1) \log \left| J_1^{-1} \right| \, dx_1 dx_2 \\
&= \int_{\mathbb{R}^2} \rho(x_1, x_2) \log |J_1| \, dx_1 dx_2 \\
&= E \log |J_1| \quad\quad\quad\quad\quad\quad\quad\quad\quad\quad\quad\quad\quad\quad (27)
\end{aligned}
$$

The conclusion follows subsequently from Equation (16).

The above theorem actually states another interesting fact that parallels what we introduced previously in §2.2 via heuristic reasoning. To see this, reconsider the mapping $\Phi : \mathbb{R}^n \to \mathbb{R}^n, x \mapsto x$. Let Φ be nonsingular and invertible. By Equation (18), the F-P operator of the joint pdf ρ can be explicitly evaluated. Accordingly, the entropy increases, as time moves from step τ to step $\tau + 1$, by

$$
\begin{aligned}
\Delta H &= -\int_{\mathbb{R}^n} \mathcal{P}\rho(x) \log \mathcal{P}\rho(x) \, dx + \int_{\mathbb{R}^n} \rho(x) \log \rho(x) \, dx \\
&= -\int_{\mathbb{R}^n} \rho \left[\Phi^{-1}(x) \right] \left| J^{-1} \right| \log \rho \left[\Phi^{-1}(x) \right] \left| J^{-1} \right| \, dx + \int_{\mathbb{R}^n} \rho(x) \log \rho(x) \, dx
\end{aligned}
$$

After some manipulation (see [48] for details), this is reduced to

$$
\Delta H = E \log |J| \quad\quad\quad\quad\quad\quad\quad\quad\quad\quad\quad\quad\quad\quad (28)
$$

This is the discrete counterpart of Equation (12), yet another remarkably concise formula. Now, if the system in question is 2-dimensional, then, as argued in §2.2, the information flow from x_2 to x_1 should be $\Delta H_1 - \Delta H_1^*$, with ΔH_1^* being the marginal entropy increase due to x_1 itself. Furthermore, if Φ_1 is nonsingular and invertible, then Equation (28) tells us it must be that

$$
\Delta H_1^* = E \log |J_1|
$$

and this is precisely what Theorem 3.2 reads.

4. Continuous Systems

For continuous systems in the form of Equation (3), we may take advantage of what we already have from the previous section to obtain the information flow. Without loss of generality, consider only the flow/transfer from x_2 to x_1, $T_{2\to1}$. We adopt the following strategy to fulfill the task:

- Discretize the continuous system in time on $[t, t + \Delta t]$, and construct a mapping Φ to take $x(t)$ to $x(t + \Delta t)$;
- Freeze x_2 in Φ throughout $[t, t + \Delta t]$ to obtain a modified mapping $\Phi_{\not{2}}$;
- Compute the marginal entropy change ΔH_1 as Φ steers the system from t to $t + \Delta t$;
- Derive the marginal entropy change $\Delta H_{1\not{2}}$ as $\Phi_{\not{2}}$ steers the modified system from t to $t + \Delta t$;
- Take the limit

$$
T_{2\to1} = \lim_{\Delta t \to 0} \frac{\Delta H_1 - \Delta H_{1\not{2}}}{\Delta t}
$$

to arrive at the desiderata.

4.1. Discretization of the Continuous System

As the first step, construct out of Equation (3) an n-dimensional discrete system, which steers $x(t) = (x_1, x_2, ..., x_n)$ to $x(t + \Delta t)$. To avoid any confusion that may arise, $x(t + \Delta t)$ will be denoted as $\boldsymbol{y} = (y_1, y_2, ..., y_n)$ hereafter. Discretization of Equation (3) results in a mapping, to the first order of Δt, $\Phi = (\Phi_1, \Phi_2, ..., \Phi_n): \mathbb{R}^n \to \mathbb{R}^n, \boldsymbol{x} \mapsto \boldsymbol{y}$:

$$
\Phi : \begin{cases}
y_1 = & x_1 + \Delta t \cdot F_1(\boldsymbol{x}) \\
y_2 = & x_2 + \Delta t \cdot F_2(\boldsymbol{x}) \\
\vdots & \vdots \\
y_n = & x_n + \Delta t \cdot F_n(\boldsymbol{x})
\end{cases}
\tag{29}
$$

Clearly, this mapping is always invertible so long as Δt is small enough. In fact, we have

$$
\Phi^{-1} : \begin{cases}
x_1 = & y_1 - \Delta t \cdot F_1(\boldsymbol{y}) + O(\Delta t^2) \\
x_2 = & y_2 - \Delta t \cdot F_2(\boldsymbol{y}) + O(\Delta t^2) \\
\vdots & \vdots \\
x_n = & y_n - \Delta t \cdot F_n(\boldsymbol{y}) + O(\Delta t^2)
\end{cases}
\tag{30}
$$

to the first order of Δt. Furthermore, its Jacobian J is

$$
\begin{aligned}
J & = \det \left[\frac{\partial(y_1, y_2, ..., y_n)}{\partial(x_1, x_2, ..., x_n)} \right] \\
& = \prod_i \left(1 + \Delta t \frac{\partial F_i}{\partial x_i} \right) + O(\Delta t^2) \\
& = 1 + \Delta t \sum_{i=1}^{n} \frac{\partial F_i}{\partial x_i} + O(\Delta t^2)
\end{aligned}
\tag{31}
$$

Likewise, it is easy to get

$$
\begin{aligned}
J^{-1} = & = \det \left[\frac{\partial(x_1, x_2, ..., x_n)}{\partial(y_1, y_2, ..., y_n)} \right] \\
& = 1 - \Delta t \sum_{i=1}^{n} \frac{\partial F_i}{\partial x_i} + O(\Delta t^2)
\end{aligned}
\tag{32}
$$

This makes it possible to evaluate the F-P operator associated with Φ. By Equation (18),

$$
\begin{aligned}
\mathcal{P}\rho(y_1, ..., y_n) & = \rho \left(\Phi^{-1}(y_1, ... y_n) \right) \left| J^{-1} \right| \\
& = \rho(x_1, x_2, ..., x_n) \cdot |1 - \Delta t \nabla \cdot \boldsymbol{F}| + O(\Delta t^2)
\end{aligned}
\tag{33}
$$

Here $\nabla \cdot \boldsymbol{F} = \sum_i \frac{\partial F_i}{\partial x_i}$; we have suppressed its dependence on $\boldsymbol{x}$ to simplify the notation.

As an aside, the explicit evaluation (31), and subsequently (32) and (33), actually can be utilized to arrive at the important entropy evolution law (12) without invoking any assumptions. To see this, recall that $\Delta H = E \log |J|$ by Equation (28). Let Δt go to zero to get

$$
\frac{dH}{dt} = \lim_{\Delta t \to 0} \frac{\Delta H}{\Delta t} = E \lim_{\Delta t \to 0} \frac{1}{\Delta t} \log \left(1 + \Delta t \nabla \cdot \boldsymbol{F} + O(\Delta t^2) \right)
$$

which is the very result $E(\nabla \cdot \boldsymbol{F})$, just as one may expect.

4.2. Information Flow

To compute the information flow $T_{2 \to 1}$, we need to know dH_1/dt and $dH_{1\not{2}}/dt$. The former is easy to find from the Liouville equation associated with Equation (3), *i.e.,*

$$
\frac{\partial \rho}{\partial t} + \frac{\partial(F_1 \rho)}{\partial x_1} + \frac{\partial(F_2 \rho)}{\partial x_2} + ... + \frac{\partial(F_n \rho)}{\partial x_n} = 0
\tag{34}
$$

following the same derivation as that in §2.2:

$$\frac{dH_1}{dt} = \int_{\mathbb{R}^n} \log \rho_1 \frac{\partial (F_1 \rho)}{\partial x_1} \, dx \qquad (35)$$

The challenge lies in the evaluation of $dH_{1\tilde{2}}/dt$. We summarize the result in the following proposition:

Proposition 4.1. *For the dynamical system* (3), *the rate of change of the marginal entropy of* x_1 *with the effect of* x_2 *instantaneously excluded is:*

$$\frac{dH_{1\tilde{2}}}{dt} = \int_{\mathbb{R}^n} (1 + \log \rho_1) \cdot \frac{\partial (F_1 \rho_{\tilde{2}})}{\partial x_1} \cdot \Theta_{2|1} \, dx +$$
$$\int_{\mathbb{R}^n} \rho_1 \log \rho_1 \cdot F_1 \cdot \frac{\partial (\rho / \rho_{\tilde{2}})}{\partial x_1} \cdot \rho_{1\tilde{2}} \, dx \qquad (36)$$

where

$$\theta_{2|1} = \theta_{2|1}(x_1, x_2, x_3, ..., x_n) = \frac{\rho}{\rho_{\tilde{2}}} \rho_{1\tilde{2}} \qquad (37)$$

$$\Theta_{2|1} = \int_{\Omega_{\mathbb{R}^{n-2}}} \theta_{2|1}(\mathbf{x}) \, dx_3 ... dx_n \qquad (38)$$

and $\rho_{\tilde{2}} = \int_{\mathbb{R}} \rho \, dx_2$, $\rho_{1\tilde{2}} = \int_{\mathbb{R}^2} \rho \, dx_1 dx_2$ *are the densities after marginalized with* x_2 *and* (x_1, x_2), *respectively.*

The proof is rather technically involved; for details, see [51], section 5.

With the above result, subtract $dH_{1\tilde{2}}/dt$ from dH_1/dt and one obtains the flow rate from x_2 to x_1. Likewise, the information flow between any component pair (x_i, x_j), $i, j = 1, 2, ..., n$; $i \neq j$, can be obtained henceforth.

Theorem 4.1. *For the dynamical system* (3), *the rate of information flow from* x_j *to* x_i *is*

$$T_{j \to i} = \int_{\Omega} (1 + \log \rho_i) \left(\frac{\partial (F_i \rho)}{\partial x_i} - \frac{\partial (F_i \rho_{\tilde{j}})}{\partial x_i} \cdot \Theta_{j|i} \right) dx$$
$$+ \int_{\Omega} \frac{\partial (F_i \rho_i \log \rho_i)}{\partial x_i} \cdot \theta_{j|i} \, dx \qquad (39)$$

where

$$\theta_{j|i} = \theta_{j|i}(\mathbf{x}) = \frac{\rho}{\rho_{\tilde{i}}} \rho_{\tilde{i}\tilde{j}} \qquad (40)$$

$$\rho_{\tilde{i}} = \int_{\mathbb{R}} \rho(\mathbf{x}) \, dx_i \qquad (41)$$

$$\rho_{\tilde{i}\tilde{j}} = \int_{\mathbb{R}^2} \rho(\mathbf{x}) \, dx_i dx_j \qquad (42)$$

$$\Theta_{j|i} = \Theta_{j|i}(x_i, x_j) = \int_{\mathbb{R}^{n-2}} \theta_{j|i}(\mathbf{x}) \prod_{v \neq i, j} dx_v \qquad (43)$$

In this formula, $\Theta_{j|i}$ reminds one of the conditional density x_j on x_i, and, if $n = 2$, it is indeed so. We may therefore call it the "generalized conditional density" of x_j on x_i.

4.3. Properties

Recall that, as we argue in §2.2 based on the entropy evolution law (12), the time rate of change of the marginal entropy of a component, say x_1, due to its own reason, is $dH_1^*/dt = E(\partial F_1/\partial x_1)$. Since for a 2D system, dH_1^*/dt is precisely $dH_{1\tilde{2}}/dt$, we expect that the above formalism (36) or (39) verifies this result.

Theorem 4.2. *If the system* (3) *has a dimensionality 2, then*

$$\frac{dH_{1\tilde{2}}}{dt} = E\left(\frac{\partial F_1}{\partial x_1}\right) \tag{44}$$

and hence the rate of information flow from x_2 to x_1 is

$$T_{2\to 1} = -E\left[\frac{1}{\rho_1}\frac{\partial(F_1\rho_1)}{\partial x_1}\right] \tag{45}$$

What makes a 2D system so special is that, when $n = 2$, $\rho_{\tilde{2}} = \rho_1$, and $\Theta_{2|1}$ is just the conditional distribution of x_2 given x_1, $\rho/\rho_1 = \rho(x_2|x_1)$. Equation (36) can thereby be greatly simplified:

$$
\begin{aligned}
\frac{dH_{1\tilde{2}}}{dt} &= \int_{\mathbb{R}^n}(1+\log\rho_1)\frac{\partial F_1\rho_1}{\partial x_1}\cdot\frac{\rho}{\rho_1}\,dx + \int_{\mathbb{R}^n}\rho_1\log\rho_1\cdot F_1\cdot\frac{\partial\rho(x_2|x_1)}{\partial x_1}\,dx \\
&= \int_{\mathbb{R}^n}\frac{\partial(F_1\rho_1)}{\partial x_1}\frac{\rho}{\rho_1}\,dx + \int_{\mathbb{R}^n}\log\rho_1\cdot\frac{\partial(F_1\rho)}{\partial x_1}\,dx \\
&= \int_{\mathbb{R}^n}\rho\left(\frac{\partial F_1}{\partial x_1}\right)dx = E\left(\frac{\partial F_1}{\partial x_1}\right)
\end{aligned}
\tag{46}
$$

Subtract this from what has been obtained above for dH_1/dt, and we get an information flow just as that in Equation (14) via heuristic argument.

As in the discrete case, one important property that $T_{j\to i}$ must possess is transfer asymmetry, which has been emphasized previously, particularly by Schreiber [47]. The following is a concretization of the argument.

Theorem 4.3. (Causality) *For system* (3), *if F_i is independent of x_j, then $T_{j\to i} = 0$; in the mean time, $T_{i\to j}$ need not vanish, unless F_j has no dependence on x_i.*

Look at the right-hand side of the formula (39). Given that $(1+\log\rho_i)$ and $\rho_{\tilde{j}}$, as well as F_i (by assumption), are independent of x_j, the integration with respect to x_j can be taken within the multiple integrals. Consider the second integral first. All the variables except $\theta_{j|i}$ have dependence on x_j. But $\int\theta_{j|i}dx_j = 1$, so the whole term is equal to $\int_{\mathbb{R}^{n-1}}\frac{\partial(F_i\rho_i\log\rho_i)}{\partial x_i}dx_1...dx_{j-1}dx_{j+1}...dx_n$ which vanishes by the assumption of compact support. For the first integral, move the integration with respect to x_j into the parentheses, as the factor outside has nothing to do with x_j. This integration yields

$$
\begin{aligned}
&\int_{\mathbb{R}}\frac{\partial(F_i\rho)}{\partial x_i}dx_j - \int_{\mathbb{R}}\frac{\partial(F_i\rho_{\tilde{j}})}{\partial x_i}\cdot\Theta_{j|i}dx_j \\
&= \int_{\mathbb{R}^{n-1}}\left[\frac{\partial}{\partial x_i}\left(F_i\int\rho\,dx_j\right) - \frac{\partial}{\partial x_i}(F_i\rho_{\tilde{j}})\cdot\int\Theta_{j|i}dx_j\right]dx_1...dx_{j-1}dx_{j+1}...dx_n \\
&= 0
\end{aligned}
$$

because $\int\rho\,dx_j = \rho_{\tilde{j}}$ and $\int\Theta_{j|i}dx_j = 1$. For all that account, both the two integrals on the right-hand side of Equation (39) vanish, leaving a zero flow of information from x_j to x_i. Notice that this vanishing $T_{j\to i}$ gives no hint on the flow in the opposite direction. In other words, this kind of flow or transfer is not symmetric, reflecting the causal relation between the component pair. As Theorem 3.1 is for discrete systems, Theorem 4.3 is the *property of causality* for continuous systems.

5. Stochastic Systems

So far, all the systems considered are deterministic. In this section we turn to systems with stochasticity included. Consider the stochastic counterpart of Equation (3)

$$dx = F(x, t)dt + B(x, t)dw \tag{47}$$

where w is a vector of standard Wiener processes, and $B = (b_{ij})$ the matrix of perturbation amplitudes. In this section, we limit our discussion to 2D systems, and hence have only two flows/transfers to discuss. Without loss of generality, consider only $T_{2 \to 1}$, *i.e.*, the rate of flow/transfer from x_2 to x_1.

As before, we first need to find the time rate of change of H_1, the marginal entropy of x_1. This can be easily derived from the density evolution equation corresponding to Equation (47), *i.e.*, the Fokker-Planck equation:

$$\frac{\partial \rho}{\partial t} + \frac{\partial (F_1 \rho)}{\partial x_1} + \frac{\partial (F_2 \rho)}{\partial x_2} = \frac{1}{2} \sum_{i,j=1}^{2} \frac{\partial^2 (g_{ij} \rho)}{\partial x_i \partial x_j} \tag{48}$$

where $g_{ij} = g_{ji} = \sum_{k=1}^{2} b_{ik} b_{jk}$, $i, j = 1, 2$. This integrated over $\mathbb{R}$ with respect to x_2 gives the evolution of ρ_1:

$$\frac{\partial \rho_1}{\partial t} + \int_{\mathbb{R}} \frac{\partial (F_1 \rho)}{\partial x_1} dx_2 = \frac{1}{2} \int_{\mathbb{R}} \frac{\partial^2 (g_{11} \rho)}{\partial x_1^2} dx_2 \tag{49}$$

Multiply (49) by $-(1 + \log \rho_1)$, and integrate with respect to x_1 over $\mathbb{R}$. After some manipulation, one obtains, using the compact support assumption,

$$\frac{dH_1}{dt} = -E \left(F_1 \frac{\partial \log \rho_1}{\partial x_1} \right) - \frac{1}{2} E \left(g_{11} \frac{\partial^2 \log \rho_1}{\partial x_1^2} \right) \tag{50}$$

where E is the mathematical expectation with respect to ρ.

Again, the key to the formalism is the finding of $dH_{1 \backslash 2}/dt$. For stochastic systems, this could be a challenging task. The major challenge is that we cannot obtain an F-P operator as nice as that in the previous section for the map resulting from discretization. In early days, Majda and Harlim [53] have tried our heuristic argument in §2.2 to consider a special system modeling the atmosphere–ocean interaction, which is in the form

$$dx_1 = F_1(x_1, x_2)dt$$
$$dx_2 = F_2(x_1, x_2)dt + b_{22}dw_2$$

Their purpose is to find $T_{2 \to 1}$ namely the information transfer from x_2 to x_1. In this case, since the governing equation for x_1 is deterministic, the result is precisely the same as that of LK05, which is shown in in §2.2. The problem here is that the approach cannot be extended even to finding $T_{1 \to 2}$, since the nice law on which the argument is based, *i.e.*, Equation (12), does not hold for stochastic processes.

Liang (2008) [54] adopted a different approach to give this problem a solution. As in the previous section, the general strategy is also to discretize the system in time, modify the discretized system with x_2 frozen as a parameter on an interval $[t, t + \Delta t]$, and then let Δt go to zero and take the limit. But this time no operator analogous to the F-P operator is sought; instead, we discretize the Fokker–Planck equation and expand $x_{1 \backslash 2(t + \Delta t)}$, namely the first component at $t + \Delta t$ with x_2 frozen at t, using the Euler–Bernstein approximation. The complete derivation is beyond the scope of this review; the reader is referred to [54] for details. In the following, the final result is supplied in the form of a proposition.

Proposition 5.1. *For the 2D stochastic system* (47), *the time change of the marginal entropy of x_1 with the contribution from x_2 excluded is*

$$\frac{dH_{1\not2}}{dt} = E\left(\frac{\partial F_1}{\partial x_1}\right) - \frac{1}{2}E\left(g_{11}\frac{\partial^2 \log \rho_1}{\partial x_1^2}\right) - \frac{1}{2}E\left(\frac{1}{\rho_1}\frac{\partial^2(g_{11}\rho_1)}{\partial x_1^2}\right) \tag{51}$$

In the equation, the second and the third terms on the right hand side are from the stochastic perturbation. The first term, as one may recall, is precisely the result of Theorem 4.2. The heuristic argument for 2D systems in Equation (13) is successfully recovered here. With this the rate of information flow can be easily obtained by subtracting $dH_{1\not2}/dt$ from dH_1/dt.

Theorem 5.1. *For the 2D stochastic system* (47), *the rate of information flow from x_2 to x_1 is*

$$T_{2\to1} = -E\left(\frac{1}{\rho_1}\frac{\partial(F_1\rho_1)}{\partial x_1}\right) + \frac{1}{2}E\left(\frac{1}{\rho_1}\frac{\partial^2(g_{11}\rho_1)}{\partial x_1^2}\right) \tag{52}$$

where E is the expectation with respect to $\rho(x_1, x_2)$.

It has been a routine to check for the obtained flow the property of causality or asymmetry. Here in Equation (52), the first term on the right hand side is from the deterministic part of the system, which has been checked before. For the second term, if b_{11}, b_{12}, and hence $g_{11} = \sum_k b_{1k}b_{1k}$ have no dependence on x_2, then the integration with respect to x_2 can be taken inside with ρ/ρ_1 or $\rho(x_2|x_1)$, and results in 1. The remaining part is in a divergence form, which, by the assumption of compact support, gives a zero contribution from the stochastic perturbation. We therefore have the following theorem:

Theorem 5.2. *If, in the stochastic system* (47), *the evolution of x_1 is independent of x_2, then $T_{2\to1} = 0$.*

The above argument actually has more implications. Suppose $B = (b_{ij})$ are independent of x, i.e., the noises are uncorrelated with the state variables. This model is indeed of interest, as in the real world, a large portion of noises are additive; in other words, b_{ij}, and hence g_{ij}, are constant more often than not. In this case, no matter what the vector field F is, by the above argument the resulting information flows within the system will involve no contribution from the stochastic perturbation. That is to say,

Theorem 5.3. *Within a stochastic system, if the noise is additive, then the information flows are the same in form as that of the corresponding deterministic system.*

This theorem shows that, if only information flows are considered, a stochastic system with additive noise functions just like deterministic. Of course, the resemblance is limited to the form of formula; the marginal density ρ_1 in Equation (52) already takes into account the effect of stochasticity, as can be seen from the integrated Fokker–Planck Equation (49). A more appropriate statement might be that, for this case, stochasticity is disguised within the formula of information flow.

6. Applications

Since its establishment, the formalism of information flow has been applied with a variety of dynamical system problems. In the following we give a brief description of these applications.

6.1. Baker Transformation

The baker transformation as a prototype of an area-conserving chaotic map is one of the most studied discrete dynamical systems. Topologically it is conjugate to another well-studied system, the horseshoe map, and has been be used to model the diffusion process in real physical world.

The baker transformation mimics the kneading of dough: first the dough is compressed, then cut in half; the two halves are stacked on one another, compressed, and so forth. Formally, it is defined as a mapping on the unit square $\Omega = [0,1] \times [0,1]$, $\Phi : \Omega \to \Omega$,

$$\Phi(x_1, x_2) = \begin{cases} (2x_1, \frac{x_2}{2}), & 0 \le x_1 \le \frac{1}{2},\, 0 \le x_2 \le 1 \\ (2x_1 - 1, \frac{1}{2}x_2 + \frac{1}{2}), & \frac{1}{2} < x_1 \le 1,\, 0 \le x_2 \le 1 \end{cases} \tag{53}$$

with a Jacobian $J = \det\left[\frac{\partial(\Phi_1(x), \Phi_2(x))}{\partial(x_1, x_2)} \right] = 1$. This is the area-conserving property, which, by Equation (28) yields $\Delta H = E \log |J| = 0$; that is to say, the entropy is also conserved. The nonvanishing Jacobian implies that it is invertible; in fact, it has an inverse

$$\Phi^{-1}(x_1, x_2) = \begin{cases} (\frac{x_1}{2}, 2x_2), & 0 \le x_2 \le \frac{1}{2},\, 0 \le x_1 \le 1 \\ (\frac{x_1+1}{2}, 2x_2 - 1), & \frac{1}{2} \le x_2 \le 1,\, 0 \le x_1 \le 1 \end{cases} \tag{54}$$

Thus the F-P operator $\mathcal{P}$ can be easily found

$$\mathcal{P}\rho(x_1, x_2) = \rho\left[\Phi^{-1}(x_1, x_2)\right] \cdot \left|J^{-1}\right| = \begin{cases} \rho(\frac{x_1}{2}, 2x_2), & 0 \le x_2 < \frac{1}{2} \\ \rho(\frac{1+x_1}{2}, 2x_2 - 1), & \frac{1}{2} \le x_2 \le 1 \end{cases} \tag{55}$$

First compute $T_{2 \to 1}$, the information flow from x_2 to x_1. Let ρ_1 be the marginal density of x_1 at time step τ. Taking integration of Equation (55) with respect to x_2, one obtains the marginal density of x_1 at $\tau + 1$

$$\begin{aligned} (\mathcal{P}\rho)_1(x_1) &= \int_0^{1/2} \rho(\frac{x_1}{2}, 2x_2)\, dx_2 + \int_{1/2}^1 \rho(\frac{x_1+1}{2}, 2x_2 - 1)\, dx_2 \\ &= \frac{1}{2} \int_0^1 \left[\rho\left(\frac{x_1}{2}, x_2\right) + \rho\left(\frac{x_1+1}{2}, x_2\right) \right] dx_2 \\ &= \frac{1}{2} \left[\rho_1\left(\frac{x_1}{2}\right) + \rho_1\left(\frac{x_1+1}{2}\right) \right] \end{aligned} \tag{56}$$

One may also compute the marginal entropy $H_1(\tau + 1)$, which is an entropy functional of $(\mathcal{P}\rho)_1$. However, here it is not necessary, as will soon become clear.

If, on the other hand, x_2 is frozen as a parameter, the transformation (53) then reduces to a dyadic mapping in the stretching direction, $\Phi_1 : [0,1] \to [0,1]$, $\Phi_1(x_1) = 2x_1 \pmod 1$. For any $0 < x_1 < 1$, The counterimage of $[0, x_1]$ is

$$\Phi^{-1}([0, x_1]) = \left[0, \frac{x_1}{2}\right] \cup \left[\frac{1}{2}, \frac{1+x_1}{2}\right]$$

So

$$\begin{aligned} (\mathcal{P}_2\rho)_1(x_1) &= \frac{\partial}{\partial x_1} \int_{\Phi^{-1}([0,x_1])} \rho(s)\, ds \\ &= \frac{\partial}{\partial x_1} \int_0^{x_1/2} \rho(s)\, ds + \frac{\partial}{\partial x_1} \int_{1/2}^{(1+x_1)/2} \rho(s)\, ds \\ &= \frac{1}{2} \left[\rho\left(\frac{x_1}{2}\right) + \rho\left(\frac{1+x_1}{2}\right) \right] \end{aligned}$$

Two observations: (1) This result is exactly the same as Equation (56), *i.e.*, $(\mathcal{P}_2\rho)_1$ is equal to $(\mathcal{P}\rho)_1$. (2) The resulting $(\mathcal{P}_2\rho)_1$ has no dependence on the parameter x_2. The latter helps to simplify the computation of $H_{1\nmid2}(\tau + 1)$ in Equation (22): Now the integration with respect to x_2 can be taken inside, giving $\int \rho(x_2|x_1)dx_2 = 1$. So $H_{1\,\underline{2}}(\tau+1)$ is precisely the entropy functional of $(\mathcal{P}_{\underline{2}}\rho)_1$. But $(\mathcal{P}_{\underline{2}}\rho)_1 = (\mathcal{P}\rho)_1$ by observation (1). Thus $H_1(\tau+1) = H_{1\,\underline{2}}(\tau+1)$, leading to a flow/transfer

$$T_{2\to1} = 0 \qquad (57)$$

The information flow in the opposite direction is different. As above, first compute the marginal density

$$(\mathcal{P}\rho)_2(x_2) = \int_0^1 \mathcal{P}\rho(x_1, x_2)\, dx_1 = \begin{cases} \int_0^1 \rho\left(\frac{x_1}{2}, 2x_2\right) dx_1, & 0 \le x_2 < \frac{1}{2} \\ \int_0^1 \rho\left(\frac{x_1+1}{2}, 2x_2 - 1\right) dx_1, & \frac{1}{2} \le x_2 \le 1 \end{cases} \qquad (58)$$

The marginal entropy increase of x_2 is then

$$\begin{aligned} \Delta H_2 &= -\int_0^1 \int_0^1 \mathcal{P}\rho(x_1, x_2) \cdot \left[\log\left(\int_0^1 \mathcal{P}\rho(\lambda, x_2)d\lambda\right)\right] dx_1 dx_2 \\ &+ \int_0^1 \int_0^1 \rho(x_1, x_2) \cdot \left[\log\left(\int_0^1 \rho(\lambda, x_2)d\lambda\right)\right] dx_1 dx_2, \end{aligned} \qquad (59)$$

which is reduced to, after some algebraic manipulation,

$$\Delta H_2 = -\log 2 + (I + II) \qquad (60)$$

where

$$I = \int_0^1 \int_0^{1/2} \rho(x_1, x_2) \cdot \left[\log \frac{\int_0^1 \rho(\lambda, x_2)d\lambda}{\int_0^{1/2} \rho(\lambda, x_2)d\lambda}\right] dx_1 dx_2 \qquad (61)$$

$$II = \int_0^1 \int_{1/2}^1 \rho(x_1, x_2) \cdot \left[\log \frac{\int_0^1 \rho(\lambda, x_2)d\lambda}{\int_{1/2}^1 \rho(\lambda, x_2)d\lambda}\right] dx_1 dx_2 \qquad (62)$$

To compute $H_{2\,\underline{1}}$, freeze x_1. The transformation is invertible and the Jacobian J_2 is equal to a constant $\frac{1}{2}$. By Theorem 3.2,

$$\Delta H_{2\,\underline{1}} = E \log \frac{1}{2} = -\log 2 \qquad (63)$$

So,

$$T_{1\to2} = \Delta H_2 - \Delta H_{2\,\underline{1}} = I + II \qquad (64)$$

In the expressions for I and II, since both ρ and the terms within the brackets are nonnegative, $I + II \ge 0$. Furthermore, the two brackets cannot vanish simultaneously, hence $I + II > 0$. By Equation (64) $T_{1\to2}$ is strictly positive; in other words, there is always information flowing from x_1 to x_2.

To summarize, the baker transformation transfers information asymmetrically between the two directions x_1 and x_2. As the baker stretches the dough, and folds back on top the other, information flows continuously from the stretching direction x_1 to the folding direction x_2 ($T_{1\to2} > 0$), while no transfer occurs in the opposite direction ($T_{2\to1} = 0$). These results are schematically illustrated in Figure 2; they are in agreement with what one would observe in daily life, as described in the beginning of this review.

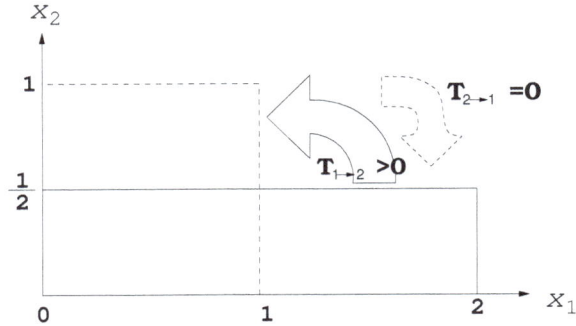

Figure 2. Illustration of the unidirectional information flow within the baker transformation.

6.2. Hénon Map

The Hénon map is another most studied discrete dynamical systems that exhibit chaotic behavior. Introduced by Michel Hénon as a simplified Poincaré section of the Lorenz system, it is a mapping $\Phi = (\Phi_1, \Phi_2) : \mathbb{R}^2 \mapsto \mathbb{R}^2$ defined such that

$$\begin{cases} \Phi_1(x_1, x_2) = 1 + x_2 - ax_1^2 \\ \Phi_2(x_1, x_2) = bx_1 \end{cases} \tag{65}$$

with $a > 0$, $b > 0$. When $a = 1.4$, $b = 0.3$, the map is termed "canonical," for which initially a point will either diverge to infinity, or approach an invariant set known as the Hénon strange attractor. Shown in Figure 3 is the attractor.

Like the baker transformation, the Hénon map is invertible, with an inverse

$$\Phi^{-1}(x_1, x_2) = \left(\frac{x_2}{b}, \ x_1 - 1 + \frac{a}{b^2} x_2^2 \right) \tag{66}$$

The F-P operator thus can be easily found from Equation (18):

$$\begin{aligned} \mathcal{P}\rho(x_1, x_2) &= \rho(\Phi^{-1}(x_1, x_2)) |J^{-1}| \\ &= \frac{1}{b} \cdot \rho \left(\frac{x_2}{b}, \ x_1 - 1 + \frac{a}{b^2} x_2^2 \right) \end{aligned} \tag{67}$$

In the following, we compute the flows/transfers between x_1 and x_2.

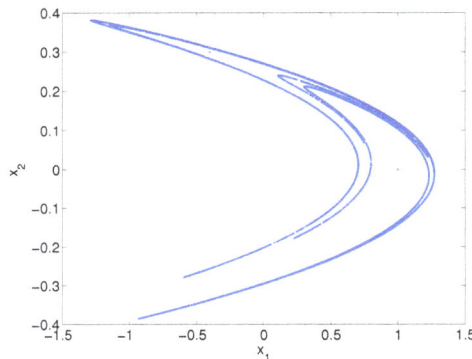

Figure 3. A trajectory of the canonical Hénon map ($a = 1.4$, $b = 0.3$) starting at $(x_1, x_2) = (1, 0)$.

First, consider $T_{2 \to 1}$, *i.e.*, the flow from the linear component x_2 to the quadratic component x_1. By Equation (23), we need to find the marginal density of x_1 at step $\tau + 1$ with and without the effect of x_2, *i.e.*, $(\mathcal{P}\rho)_1$ and $(\mathcal{P}\rho)_{1\tilde{2}}$. With the F-P operator obtained above, $(\mathcal{P}\rho)_1$ is

$$
\begin{aligned}
(\mathcal{P}\rho)_1(x_1) &= \int_{\mathbb{R}} \mathcal{P}\rho(x_1, x_2)\, dx_2 \\
&= \int_{\mathbb{R}} \frac{1}{b} \cdot \rho\left(\frac{x_2}{b}, x_1 - 1 + \frac{a}{b}x_2^2\right) dx_2 \\
&= \int_{\mathbb{R}} \rho(\eta, x_1 - 1 + a\eta^2)\, d\eta \qquad (x_2/b \equiv \eta)
\end{aligned}
$$

If $a = 0$, this integral would be equal to $\rho_2(x_1 - 1)$. Note it is the marginal density of x_2, but the argument is $x_1 - 1$. But here $a > 0$, the integration is taken along a parabolic curve rather than a straight line. Still the final result will be related to the marginal density of x_2; we may as well write it $\tilde{\rho}_2(x_1)$, that is

$$
(\mathcal{P}\rho)_1(x_1) = \tilde{\rho}_2(x_1) \tag{68}
$$

Again, notice that the argument is x_1.

To compute $(\mathcal{P}_{\tilde{2}}\rho)_1$, let

$$
y_1 \equiv \Phi_1(x_1) = 1 + x_2 - ax_1^2
$$

following our convention to distinguish variables at different steps. Modify the system so that x_2 is now a parameter. As before, we need to find the counterimage of $(-\infty, y_1]$ under the transformation with x_2 frozen:

$$
\Phi_1^{-1}((-\infty, y_1]) = \left(-\infty, \ -\sqrt{(1 + x_2 - y_1)/a}\right] \cup \left[\sqrt{(1 + x_2 - y_1)/a}, \ \infty\right)
$$

Therefore,

$$
\begin{aligned}
(\mathcal{P}_{\tilde{2}}\rho)_1(y_1) &= \frac{d}{dy_1} \int_{\Phi_1^{-1}((-\infty, y_1])} \rho_1(s)\, ds \\
&= \frac{d}{dy_1} \int_{-\infty}^{-\sqrt{(1+x_2-y_1)/a}} \rho_1(s)\, ds + \frac{d}{dy_1} \int_{\sqrt{(1+x_2-y_1)/a}}^{\infty} \rho_1(s)\, ds \\
&= \frac{1}{2\sqrt{a(1+x_2-y_1)}} \left[\rho_1\left(-\sqrt{(1+x_2-y_1)/a}\right) + \rho_1\left(\sqrt{(1+x_2-y_1)/a}\right)\right] \\
&\qquad\qquad\qquad\qquad\qquad\qquad\qquad (y_1 < 1 + x_2) \\
&= \frac{1}{2a|x_1|} \left[\rho_1(-x_1) + \rho_1(x_1)\right]. \qquad \text{(recall } y_1 = 1 + x_2 - ax_1^2)
\end{aligned}
$$

Denote the average of $\rho_1(-x_1)$ and $\rho_1(x_1)$ as $\bar{\rho}_1(x_1)$ to make an even function of x_1. Then $(\mathcal{P}_{\tilde{2}}\rho)_1$ is simply

$$
(\mathcal{P}_{\tilde{2}}\rho)_1(y_1) = \frac{\bar{\rho}_1(x_1)}{a|x_1|} \tag{69}
$$

Note that the parameter x_2 does not appear in the arguments. Furthermore, $J_1 = \det\left(\frac{\partial \Phi_1}{\partial x_1}\right) = -2ax_1$. Substitute all the above into Equation (23) to get

$$
T_{2 \to 1} = -\int_{\mathbb{R}} (\mathcal{P}\rho)_1(x_1) \cdot \log(\mathcal{P}\rho)_1(x_1)\, dx_1
$$

$$+ \int_{\mathbb{R}^2} (\mathcal{P}_{\mathfrak{D}}\rho)_1(y_1) \log(\mathcal{P}\rho)_{1\mathfrak{D}}(y_1) \cdot \rho(x_2|x_1) \cdot |J_1| \, dx_1 dx_2$$

$$= - \int_{\mathbb{R}} \tilde{\rho}_2(x_1) \log \tilde{\rho}_2(x_1) \, dx_1$$

$$+ \int_{\mathbb{R}} \frac{\tilde{\rho}_1(x_1)}{a|x_1|} \log \frac{\tilde{\rho}_1(x_1)}{a|x_1|} \cdot |-2ax_1| \cdot \left[\int_{\mathbb{R}} \rho(x_2|x_1) dx_2 \right] dx_1$$

The taking of the integration with respect to x_2 inside the integral is legal since all the terms except the conditional density are independent of x_2. With the fact $\int_{\mathbb{R}} \rho(x_2|x_1) dx_2 = 1$, and the introduction of notations $\tilde{H}$ and $\bar{H}$ for the entropy functionals of $\tilde{\rho}$ and $\bar{\rho}$, respectively, we have

$$T_{2 \to 1} = \tilde{H}_2 - 2\bar{H}_1 - \log|ax_1| \tag{70}$$

Next, consider $T_{1 \to 2}$, the flow from the quadratic component to the linear component. As a common practice, one may start off by computing $(\mathcal{P}\rho)_2$ and $(\mathcal{P}_{\backslash 1}\rho)_2$. However, in this case, things can be much simplified. Observe that, for the modified system with x_1 frozen as a parameter, the Jacobian of the transformation $J_2 = \det \left[\frac{\partial \Phi_2}{\partial x_2} \right] = 0$. So, by Equation (24),

$$\begin{aligned} T_{1 \to 2} &= - \int_{\mathbb{R}} (\mathcal{P}\rho)_2(x_2) \cdot \log(\mathcal{P}\rho)_2(x_2) \, dx_2 \\ &\quad + \int_{\mathbb{R}} (\mathcal{P}_{\backslash 1}\rho)_2(y_2) \cdot \log(\mathcal{P}_{\backslash 1}\rho)_2(y_2) \cdot \rho(x_1|x_2) \cdot |J_2| \, dx_1 dx_2, \\ &\qquad\qquad\qquad (y_2 \equiv \Phi_2(x_1, x_2)) \\ &= - \int_{\mathbb{R}} (\mathcal{P}\rho)_2(x_2) \cdot \log(\mathcal{P}\rho)_2(x_2) \, dx_2 \end{aligned}$$

with Equation (67), the marginal density

$$\begin{aligned} (\mathcal{P}\rho)_2(x_2) &= \int_{\mathbb{R}} \mathcal{P}\rho(x_1, x_2) \, dx_1 \\ &= \int_{\mathbb{R}} \frac{1}{b} \rho \left(\frac{x_2}{b}, \ x_1 - 1 + a\frac{x_2^2}{b^2} \right) dx_1 \\ &= \frac{1}{b} \int_{\mathbb{R}} \rho(y, \xi) \, d\xi = \frac{1}{b} \rho_1 \left(\frac{x_2}{b} \right) \end{aligned}$$

allowing us to arrive at an information flow from x_1 to x_2 in the amount of:

$$\begin{aligned} T_{1 \to 2} &= - \int_{\mathbb{R}} \frac{1}{b} \rho_1 \left(\frac{x_2}{b} \right) \cdot \log \left[\frac{1}{b} \rho_1 \left(\frac{x_2}{b} \right) \right] dx_2 \\ &= H_1 + \log b \end{aligned} \tag{71}$$

That is to say, the flow from x_1 to x_2 has nothing to do with x_2; it is equal to the marginal entropy of x_1, plus a correction term due to the factor b.

The simple result of Equation (71) is remarkable; particularly, if $b = 1$, the information flow from x_1 to x_2 is just the entropy of x_1. This is precisely what what one would expect of the mapping component $\Phi_2(x_1, x_2) = bx_1$ in Equation (65). While the information flow is interesting *per se*, it also serves as an excellent example for the verification of our formalism.

6.3. Truncated Burgers–Hopf System

In this section, we examine a more complicated system, the Truncated Burgers–Hopf system (TBS hereafter). Originally introduced by Majda and Timofeyev [55] as a prototype of climate modeling, the TBS results from a Galerkin truncation of the Fourier expansion of the inviscid Burgers' equation, *i.e.*,

$$\frac{\partial u}{\partial t} + u \frac{\partial u}{\partial x} = 0 \tag{72}$$

to the nth order. Liang and Kleeman [51] examined such a system with two Fourier modes retained, which is governed by 4 ordinary differential equations:

$$\frac{dx_1}{dt} = F_1(x) = x_1 x_4 - x_3 x_2 \tag{73}$$

$$\frac{dx_2}{dt} = F_2(x) = -x_1 x_3 - x_2 x_4 \tag{74}$$

$$\frac{dx_3}{dt} = F_3(x) = 2x_1 x_2 \tag{75}$$

$$\frac{dx_4}{dt} = F_4(x) = -x_1^2 + x_2^2 \tag{76}$$

Despite its simplicity, the system is intrinsically chaotic, with a strange attractor lying within

$$[-24.8, 24.6] \times [-25.0, 24.5] \times [-22.3, 21.9] \times [-23.7, 23.7]$$

Shown in Figure 4 are its projections onto the x_1-x_2-x_4 and x_1-x_3-x_4 subspaces, respectively.

Finding the information flows within the TBS system turns out to be a challenge in computation, since the Liouville equation corresponding to Equations (73)–(76) is a four-dimensional partial differential equation. In [51], Liang and Kleeman adopt a strategy of ensemble prediction to reduce the computation to an acceptable level. This is summarized in the following steps:

1. Initialize the joint density of (x_1, x_2, x_3, x_4) with some distribution ρ_0; make random draws according to ρ_0 to form an ensemble. The ensemble should be large enough to resolve adequately the sample space.

2. Discretize the sample space into "bins."

3. Do ensemble prediction for the system (73)–(74).

4. At each step, estimate the probability density function ρ by counting the bins.

5. Plug the estimated ρ back to Equation (39) to compute the rates of information flow at that step.

Notice that the invariant attractor in Figure 4 allows us to perform the computation on a compact subspace of $\mathbb{R}^4$. Denote by $[-d,d]^4$ the Cartesian product $[-d,d] \times [-d,d] \times [-d,d] \times [-d,d]$. Obviously, $[-30,30]^4$ is large enough to cover the whole attractor, and hence can be taken as the sample space. Liang and Kleeman [51] discretize this space into 30^4 bins. With a Gaussian initial distribution $N(\boldsymbol{\mu}, \boldsymbol{\Sigma})$, where

$$\boldsymbol{\mu} = \begin{bmatrix} \mu_1 \\ \mu_2 \\ \mu_3 \\ \mu_4 \end{bmatrix}, \qquad \boldsymbol{\Sigma} = \begin{bmatrix} \sigma_1^2 & 0 & 0 & 0 \\ 0 & \sigma_2^2 & 0 & 0 \\ 0 & 0 & \sigma_3^2 & 0 \\ 0 & 0 & 0 & \sigma_4^2 \end{bmatrix}$$

they generate an ensemble of 2,560,000 members, each steered independently under the system (73)–(76). The details about the sample space discretization, probability estimation, *etc.*, are referred to [51]. Shown in the following are only the major results.

Between the four components of the TBS system, pairwise there are 12 information flows, namely,

$$\begin{array}{ccc} T_{2\to1}, & T_{3\to1}, & T_{4\to1} \\ T_{1\to2}, & T_{3\to2}, & T_{4\to2} \\ T_{1\to3}, & T_{2\to3}, & T_{4\to3} \\ T_{1\to4}, & T_{2\to4}, & T_{3\to4} \end{array}$$

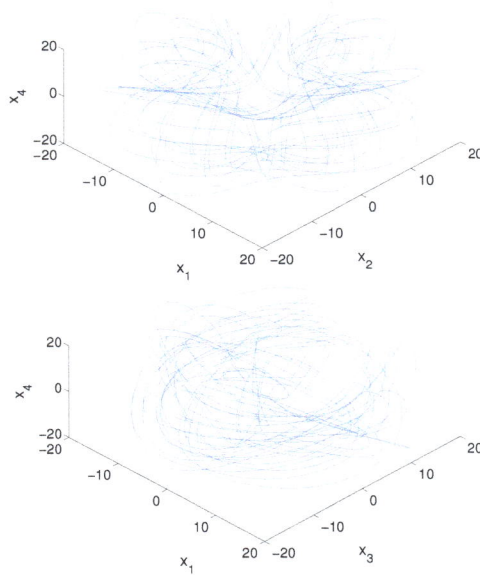

Figure 4. The invariant attractor of the truncated Burgers–Hopf system (73)–(76). Shown here is the trajectory segment for $2 \leq t \leq 20$ starting at $(40, 40, 40, 40)$. (a) and (b) are the 3-dimensional projections onto the subspaces x_1-x_2-x_3 and x_2-x_3-x_4, respectively.

To compute these flows, Liang and Kleeman [51] have tried different parameters μ and σ_k^2 $(k = 1, 2, 3, 4)$, but found the final results are the same after $t = 2$ when the trajectories are attracted into the invariant set. It therefore suffices to show the result of just one experiment: $\mu_k = 9$ and $\sigma_k^2 = 9$, $k = 1, 2, 3, 4$.

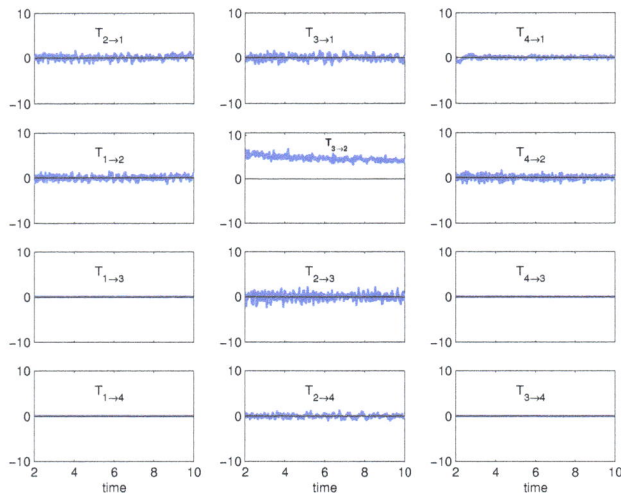

Figure 5. Information flows within the 4D truncated Burgers-Hopf system. The series prior to $t = 2$ are not shown because some trajectories have not entered the attractor by that time.

Plotted in Figure 5 are the 12 flow rates. First observe that $T_{3 \to 4} = T_{4 \to 3} = 0$. This is easy to understand, as both F_3 and F_4 in Equations (75) and (76) have no dependence on x_3 nor on x_4, implying a zero flow in either direction between the pair (x_3, x_4) by the property of causality. What makes the result remarkable is, besides $T_{3 \to 4}$ and $T_{4 \to 3}$, essentially all the flows, except $T_{3 \to 2}$, are negligible, although obvious oscillations are found for $T_{2 \to 1}$, $T_{3 \to 1}$, $T_{1 \to 2}$, $T_{4 \to 1}$, $T_{2 \to 3}$, and $T_{2 \to 4}$. The only significant flow, *i.e.*, $T_{3 \to 2}$, means that, within the TBS system, it is the fine component that causes an increase in uncertainty in a coarse component but not conversely. Originally the TBS was introduced by Majda and Timofeyev [55] to test their stochastic closure scheme that models the unresolved high Fourier modes. Since additive noises are independent of the state variables, information can only be transferred from the former to the latter. The transfer asymmetry observed here is thus reflected in the scheme.

6.4. Langevin Equation

Most of the applications of information flow/transfer are expected with stochastic systems. Here we illustrate this with a simple 2D system, which has been studied in reference [54] for the validation of Equation (52):

$$dx = Ax\,dt + B\,dw \tag{77}$$

where $A = (a_{ij})$ and $B = (b_{ij})$ are 2×2 constant matrices. This is the linear version of Equation (47). Linear systems are particular in that, if initialized with a normally distributed ensemble, then the distribution of the variables will be a Gaussian subsequently (e.g., [56]). This greatly simplifies the computation which, as we have seen in the previous subsection, is often a formidable task. Let $x \sim N(\mu, \Sigma)$. Here $\mu = \begin{pmatrix} \mu_1 \\ \mu_2 \end{pmatrix}$ is the mean vector, and $\Sigma = \begin{pmatrix} \sigma_1^2 & \sigma_{12} \\ \sigma_{21} & \sigma_2^2 \end{pmatrix}$ the covariance matrix; they evolve as

$$d\mu/dt = A\,\mu \tag{78a}$$
$$d\Sigma/dt = A\,\Sigma + \Sigma\,A^T + B\,B^T \tag{78b}$$

(BB^T is the matrix (g_{ij}) we have seen in Section 5), which determine the joint density of x:

$$\rho(x) = \frac{1}{2\pi (\det \Sigma)^{1/2}} e^{-\frac{1}{2}(x-\mu)^T \Sigma^{-1}(x-\mu)} \tag{79}$$

By Theorem 5.1, the rates of information flow thus can be accurately computed.

Several sets of parameters have been chosen in [54] to study the model behavior. Here we just look at one such choice: $B = \begin{pmatrix} 1 & 1 \\ 1 & 1 \end{pmatrix}$, $A = \begin{pmatrix} -0.5 & 0.1 \\ 0 & -0.5 \end{pmatrix}$. Its corresponding mean and covariance approach to an equilibrium: $\mu(\infty) = \begin{pmatrix} 0 \\ 0 \end{pmatrix}$, $\Sigma(\infty) = \begin{pmatrix} 2.44 & 2.2 \\ 2.2 & 2 \end{pmatrix}$. Shown in Figure 6 are the time evolutions of μ and Σ initialized with $\mu(0) = \begin{pmatrix} 1 \\ 2 \end{pmatrix}$ and $\Sigma(0) = \begin{pmatrix} 9 & 0 \\ 0 & 9 \end{pmatrix}$, and a sample path of x starting from $(1, 2)$. The computed rates of information flow, $T_{2 \to 1}$ and $T_{1 \to 2}$, are plotted in Figure 7a and b. As time moves on, $T_{2 \to 1}$ increases monotonically and eventually approaches a constant; on the other hand, $T_{1 \to 2}$ vanishes throughout. While this is within one's expectations, since $dx_2 = -0.5x_2\,dt + dw_1 + dw2$ has no dependence on x_1 and hence there should be no transfer of information from x_1 to x_2, it is interesting to observe that, in contrast, the typical paths of x_1 and x_2 could be highly correlated, as shown in Figure 6c. In other words, for two highly correlated time series, say $x_1(t)$ and $x_2(t)$, one series may have nothing to do with the other. This is a good example illustrating how information flow extends the classical notion of correlation analysis, and how it may be potentially utilized to identify the causal relation between complex dynamical events.

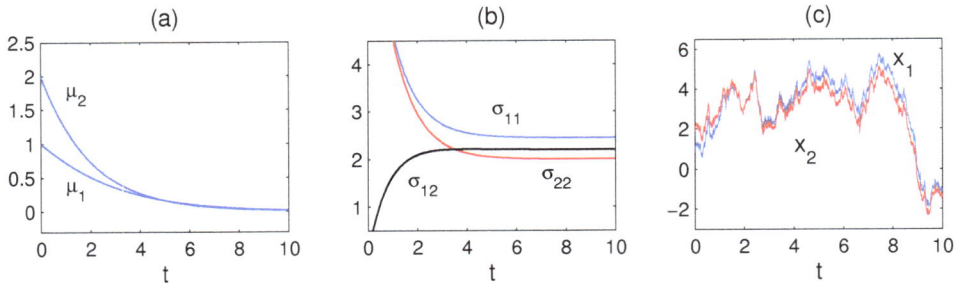

Figure 6. A solution of Equation (78), the model examined in [54], with $a_{21} = 0$ and initial conditions as shown in the text: (a) μ; (b) Σ; and (c) a sample path starting from (1,2).

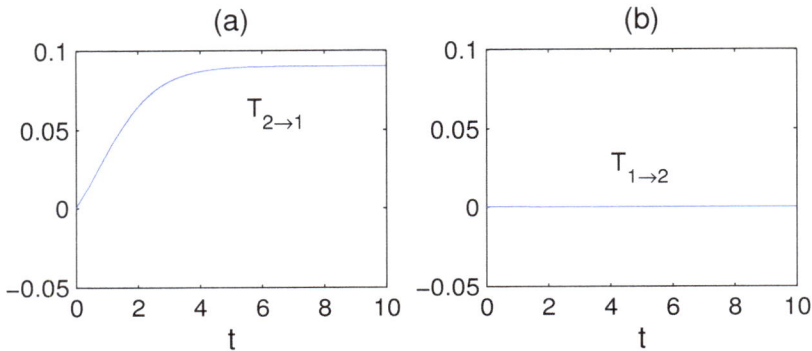

Figure 7. The computed rates of information flow for the system (77): (a) $T_{2\to1}$, (b) $T_{1\to2}$.

7. Summary

The past decades have seen a surge of interest in information flow (or information transfer, as it is sometimes called) in different fields of scientific research, mostly in the appearance of some empirical/half-empirical form. We have shown that, given a dynamical system, deterministic or stochastic, this important notion can actually be formulated on a rigorous footing, with flow measures explicitly derived. The general results are summarized in the theorems in Sections 3, 4 and 5. For two-dimensional systems, the result is fairly tight. In fact, if writing such a system as

$$\begin{cases} dx_1 = F_1(\boldsymbol{x}, t)dt + b_{11}(\boldsymbol{x}, t)dw_1 + b_{12}(\boldsymbol{x}, t)dw_2 \\ dx_2 = F_2(\boldsymbol{x}, t)dt + b_{21}(\boldsymbol{x}, t)dw_1 + b_{22}(\boldsymbol{x}, t)dw_2 \end{cases}$$

where (w_1, w_2) are standard Wiener processes, we have a rate of information flowing from x_2 to x_1,

$$T_{2\to1} = -E\left(F_1 \frac{\partial \log \rho_1}{\partial x_1}\right) - E\left(\frac{\partial F_1}{\partial x_1}\right) + \frac{1}{2}E\left(\frac{1}{\rho_1} \frac{\partial^2 (g_{11}\rho_1)}{\partial x_1^2}\right)$$

This is an alternative expression of that in Theorem 5.1; $T_{1\to2}$ can be obtained by switching the subscripts 1 and 2. In the formula, $g_{11} = \sum_k b_{1k}^2$, ρ_1 is the marginal density of x_1, and E stands for mathematical expectation with respect to ρ, *i.e.*, the joint probability density. On the right-hand side, the third term is contributed by the Brownian notion; if the system is deterministic, this term vanishes. In the remaining two terms, the first is the tendency of H_1, namely the marginal entropy of x_1; the second can be interpreted as the rate of H_1 increase on x_1 its own, thanks to the law of entropy production (12) [49], which we restate here:

For an n-dimensional system $\frac{dx}{dt} = \mathbf{F}(\mathbf{x}, t)$, its joint entropy H evolves as $\frac{dH}{dt} = E(\nabla \cdot \mathbf{F})$

This interpretation lies at the core of all the theories along this line. It illustrates that the marginal entropy increase of a component, say, x_1, is due to two different mechanisms: the information transferred from some component, say, x_2, and the marginal entropy increase associated with a system without taking x_2 into account. On this ground, the formalism is henceforth established, with respect to discrete mappings, continuous flows, and stochastic systems, respectively. Correspondingly, the resulting measures are summarized in Equations (24), (39) and (52).

The above-obtained measures possess several interesting properties, some of which one may expect based on daily life experiences. The first one is a property of flow/transfer asymmetry, which has been set as the basic requirement for the identification of causal relations between dynamical events. The information flowing from one event to another event, denoted respectively as x_2 and x_1, may yield no clue about its counterpart in the opposite direction, *i.e.*, the flow/transfer from x_1 to x_2. The second says that, if the evolution of x_1 is independent of x_2, then the flow from x_2 to x_1 is zero. The third one is about the role of stochasticity, which asserts that, if the stochastic perturbation to the receiving component does not rely on the given component, the flow measure then has a form same as that for the corresponding deterministic system. As a direct corollary, when the noise is additive, then in terms of information flow, the stochastic system functions in a deterministic manner.

The formalism has been put to application with benchmark dynamical systems. In the context of the baker transformation, it is found that there is always information flowing from the stretching direction to the folding direction, while no flow exists conversely. This is in agreement with what one would observe in kneading dough. Application to the Hénon map also yields a result just as expected on physical grounds. In a more complex case, the formalism has been applied to the study of the scale–scale interaction and information flow between the first two modes of the chaotic truncated Burgers equation. Surprisingly, all the twelve flows are essentially zero, save for one strong flow from the high-frequency mode to the low-frequency mode. This demonstrates that the route of information flow within a dynamical system, albeit seemingly complex, could be simple. In another application, we test how one may control the information flow by tuning the coefficients in a two-dimensional Langevin system. A remarkable observation is that, for two highly correlated time series, there could be no transfer from one certain series, say x_2, to the other (x_1). That is to say, the evolution of x_1 may have nothing to do with x_2, even though x_1 and x_2 are highly correlated. Information flow/transfer analysis thus extends the traditional notion of correlation analysis and/or mutual information analysis by providing a quantitative measure of causality between dynamical events, and this quantification is based firmly on a rigorous mathematical and physical footing.

The above applications are mostly with idealized systems; this is, to a large extent, intended for the validation of the obtained flow measures. Next, we would extend the results to more complex systems, and develop important applications to realistic problems in different disciplines, as envisioned in the beginning of this paper. The scale–scale information flow within the Burgers–Hopf system in § 6.3, for example, may be extended to the flow between *scale windows*. By a scale window we mean, loosely, a subspace with a range of scales included (cf. [57]). In atmosphere–ocean science, important phenomena are usually defined on scale windows, rather than on individual scales (e.g., [58]). As discussed in [53], the dynamical core of the atmosphere and ocean general circulation models is essentially a quadratically nonlinear system, with the linear and nonlinear operators possessing certain symmetry resulting from some conservation properties (such as energy conservation). Majda and Harlim [53] argue that the state space may be decomposed into a direct sum of scale windows which inherit evolution properties from the quadratic system, and then information flow/transfer may be investigated between these windows. Intriguing as this conceptual model might be, there still exist some theoretical difficulties. For example, the governing equation for a window may be problem-specific; there may not be such governing equations as simply written as those like Equation (3) for individual components. Hence one may need to seek new ways to the derivation of the information flow formula. Nonetheless, central at the problem is still the aforementioned classification of mechanisms that govern the marginal entropy evolution; we are expecting new breakthroughs along this line of development.

The formalism we have presented thus far is with respect to Shannon entropy, or absolute entropy as one may choose to refer to it. In many cases, such as in the El Niño case where predictability is concerned, this may need to be modified, since the predictability of a dynamical system is measured by relative entropy. Relative entropy is also called Kullback–Leibler divergence; it is defined as

$$D(\rho\|q) = E_\rho \left[\log \left(\frac{\rho}{q} \right) \right]$$

i.e., the expectation of the logarithmic difference between a probability ρ and another reference probability q, where the expectation is with respect to ρ. Roughly it may be interpreted as the "distance" between ρ and q, though it does not satisfy all the axioms for a distance functional. Therefore, for a system, if letting the reference density be the initial distribution, its relative entropy at a time t informs how much additional information is added (rather than how much information it has). This provides a natural choice for the measure of the utility of a prediction, as pointed out by Kleeman (2002) [59]. Kleeman also argues in favor of relative entropy because of its appealing properties, such as nonnegativity and invariance under nonlinear transformations [60]. Besides, in the context of a Markov chain, it has been proved that it always decreases monotonically with time, a property usually referred to as the generalized second law of thermodynamics (e.g., [60,61]). The concept of relative entropy is now a well-accepted measure of predictability (e.g., [59,62]). When predictability problems (such as those problems in atmosphere-ocean science and financial economics as mentioned in the introduction) are dealt with, it is necessary to extend the current formalism to one with respect to the relative entropy functional. For all the dynamical system settings in this review, the extension should be straightforward.

Acknowledgments: This study was supported by the National Science Foundation of China (NSFC) under Grant No. 41276032 to NUIST, by Jiangsu Provincial Government through the "Jiangsu Specially-Appointed Professor Program" (Jiangsu Chair Professorship), and by the Ministry of Finance of China through the Basic Research Funding to China Institute for Advanced Study.

References

1. Baptista, M.S.; Garcia, S.P.; Dana S.K.; Kurths, J. Transmission of information and synchronization in a pair of coupled chaotic circuits: An experimental overview. *Eur. Phys. J.-Spec. Top.* **2008**, *165*, 119–128.
2. Baptista, M.D.S.; Kakmeni, F.M.; Grebogi, C. Combined effect of chemical and electrical synapses in Hindmarsh-Rose neural networks on synchronization and the rate of information. *Phys. Rev. E* **2010**, *82*, 036203.
3. Bear, M.F.; Connors, B.W.; Paradiso, M.A. *Neuroscience: Exploring the Brain*; 3rd ed.; Lippincott Williams & Wilkins: Baltimore, MD, USA, 2007; p. 857.
4. Vakorin, V.A.; MiAiA, B.; Krakovska, O.; McIntosh, A.R. Empirical and theoretical aspects of generation and transfer of information in a neuromagnetic source network. *Front. Syst. Neurosci.* **2011**, *5*, 96.
5. Ay, N.; Polani, D. Information flows in causal networks. *Advs. Complex Syst.* **2008**, *11*, doi: 10.1142/S0219525908001465.

6.   Peruani, F.; Tabourier, L. Directedness of information flow in mobile phone communication networks. *PLoS One* **2011**, *6*, e28860.

7.   Sommerlade, L.; Amtage, F.; Lapp, O.; Hellwig, B.; Licking, C.H.; Timmer, J.; Schelter, B. On the estimation of the direction of information flow in networks of dynamical systems. *J. Neurosci. Methods* **2011**, *196*, 182–189.

8.   Donner, R.; Barbosa, S.; Kurths, J.; Marwan, N. Understanding the earth as a complex system-recent advances in data analysis and modelling in earth sciences. *Eur. Phys. J.* **2009**, *174*, 1–9.

9.   Kleeman, R. Information flow in ensemble weather prediction. *J. Atmos. Sci.* **2007**, *64*, 1005–1016.

10.   Materassi, M.; Ciraolo, L.; Consolini, G.; Smith, N. Predictive space weather: An information theory approach. *Adv. Space Res.* **2011**, *47*, 877–885.

11.   Tribbia, J.J. Waves, Information and Local Predictability. In Proceedings of the Workshop on Mathematical Issues and Challenges in Data Assimilation for Geophysical Systems: Interdisciplinary Perspectives, IPAM, UCLA, 22–25 February 2005.

12.   Chen, C.R.; Lung, P.P.; Tay, N.S.P. Information flow between the stock and option markets: Where do informed traders trade? *Rev. Financ. Econ.* **2005**, *14*, 1–23.

13.   Lee, S.S. Jumps and information flow in financial markets. *Rev. Financ. Stud.* **2012**, *25*, 439–479.

14.   Sommerlade, L.; Eichler, M.; Jachan, M.; Henschel, K.; Timmer, J.; Schelter, B. Estimating causal dependencies in networks of nonlinear stochastic dynamical systems. *Phys. Rev. E* **2009**, *80*, 051128.

15.   Zhao, K.; Karsai, M.; Bianconi, G. Entropy of dynamical social networks. *PLoS One* **2011**, doi:10.1371/journal.pone.0028116.

16.   Cane, M.A. The evolution of El Niño, past and future. *Earth Planet. Sci. Lett.* **2004**, *164*, 1–10.

17.   Jin, F.-F. An equatorial ocean recharge paradigm for ENSO. Part I: conceptual model. *J. Atmos. Sci.* **1997**, *54*, 811–829.

18.   Philander, S.G. *El Niño, La Niña, and the Southern Oscillation*; Academic Press: San Diego, CA, USA, 1990.

19.   Ghil, M.; Chekroun, M.D.; Simonnet, E. Climate dynamics and fluid mechanics: Natural variability and related uncertainties. *Physica D* **2008**, *237*, 2111–2126.

20.   Mu, M.; Xu, H.; Duan, W. A kind of initial errors related to "spring predictability barrier" for El Niño events in Zebiak-Cane model. *Geophys. Res. Lett.* **2007**, *34*, L03709, doi:10.1029/2006GL027412.

21.   Zebiak, S.E.; Cane, M.A. A model El Niño-Southern Oscillation. *Mon. Wea. Rev.* **1987**, *115*, 2262–2278.

22.   Chen, D.; Cane, M.A. El Niño prediction and predictability. *J. Comput. Phys.* **2008**, *227*, 3625–3640.

23.   Mayhew, S.; Sarin, A.; Shastri, K. The allocation of informed trading across related markets: An analysis of the impact of changes in equity-option margin requirements. *J. Financ.* **1995**, *50*, 1635–1654.

24.   Goldenfield, N.; Woese, C. Life is physics: Evolution as a collective phenomenon far from equilibrium. *Ann. Rev. Condens. Matt. Phys.* **2011**, *2*, 375–399.

25.   Küppers, B. *Information and the Origin of Life*; MIT Press: Cambridge, UK, 1990.

26.   Murray, J.D. *Mathematical Biology*; Springer-Verlag: Berlin, Germany, 2000.

27.   Allahverdyan, A.E.; Janzing, D.; Mahler, G. Thermodynamic efficiency of information and heat flow. *J. Stat. Mech.* **2009**, *PO9011*, doi:10.1088/1742-5468/2009/09/P09011.

28.   Crutchfield, J.P.; Shalizi, C.R. Thermodynamic depth of causal states: Objective complexity via minimal representation. *Phys. Rev. E* **1999**, *59*, 275–283.

29.   Davies, P.C.W. The Physics of Downward Causation. In *The Re-emergence of Emergence*; Clayton, P., Davies, P.C.W., Eds.; Oxford University Press: Oxford, UK, 2006; pp. 35–52.

30.   Ellis, G.F.R. Top-down causation and emergence: Some comments on mechanisms. *J. R. Soc. Interface* **2012**, *2*, 126–140.

31.   Okasha, S. Emergence, hierarchy and top-down causation in evolutionary biology. *J. R. Soc. Interface* **2012**, *2*, 49–54.

32.   Walker, S.I.; Cisneros, L.; Davies, P.C.W. Evolutionary transitions and top-down causation. arXiv:1207.4808v1 [nlin.AO], **2012**.

33.   Wu, B.; Zhou, D.; Fu, F.; Luo, Q.; Wang, L.; Traulsen, A. Evolution of cooperation on stochastic dynamical networks. *PLoS One* **2010**, *5*, e11187.

34.   Pope, S. *Turbulent Flows*; 8th ed.; Cambridge University Press: Cambridge, UK, 2011.

35.   Faes, L.; Nollo, G.; Erla, S.; Papadelis, C.; Braun, C.; Porta, A. Detecting Nonlinear Causal Interactions between Dynamical Systems by Non-uniform Embedding of Multiple Time Series. In Proceedings of

the Engineering in Medicine and Biology Society, Buenos Aires, Argentina, 31 August–4 September 2010; pp. 102–105.

36. Kantz, H.; Shreiber, T. *Nonlinear Time Series Analysis*; Cambridge University Press: Cambridge, UK, 2004.

37. Schindler-Hlavackova, K.; Palus, M.; Vejmelka, M.; Bhattacharya, J. Causality detection based on information-theoretic approach in time series analysis. *Phys. Rep.* **2007**, *441*, 1–46.

38. Granger, C. Investigating causal relations by econometric models and cross-spectral methods. *Econometrica* **1969**, *37*, 424–438.

39. McWilliams, J.C. The emergence of isolated, coherent vortices in turbulence flows. *J. Fluid Mech.* **1984**, *146*, 21–43.

40. Salmon, R. *Lectures on Geophysical Fluid Dynamics*; Oxford University Press: Oxford, UK, 1998; p. 378.

41. Bar-Yam, Y. *Dynamics of Complex Systems*; Addison-Welsley Press: Reading, MA, USA, 1997; p. 864.

42. Crutchfield, J.P. The calculi of emergence: computation, dynamics, and induction induction. "Special issue on the Proceedings of the Oji International Seminar: Complex Systems-From Complex Dynamics to Artifical Reality". *Physica D* **1994**, *75*, 11–54.

43. Goldstein, J. Emergence as a construct: History and issues. *Emerg. Complex. Org.* **1999**, *1*, 49–72.

44. Corning, P.A. The re-emergence of emergence: A venerable concept in search of a theory. *Complexity* **2002**, *7*, 18–30.

45. Vastano, J.A.; Swinney, H.L. Information transport in sptiotemporal systems. *Phys. Rev. Lett.* **1988**, *60*, 1773–1776.

46. Kaiser, A.; Schreiber, T. Information transfer in continuous processes. *Physica D* **2002** *166*, 43–62.

47. Schreiber, T. Measuring information transfer. *Phys. Rev. Lett.* **2000**, *85*, 461.

48. Liang, X.S.; Kleeman, R. A rigorous formalism of information transfer between dynamical system components. I. Discrete mapping. *Physica D* **2007**, *231*, 1–9.

49. Liang, X.S.; Kleeman, R. Information transfer between dynamical system components. *Phys. Rev. Lett.* **2005**, *95*, 244101.

50. Lasota, A.; Mackey, M.C. *Chaos, Fractals, and Noise: Stochastic Aspects of Dynamics*; Springer: New York, NY, USA, 1994.

51. Liang, X.S.; Kleeman, R. A rigorous formalism of information transfer between dynamical system components. II. Continuous flow. *Physica D* **2007**, *227*, 173–182.

52. Liang, X.S. Uncertainty generation in deterministic fluid flows: Theory and applications with an atmospheric stability model. *Dyn. Atmos. Oceans* **2011**, *52*, 51–79.

53. Majda, A.J.; Harlim, J. Information flow between subspaces of complex dynamical systems. *Proc. Natl. Acad. Sci. USA* **2007**, *104*, 9558–9563.

54. Liang, X.S. Information flow within stochastic dynamical systems. *Phys. Rev. E* **2008**, *78*, 031113.

55. Majda, A.J.; Timofeyev, I. Remarkable statistical behavior for truncated Burgers-Hopf dynamics. *Proc. Natl. Acad. Sci. USA* **2000**, *97*, 12413–12417.

56. Gardiner, C.W. *Handbook of Stochastic Methods for Physics, Chemistry, and the Natural Sciences*; Springer-Verlag: Berlin/Heidelberg, Germany, 1985.

57. Liang, X.S.; Anderson, D.G.M.A. Multiscale window transform. *SIAM J. Multiscale Model. Simul.* **2007**, *6*, 437–467.

58. Liang, X.S.; Robinson, A.R. Multiscale processes and nonlinear dynamics of the circulation and upwelling events off Monterey Bay. *J. Phys. Oceanogr.* **2009**, *39*, 290–313.

59. Kleeman, R. Measuring dynamical prediction utility using relative entropy. *J. Atmos. Sci.* **2002**, *59*, 2057–2072.

60. Cover, T.M.; Thomas, J.A. *Elements of Information Theory*; Wiley: New York, NY, USA, 1991.

61. Ao, P. Emerging of stochastic dynamical equalities and steady state thermodynamics from Darwinian dynamics. *Commun. Theor. Phys.* **2008**, *49*, 1073–1090.

62. Tang, Y.; Deng, Z.; Zhou, X.; Cheng, Y.; Chen, D. Interdecadal variation of ENSO predictability in multiple models. *J. Clim.* **2008**, *21*, 4811–4832.

MDPI

St. Alban-Anlage 66

4052 Basel

Switzerland

Tel. +41 61 683 77 34

Fax +41 61 302 89 18

www.mdpi.com

Entropy Editorial Office

E-mail: entropy@mdpi.com

www.mdpi.com/journal/entropy

www.ingramcontent.com/pod-product-compliance
Lightning Source LLC
Chambersburg PA
CBHW051713210326
41597CB00032B/5464